D1282339

MSU
LIBRARIES

Advances in

Insect Physiology

Volume 18

Advances in Insect Physiology

edited by

M. J. BERRIDGE
J. E. TREHERNE
and V. B. WIGGLESWORTH

Department of Zoology, The University
Cambridge, England

Volume 18

1985

ACADEMIC PRESS
(Harcourt Brace Jovanovich, Publishers)
London Orlando San Diego New York
Toronto Montreal Sydney Tokyo

ACADEMIC PRESS INC. (LONDON) LTD.
24–28 Oval Road
LONDON NW1 7DX

United States Edition published by
ACADEMIC PRESS, INC.
Orlando, Florida 32887

ISSN: 0065-2806

ISBN: 0-12-024218-4

PRINTED IN THE UNITED STATES OF AMERICA

85 86 87 88 9 8 7 6 5 4 3 2 1

Contents

Contributors

Athula B. Attygalle[*]

Department of Chemistry, University of Keele, Staffordshire ST5 5BG, England

D. Graham

Facultät Biologie, Universität Kaiserslautern, Kaiserslautern, Federal Republic of Germany

John A. Kiger, Jr.

Department of Genetics, University of California, Davis, California 95616, USA

E. David Morgan

Department of Chemistry, University of Keele, Staffordshire ST5 5BG, England

H. Frederik Nijhout

Department of Zoology, Duke University, Durham, North Carolina 27706, USA

Helen K. Salz

Department of Biology, Princeton University, Princeton, New Jersey 08544, USA

[*]*Present address: Institute for Organic Chemistry II, University of Erlangen-Nürnberg, D-8520 Erlangen, Federal Republic of Germany.*

Melody V. S. Siegler

Department of Zoology, University of Cambridge, Cambridge CB2 3EJ, England

Barbara Stay

Department of Zoology, University of Iowa, Iowa City, Iowa 52242, USA

Stephen S. Tobe

Department of Zoology, University of Toronto, Toronto, Ontario, Canada M5S 1A1

Ant Trail Pheromones

Athula B. Attygalle* and E. David Morgan

Department of Chemistry, University of Keele, Staffordshire, England

1 Introduction

Social insects (bees, wasps, ants, and termites) utilize an array of pheromones to maintain the high level of organization in their colonies. Many species of ants and termites that are essentially wingless lay terrestrial odor trails leading to food sources or nesting sites. According to Wilson (1971), the odor trail system is the most elaborate of all the known forms of chemical communication. Sudd (1959) has defined trail laying as a field activity in

* Present address: Institute for Organic Chemistry II, University of Erlangen-Nürnberg, Erlangen, West Germany.

ADVANCES IN INSECT PHYSIOLOGY, VOL. 18

which an insect marks a route with scent or odor traces such that other insects of the same community are able to follow it. Essentially, a foraging worker returning from a food source lays a more or less continuous narrow band of chemicals on the substratum as it returns to the nest ("recruitment trails"); these trails may excite other workers to follow the chemical scent to the food, where they in turn feed and, as they return to the nest, reinforce the chemical deposit. Workers returning from visiting an exhausted food source do not reinforce the trail, so that eventually it evaporates and the signal is obliterated. The ants can regulate the amount of trail substance they lay on a trail (Hangartner, 1970). Short-range trail pheromones, laid with footprints in the vicinity of the hive or nest, are also known for bees and wasps (Butler *et al.*, 1969; Lindauer and Kerr, 1958). Trail pheromones can also facilitate migration of the colony to a new site ("emigration trails"). Some long-lasting trails may serve as chemical cues in home range orientation and help in marking home territories.

The chemical stimuli of this kind that trigger an immediate, reversible and specific change in the behavior of the recipient are called releaser pheromones. Here is an example of a pheromone type, allowing close and detailed examination, that is very different from those best known among insects, such as the sexual attractants of Lepidoptera or the aggregating pheromones of Coleoptera. Ant trail pheromones are the releaser pheromones we best understand today, because they have received intensive study in recent years and new methods of experimentation have yielded detailed knowledge of their chemical nature, their glandular origins, and their specificity. These aspects are the subjects of this review, which is the first comprehensive review of ant trail pheromones in recent years.

2 Earlier observations

As early as the eighteenth century, Bonnet (1779) had observed that some ants use trails to recruit workers of the same species to a food source. Forel (1886, 1908) proposed that ants employ their antenna to follow trails. Similar observations of recruitment trails of ants were made by Eidmann (1927), Hingston (1928), and Santschi (1923). Goetsch (1934) demonstrated that artificial trails could be laid with the tip of the gaster of a freshly killed ant. The chemical nature of the ant trails had been observed by McGregor (1948), Carthy (1950, 1951a,b, 1952), and Vowles (1955). Carthy was one of the first to conduct experimental studies on trail laying. Some general reviews of insect pheromones cover earlier work on ant trail pheromones (Gabba and Pavan, 1970; Blum and Brand, 1972; Blum, 1974a). More recent works by

Hölldobler (1977), Parry and Morgan (1979), and Dumpert (1981) are further useful sources of information. A review by Hölldobler (1978) concentrated on the ethological aspects of chemical communications in ants.

3 Glandular origins

In the termites, a gland on the ventral surface of the abdomen serves as the source of the trail pheromones, but in ants pheromones can arise from a number of glandular sources including the hind gut, rectal gland, poison gland, tarsal gland, Dufour gland, Pavan's gland, tibial gland, sternal gland, and anal (pygidial?) gland. The location of the general exocrine glands of a typical ant is shown in Fig. 1. The trail is deposited on the ground by either

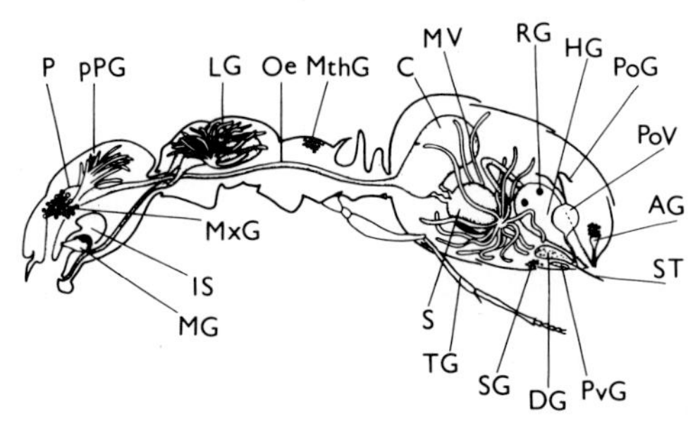

Fig. 1. Location of exocrine glands and the intestinal tract of a typical ant. P, Pharynx; pPG, postpharyngeal gland; LG, labial gland; Oe, esophagus; MthG, metathoracic or metapleural gland; C, crop; MV, Malpighi vessels; RG, rectal gland; HG, hind gut; PoG, poison gland; PoV, poison vesicle; AG, anal gland; ST, sting; PvG, Pavan gland; DG, Dufour gland; SG, sternal gland; TG, tibial gland; S, stomach; MxG, maxilliary gland; IS, infrabuccal sac; MG, mandibular gland. (After Dumpert, 1981, adapted.)

the sting, anus, abdominal sternum, or tarsi of the hind legs, as the ant moves along. The glandular origins are usually determined by laying artificial trails with extracts of the various glands in a solvent such as ether, hexane, or acetone. Wilson (1959) provided the first bioassay methods to examine trail-following behavior. A number of bioassay procedures are now available to test trail-following activity (Topoff *et al.*, 1972; Van Vorhis Key *et al.*, 1981). The method of Pasteels and Verhaeghe (1974), in which a trail is laid on a circular track, has been used popularly by many investigators. It is a quantitative method, and is therefore particularly useful in measuring the activities of different samples. Ritter and Persoons (1977) have discussed the problems associated with the choice of bioassays for establishing the authenticity of natural trail pheromones. Table 1 summarizes the reported glandular sources of the trail pheromones of some genera of Formicidae as found by bioassay of glandular extracts.

TABLE 1
Glandular sources of trail pheromones in some subfamilies of Formicidae

Subfamily, Genus	Glandular source	Reference
Ponerinae		
Pachycondyla (= *Termitopone*)	Hind gut	Blum (1966)
	Pygidial gland	Hölldobler and Engel (1978); Hölldobler and Traniello (1980a)
Leptogenys	Poison gland + pygidial gland (?)	Fletcher (1971); Maschwitz and Schönegge (1977)
Onychomyrmex	Sternal gland	Hölldobler *et al.* (1982)
Paltothyreus	Sternal gland	Hölldobler (1982)
Megaponera	Poison gland	Longhurst *et al.* (1979)
Ecitoninae		
Eciton	Hind gut or pygidial gland	Blum and Portocarrero (1964); Hölldobler and Engel (1978)
Neivamyrmex	Hind gut	Watkins (1964)
Myrmicinae		
Acromyrmex	Poison gland	Blum *et al.* (1964)
Atta	Poison gland	Blum *et al.* (1964); Moser and Blum (1963)
Crematogaster	Tibial gland applied with hind legs	Leuthold (1968); Fletcher and Brand (1968)
Cyphomyrmex	Poison gland	Blum *et al.* (1964)
Huberia	Poison gland	Blum (1966)
Leptothorax	Poison gland[a]	Möglich *et al.* (1974); Möglich (1979)
Manica	Poison gland	Blum (1974b)
Monomorium	Dufour gland	Blum (1966); Ritter *et al.* (1977a)
Myrmica	Poison gland	Blum (1974b); Cammaerts-Tricot (1974a, b)
Novomessor	Poison gland	Hölldobler *et al.* (1978)
Orectognathus	Pygidial gland + poison gland	Hölldobler (1981, 1982)
Pheidole	Dufour gland	Wilson (1963)
Pogonomyrmex	Dufour gland	Hölldobler (1971a, 1974)
	Poison gland	Blum (1974b); Hölldobler and Wilson (1970)
Pristomyrmex	?	Hayashi and Komae (1977)
Sericomyrmex	Poison gland	Blum and Portocarrero (1966)
Solenopsis	Dufour gland	Wilson (1962a)
Tetramorium	Poison gland	Blum and Ross (1965)
Trachymyrmex	Poison gland	Blum and Portocarrero (1966)
Veromessor	Poison gland	Blum (1974b)
Dolichoderinae		
Iridomyrmex	Pavan's gland	Wilson and Pavan (1959); Robertson *et al.* (1980)

(*continued*)

TABLE 1 (*continued*)

Subfamily, Genus	Glandular source	Reference
Monacis	Pavan's gland	Wilson and Pavan (1959)
Tapinoma	Pavan's gland	Wilson (1965a)
Formicinae		
Acanthomyops	Hind gut	Hangartner (1969a)
Camponotus	Hind gut	Hölldobler (1971b); Hölldobler *et al.* (1974)
Formica	Hind gut	Möglich and Hölldobler (1975)
Lasius	Hind gut	Hangartner (1967, 1969b); Hangartner and Bernstein (1964)
Myrmelachista	Hind gut	Blum and Wilson (1964)
Oecophylla	Rectal gland	Hölldobler and Wilson (1978)
Paratrechina	Hind gut	Blum and Wilson (1964)
Aneuretinae		
Aneuretus	Sternal gland	Traniello and Jayasuriya (1981a, b)

[a] The poison gland produces a tandem calling pheromone.

4 Chemistry of trail pheromones

4.1 COMPOSITION

In recent years considerable advances has been made in chemically identifying the substances that evoke trail-following behavior. Methyl 4-methylpyrrole-2-carboxylate (**I**), a poison gland substance of the myrmicine ant *Atta texana*, was the first ant trail substance to be identified (Tumlinson *et al.*, 1971, 1972). The pyrrole (**I**) was synthesized by Sonnet (1972) and the product was shown to be identical with the natural substance. The compound has a very high behavioral efficiency and the detection threshold is as low as 80 fg/cm. The same compound was subsequently demonstrated to be active for *Atta cephalotes* (Riley *et al.*, 1974) and *Acromyrmex octospinosus* (Robinson *et al.*, 1974; Cross *et al.*, 1982).

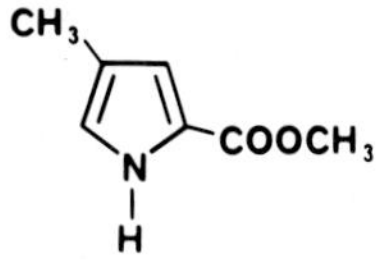

CH_3, $COOCH_3$, N, H

[I]

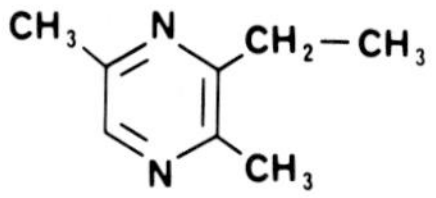

CH_3, N, CH_2-CH_3, N, CH_3

[II]

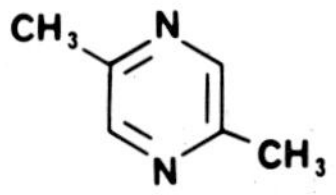

CH_3, N, N, CH_3

[III]

Atta sexdens, a species related to *A. texana* and *A. cephalotes*, did not follow trails made of synthetic methyl 4-methylpyrrole-2-carboxylate. Subsequently another compound, 3-ethyl-2,5-dimethylpyrazine (**II**), was identified as the major component of the trail pheromone of *A. sexdens rubropilosa* (Cross *et al.*, 1979). Cammaerts *et al.* (1977) first isolated the trail pheromone of *Myrmica rubra* from its poison gland. Later, Evershed *et al.* (1981, 1982) found the trail pheromone of eight species of *Myrmica* to consist of a single substance, 3-ethyl-2,5-dimethylpyrazine (**II**). Recently a further pyrazine, 2,5-dimethylpyrazine (**III**), together with 3-ethyl-2,5-dimethylpyrazine have been identified as the trail pheromone components of *Tetramorium caespitum* (Attygalle and Morgan, 1983). A concentration of 90 and 40 pg/cm trail of the two respective synthetic pyrazines released trail following equivalent to that released by an artificial trail made of one poison gland.

Among the poison gland constituents of the Pharaoh's ant are two substances, 5-methyl-3-butyloctahydroindolizine (monomorine I) (**IV**) and 2-(5′-hexenyl)-5-pentylpyrrolidine (monomorine III) (**V**), that are able to attract worker ants and that show some activity in trail following tests (Ritter *et al.*, 1973, 1975, 1977a). Sonnet and Oliver (1975) have reported a synthesis for monomorine I.

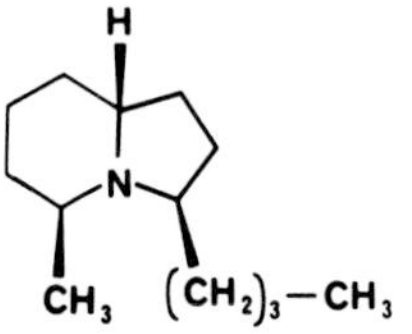

H N CH_3 $(CH_2)_3-CH_3$

[IV]

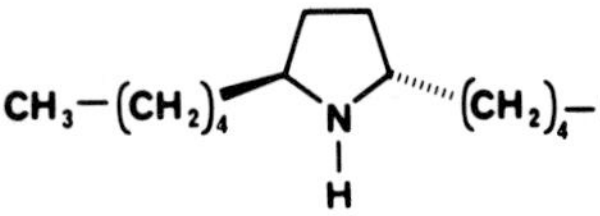

$CH_3-(CH_2)_4$ N H $(CH_2)_4-CH=CH_2$

[V]

However, the much more active true trail pheromone of the Pharaoh's ant is found in the Dufour gland in trace quantities. It was later identified as (+)-(3*S*,4*R*)-3,4,7,11-tetramethyltrideca-6*E*,10*Z*-dienal (faranal) (Ritter *et al.*, 1977a,b) (**VI**).

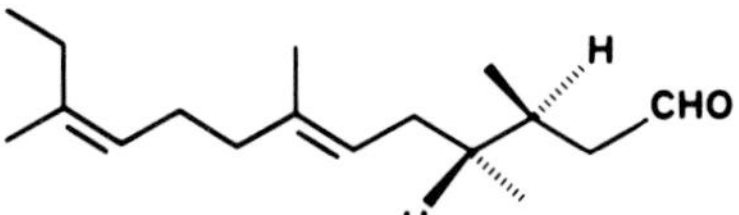

H CHO H

[VI]

The stereochemistry and geometry of faranal have been confirmed by stereospecific synthesis by several groups (Knight and Ojhara, 1981, 1983; Kobayashi *et al.*, 1980; Mori and Ueda, 1981, 1982; Baker *et al.*, 1981, 1983). Faranal has an interesting structural relationship to juvenile hormones and the farnesene homologs found in *Myrmica* species (Morgan and Wadhams, 1972; Attygalle and Morgan, 1982). Walsh *et al.* (1965) first demonstrated that the trail pheromone of the red imported fire ant *Solenopsis invicta*

(=*saevissima*) can be purified by gas chromatography. This trail pheromone is certainly multicomponent but the composition is controversial. Williams *et al.* (1981a,b) reported it to be (2*Z*,4*Z*,6*Z*)-3,7,11-trimethyl-2,4,6,10-dodecatetraene (*Z*,*Z*,*Z*-allofarnesene) (**VII**). They synthesized the eight isomers of allofarnesene and found the *Z*,*Z*,*Z* isomer to have identical chromatographic and spectral properties to the ant material from the Dufour gland of *S. invicta*. The synthetic substance showed optimal activity at the 100–500 fg/cm level.

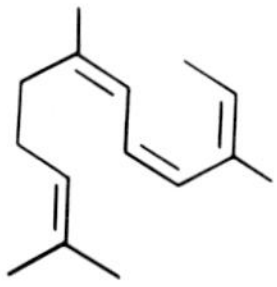

[VII]

According to Vander Meer *et al.* (1981), *S. invicta* trail pheromone is multicomponent, and four components were identified as (3*Z*,6*E*)-3,7,11-trimethyldodeca-1,3,6,10-tetraene (*Z*,*E*-α-farnesene) (**VIII**), (3*E*,6*E*)-3,7,11-trimethyldodeca-1,3,6,10-tetraene (*E*,*E*-α-farnesene) (**IX**), (3*Z*,6*Z*)-3,4,7,11-tetramethyldodeca-1,3,6,10-tetraene (*Z*,*Z*-homofarnesene) (**X**), and (3*Z*,6*E*)-3,4,7,11-tetramethyldodeca-1,3,6,10-tetraene (*Z*,*E*-homofarnesene) (**XI**). The two farnesenes were synthesized by dehydrating nerolidol and showed activity at a pheromonal level. *Z*,*E*-α-Farnesene showed significant trail-following activity at the 100 fg/cm level. The two homofarnesenes are not yet synthesized.

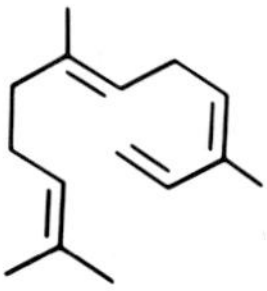

[VIII] 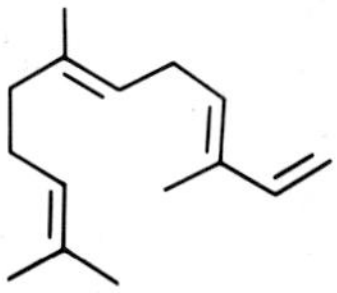 [IX]

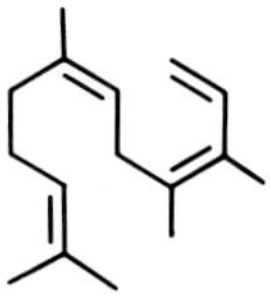

[X] 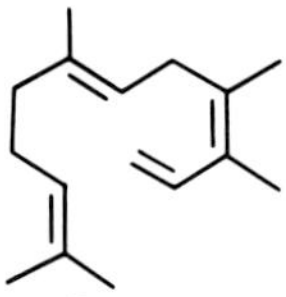 [XI]

A preliminary study had been done on the trail pheromones of few other species of *Solenopsis*. Barlin *et al.* (1976b) reported that the main trail pheromone of *S. richteri* possesses a molecular weight of 218 and an empirical formula of $C_{16}H_{26}$. They assumed the trail pheromones of *S. xyloni* and *S. geminata* are of a similar chemical type and suggested the empirical formula $C_{17}H_{28}$.

Hexanoic, heptanoic, octanoic, nonanoic, decanoic, and dodecanoic acids were reported by Huwyler *et al.* (1973, 1975) as components of the trail pheromone isolated from the hind gut of the formicine ant *Lasius fuliginosus*. It was also reported that the active material is composed of an acidic and a nonacidic fraction. The former appears to account for the greater part of the total activity. Although the absence of pentanoic, undecanoic, and tridecanoic acids has been tested, it is unknown whether any other lower and higher homologs of the fatty acid series are present in the rectal fluid. In fact, Hangartner (1967) had observed that the activity of an aqueous trail pheromone extract disappeared to a large extent upon the addition of alkali, and reappeared at the original level when it was reacidified. Commercial samples of the aforementioned six fatty acids were all found to elicit trail-following behavior in *L. fuliginosus* workers, when tested individually for activity, but the activity toward an appropriate mixture of the acids has not been examined. It is interesting to note that the trail pheromone of *Lasius niger*, isolated from the rectal fluid, is nonacidic and can be recovered from the GC effluent, although no corresponding peak can be observed (Huwyler *et al.*, 1975).

Nine fatty acids, similar to those found in *L. fuliginosus*, have been reported as the components of the trail pheromone of the myrmicine ant *Pristomyrmex pungens* (Hayashi and Komae, 1977). This mixture of saturated and unsaturated fatty acids of the C_{14} to C_{20} range falls out of line when compared with the chemical structures described as the trail pheromone components of the other myrmicine ants (Table 2). Out of the nine fatty acids, three were saturated and identified as tetradecanoic acid, hexadecanoic acid, and octadecanoic acid. The remaining five unsaturated acids have been only partially identified. The positions and the geometries of the double bonds remain unknown. Nevertheless the query whether they are in fact the true trail pheromone components remains, because the activity of the synthetic analogs has not been reported. Furthermore, the glandular origin of the pheromone remains unknown.

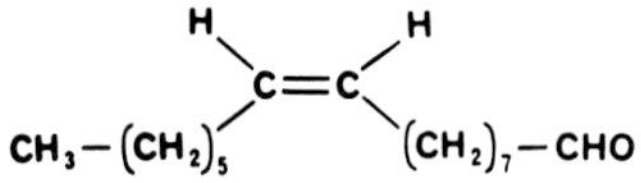

$$CH_3-(CH_2)_5-CH{=}CH-(CH_2)_7-CHO$$

[XII]

In all reported dolichoderine ants the trail pheromones are derived exclusively from the Pavan's gland (Wilson and Pavan, 1959). In *Iridomyrmex humilis* (*Z*)-9-hexadecenal (**XII**), a Pavan's gland constituent, is indicated as a trail pheromone component by behavioral evidence (Cavill *et al.*, 1979, 1980; Van Vorhis Key and Baker, 1982a,c). Although high concentrations of (*Z*)-9-hexadecenal alone elicit intense trail following by recruited workers, the true trail pheromone is considered to be multicomponent. In fact, gaster extract trails containing 100 times less (*Z*)-9-hexadecenal were comparable in activity to the synthetic trails.

4.2 CONCENTRATION

Besides the exact chemical composition, the correct physiological concentration is a critical factor in the determination of trail-following activity of synthetic mixtures. Van Vorhis Key *et al.* (1981) demonstrated for *Iridomyrmex humilis*, using extracts of whole gasters, that optimum activity was found in response to trails containing 0.1–1.0 ant equivalents per 50 cm. The activity dropped when the concentrations were lower or higher than the optimal concentrations. When the concentration was increased to 5 ant equivalents per 50 cm trail not only the trail-following activity decreased but also the mean lateral distance from the trail at which ants exhibited trail following increased. A concentration of 0.2 ng/cm of (*Z*)-9-hexadecenal could evoke the same trail-following activity as 0.002 ant equivalents per 1 cm on an artificial trail (Van Vorhis Key and Baker, 1982b).

Cross *et al.* (1979) have found for *Atta sexdens rubroliposa* that the ants showed no trail-following response when concentrations above 10^3 ng per 50 cm trail of the pyrazine (**II**) was used for bioassay. Similar results have been obtained in the studies on *Myrmica rubra* (Evershed *et al.*, 1982) and *Tetramorium caespitum* (Attygalle and Morgan, 1984). All these results confirm the importance of testing the synthetic substances at the correct physiological concentration when used for trail bioassay.

5 Stereobiology of trail pheromones

5.1 MULTICOMPONENT PHEROMONES

The pheromonal transmission of information in most insects is now accepted to be through multicomponent pheromones consisting of several stimulus compounds (Silverstein and Young, 1976). In many examples, the precise qualitative blend of the components gives the species specifity to the pheromonal signal. Similar examples are seen in ant trail pheromones. The pyrrole (**I**) is only one component in the trail pheromone of *Atta texana* because Tumlinson *et al.* (1972) have isolated at least four other active fractions, but the nature of the other constituents remains unknown. Moser and Silverstein (1967) have shown the presence of a nonvolatile component, besides the volatile component, in the trail-marking substance of *A. texana.* The trail pheromones of *Lasius fuliginosus*, *Pristomyrmex pungens*, and *Iridomyrmex humilis* are known to be multicomponent but complete quantitative evaluations have not been performed. In *Solenopsis invicta* *Z,E*-α-farnesene (**VIII**) was the most active component in the trail pheromone mixture and was capable of reproducing the trail-following activity equivalent to that of a Dufour gland extract. The other three components (**IX, X, XI**) were 10–100 times less active (Vander Meer, 1983). It is interesting to note that the four components (**VIII–XI**) were able to duplicate the recruitment response of a Dufour gland better when heptadecane was added to the mixture. Heptadecane is a Dufour gland constituent which by itself is inactive.

A complete identification of a multicomponent trail pheromone has been done only in the case of *Tetramorium caespitum*, in which a 30:70 mixture of 3-ethyl-2,5-dimethylpyrazine (**II**) and 2,5-dimethylpyrazine (**III**) constitute the synergistic mixture (Attygalle and Morgan, 1983, 1984). The trail-following activity of *Myrmica rubra* and the seven other reported species of *Myrmica* is exceptional because it is evoked only by the single substance 3-ethyl-2,5-dimethylpyrazine (**II**) (Evershed *et al.*, 1981, 1982). Table 2 summarizes all the chemical substances characterized as trail pheromone components in ants.

TABLE 2
Chemical substances identified in the trail pheromones of ants (Formicidae)

Compound	Species	Reference
Methyl 4-methylpyrrole-2-carboxylate (attalure; **I**)	*Atta cephalotes*	Riley *et al.* (1974)
	Atta texana	Tumlinson *et al.* (1971, 1972)
	Acromyrmex octospinosus	Robinson *et al.* (1974)
		Cross *et al.* (1982)
Hexanoic acid	*Lasius fuliginosus*	Huwyler *et al.* (1973, 1975)
Heptanoic acid	*L. fuliginosus*	Huwyler *et al.* (1973, 1975)
Nonanoic acid	*L. fuliginosus*	Huwyler *et al.* (1973, 1975)
Decanoic acid	*L. fuliginosus*	Huwyler *et al.* (1973, 1975)
3-Ethyl-2,5-dimethylpyrazine (**II**)	*Atta sexdens rubropilosa*	Cross *et al.* (1979)
	Atta sexdens sexdens	Evershed and Morgan (1983)
	Manica rubida	Attygalle *et al.* (1985).
	Myrmica sp.	Evershed *et al.* (1981, 1982)
	Tetramorium caespitum	Attygalle and Morgan (1983)
2,5-Dimethylpyrazine (**III**)	*T. caespitum*	Attygalle and Morgan (1983)

(*continued*)

TABLE 2 *(continued)*

Compound	Species	Reference
(+)-(3*S*,4*R*)-3,4,7,11-Tetramethyltrideca-6*E*,10*Z*-dienal (faranal; **VI**)[a]	*Monomorium pharaonis*	Ritter *et al.* (1977a, b)
5-Methyl-3-butyloctahydroindolizine	*M. pharaonis*	Ritter *et al.* (1973, 1977a)
trans-2-Pentyl-5-(5′-hexenyl)-pyrrolidine (monomorine III)	*M. pharaonis*	Ritter *et al.* (1973, 1977a)
(3*Z*,6*E*)-3,7,11-Trimethyldodeca-1,3,6,10-tetraene (*Z*,*E*-α-farnesene)	*Solenopsis invicta*	Vander Meer *et al.* (1981)
(3*E*,6*E*)-3,7,11-Trimethyldodeca-1,3,6,10-tetraene (*E*,*E*-α-farnesene)	*S. invicta*	Vander Meer *et al.* 1981)
(3*Z*,6*Z*)-3,4,7,11-Tetramethyldodeca-1,3,6,10-tetraene (*Z*,*Z*-homofarnesene)	*S. invicta*	Vander Meer *et al.* (1981)
(3*Z*,6*E*)-3,4,7,11-Tetramethyldodeca-1,3,6,10-tetraene (*Z*,*E*-homofarnesene)	*S. invicta*	Vander Meer *et al.* (1981)
(2*Z*,4*Z*,6*Z*)-3,7,11-Trimethyldodeca-2,4,6,10-tetraene (*Z*,*Z*,*Z*-allofarnesene)	*S. invicta*	Williams *et al.* (1981a, b)
Tetradecanoic acid[b]	*Pristomyrmex pungens*	Hayashi and Komae (1977)
Hexadecanoic acid[b]	*P. pungens*	Hayashi and Komae (1977)
Hexadecenoic acid[b]	*P. pungens*	Hayashi and Komae (1977)
Octadecanoic acid[b]	*P. pungens*	Hayashi and Komae (1977)
Octadecenoic acid[b]	*P. pungens*	Hayashi and Komae (1977)
Octadecadienoic acid[b]	*P. pungens*	Hayashi and Komae (1977)
Octadecatrienoic acid[b]	*P. pungens*	Hayashi and Komae (1977)
Eicosatetraenoic acid[b]	*P. pungens*	Hayashi and Komae (1977)
Eicosapentenoic acid[b]	*P. pungens*	Hayashi and Komae (1977)
Z-9-Hexadecenal	*Iridomyrmex humilis*	Cavill *et al.* (1979); Van Vorhis Key and Baker (1982a)

[a] Faranal is the true trail pheromone; monomorine I and III may act as attractants.
[b] Compound isolated from the insect, but activity response toward a synthetic sample not reported.

5.2 ACTIVITY OF CONGENERS

Pheromonal activity shown by structurally related compounds (or congeners, used in its chemical sense), isomers, and optical antipodes helps us in understanding the stereochemical requirements needed to manifest trail-following activity. Sonnet and Moser (1972, 1973) have conducted trail-following activity studies on *Atta texana* using several synthetic analogs of methyl 4-methylpyrrole-2-carboxylate (**I**). They recognized the importance of the 2,4-substitution pattern and the pyrrolic nitrogen. The substituent on position 4 cannot be removed but can change its size and polarity. For example, the methyl group at position 4 can be replaced by a chlorine atom (approximately of the same size) without the loss of activity. The substitution on position 2 cannot be enlarged but can be smaller with similar properties. For example, a carboethoxy group at position 2 is inactive but an acetyl group showed some activity. Caputo *et al.* (1979), by calculating the charge density on the pyrrolic nitrogen for a variety of analogs, have suggested that the receptor requires the charge to have a precise value, in a compound with the correct steric properties, to show chemorecognition.

A similar study has been done on *Tetramorium caespitum.* The 2,5-substitution on the pyrazine ring is important because 2,3- and 2,6-dimethylpyrazine were inactive (Attygalle and Morgan, 1984). Furthermore, when the ethyl group at position 3 of the pyrazine (**II**) was replaced by a methyl group, the compound could still show some activity.

On studies of faranal (**VI**), the trail pheromone of *Monomorium pharaonis,* Kobayashi *et al.* (1980) found that the 3 epimer (3*R*, 4*R*) also shows a weak pheromone activity when tested separately, though the preference of the ants to the (3*S*, 4*R*) enantiomer is unambiguous when tested in a choice test. The 3*R* enantiomer does not interfere with the activity of 3*S* since ants follow a trail made of a mixture. Koyama *et al.* (1983) have studied the importance of the 4*R* configuration. Their results show the geometry of the C-10 double bond is not important. The trail releasing activity of the cis isomer is comparable to the trans isomer. Furthermore it is not very important for the trail-releasing activity whether the substituent at C-11 is an ethyl or a methyl, in contrast to the 7-methyl which cannot be replaced by an ethyl group without losing the activity.

In the case of *Solenopsis invicta,* Williams *et al.* (1981b) synthesized the geometrical isomers of *Z,Z,Z*-allofarnesene and found all isomers with *Z*-4 configuration to show biological activity. On the other hand, Vander Meer (1983) has reported that the four unnatural isomers (*Z,Z*- and *E,Z*-α-farnesene and *E*- and *Z*-β-farnesene) were several thousand times less active than the two natural *Z,E*- and *E,E*-α-farnesenes when tested on *S. invicta.*

In *Iridomyrmex humilis* the geometry of the C-9 double bond of hexadecenal is important for the activity. (*E*)-9-Hexadecenal could evoke only insignificant trail following when compared to the natural isomer (*Z*)-9-hexadecenal (**XII**) (Van Vorhis Key and Baker, 1982b). Furthermore, the congeners, (*Z*)-7-tetradecenyl formate, (*E*)-7-tetradecenyl formate, and tetradecyl formate, also did not exhibit any significant activity.

We can conclude that some small changes in structure of a pheromone molecule can be tolerated and the altered substance is also active, but other or more drastic changes completely destroy activity. Furthermore, the presence of other nonactive components appear not to interfere or inhibit the activity of the active substance. It is frequently found among the Myrmicinae that one species will follow artificial trails made with the glandular contents of another species that happen to contain among its compounds the trail pheromone of the first, regardless whether that compound is active or inactive for the second. The study of the sensitivity of the receptors of the insect to the exact shape (or stereochemistry) of a pheromone molecule is called stereobiology.

6 Source and specificity of ant trail pheromones

Examinations of the behavior response of one ant species to an artificial trail made from the trail substances of another species may help in the identification of the trail pheromones and also in the understanding of their phylogeny. Such cross-examinations are called transposition studies. Initially, the trail pheromones were considered to be highly species specific, and the concept appeared incidentally to provide a simple method to distinguish one species from another within a given genus (Wilson and Pavan, 1959; Wilson, 1963). However this expectation was short lived because the existence of a wide variability in specificity in the trail-following behavior has become apparent recently. In fact, it is not surprising, as ants do not use a single specialized gland for the synthesis of the trail pheromones but a number of glands, which also have other functions (Table 1). It seems plausible that the chemical recruitment mechanism was derived from a more primitive tandem calling behavior (similar to that used by *Leptothorax*; Möglich *et al.*, 1974) by the conversion of a gland normally used for other functions into a social organ. Such an evolutionary process may be common to almost all subfamilies of ants, except perhaps the dolichoderine ants which utilize Pavan's gland, a gland with no other known function. Maschwitz (1975) has proposed a scheme by which the chemical recruitment system of ants has evolved. Tandem running is a very primitive recruitment method in which only one ant is recruited at a time. The recruited ant follows closely behind the leader ant. The chemicals that facilitate this behavior are called tandem running pheromones. They can be considered as chemical ancestors of the trail pheromones that bring mass recruitment. In the examples quoted here a gradation of behavior, from primitive tandem running to more sophisticated mass recruitment, can be seen and it is likely the evolution may have taken place on similar lines. Gabba and Pavan (1970) have reviewed the results from transposition studies up to 1970.

6.1 DOLICHODERINE ANTS

The trail pheromones of species of four dolichoderine genera, *Iridomyrmex*, *Tapinoma*, *Liometopum*, and *Monacis*, were demonstrated by transposition studies to be highly species specific (Wilson and Pavan, 1959). The trail pheromones of the dolichoderine ants are produced in the Pavan's gland and secreted via the posterior border of the sixth sternite. According to Blum (1974a) such a great specificity of dolichoderine trail pheromones may be possible because Pavan's gland, a unique organ for the synthesis and dispensing of trail pheromones, has been selected for the production of particular metabolic end products which serve as trail pheromones. Presumably, highly species-specific trail pheromones, similar to those found in the dolichoderine ants, would be much more difficult to achieve in those subfamilies that utilize trace constituents of the poison gland, the Dufour gland, or the hind gut as trail pheromones.

6.2 MYRMICINE ANTS

Three different glands, the poison gland, Dufour gland, or tibial gland, may serve as the source of the trail pheromones of myrmicine ants. The tibial glands and the elaborate trail-laying mechanisms appear to be unique only to the genus *Crematogaster* (Leuthold, 1968; Fletcher and Brand, 1968). In the genus *Leptothorax* a very primitive form of recruitment is observed. The recruiting workers invite nestmates to tandem following by extruding the sting and releasing a secretion from the poison gland, but the sting is not dragged over the surface as found in most species which lay chemical trails with sting gland secretion (Möglich, 1979). Given that trail-following behavior has arisen several times separately and the myrmicine ants are the largest and the most diverse group of ants, it is not surprising to find that they utilize a number of different glands for the production of their respective trail substances.

In contrast to dolichoderine ants, the trail substance of myrmicines investigated so far show a marked absence of species specificity in transposition studies conducted in the laboratory. However, field observations indicate that the natural trails are considerably more specific. The wide interest on leaf-cutting ants has provided a considerable amount of information about the variability of specificity in their trail-following behavior. The information available from transposition studies is summarized in Table 3. The trail-following behavior of some leaf-cutting ants toward two synthetic trail substances, methyl 4-methylpyrrole-2-carboxylate (**I**) and 3-ethyl-2,5-dimethylpyrazine (**II**), is summarized in Table 4. The workers of *Atta sexdens* followed synthetic trails of the pyrazine (**II**) only, not the pyrrole (**I**), and they

TABLE 3
Responses of some myrmicine ants to artificial trails laid from their poison glands[a]

	Test species						
Source species	*A. texana*	*A. cephalotes*	*A. sexdens*	*A. octospinosus*	*T. septentrionalis*	*S. urichi*	*D. armigerum*
Atta texana	+ + +[2,3]	+ + +[1]	+ + +[1]	+ + +[1]	+ + +[2,3]	0[3]	
Atta cephalotes	+ + +[1]	+ + +[1]	+ + +[1]	+ + +[1]			
Atta sexdens	+ + +[1]	+ + +[1]	+ + +[1]	+ + +[1]			
Acromyrmex octospinosus	+ + +[1]	+ + +[1]	0[1]	+ + +[1]			
Trachymyrmex septentrionalis	+ + +[2,3]				+ + +[2,3]	0[3]	
Sericomyrmex urichi	0?[3]				0?[3]	+ + +[3]	
Daceton armigerum	+ + +[3]	+ + +[3]	+ + +[3]	+ + +[3]	+ + +[3]	0[3]	0[3]

[a] Response code: 0, no trail-following activity; + + +, high trail-following activity. Reference code: 1, Robinson *et al.* (1974); 2, Blum and Ross (1965); 3, Blum and Portocarrero (1966).

did not follow trails made of the poison gland of *Acromyrmex*; therefore it can be expected that the pyrazine (**II**) is absent from the poison gland of *Acromyrmex*. The recent chemical investigation of the poison gland of *Acromyrmex octospinosus* by Evershed and Morgan (1983) demonstrated the absence of the pyrazine (**II**) at levels that would show activity. The poison glands of *Atta cephalotes*, *A. sexdens sexdens*, and *A. sexdens rubropilosa* contain both the pyrrole (**I**) and pyrazine (**II**) (Evershed and Morgan, 1983). Hence it becomes clear why *A. sexdens* would follow the poison gland trails of *A. cephalotes*, which also contain the pyrazine (**II**), but not a synthetic trail of the pyrrole (**I**) only. A number of nonattine ants showed no response to artificial trails of pyrrole (**I**) (Robinson *et al.*, 1974).

Sericomyrmex apparently differs from the genera representing the mainstream of attine ants, as it does not follow trails generated from the poison glands of *Atta texana* or *Trachymyrmex septentrionalis* (Table 3). Furthermore, the poison gland contents of *Sericomyrmex urichi* induces feeble or no trail-following behavior on *A. texana* or *T. septentrionalis* workers (Blum and Portocarrero, 1966). Hence the pheromone which releases the trail-following behavior in *S. urichi* appears species specific and is expected to be different from pyrrole (**I**) or pyrazine (**II**). An interesting discovery, which indicates how the poison gland substances have evolved to assume the secondary

TABLE 4
Responses of some myrmicine ants to artificial trails laid with three synthetic substances[a]

Test species	Test substance			Reference
	N H COOCH$_3$	N N	N N	
Atta texana	+ + +			1
Atta cephalotes	+ + +	+ +		1,2
Atta colombica	+ + +			1
Atta laevigata	+ +			1
Acromyrmex octospinosus	+ + +	+ +		1,2
Acromyrmex versicolor	+ + +			1
Atta sexdens sexdens	0	+ + +		1,2
Atta sexdens rubropilosa	0	+ + +		1,2
Trachymyrmex septentrionalis	+ + +			1
Trachymyrmex urichi	+ + +			1
Cyphomyrmex rimosus	+ + +			1
Apterostigma collare	+ + +			1
Myrmica sp.	0	+ + +	0	3
Tetramorium caespitum	0	+ +	+ + +	4
Manica rubida		+ + +	0	5

[a] Response code: 0, no trail-following activity; +, detected but not followed convincingly; + +, trail following by most ants, but a few with hesitation; + + +, natural following. References: 1, Robinson *et al.* (1974); 2, Cross *et al.* (1979); 3, Evershed *et al.* (1982); 4, Attygalle and Morgan (1983); 5, Attygalle *et al.* (1985).

function of trail releasing, was reported from a non-trail-laying myrmicine, *Daceton armigerum* (Blum and Portocarreo, 1966). A trail laid with the poison gland of *D. armigerum* was not followed by the workers of *D. armigerum* itself, but, surprisingly, it was strongly followed by *T. septentrionalis, A. texana, A. cephalotes,* and *Acromyrmex coronatus.* These results demonstrate that though the trail-following behavior is not evolved yet in the primitive *Daceton*, its venom contains pheromone components [most probably the pyrrole (**I**) and the pyrazine (**II**)] utilized by some other more advanced ant species as trail substances. Furthermore, *S. urichi* did not follow a trail made of *Daceton* venom, as one might expect (Blum and Portocarrero, 1966). Although Daceton does not show trail-following behavior, another species of Dacetini, *Orectognathus versicolor*, has been shown to lay recruitment trails during nest migration. Hölldobler (1981) was able to lay artificial trails with poison gland extracts of *O. versicolor*. Furthermore, the pygidial glands also appear to play a role in the trail communication behavior of *Orectognathus* (Hölldobler, 1981).

Similarly, although no one has verified it experimentally, one can expect *Solenopsis invicta* to follow Dufour gland extracts of *Myrmica rubra*, *Myrmica scabrinodis*, and several other related species of *Myrmica* because the major component of *S. invicta* trail pheromone, *Z*,*E*-α-farnesene (Vander Meer *et al.*, 1981), is found in the Dufour gland of these *Myrmica* species (Morgan and Wadhams, 1972; Morgan *et al.*, 1979; Attygalle and Morgan, 1982; Attygalle *et al.*, 1983). However, *Z*,*E*-α-farnesene does not evoke any trail-following activity on *Myrmica* ants, whose own trail substances originate from the poison gland.

Further evidence on the variability of specificity in the trail-following behavior of myrmicine ants, as summarized in Table 5, was found by Blum and Ross (1965) in their study on *Tetramorium*. The source of odor trail pheromone of *Tetramorium* ants is the poison gland and they do not follow the artificial trails laid with either the hind gut or the Dufour gland. The trails are completely species specific between *Tetramorium guineense* and *Tetramorium caespitum* (Table 5) but in contrast, *T. guineense* trails are well

TABLE 5
Responses of some myrmicine ants to artificial trails laid from the glandular sources of their odor trail pheromones[a]

	Test species					
Source species	*T. guineense*	*T. caespitum*	*A. texana*	*T. septentrionalis*	*S. saevissima*	*M. ruginodis*
Tetramorium guineense	+++[1]	0[1]	+++[1]	++[1]	0[1]	
Tetramorium caespitum	0[1]	+++[1,2]	0[1]	0[1]	0[1]	+++[2]
Atta texana	+++[1]	0[1]	+++[1]	+++[1]	0[1]	
Trachymyrmex septentrionalis	+++[1]	0[1]	+++[1]	+++[1]	0[1]	
Solenopsis saevissima	0[1]	0[1]	0[1]	0[1]	+++[1]	
Myrmica ruginodis		++[2]				+++[2]

[a] Response code: 0, no trail-following activity; +, detected but not followed convincingly; ++, trail following by most ants, but a few with hesitation; +++, natural following. References: 1, Blum and Ross (1965); 2, Attygalle and Morgan (1984).

followed by two species not so closely related, *A. texana* and *T. septentrionalis*, and vice versa. Furthermore, *T. caespitum* trails are not followed by the two attine species. Hence it appears that the trails of *T. guineense* may contain methyl 4-methylpyrrole-2-carboxylate (**I**). Recently, Attygalle and Morgan (1984) found that *T. caespitum* and *Myrmica ruginodis* would follow artificial trails made from each other's poison glands. This can be expected because of the presence of 3-ethyl-2,5-dimethylpyrazine (**II**) in both poison glands. However, *T. caespitum* will not follow a *M. ruginodis* trail as well as their own trail, because it needs the further component 2,5-dimethylpyrazine (III) for complete activity, which is absent in the poison gland of *M. ruginodis* (Attygalle and Morgan, 1984). Group recruitment in *T. caespitum* has been investigated by Verhaeghe (1977).

The least species specificity in the trail-following behavior is demonstrated by the genus *Myrmica*. With the exception of *Myrmica monticola*, 13 other *Myrmica* species almost equally follow trails made of each other's poison glands (Blum, 1974b; Evershed *et al*., 1982). It is interesting to note that *Veromessor pergandei*, *Pogonomyrmex badius*, and three species of *Manica* also follow each other's trails and those of *Myrmica* (Blum, 1974b). The poison gland extracts of *Aphaenogaster fulva*, *Novomessor cockerelli*, and *Veromessor pergandei* do not evoke any positive trail-following behavior in *Myrmica* or *Manica* workers.

The chemical nature of the trails of *Monomorium pharaonis* was first recognized by Sudd (1960). According to Hölldobler (1973), the trail pheromone of *Monomorium* originates from the Dufour gland, although Blum (1966) had previously reported the source as the poison gland. The trail of *Monomorium floricola*, *Moromorium minimum*, and *Monomorium striate* are strictly species specific, although the trail pheromone of *M. pharaonis* produced trail-following responses in workers of *M. minimum* as well as in workers of its own species (Hölldobler, 1973).

In a similar study on two species of *Novomessor*, Hölldobler *et al.* (1978) found that *N. albisetosus* also followed artificial trails laid with the poison gland extracts of *N. cockerelli*, but *N. cockerelli* did not follow the poison gland extracts of *N. albisetosus*. Neither species responded to poison gland trails of several species of *Pogonomyrmex*.

According to Barlin *et al.* (1976a), among the fire ants, *Solenopsis richteri* and *S. invicta* follow each other's artificial trails, laid separately from their Dufour gland extracts. But their Dufour gland contents were found to be different from each other on gas chromatographic examination. On the other hand *Solenopsis geminata* and *Solenopsis xyloni* appear to have a common trail pheromone as they follow each other's trails, and the Dufour gland contents are also similar to each other. The Dufour gland contents of *S.*

TABLE 6
Species specificity of trails laid with Dufour gland extracts of four species of *Solenopsis*[a]

	Test species			
Source species	*S. richteri*	*S. invicta*	*S. geminata*	*S. xyloni*
S. richteri	+ + +[2,3]	+ + +[2,3]	0[2]	+[2]
S. invicta	+ + +[2,3]	+ + +[1,2,3]	0[2]; + + +[1,3]	+[2]
S. geminata	+[2]	0[1,2]	+ + +[1,2,3]	+ + +[2,3]
S. xyloni	+ + +[2]; 0[3]	0[2]; + + +[1]	+ + +[1,2,3]	+ + +[1,2,3]

[a] Response code: 0, no trail-following activity; +, detected but not followed convincingly; + +, trail following by most ants, but a few with hesitation; + + +, natural following. References: 1, Wilson (1962a); 2, Barlin *et al.* (1976b); 3, Jouvenaz *et al.* (1978).

geminata produce virtually no response on the workers of *S. invicta* and *S. richteri.* The results are summarized in Table 6. These results are somewhat different from those reported previously by Wilson (1962a–c) on fire ant trails, but Jouvenaz *et al.* (1978) obtained similar results using purified whole ant extracts. Furthermore the Dufour gland contents of the dolichoderine ant *Monacis bispinosa* can release strong trail following activity in *S. invicta* (= *saevissima*) workers. However, *M. bispinosa* produces its own trail pheromone in the Pavan's gland (Wilson, 1962b). This indicates that trail releasing by the constituents of the Dufour gland has evolved as a secondary function. Hangartner (1969c) has demonstrated that workers of *Solenopsis* are able to adjust the amounts of their trail pheromone secretion according to the quality of the food source.

In the genus *Leptothorax* some specificity studies have been done on its recruitment pheromone from the poison gland (Möglich, 1979). The tandem followers of *L. acervorum* could be led with pheromone preparations from *L. muscorum* and from conspecifics, but not with poison gland secretions of *L. nylanderi.* Conversely, tandem followers of *L. nylanderi* (subgenus *Leptothorax*) do not respond to the poison gland secretions of *L. acervorum* and *L. muscorum* (both subgenus *Mycothorax*).

The trail pheromones of many myrmicine species are nitrogen-containing compounds from the poison gland. The raw materials of the proteinaceous myrmicine venom are amino acids, and the pheromone substances appear to be metabolic by-products of the amino acids. The pyrrole (**I**) can be seen as a derivative of leucine; the pyrazines (**II**) and (**III**) can be plausibly derived from threonine via aminoacetone (Morgan, 1984). In those species in which the venom consists of alkaloids, compounds of quite different biosynthetic origin in the Dufour gland appear to have been adopted as trail pheromones.

6.3 FORMICINE ANTS

The hind gut or the rectal gland is the source of the formicine trail pheromones. Hangartner (1967, 1969a) used artificial trails made of hind gut contents and demonstrated that the trail substance of *Lasius fuliginosus* did not evoke any response among other *Lasius* species, such as *L. emerginatus*, *L. niger*, and *L. flavus*. On the contrary, *L. fuliginosus* was able to follow the trails of *L. emerginatus* and *L. niger*. *Lasius flavus*, an underground species, produces no trail substance, or only a minimally effective one. The workers of *L. fuliginosus* were also found to follow the trails of *Formica rufibarbis* and *Formica rufa* (Hangartner, 1967).

In *Oecophylla longinoda*, Hölldobler and Wilson (1978) were able to dye the hind gut contents, but no traces of the dye was found in the recruitment trails laid by the ants. An invagination of the lower rear surface of the rectal sac, with a strongly developed glandular epithelium which they call the rectal gland, is considered to be involved in recruitment communication. Whether this glandular structure is found in other formicines remains to be investigated.

Another formicine, *Camponotus pennsylvanicus*, strongly followed trails prepared from the hind guts of *Camponotus americanus*. *Camponotus pennsylvanicus* and *C. americanus* belong to the same subgenus. It was also shown that the trail pheromones of six other species of *Camponotus* of other subgenera were partially specific (Barlin *et al*., 1976b). Wilson (1965b) has observed that the natural trails of the dolichoderine ant *Azteca chatifex* were followed by the formicine *Camponotus beebei*, but not vice versa.

The studies on *Camponotus socius* (Hölldobler, 1971b) have shown the hind gut material has no immediate effect as a recruitment signal. The pioneer workers lay a trail of hindgut contents from the food source to the nest, but the trail is followed by worker ants only if they are preexcited by a "waggle" display of the recruiting ant and kept stimulated by the poison gland contents of the leader ant. Formic acid could provide this stimulation. However, similar experiments with *Camponotus sericeus* (Hölldobler *et al*., 1974) could not release any trail-following behavior. Furthermore, in *C. pennsylvanicus* (Traniello, 1977) the mechanical stimulation was not essential and the workers would follow an artificial trail made from hindgut contents. Hartwick *et al*. (1977) also investigated the trail-laying behavior of *C. pennsylvanicus*. In *Formica fusca* (Möglich and Hölldobler, 1975) the trail pheromone from the hindgut alone does not elicit a recruitment effect.

However, once the ants are stimulated by the specific recruitment signals, they will follow the trail even in the absence of a recruiting leader ant.

6.4 ECITONINE ANTS

This subfamily of ants includes the New World army ants, long placed within Dorylinae. The trails of army ants are long lasting and could be followed by workers weeks after they had been laid. The source of the trail substances in ecitonine ants is either the hind gut or the pygidial glands. In addition, Chadab and Rettenmeyer (1975) and Topoff and Mirenda (1975) have demonstrated the involvement of other secretions in the organization of "mass recruitment" in *Eciton* and *Neivamyrmex*. The trails laid by four species of *Neivamyrmex* and by *Labidus coecus* were followed by all five species, but, in general, each species preferred its own trail when presented with a choice (Watkins, 1964; Watkins *et al.*, 1967). However, a fifth species, *Neivamyrmex pilosus*, would not follow any trail other than its own. Later, Torgerson and Akre (1970) showed, utilizing five species of *Eciton*, two of *Labidus*, and a single species each of *Neivamyrmex* and *Nomamyrmex*, that all possible combinations of specificity can be encountered in this subfamily. The studies of Topoff and Lawson (1979) demonstrated that tactile stimuli can facilitate the orientation of *Neivamyrmex nigrescens* workers to weak chemical trails. Furthermore, Topoff *et al.* (1980) have shown the recruitment trails of *N. nigrescens* are qualitatively different from their exploratory trails.

6.5 PONERINE ANTS

Although the hind gut was long considered to be the source of ponerine trails, recent discoveries show trail substances can originate from the pygidial glands (Hölldobler and Engel, 1978; Hölldobler and Traniello, 1980a,b), the poison gland (Fletcher, 1971), or the newly discovered sternal gland (Hölldobler *et al.*, 1982). In *Pachycondyla obscuricornis* a very primitive form of recruitment behavior is observed. This tandem running recruitment mechanism is mediated by a pheromone from the pygidial glands. The pheromone can be extracted into solvents such as ether and acetone but nothing is known about the chemistry (Hölldobler and Traniello, 1980b). Very little information is available on transposition studies of the trails of ponerine ants. The three species of *Onychomyrmex* studied by Hölldobler *et al.*, (1982) demonstrated a partial species specificity.

6.6 ANEURETINE ANTS

This primitive subfamily is represented by the single genus *Aneuretus*, confined to limited regions of the tropical rain forests of Sri Lanka (Wilson *et al.*, 1956; Jayasuriya, 1980). Traniello and Jayasuriya (1981a,b) have described the recruitment behavior of *A. simoni*. The trail substances originate from the sternal gland and the extracts made from other glands were inactive (Traniello and Jayasuriya, 1981a). No information is available on the chemistry or transposition studies.

7 Conclusions

Studies of pheromones of other insect orders have led us to expect a high degree of species specificity in these compounds and mixtures. Indeed, in studies of other pheromones of ants (from the mandibular and Dufour glands), highly specific mixtures of chemicals appear to be encountered for each species. But with the exception of dolichoderines, trail pheromones of ants appear to be rarely species specific, and sometimes a single compound is used by a large number of species, not even closely related. This leads one to predict that the trail pheromone may be used in conjunction with other pheromones. There is evidence for *Myrmica* species, at least, that the unspecific trail pheromones are used with other, specific pheromones in recruitment and food gathering (Morgan, 1984).

Some ant species are important economic pests. Next to grasshoppers and locusts, leaf-cutting ants are said to be the group causing the greatest damage to world agriculture. Because of their largely subterranean existence, they are very difficult to control and the most effective method is the use of bait containing the insecticide Mirex. Fire ants (*Solenopsis* sp.) are an important economic pest and a public health hazard, at least in the southern United States (Apperson and Powell, 1983; Joyce, 1983). Their venom exhibits pronounced necrotic (Buffkin and Russel, 1972) and hemolytic activities (Adrouny *et al.*, 1959). Toxic baits have been used since 1957 for their control (Lofgren *et al.*, 1975), and Mirex was the most effective toxicant (Lofgren *et al.*, 1964). However, the ill effects due to indiscriminate use of aerial spraying of insecticides for fire ant eradication soon became evident. Rachel Carson ("Silent Spring," 1962) severely criticized their use for this purpose. The production and use of Mirex have been banned in the United States since 1975 because of environmental pollution and suspected carcinogenicity, but many states are seeking to introduce it because the fire ants are continuing to spread (Press, 1982).

In order to avoid environmental damage caused by persistant pesticides, novel and more effective insect control methods are being sought. The use of semiochemicals (chemical messengers) to manipulate and control the behavior of insects has gained popularity as a more acceptable means of insect control (Ritter and Persoons, 1976; Ritter *et al.*, 1977c). Therefore the investigations on ant trail pheromones not only can help in chemosystematics and taxonomy, but also can show ways to disrupt communication codes which will eventually be useful in integrated pest control (Ritter *et al.*, 1977c).

Initial laboratory experiments showed that bait particles were more easily found by *Atta sexdens* workers if the bait contained between 0.8 and 80 ng of methyl 4-methylpyrrole-2-carboxylate (**I**) (Robinson and Cherrett, 1973). Further laboratory tests by Robinson and Cherrett (1978) demonstrated that addition of pyrrole (**I**) to baits only made it easier for the ants to find them but did not increase the likelihood of the baits being picked up. Jaffe and Howse (1979) have demonstrated that although pyrrole (I) recruits ants, it cannot reproduce all aspects of natural recruitment. The field trials in Trinidad, Brazil, and Paraguay too were disappointing (Robinson *et al.*, 1982). The failures may be partly due to incomplete identification of the trail pheromone and the communication system. However, a trail pheromone need not necessarily induce bait pick-up. It is evident that a closer understanding of their communication system is important before further attempts are made to use the trail pheromones in integrated pest control.

Acknowledgments

The authors thank Miss Miho Yamakawa for her invaluable assistance in the preparation of this article.

References

Adrouny, G. A., Derbes, V. J., and Jung, R. C. (1959). Isolation of a hemolytic component of fire ant venom. *Science* **130**, 449.

Apperson, C. S., and Powell, E. E. (1983). Correlation of the red imported fire ants with reduced soybean yields in North Carolina. *J. Econ. Entomol.* **76**, 259–263.

Attygalle, A. B., and Morgan, E. D. (1982). Structures of homofarnesene and bishomofarnesene from *Myrmica* ants. *J. Chem. Soc. Perkin Trans.* **1**, 949–951.

Attygalle, A. B., and Morgan, E. D. (1983). Trail pheromone of the ant *Tetramrium caespitum* L. *Naturwissenschaften* **70**, 364–365.

Attygalle, A. B., and Morgan, E. D. (1984). Identification of the trail pheromone of the ant *Tetramorium caespitum* (Hymenoptera: Myrmicinae). *J. Chem. Ecol.* **10**, 1453–1468.

Attygalle, A. B., Evershed, R. P., Morgan, E. D., and Cammaerts, M. C. (1983). Dufour gland secretions of workers of the ants *Myrmica sulcinodis* and *Myrmica lobicornis* and comparison with six other species of *Myrmica*. *Insect Biochem.* **13**, 507–512.

Attygalle, A. B., Lancaster, V. K., and Morgan, E. D. (1985). The trail pheromone of the ant *Manica rubida*. *Actes Coll. Insectes Soc.* **2**, 159—166.

Baker, R., Billington, D. C., and Ekanayake, N. (1981). Stereoselective synthesis of (3*R*, 4*S*/3*S*, 4*R*)-(6*E*, 10*Z*)-3,4,7,11-tetramethyltrideca-6,10-dienal (faranal); the trail pheromone of the Pharaoh's ant. *J. Chem. Soc. Chem. Commun*, 1234–1235.

Baker, R., Billington, D. C., and Ekanayake, N. (1983). Stereoselective total synthesis of recemic (3*S*, 4*R*/3*R*, 4*S*)- and a diastereomeric mixture of (6*E*, 10*Z*)-3,4,7,11-tetramethyltrideca-6,10-dienal (faranal) the trail pheromone of the Pharaoh's ant. *J. Chem. Soc. Perkin Trans.* **1**, 1387–1393.

Barlin, M. R., Blum, M. S., and Brand, J. M. (1976a). Fire ant trail pheromones; analysis of species specificity after gas chromatographic fractionation. *J. Insect Physiol.* **22**, 839–844.

Barlin, M. R., Blum, M. S., and Brand, J. M. (1976b). Species-specificity studies on the trail pheromone of the carpenter ant, *Camponotus pennsylvanicus* (Hymenoptera: Formicidae). *J. Georgia Entomol. Soc.* **11**, 162–164.

Blum, M. S. (1966). The source and specificity of trail pheromones in *Termitopone*, *Monomorium* and *Huberia* and their relation to those of some other ants. *Proc. R. Entomol. Soc. London* 155–160.

Blum, M. S. (1974a). Pheromonal sociality in the Hymenoptera. *In* "Pheromones"(M. C. Birch, ed.), pp. 222–249. Elseviar, Amsterdam.

Blum, M. S. (1974b). Myrmicine trail pheromones; specificity, source and significance. *J. N. Y. Entomol. Soc.* **82**, 141–147.

Blum, M. S., and Brand, J. M. (1972). Social insect pheromones: Their chemistry and function. *Am. Zool.* **12**, 553–576.

Blum, M. S., and Portocarrero, C. A. (1964). Chemical releasers of social behavior. IV. The hindgut as the source of the odor trail pheromone in the neotropical army ant genus, *Eciton*. *Ann. Entomol. Soc. Am.* **57**, 793–794.

Blum, M. S., and Portocarrero, C. A. (1966). Chemical releasers of social behavior. X. An Attine trail substance in the venom of a non-trail laying myrmicine *Daceton armigerum* (Latreille). *Psyche* **73**, 150–155.

Blum, M. S., and Ross, G. N. (1965). Chemical releasers of social behavior. V. Source, specificity and properties of the odour trail pheromone of *Tetramorium guineense* (F.) (Formicidae: Myrmicinae). *J. Insect Physiol.* **11**, 857–868.

Blum, M. S., and Wilson, E. O. (1964). The anatomical source of trail substances in formicine ants. *Psyche* **71**, 28–31.

Blum, M. S., Moser, J. C., and Cordero, A. D. (1964). Chemical releasers of social behavior. II. Source and specificity of the odour trail substance in four Attine genera (Hymenoptera: Formicidae). *Psyche* **71**, 1–7.

Bonnet, C. (1779). "Observation XLIII. Sur un Procédé des Fourmis," Vol. 1, pp. 535–536 (1779–1783). Oeuvres d'Histoire Naturelle et de Philosophie, Neuchatel.

Buffkin, D. C., and Russell, F. F. (1972). Some chemical and pharmacological properties of the venom of the imported fire ant, *Solenopsis saevissima richteri*. *Toxicon* **10**, 526.

Butler, C. G., Fletcher, D. J. C., and Walter, D. (1969). Nest-entrance marking with pheromones by the honey bee *Apis mellifera* L. and by a wasp *Vespula vulgaris* L. *Anim. Behav.* **17**, 142–147.

Cammaerts-Tricot, M. C. (1974a). Recrutement d'ouvrières chez *Myrmica rubra*, les phéromones de l'appareil à venin. *Behavior* **50**, 111–122.

Cammaerts-Tricot, M. C. (1974b). Piste et phèromone attractive chez la fourmi *Myrmica rubra*. *J. Comp. Physiol.* **88**, 373–382.

Cammaerts-Tricot, M. C., Morgan, E. D., and Tyler, R. C. (1977). Isolation of the trail pheromone of the ant *Myrmica rubra*. *J. Insect Physiol.* **23**, 421–427.

Caputo, J. F., Caputo, R. E., and Brand, J. M. (1979). Significance of the pyrrolic nitrogen atom in receptor recognition of *Atta texana* (Buckley) (Hymenoptera: Formicidae) trail pheromone and parapheromones. *J. Chem. Ecol.* **5**, 273–278.

Carson, R. (1962). "Silent Spring," pp. 147–153. Penguin, London.

Carthy, J. D. (1950). Odour trails of *Acanthomyops fuliginosus*. *Nature* (*London*) **166**, 154.

Carthy, J. D. (1951a). The orientation of two allied species of British ant. I. Visual direction finding in *Acanthomyops* (*Lasius*) *niger*. *Behavior* **3**, 275–303.

Carthy, J. D. (1951b). The orientation of two allied species of British ant. II. Odour trail laying and following in *Acanthomyops* (*Lasius*) *fuliginosus*. *Behavior* **3**, 304–318.

Carthy, J. D. (1952). The return of ants to their nest. *Int. Congr. Entomol. 9th, Trans.* **1**, 365–369.

Cavill, G. W. K., Robertson, P. L., and Davies, N. W. (1979). An Argentine ant aggregation factor. *Experientia* **35**, 989–990.

Cavill, G. W. K., Davies, N. W., and McDonald, F. S. (1980). Characterization of aggregation factors and associated compounds from the Argentine ant *Iridomyrmex humilis*. *J. Chem. Ecol.* **6**, 371–384.

Chadab, R., and Rettenmeyer, C. (1975). Mass recruitment by army ant. *Science* **188**, 1124–1125.

Cross, J. H., Byler, R. C., Ravid, U., Silverstein, R. M., Robinson, S. W., Baker, P. M., De Oliveira, J. S., Jutsum, A. R., and Cherrett, J. M. (1979). The major component of the trail pheromone of the leaf-cutting ant, *Atta sexdens rubropilosa* Forel. *J. Chem. Ecol.* **5**, 187–203.

Cross, J. H., West, J. R., Silverstein, R. M., Jutsum, A. R., and Cherrett, J. M. (1982). Trail pheromone of the leaf-cutting ant *Acromyrmex octospinosus* (Reich) (Formicidae: Myrmicinae). *J. Chem. Ecol.* **8**, 1119–1124.

Dumpert, K. (1981). "The Social Biology of Ants." Pitman, London.

Eidmann, H. (1927). Die Sprache der Ameisen. *Rev. Zool. Russe* **7**, 39–47.

Evershed, R. P., and Morgan, E. D. (1983). The amounts of trail pheromone substances in the venom of workers of four species of Attine ants. *Insect Biochem.* **13**, 469–474.

Evershed, R. P., Morgan, E. D., and Cammaerts, M. C. (1981). Identification of the trail pheromone of the ant *Myrmica rubra* L., and related species. *Naturwissenschaften* **67**, 374–375.

Evershed, R. P., Morgan, E. D., and Cammaerts, M. C. (1982). 3-Ethyl-2,5-dimethylpyrazine, the trail pheromone from the venom gland of eight species of *Myrmica* ants. *Insect Biochem.* **12**, 383–391.

Fletcher, J. (1971). The glandular source and social functions of trail pheromones in two species of ants. (*Leptogenys*). *J. Entomol.* **A46**, 27–37.

Fletcher, D. J. C., and Brand, J. M. (1968). Source of the trail pheromone and method of trail laying in the ant *Crematogaster peringueyi*. *J. Insect Physiol.* **14**, 783–788.

Forel, A. (1886). Études myrmecologiques en 1886. *Ann. Soc. Entomol. Belg.* **30**, 131–215.

Forel, A. (1908). "The Senses of Insects." Methuen, London.

Gabba, A., and Pavan, M. (1970). Researchers on trail and alarm substances in ants. *In* "Communication by Chemical Signals" (J. W. Johnston, D. G. Moulton, and A. Turk, eds.), pp. 161–204. Appleton, New York.

Goetsch, W. (1934). Untersuchungen über die Zusammenarbeit im Ameisenstaat. *Z. Morphol. Oekol. Tiere* **28**, 319–401.

Hangartner, W. (1967). Spezifität und Inaktivierung des Spurpheromones von *Lasius fuliginosus* Latr. und Orientierung der Arbeiterinnen im Duftfeld. *Z. Vergl. Physiol.* **57**, 103–106.

Hangartner, W. (1969a). Trail laying in the subterranean ant, *Acanthomyops interjectus*. *J. Insect Physiol.* **15**, 1–4.

Hangartner, W. (1969b). Orientierung von *Lasius fuliginosus* an einer Gabelung der Geruchspur. *Insectes Soc.* **16**, 55–60.

Hangartner, W. (1969c). Structure and variability of the individual odor trail in *Solenopsis* (Formicidae). *Z. Vergl. Physiol.* **62**, 111–120.

Hangartner, W. (1970). Control of pheromone quantity in odour trails of the ant *Acanthomyops interjectus* Mayr. *Experientia* **26**, 664–665.

Hangartner, W., and Bernstein, S. (1964). Über die Geruchsspur von *Lasius fuliginosus* zwischen Nest und Futterquelle. *Experientia* **20**, 392–393.

Hartwick, E. B., Friend, W. G., and Atwood, C. E. (1977). Trail-laying behaviour of the carpenter ant, *Camponotus pennsylvanicus* (Hymenoptera: Formicidae). *Can. Entomol.* **109**, 129–136.

Hayashi, N., and Komae, H. (1977). The trial and alarm pheromones of the ant, *Pristomyrmex pungens* Mayr. *Experientia* **33**, 424–425.

Hingston, R. W. G. (1928). "Problems of Instinct and Intelligence." Arnold, London.

Hölldobler, B. (1971a). Homing in the Harvester ant *Pogonomyrmex badius. Science* **171**, 1149–1151.

Hölldobler, B. (1971b). Recruitment behavior in *Camponotus socius* (Hym. Formicidae). *Z. Vergl. Physiol.* **75**, 123–142.

Hölldobler, B. (1973). Chemische Strategie beim Nahrungserwerb der Diebameise (*Solenopsis fugax* Latr.) und der Pharaoameise (*Monomorium pharaonis* L.) *Oecologia* **11**, 371-380.

Hölldobler, B. (1974). Home range orientation and territoriality in harvesting ants. *Proc. Natl. Acad. Sci. U.S.A.* **71**, 3274–3277.

Hölldobler, B. (1977). Communication in social Hymenoptera. *In* "How Animals Communicate" (T. A. Sebeok, ed.), pp. 418–471. Indiana Univ. Press, Bloomington, Indiana.

Hölldobler, B. (1978). Ethological aspects of chemical communication in ants. *Adv. Study Behav.* **8**, 75–115.

Hölldobler, B. (1981). Trail communication in the Dacetine ant *Orectognathus versicolor* (Hymenoptera: Formicidae). *Psyche* **88**, 245–257.

Hölldobler, B. (1982). Chemical communication in ants: New exocrine glands and their behavioral function. The biology of social insects. *Proc. Int. Congr. IUSSI, 9th*, pp. 312–317.

Hölldobler, B., and Engel, H. (1978). Tergel and sternal glands in ants. *Psyche* **85**, 285–330.

Hölldobler, B., and Traniello, J. F. A. (1980a). The pygidial gland and chemical recruitment communication in *Pachycondyla* (= *Termitopone*) *laevigata. J. Chem. Ecol.* **6**, 883–893.

Hölldobler, B., and Traniello, J. (1980b). Tandem running pheromone in ponerine ants. *Naturwissenschaften* **67**, 360.

Hölldobler, B., and Wilson, E. O. (1970). Recruitment trails in the harvester ant *Pogonomyrmex badius. Psyche* **77**, 385–399.

Hölldobler, B., and Wilson, E. O. (1978). The multiple recruitment systems of the African weaver ant *Oecophylla longinoda* (Latreille) (Hymenoptera: Formicidae). *Behav. Ecol. Sociobiol.* **3**, 19–60.

Hölldobler, B., Moglich, M., and Maschwitz, U. (1974). Communication by tandem running in the ant *Camponotus sericeus. J. Comp. Physiol.* **90**, 105–127.

Hölldobler, B., Stanton, R. C., and Markl, H. (1978). Recruitment and food-retrieving behavior in *Novomessor* (Formicidae, Hymenoptera). *Behav. Ecol. Sociobiol.* **4**, 163–181.

Hölldobler, B., Engel, H., and Taylor, R. W. (1982). A new sternal gland in ants and its function in chemical communication. *Naturwissenschaften* **69**, 90-91.

Huwyler, S., Grob, K., and Viscontini, M. (1973). Identifizierung von sechs Komponenten des Spurpheromons der Ameisenart *Lasius fuliginosus. Helv. Chim. Acta* **56**, 976–977.

Huwyler, S., Grob, K., and Viscontini, M. (1975). The trail pheromone of the ant *Lasius fuliginosus*: Identification of six components. *J. Insect Physiol.* **21**, 299–304.

Jaffe, K., and Howse, P. E. (1979). The mass recruitment system of the leaf-cutting ant *Atta cephalotes. Anim. Behav.* **27**, 930–939.

Jayasuriya, A. K. (1980). The behavior and ecology of *Aneuretus simoni* Emery. B.A. thesis, Harvard University, Cambridge, Massachusetts.

Jouvenaz, D. P., Lofgren, C. S., Carlson, D. A., and Banks, W. A. (1978). Specificity of the trail pheromone of four species of fire ant, *Solenopsis* spp. *Fl. Entomol.* **61**, 244.

Joyce, J. R. (1983). Multifocal ulcerative keratoconjunctivites as a result of sting by imported fire ant. *Vet. Med. Small Anim. Clin.* **78**, 1107–1108.

Knight, D. W., and Ojhara, B. (1981). A total synthesis of ±-faranal, the true trail pheromone of Pharaoh's ant, *Monomorium pharaonis. Tetrahedron Lett.* **22**, 5101–5104.

Knight, D. W., and Ojhara, B. (1983). A total synthesis of (±)-faranal, the true trail pheromone of Pharaoh's ant, *Monomorium pharaonis. J. Chem. Soc. Perkin Trans* **1**, 955–960.

Kobayashi, M., Koyama, T., Ogura, K., Seto, S., Ritter, F. J., and Bruggemann-Rotgans, I. E. M. (1980). Bioorganic synthesis and absolute configuration of faranal. *J. Am. Chem. Soc.* **102**, 6602–6604.

Koyama, T., Matsubara, M., Ogura, K., Bruggemann, I. E. M., and Vrielink, A. (1983). Congeners of faranal and their trail pheromone activity. *Naturwissenshaften* **70**, 469–470.

Leuthold, R. H. (1968). A tibial gland scent-trail and trail-laying behavior in the ant *Crematogaster ashmeidi* Mayr. *Psyche* **75**, 233–248.

Lindauer, M., and Kerr, W. E. (1958). Die gegenseitige Verständigung bei den stachellosen Bienen. *Z. Vergl. Physiol.* **41**, 405–434.

Lofgren, C. S., Bartlet, F. J., Stringer, C. E., and Banks, W. A. (1964). Imported fire ant toxic bait studies: Further tests with granulated mirex-soybean oil bait. *J. Ecol. Entomol.* **57**, 695-698.

Lofgren, C. S., Banks, W. A., and Glancey, B. M. (1975). Biology and control of imported fire ant. *Annu. Rev. Entomol.* **20**, 1–30.

Longhurst, C., Baker, R., and Howse, P. E. (1979). Termite predation by *Megaponera foetens* (FAB) (Hymenoptera: Formicidae). *J. Chem. Ecol.* **5**, 703–719.

McGregor, E. G. (1948). Odours as a basis for oriented movement in ants. *Behavior* **1**, 267–296.

Maschwitz, U. (1975). Old and new trends in the investigation of chemical recruitment in ants. *In* "Pheromones and Defensive Secretions in Social Insects" (Ch. Noirot, P. E. Howse, and G. Le Masne, eds.), pp. 47–59. University of Dijon Press, Dijon.

Maschwitz, U., and Schönegge, P. (1977). Recruitment gland of *Leptogenys chinensis*. *Naturwissenschaften* **64**, 589–590.

Möglich, M. (1979). Tandem calling pheromone in the genus *Leptothorax* (Hymenoptera: Formicidae). Behavioral analysis of specificity. *J. Chem. Ecol.* **5**, 35–52.

Möglich, M., and Hölldobler, B. (1975). Communication and orientation during foraging and emigration in the ant *Formica fusca*. *J. Comp. Physiol.* **101**, 275–288.

Möglich, M., Maschwitz, U., and Hölldobler, B. (1974). Tandem calling: A new kind of signal in ant communication. *Science* **186**, 1046–1074.

Morgan, E. D. (1984). Chemical words and phrases in the language of pheromone for foraging and recruitment. *In* "Insect Communication" (T. Lewis, ed.), pp. 169–194. Academic Press, New York.

Morgan, E. D., and Wadhams, L. J. (1972). Chemical constituents of Dufour's gland in the ant, *Myrmica rubra*. *J. Insect Physiol.* **18**, 1125–1135.

Morgan, E. D., Parry, K., and Tyler, R. C. (1979). The chemical composition of the Dufour gland secretion of the ant *Myrmica scabrinodis*. *Insect Biochem.* **9**, 117–121.

Mori, K., and Ueda, H. (1981). Synthesis of optically active forms of faranal, the trail pheromone of Pharaoh's ant. *Tetrahedron Lett.* **22**, 461–464.

Mori, K., and Ueda, H. (1982). Synthesis of optically active forms of faranal, the trail pheromone of Pharaoh's ant. *Tetrahedron* **38**, 1227–1233.

Moser, J. C., and Blum, M. S.. (1963). Trail marking substance of the Texas leaf-cutting ant: Source and potency. *Science* **140**, 1228.

Moser, J. C., and Silverstein, R. M. (1967). Volatility of trail marking substance of the town ant. *Nature (London)* **215**, 206–207.

Parry, K., and Morgan, E. D. (1979). Pheromones of ants: A review. *Physiol. Entomol.* **4**, 161–189.

Pasteels, J. M., and Verhaeghe, J. C. (1974). Dosage biologique de la phèromone de piste chez les fourrageures et les reines de *Myrmica rubra*. *Insectes Soc.* **21**, 167–180.

Press, R. M. (1982). Battle over fire ant. *Christian Science Monitor*, 15th Dec., p. 5.

Riley, R. G., Silverstein, R. M., Carroll, B., and Carroll, R. (1974). Methyl 4-methylpyrrole-2-carboxylate, a volatile trail pheromone from the leaf cutting ant *Atta cephalotes*. *J. Insect Physiol.* **20**, 651–654.

Ritter, F. J., and Persoons, C. J. (1976). Insect pheromones as a basis for the development of more effective selective pest control agents. *Med. Chem.* **7**, 59–114.

Ritter, F. J., and Persoons, C. J. (1977). Trail pheromones and related compounds in termites and ants. *Proc. Int. Congr. IUSSI, 8th,* pp. 34–38.

Ritter, F. J., Rotgans, I. E. M., Talman, E., Verwiel, P. E. J., and Stein, F. (1973). 5-Methyl-3-butyloctahydroindolizine, a novel type of pheromone attractive to Pharaoh's ant. *Monomorium pharaonis. Experientia* **29**, 530–531.

Ritter, F. J., Brüggemann-Rotgans, I. E. M., Verkuil, E., and Persoons, C. J. (1975). *In* "Pheromones and Defensive Secretions in Social Insects" (Ch. Noirot, P. E. Howse, and G. Le Masne, eds.), pp. 99–103. Univ. of Dijon Press, Dijon.

Ritter, F. J., Brüggemann-Rotgans, I. E. M., Verwiel, P. E. J., Talman, E., Stein, F., La Brijn, J., and Persoons, C. J. (1977a). Faranal, a trail pheromone from the Dufour's gland of the Pharaoh's ant, structurally related to juvenile hormone. *Proc. Int. Congr. IUSSI, 8th,* pp. 41–43.

Ritter, F. J., Brüggemann-Rotgans, I. E. M., Verwiel, P. E. J., Persoons, C. J., and Talman, E. (1977b). Trail pheromone of the Pharaoh's ant *Monomorium pharaonis*; isolation and identification of faranal, a terpenoid related to juvenile hormone II. *Tetrahedron Lett.* 2617-2618

Ritter, F. J., Brüggemann-Rotgans, I. E. M., Persoons, C. J., Talman, E., van Osten, A. M., and Verwiel, P. E. J. (1977c). Evaluation of social insect pheromones in pest control with special reference to subterranean termites and Pharaoh's ants. *In* "Crop Protection Agent: Their Biological Evaluation" (N. R. McFarlane, ed.), pp. 195–216. Academic Press, New York.

Robertson, P. L., Dudzinski, M. L., and Orton, C. J. (1980). Exocrine gland involvement in trailing behavior in the Argentine ant (Formicidae: Dolichoderine). *Anim. Behav.* **28**, 1255–1273.

Robinson, S. W., and Cherrett, J. M. (1973). Studies on the use of leaf-cutting ant trail pheromones as attractant in bait. *Proc. Int. Congr. IUSSI, 7th,* pp. 332–338.

Robinson, S. W., and Cherrett, J. M. (1978). The possible use of methyl 4-methylpyrrole-2-carboxylate, an ant trail pheromone, as a component of an improved bait for leaf-cutting ant (Hymenoptera: Formicidae) control. *Bull. Entomol. Res.* **68**, 159–170.

Robinson, S. W., Moser, J. C., Blum, M. S., and Amante, E. (1974). Laboratory investigations of the trail-following responses of four species of leaf-cutting ants with notes on the specificity of a trail pheromone of *Atta texana* (Buckley). *Insectes Soc.* **21**, 87–94.

Robinson, S. W., Jutsum, A. R., Cherrett, J. M., and Quinlan, R. J. (1982). Field evaluation of methyl 4-methylpyrrole-2-carboxylate, an ant trail pheromone, as a component of baits for leaf-cutting ant (Hymenoptera: Formicidae) control. *Bull. Entomol. Res.* **72**, 345–356.

Santschi, F. (1923). Les différentes orientations chez les fourmis. *Rev. Zool. Afr.* **11**, 111–143.

Silverstein, R. M., and Young, J. C. (1976). Insects generally use multicomponent pheromones. *ACS Symp. Ser.* (23), 1–29.

Sonnet, P. E. (1972). Synthesis of the trail marker of the Texas leaf-cutting ant *Atta texana* (Buckley). *J. Med. Chem.* **15**, 97–98.

Sonnet, P. E., and Moser, J. C. (1972). Synthetic analogs of the trail pheromone of the leaf-cutting ant, *Atta texana* (Buckley). *Agric. Food Chem.* **20**, 1191–1194.

Sonnet, P. E., and Moser, J. C. (1973). Trail pheromone: Responses of the Texas leafcutting ant, *Atta texana* to selected halo- and cyanopyrrole-2-aldehyde, ketones, and esters. *Environ. Entomol.* **2**, 851–854.

Sonnet, P. E., and Oliver, J. E. (1975). Synthesis of insect trail pheromones: The isomeric 3-butyl-5-methyloctahydroindolizines. *J. Heterocycl. Chem.* **12**, 289–294.

Sudd, J. H. (1959). Interaction between ants on a scent trail. *Nature* (*London*) **183**, 1588.

Sudd, J. H. (1960). The foraging method of Pharaoh's ants, *Monomorium pharaonis* (L.) *Anim. Behav.* **8**, 67–75.

Topoff, H., and Lawson, K. (1979). Orientation of the army ant *Neivamyrmex nigrescens*, integration of chemical and tactile information. *Anin. Behav.* **27**, 429–433.

Topoff, H., and Mirenda, J. (1975). Trail-following by the army ant *Neivamyrmex nigrescens*: Responses by workers to volatile odors. *Ann. Entomol Soc. Am.* **68**, 1044–1046.

Topoff, H., Boshes, M., and Trakimas, W. (1972). A comparison of trail following between callow and adult workers of the army ant, *Neivamyrmex nigrescens* (Formicidae). *Anim. Behav.* **20**, 361–366.

Topoff, H., Mirenda, J., Droual, R., and Henrick, S. (1980). Behavioural ecology of mass recruitment in the army ant *Neivamyrmex nigrescens. Anim. Behav.* **28**, 779–789.

Torgerson, R. L., and Akre, R. D. (1970). The persistence of army ant chemical trails and their significance in the Ecitonine-Ecitophile association (Formicidae: Ecitonini). *Melanderia* **5**, 1–28.

Traniello, J. F. A. (1977). Recruitment behavior, orientation and the organization of foraging in the carpenter ant *Camponotus pennsylvanicus* De Geer (Hymenoptera: Formicidae). *Behav. Ecol. Sociobiol.* **2**, 61–79.

Traniello, J., and Jayasuriya, A. (1981a). The sternal gland and recruitment communication in the primitive ant *Aneuretus simoni. Experientia* **37**, 46–47.

Traniello, J. F. A., and Jayasuriya, A. K. (1981b). Chemical communication in the primitive ant *Aneuretus simoni*: The role of the sternal and pygidial glands. *J. Chem. Ecol.* **7**, 1023–1033.

Tumlinson, J. H., Silverstein, R. M., Moser, J. C., Brownlee, R. G., and Ruth, J. M. (1971). Identification of the trail pheromone of a leaf-cutting ant, *Atta texana. Nature (London)* **234**, 348–349.

Tumlinson, J. H., Moser, J. C., Silverstein, R. M., Brownlee, R. G., and Ruth, J. M. (1972). A volatile trail pheromone of the leaf-cutting ant, *Atta texana. J. Insect Physiol.* **18**, 809–814.

Vander Meer, R. K. (1983). Semiochemicals and the red imported fire ant (*Solenopsis invicta* Buren) (Hymenoptera; Formicidae). *Fl. Entomol.* **66**, 139–161.

Vander Meer, R. K., Williams, F. D., and Lofgren, C. S. (1981). Hydrocarbon components of the trail pheromone of the red imported fire ant, *Solenopsis invicta. Tetrahedron Lett.* 1651–1654.

Van Vorhis Key S. E., and Baker, T. C. (1982a). Trail-following responses of the Argentine ant, *Iridomyrmex humilis* (Mayr), to a synthetic trail pheromone component and analogs. *J. Chem. Ecol.* **8**, 3–14.

Van Vorhis Key, S. E., and Baker, T. C. (1982b). Specificity of laboratory trail-following by Argentine ant, *Iridomyrmex humilis* (Mayr) to (*Z*)-9-hexadecenal, analogs and gaster extracts. *J. Chem. Ecol.* **8**, 1057–1063.

Van Vorhis Key, S. E., and Baker, T. C. (1982c). Trail pheromone conditioned anemotaxis by the Argentine ant, *Iridomyrmex humilis. Entomol. Exp. Appl.* **32**, 232.

Van Vorhis Key, S. E., Gaston, L. K., and Baker, T. C. (1981). Effects of gaster extract trail concentration on the trail following behavior of the Argentine ant *Iridomyrmex humilis* (Mayr). *J. Insect. Physiol.* **27**, 363–370.

Verhaeghe, J. C. (1977). Group recruitment in *Tetramorium caespitum. Proc. Int. Congr. IUSSI, 8th*, pp. 67–68.

Vowles, D. M. (1955). The foraging of ants. *Br. J. Anim. Behav.* **3**, 1–13.

Walsh, C. T., Law, J. H., and Wilson, E. O. (1965). Purification of the fire ant trail substance. *Nature (London)* **207**, 320–321.

Watkins, J. F. (1964). Laboratory experiments on the trail following of army ants of the genus *Neivamyrmex* (Formicidae: Dorylinae). *J. Kansas Entomol. Soc.* **37**, 22–28.

Watkins, J. F., Cole, T. W., and Baldridge, R. S. (1967). Laboratory studies on inter species trail following and trail preference of army ants (Dorylinae). *J. Kansas Entomol. Soc.* **40**, 146–151.

Williams, H. J., Strand, M. R., and Vinson, S. B. (1981a). Trail pheromone of the red imported fire ant *Solenopsis invicta* (Buren). *Experientia* **37**, 1159–1160.

Williams, H. J., Strand, M. R., and Vinson, S. B. (1981b). Synthesis and purification of the allofarnesenes. *Tetrahedron* **37**, 2763–2767.

Wilson, E. O. (1959). Source and possible nature of the odor trail of fire ants. *Science* **129**, 643–644.

Wilson, E. O. (1962a). Chemical communication among workers of the fire ant *Solenopsis saevissima* (Fr. Smith). I. The organization of mass foraging. *Anim. Behav.* **10**, 134–147.

Wilson, E. O. (1962b). Chemical communication among workers of the fire ant *Solenopsis saevissima* (Fr. Smith). II. An information analysis of the odour trail. *Anim. Behav.* **10**, 148–158.

Wilson, E. O. (1962c). Chemical communication among workers of the fire ant *Solenopsis saevissima* (Fr. Smith). III. The experimental induction of social responses. *Anim. Behav.* **10**, 159–164.

Wilson, E. O. (1963). The social biology of ants. *Ann J. Rev. Entomol.* **8**, 345–368.

Wilson, E. O. (1965a). Chemical communication in the social insects. *Science* **149**, 1064–1071.

Wilson, E. O. (1965b). Trail sharing in ants. *Psyche* **72**, 2–7.

Wilson, E. O. (1971). "The Insect Societies." Harvard Univ. Press, Cambridge, Mass.

Wilson, E. O., and Pavan, M. (1959). Glandular sources and specificity of some chemical releasers of social behavior in dolichoderine ants. *Psyche* **66**, 70–76.

Wilson, E. O., Eisner, T., Wheeler, G. C., and Wheeler, J. (1956). *Aneuretus simoni* Emery, a major link in ant evolution. *Bull. Mus. Comp. Zool. Harvard* **115**, 81–99.

Pattern and Control of Walking in Insects

D. Graham

Facultät Biologie, Universität Kaiserslautern, Kaiserslautern, Federal Republic of Germany

ADVANCES IN INSECT PHYSIOLOGY, VOL. 18

1 Introduction

The study of walking locomotion may be approached from several different points of view and at a variety of levels. This review considers the task from a behavioral context and moves inward from a description of the coordination of leg movements in insects, via a consideration of the muscle activity producing individual leg movements, to an examination first of the nervous control of the overall system and then of the sensory inputs to the individual legs and their influence on different parts of the leg movement. Finally, various models of walking control are described and the current state of research is discussed. I have attempted to broaden the normal horizontal surface context of walking to include the consideration of uphill and downhill walking and to a very limited extent the climbing capabilities of insects which have probably been of considerable importance in the evolution of the insect walking system.

I tend from my background in physics and bioengineering to consider the insect as a walking machine and hope that this review will be of use not only to the insect behavioral physiologist as a review of the present status of the field, but also to those physicists, mathematicians, and engineers who are currently engaged in the study of the design principles of legged vehicles. Current models of insect walking behavior can be significantly improved by the rigor of these sciences and I suspect that during the coming years these disciplines will have much to offer in this exciting task, the role of the biologist being to examine the tactics and strategy of walking behavior in a successful walking machine in a manner that is not only dedicated to the elucidation of neurological principles of control but also to the wider context of understanding how a walking machine works as a general purpose transportation system.

In recent years several reviews have appeared that cover walking and other forms of locomotion in arthropods and mammals and that emphasize the similarity between different forms of locomotion (Hoyle, 1976; Grillner, 1977; Moffett, 1977; Bowerman, 1977; Delcomyn, 1980, 1981). There is now sufficient walking data from several insects to examine their behavior in considerable detail and comment on the similarities and differences which appear to exist between the insects and other animals in the design of their walking system.

Historically, interest in insect walking reaches back at least to the seventeenth century (and I am sure the ancient Greeks could not have resisted a little speculation), but it is only in the last three decades that intensive quantitative analysis has been applied to this field. Before this period research was often anecdotal and, with a few important exceptions, little use was made of film or other high-speed analysis techniques. Since the late 1950s the subject has demanded quantitative assessment and improvements in technology have greatly broadened the range of walking parameters amenable to measurement.

The film analysis of cockroach walking (Hughes, 1952), the shadow technique for the stick insect walking on a treadmill (Wendler, 1964), and the direct recording of muscle activity in free-walking locusts (Hoyle, 1964) showed the direction for future research and defined a new term—*neuroethology*, the study of the neural basis of behavior. While the work of Hughes firmly established the basic parameters of walking for an insect and Hoyle showed what might become possible in the future in terms of neuronal analysis, the paper of Wendler (1966) quantitatively established that at least two different patterns of walking behavior could exist in an insect and these modes of walking depended upon whether or not the middle leg moved and contributed to the walk (Fig. 1).

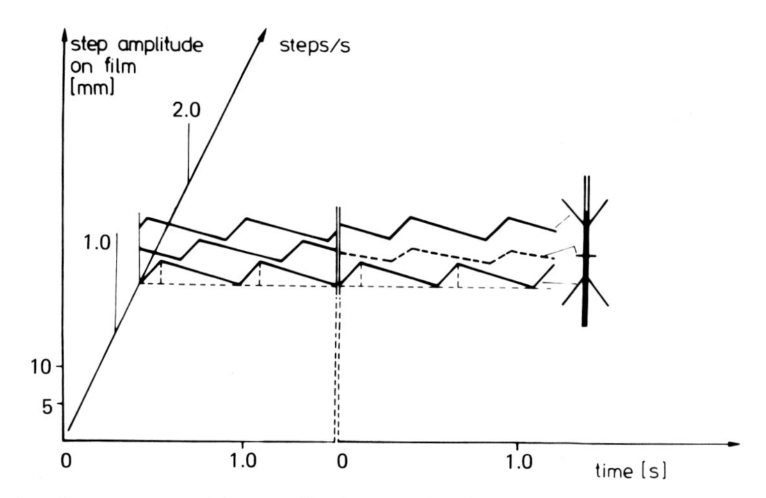

Fig. 1. Step pattern and leg coordination as a function of the step frequency. The phase relationships between ipsilateral legs before (left) and after (right) autotomization of both middle legs. The movement of the middle leg stump with decreased amplitude is indicated by heavy dashed line. (From Wendler, 1978b.)

The search for a sense organ responsible for such a change was unsuccessful but, by means of an elegant gravitational experiment, it was made clear that the lack of success did not mean that the tested organs might not be involved. The Wendler article indicated that the system depended heavily on some input from the middle legs which provided the normal phase-locked synchronous coordination between the front and hind legs. These legs showed some independence of action when the middle legs stood on a support or were removed. This is illustrated in Fig. 2, which shows gliding coordination in the bilateral, middle-leg amputee. It can be seen that the front and hind legs tend to alternate in this mode of walking but the front legs step slightly earlier in each hind-leg cycle, showing that they have a slightly higher mean frequency than the hind legs.

Wilson (1966) reviewed the earlier work on insect walking, collecting together the basic features of the temporal patterns of walking, and carried out numerous unpublished experiments of his own. His conclusion, based on the way in which step patterns change with the speed of walking, was that walking could best be described as a cooperative interaction between individual leg oscillators. A model was proposed in which right and left oscillators of the same segment always stepped in exact alternation. Along the body axis a relatively constant interval or lag was proposed which would produce separated metachronal waves at low step frequency and overlap at high speeds giving the symmetrical tripod gait in which the front and hind legs on one side swing forward at the same time as the middle leg on the other side. In this gait the animal steps from one tripod of support to another (Fig. 3e). Wilson pointed out that there might be exceptions to this simple hypothesis. Indeed, examples of slow stepping in the hind legs of some long-legged grasshoppers and other temporal patterns in which legs appeared to step in diagonal pairs, rather than three at once, were noted. However, the

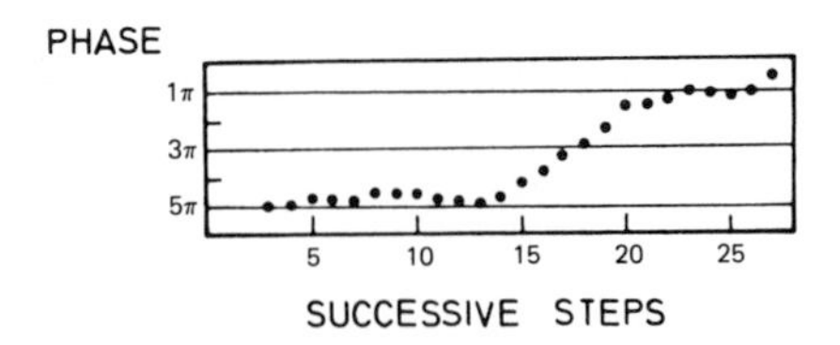

Fig. 2. Phase relationship between right fore and hind legs of a stick insect which has autotomized both middle legs. Even π legs are in phase; odd π legs alternate. At the beginning the legs alternate. From the fifteenth step on, the front leg adopts a slightly higher stepping frequency, so that the phase relationship changes progressively (gliding). Fixed alternation recurs from the twentieth to the twenty-sixth step. Phase is defined as the time interval between the same, successive event in two different legs divided by the step period of the reference leg. If two legs are in exact alternation the relative phase is π radians, 180°, or 0.5 as a decimal fraction of the step period. (From Wendler, 1978b.)

majority of the earlier work of von Buddenbrock (1921), Bethe (1930), ten Cate (1941), von Holst (1943), and especially the observations of Hughes (1952) and Wendler (1964) provided an excellent basis for the conceptual model of Wilson.

This article will consider how these earlier ideas have been put to the test and the extent to which our view of the functional organization of insect walking has changed in the intervening period. To begin with, let us define the basic mechanical requirements of the insect or any other self-supporting walking system. When an upright animal which holds its body clear of the ground walks on a horizontal surface the legs have to perform four functions. They must support the body, while holding it clear of small ground obstacles that would impede its progress. The legs must also stabilize the animal both along and across the body axis so that it does not fold and collapse like a card house. To achieve movement in the required direction the legs must provide leverage using those legs in contact with the ground. Finally, the legs must lift themselves from the ground in recovery strokes which place each leg in a favorable position to exert the next power stroke.

The study of walking behavior in arthropods has tended to consider these requirements in the reverse order, concentrating on the power stroke and recovery rather than stability and support. Coordination is not mentioned as a fundamental requirement because its importance appears to depend upon how many legs are available, their separation and orientation with respect to each other, and the load they have to support. As an example, the numerous tube feet of the starfish can operate independently with only a loose cooperation between neighboring feet, while the millipedes must observe strict coordination merely because the legs are close packed along the body (Franklin, 1984). As the number of legs is reduced the considerations of support begin to dominate the strategy of movement but where this constraint is relaxed, as in water-supported walkers such as the decapod crustacea, the coordination of recovery strokes can be relaxed into a hazy alternating pattern. When the number of legs is reduced to six and the full body weight must be supported at all orientations of the walk surface to the gravity vector, then coordinated recovery and body stabilization relative to the surface become vital considerations. If the number of legs is reduced further, then parts of the walk can no longer be considered to be in static equilibrium and it becomes necessary to provide a mechanism for dynamically stabilizing the body. This was probably beyond the capabilities of the arthropod nervous system and required the evolution of the cerebellum.

The insect, when walking upright, is always in static equilibrium and the leg system appears to be designed to maintain the body in space with a suitable orientation relative to the substrate. The substrate is grasped by the sticky pads and claws of the tarsus and the legs provide the forces which

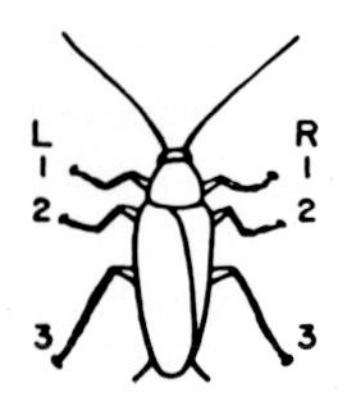
L
1
2
3
R
1
2
3

a

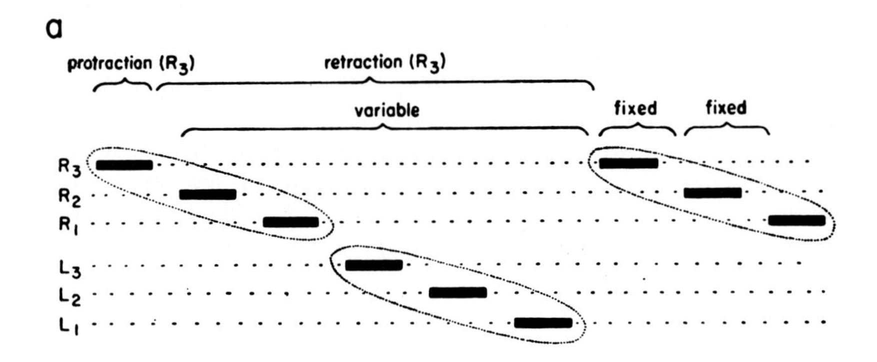
protraction (R_3)
retraction (R_3)
variable
fixed
fixed
R_3
R_2
R_1
L_3
L_2
L_1

b

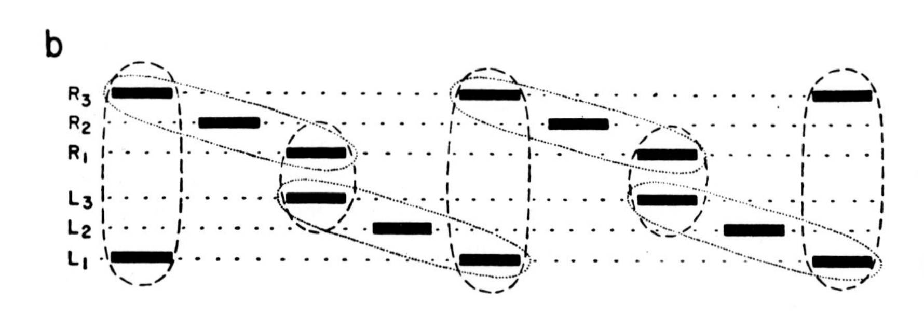
R_3
R_2
R_1
L_3
L_2
L_1

c

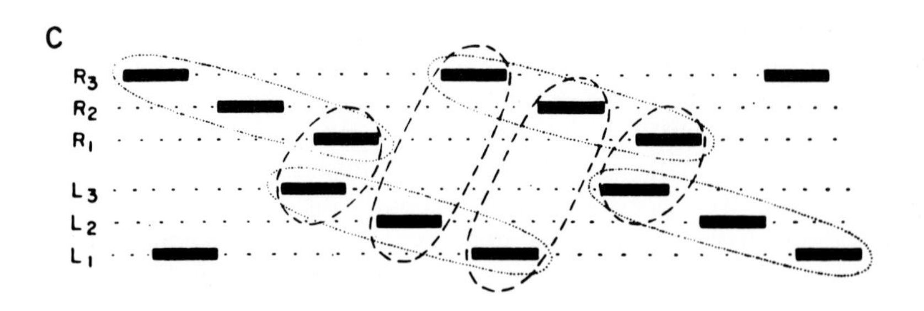
R_3
R_2
R_1
L_3
L_2
L_1

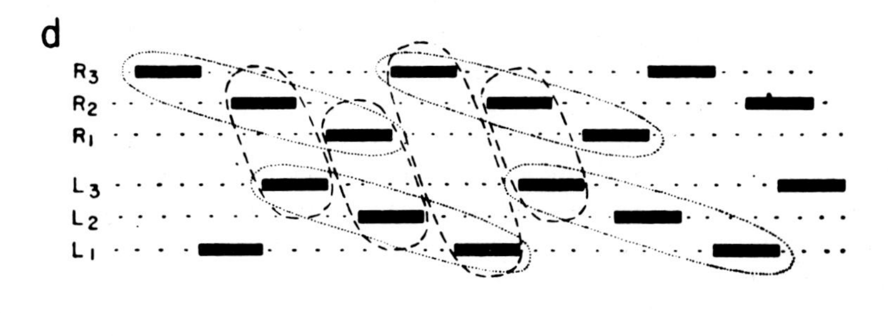

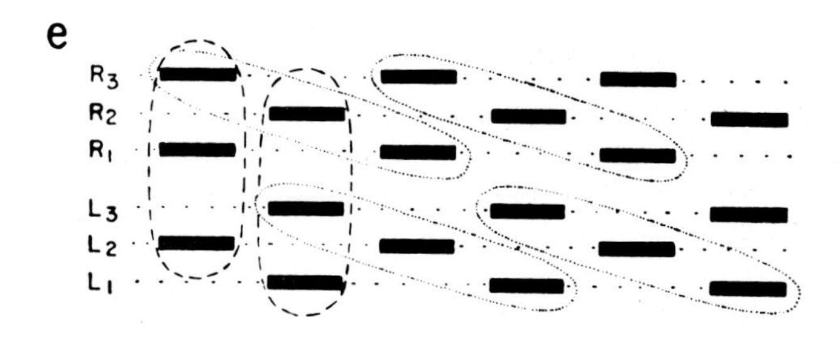

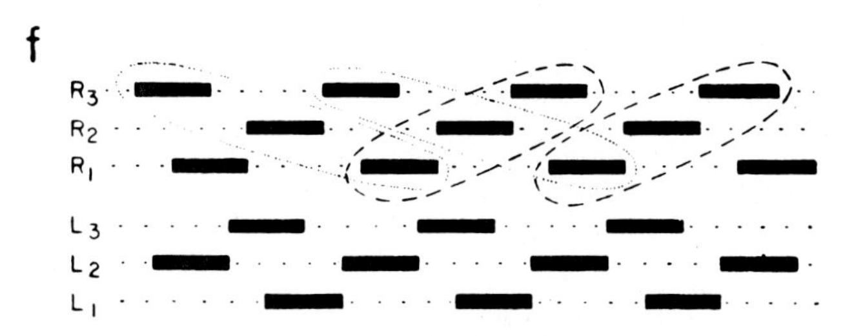

Fig. 3. Diagram illustrating an hypothesis relating the various six-legged gaits. The legs are numbered as in the drawing. The horizontal axis represents time. The solid bars indicate protraction (leg off ground and moving forward relative to the body and ground). For the purpose of making the simplest presentation, protraction time and interstep intervals between legs 3 to 2 and 2 to 1 are held constant and frequency changes are accomplished by varying only the interval between stepping in leg 1 and leg 3. Legs on opposite sides in the same segment are held in strict antiphase. Dotted enclosures indicate fixed basic sequences of steps of the legs of one side. In (a), the lowest frequency pattern is shown. Each leg steps by itself. In (b), the basic sequences overlap and some legs step in pairs (dashed enclosures). Removal of the middle legs, R2 and L2, would result in a diagonal stepping. In (c) and (d), the sequences overlap more. In order to have the sides alternate strictly, legs cannot step in exact pairs. Dashed enclosures indicate nearest temporal neighbor pairs. In (e), further overlap results in the tripod gait. Even the sequences on one side overlap so that legs 1 and 3 are synchronous. In (f), at the highest frequency the sequence beginning with leg 3 is started before the previous sequence has ended so that there is apparent reversal of the direction of the stepping sequence (dashed enclosures). (From Wilson, 1966.)

counteract gravity and propel the animal in the required direction. In addition, the small number of legs requires that recovery strokes be organized in such a way that at least three legs maintain contact with substrate to hold the body in a suitable walking orientation. In the literature, little mention is made of the vertical climbing and suspended walking capabilities of many insects and walking is viewed as the problem of supporting the body above its standing legs on a horizontal surface. This limited view persists but is giving way to a wider view as experimental techniques improve.

2 Free walking on horizontal surfaces

Surprisingly, little is known of the general behavioral context of walking. Apart from a few studies on the frequency of occurrence of walks (Moorhouse, 1971; Moorhouse *et al.*, 1978) and their relationship to circadian rhythms (Brady, 1969; Godden, 1973) there is almost no information on the distribution of speeds of walking, terrain parameters, the incidence of climbing, running and scrambling, or the relationship between walking and other behavior such as prey capture, mating, and feeding in the wild. This would appear to be a major gap in our understanding which makes it difficult to estimate the importance to the animal of different aspects of its design and structural organization. Research has concentrated primarily on horizontal smooth surface walking in an upright position and considerable quantitative information is now available for several insects, although they represent only a small fraction of the insect kingdom.

2.1 PARAMETERS OF WALKING BEHAVIOR

Taking a step involves the relative movement of the segments of the leg and the movement of the leg as a whole unit. Pattern is apparent at all levels, as the process is a cyclic or repetitive one.

The cycle of movement in one leg is a function of the exact structure and orientation of the joints but most walking behavior can readily be resolved into a recovery, protraction, or swing phase in which the leg is lifted clear of the ground and swings forward to place itself in contact with the ground. This forward position relative to the body, at the moment of touchdown, is termed the anterior extreme position or AEP of the leg (Baessler, 1972a). The leg in this position may some times pause before starting its rearward movement (Graham, 1979b). After such a pause the leg then moves to the rear in the support, retraction, or stance phase until it reaches the posterior extreme position or PEP where it may pause again (Burns, 1973), detaching the tarsal pads and claws from the surface, in preparation for lifting the leg in the next recovery stroke.

The orientation and action of the leg vary considerably in different insects. In ants, flies, and many insect nymphs the body is held well clear of the walk surface and the plane containing the coxa, femur, and tibia is approximately perpendicular to the walk surface. In the larger insects the abdomen is often dragged behind the legs. In the cockroach and beetles the plane of leg action may be rotated until it is almost parallel to the walk surface and most of the leg musculature contributes directly to the propulsive stroke. Thus the manner in which particular muscles are used in propelling, supporting, or stabilizing the body may differ greatly from one preparation to another and must be remembered when attempting to compare function and behavior between species.

2.2 PATTERN OF LEG RECOVERY

The pattern of leg recovery varies from species to species but all insects so far examined fit within a spectrum of metachronal patterns in which the legs of one side are recovered in the order of hind, middle, front. This is particularly apparent in slow-walking cockroaches, as reported by Hughes (1952), in which all the legs on one side are recovered in a metachronal sequence, followed by all the legs on the other side (similar to Fig. 3a). In faster walking cockroaches and most other insects at all speeds the sequences of recovery, on opposite sides, overlap (Figs. 4 and 5).

On the same side the sequences may be well separated in time as described above for the slow-stepping cockroach but are usually separated by a much shorter interval. This is typical of the slow walking behavior of adult stick insects (Graham, 1972), locusts (Burns, 1973), and mantids (Thomson, private communication). At high speeds, the sequences just overlap, producing simultaneous recovery of front and hind legs on the same side. Legs on opposite sides, in the same segment, tend to step in exact alternation producing the tripod stepping pattern which is typical of the cockroach (Figs. 4a and 5a).

In this condition the front and hind legs on one side lift at the same time as the middle leg on the other side. However, Wendler (1964) has shown that overlap can exceed the tripod condition at the highest speeds of walking on a rather heavy wheel (similar to Fig. 3). Excessive overlap has also been reported for free-walking stick insects (Graham, 1972). These patterns are similar to those proposed by Wilson with one exception. Legs of the same segment only step in exact alternation at the highest speeds of walking in these insects. At lower speeds the time interval (lag) between the recovery of legs on the same body segment does not equal half the step period. The more

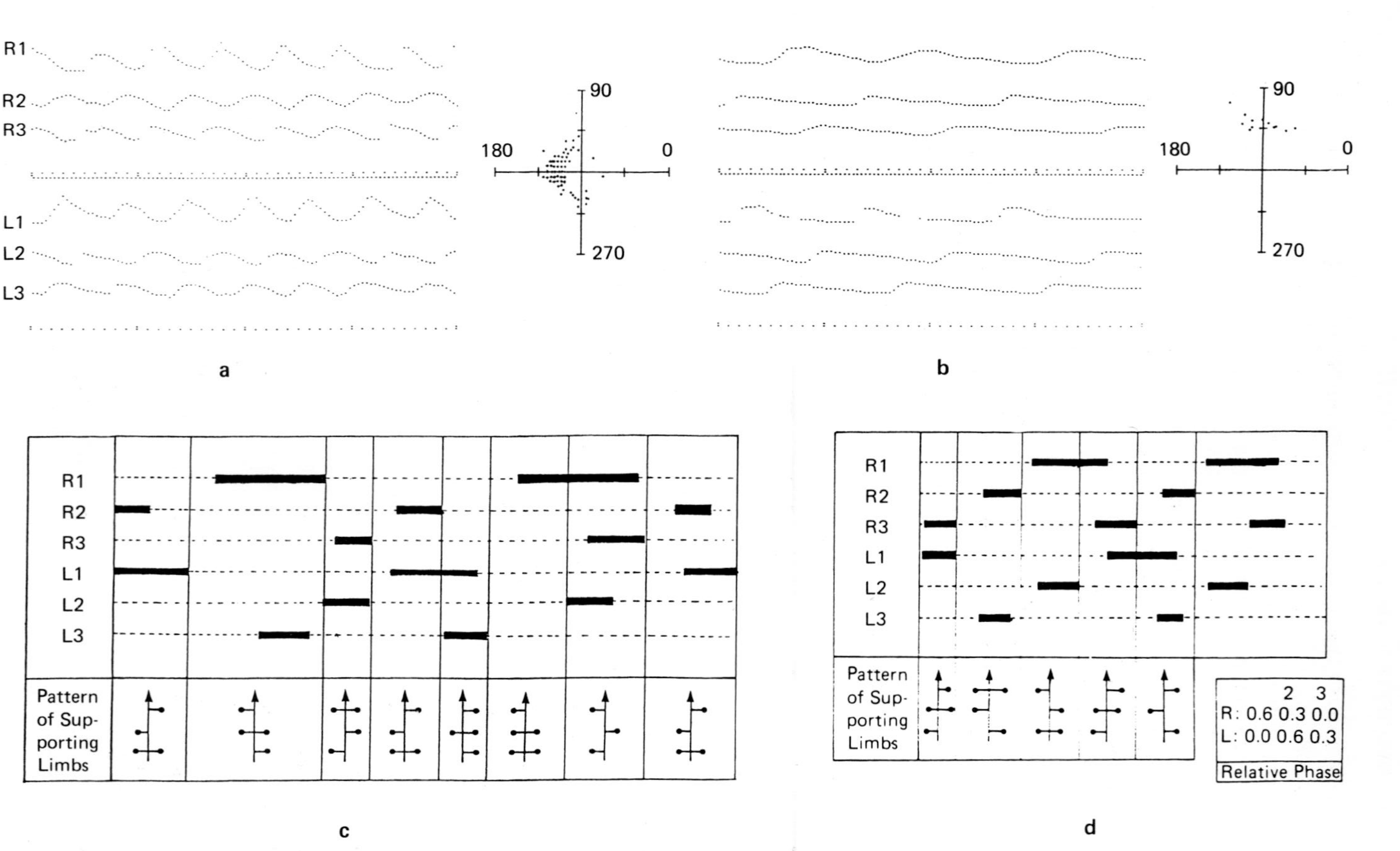

Fig. 4. Step patterns for a stick insect walking fast (a) and slow (b). Time marks are 100 ms and forward movement or protraction is measured upward. The associated polar plot gives the phase angle for R3 and L3 and the step period on the radius. Two slow walking step patterns for an adult mantis are shown in (c) and (d) (Thomson, 1985). The black bar represents a protraction stroke. These demonstrate the two alternative right and left

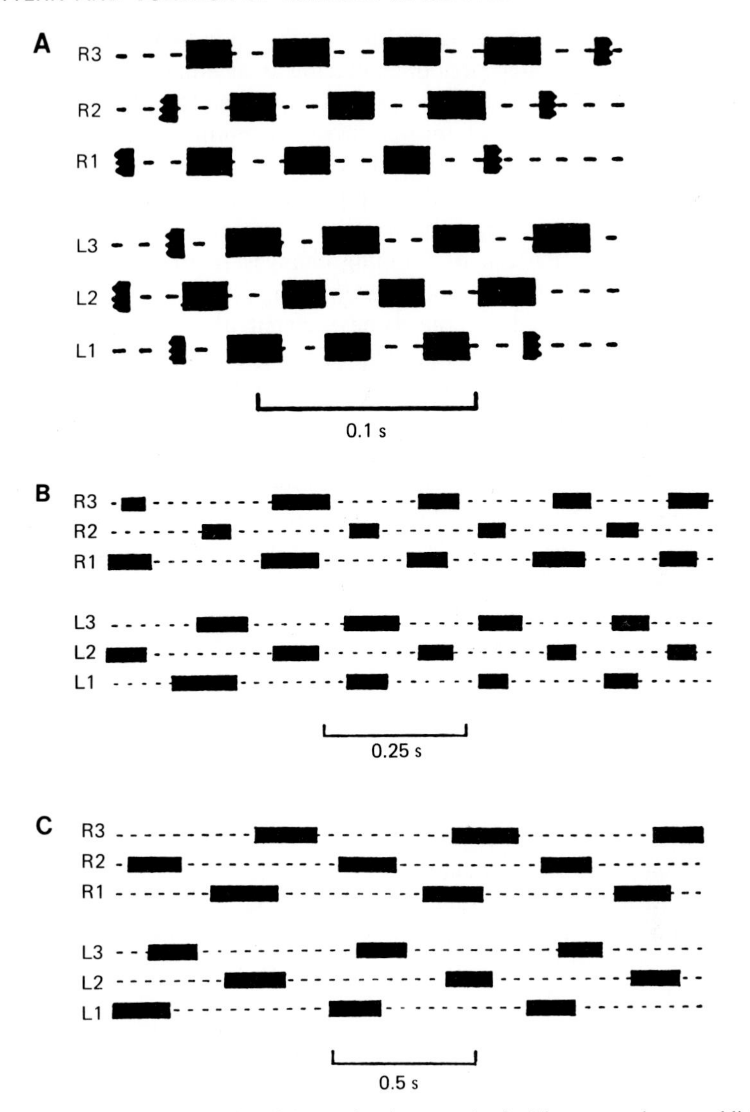

Fig. 5. Stepping patterns of rapidly moving intact animals. The conventions established by Wilson (1966) have been followed. The legs on the right (R) and left (L) sides of the body are numbered from front to rear. Each row represents the movements of the indicated leg, the solid bars representing protraction, dotted lines retraction. The diagram reads from left to right. Notice the simultaneous movements of the two sets of three legs, R3, R1, L2 and L3, L1, R2, each of which form a triangle. (A) Frequency of leg movement about 22 Hz. (B) The alternating movements of the two triangles of legs are still maintained at this low frequency of leg movement (4 Hz). Protraction durations have now become significantly shorter than retraction, resulting in periods during which all six legs are on the ground at once. (C) The alternating triangle pattern has begun to break down at this frequency of leg movement (1.5 Hz) and metachronal sequences are becoming separated in time. (From Delcomyn, 1971a.)

general rule is that the lag along and across the body is relatively constant and is less than half the step period, producing an asymmetrical step pattern (Fig. 4b, c, d). These asymmetrical patterns have been found in stick insects (Graham, 1972), mantids (Thomson, private communication), and some slow-walking cockroaches (Spirito and Mushrush, 1979).

As the stick insect decelerates, the step period increases more rapidly than the lag between adjacent legs and the exact alternation of legs in the same segment is lost. The pattern of stepping changes so that legs tend to be recovered in pairs rather than three at a time (Figs. 4b and 10). The phase or relative timing of the legs in the same segment alters from 180° or 0.5 (normalized to unity) to an asymmetrical value. Instead of the right leg lifting halfway through the left leg step, the right lifts after only one-quarter of the left leg step (phase R3: L3 = 0.25). If we measured left on right this would give a phase of 0.75 for the same sequence of steps (Fig. 4b and c).

In stick insects this asymmetry is associated with a slight curvature of the straight walking track. The asymmetry can be of either handedness, giving slight left or right curvature to the track. This observation suggested that a particular asymmetry may be caused by a difference in the power output to the right and left sides of the insect and formed the basis for a quantitative reexamination of Wilson's proposal for the control of leg recovery in insects (Graham, 1972, 1977a).

2.3 DEPENDENCE OF PARAMETERS ON STEP FREQUENCY

The important feature of Wilson's analysis of walking behavior was the recognition that some temporal features changed with step frequency while others remained relatively unchanged. He noticed that if the time interval between the protractions of adjacent legs on the same side is approximately constant, while the retraction time changes with the step frequency, a whole range of different step patterns are created (Fig. 3). This assumption has been partly confirmed in the adult stick insect (Wendler, 1964; Graham, 1972), although it is clear that in other insects the time interval between the recovery of adjacent legs on the same side, now usually referred to as the lag, can change with step frequency at much the same rate as the retraction time. In these animals the pattern of stepping remains constant over a wide range of step frequency. The best examples are the cockroach (Delcomyn, 1971a) and the stick insect nymph (Graham, 1972) (see Fig. 6). While such disparities might suggest a fundamental difference in mechanism, the appearance of both gaits in the stick insect nymph (Graham, 1972) indicated that the two patterns of behavior should be reconciled within one structure (Fig. 6).

Another apparent difference between species is the dependence of the recovery time upon step frequency or step period. In the earlier work

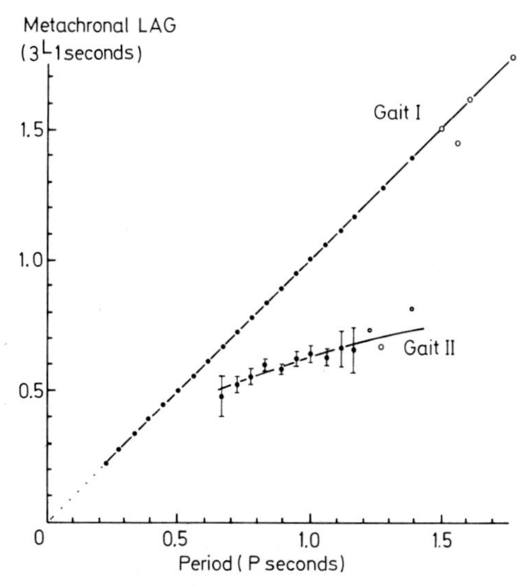

Fig. 6. Metachronal lag (${}_3L_1$) versus period (P) for 10 first instar stick insects. Closed circles represent the mean value of ${}_3L_1$ for a given mean period and open circles single observations. Error bars denote the error of the mean where this is larger than the symbol. Metachronal lag (${}_3L_1$) is defined as the sum of the lag between hind and middle leg (${}_3L_2$) and middle and front legs (${}_2L_1$) in the metachronal wave of protractions and each lag is measured from the protraction onset of one leg to that of the leg in front. (From Graham, 1972.)

considerable emphasis was placed on the ratio of the time required for a recovery stroke or protraction (p) to that required for retraction (r). This quantity is related to the duty factor, which represents the fraction of time for which the leg can be active in propelling the body. However, the p/r ratio is a rather unsatisfactory parameter because it is not clear, when it is constant, whether p and r are constant or varying at the same rate. This parameter is now rarely used and has been replaced by consideration of the dependence of protraction or retraction duration on step period (the latter ratio is defined as the duty factor; English, 1979). The choice of period or frequency as a variable depends upon the system under examination. In the cockroach, in which step frequency covers the range from 0.5 to 20 Hz, frequency is the obvious choice. In the stick insect the step frequency range is only 0.5–4 Hz and period can be used rather than frequency to provide a direct comparison with lag, protraction duration, and other temporal parameters of the walk.

In the cockroach and stick insect nymph the time taken to recover the leg varies with the step period. When stepping at maximum speed the protraction duration (t_p) is short (cockroach $t_p = 20$ ms; stick insect nymph $t_p = 60$ ms). As the speed decreases the recovery time lengthens (cockroach $t_p = 160$ ms; stick insect nymph $t_p = 400$ ms). The speed of locomotion can be assumed to be directly related to the force generated in the retraction stroke, and corresponding forces are developed in the protraction or recovery musculature. Thus, the energy expended in moving the leg to a new position of support is minimized during slow walking in these animals. The results for the adult stick insect are quite different (gait II, Fig. 7) and show a constant recovery time for the legs in free walking or on the Wendler treadwheel. The protraction time is a minimum (100 ± 20 ms) regardless of the walking speed.

With respect to spatial parameters, individual insects can show considerable variability in the mean range of movement during a walk, but no reports have appeared of any strong dependence of stroke amplitude or the extreme positions on step frequency in the cockroach or stick insect. The leg steps of an adult stick insect are shown in Fig. 8 (Epstein and Graham, 1983), where it can be seen that the range of movement depends upon the segment and a middle leg may operate forward or behind the most typical range, indicating that the AEP and PEP may vary considerably from step to step. Such variations do not appear to take place at random and their role in both coordination and the relative timing of parts of the leg cycle is probably of considerable importance.

Several insects do not always obey the general rules proposed by Wilson (1966). They are marked by extreme specialization of the hind legs into a jumping organ. The locust and grasshopper have evolved long back legs with powerful musculature to project themselves from danger. In walking at high

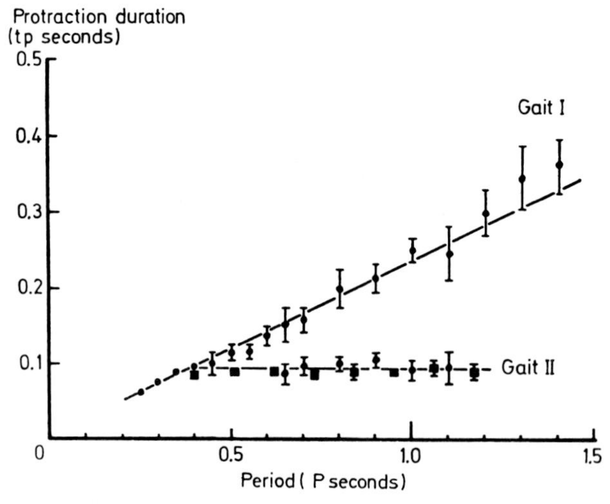

Fig. 7. Protraction time t_p as a function of period P for the mesothoracic legs of first instar and adult insects. Closed circles are average values of t_p for a first instar insect. Closed squares are average values of t_p for an adult insect. Error bars denote the error of the mean where this is larger than the symbol. (From Graham, 1972.)

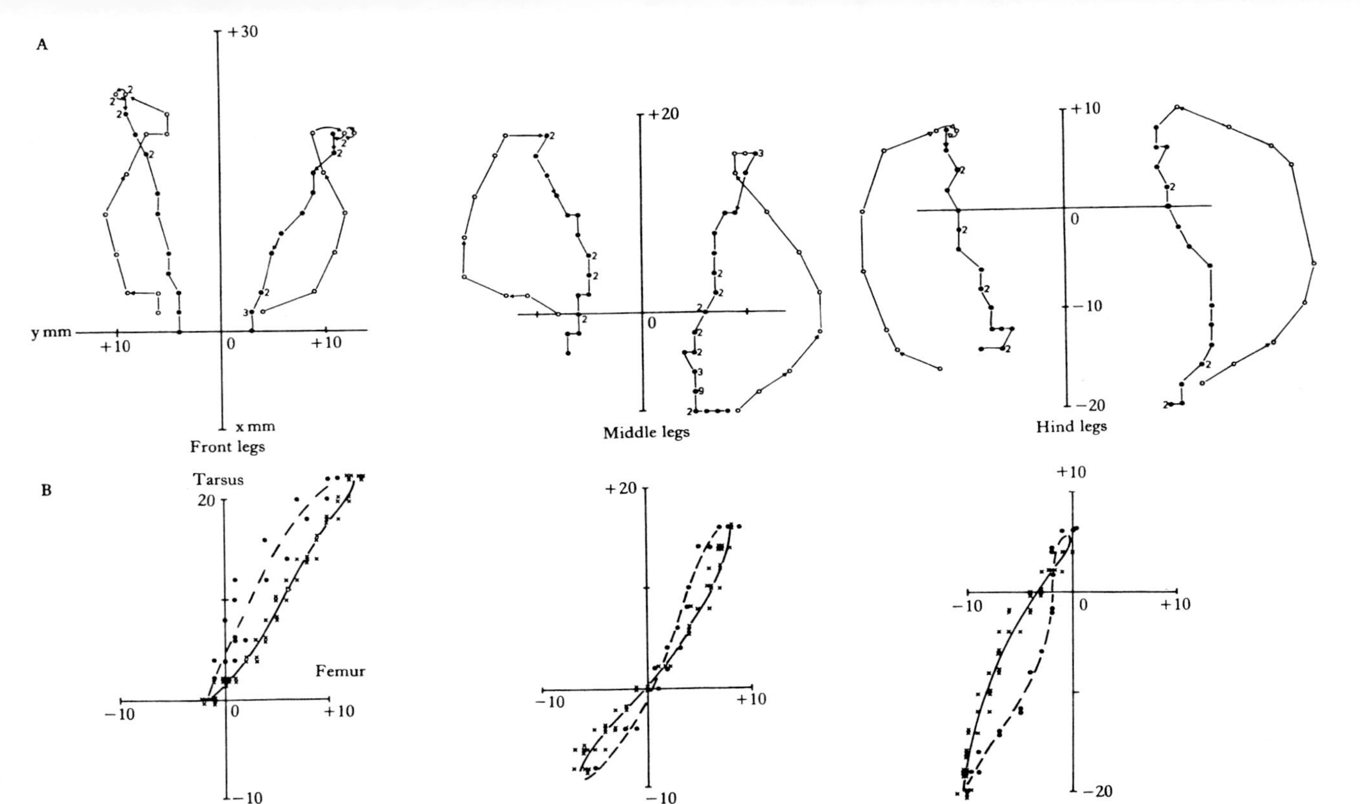

Fig. 8. The trajectory of the tarsus in the horizontal (xy) plane is shown for front, middle, and hind legs in (A). Stance positions are shown by the filled circles and swing positions by open circles. Numbers denote that the leg is stationary for several film frames. Arrows show the sequence of position changes. Plots on right and left sides of the x axis merely show examples of two different steps. In (B) a comparison of the femur tip (x mm) and tarsus position (y mm) is shown. Stance positions are shown by crosses and swing positions by closed circles. This figure illustrates the smaller range of movement at the femur tip caused by rotation about the long axis of the femur. (From Epstein and Graham, 1983.)

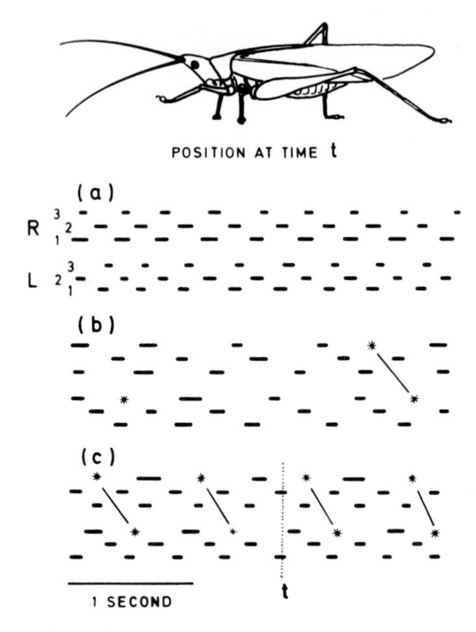

Fig. 9. Side elevation of *N. robustus* walking in the systematic absence mode at the moment shown by the dotted line in step pattern (c). (a) Step pattern of "complete" walk (no absences). Black bar denotes the duration of the protraction movement in which the leg is lifted and moves forward. (c) Same animal some minutes later shows occasional "isolated" and "paired" absences marked by asterisks. (c) Fully coordinated systematic absence walk. Note asymmetry of rear leg protractions. R denotes right legs and the legs are numbered from 1 to 3 from the front of the animal. (From Graham, 1978a.)

step frequency these legs are used quite normally and step with a similar amplitude and frequency to the other legs. After prolonged walks the large hind legs may begin to make steps of twice or in exceptional cases three times the normal amplitude (Graham, 1978a) (see Fig. 9). The doubling or tripling of the hind-leg amplitude is demonstrated in the step pattern by the absence of one or two protraction strokes compared with the middle and front legs. These absences appear irregularly in fast walks or systematically in the lower speed walks. The recovery strokes of the hind legs are still coordinated with the anterior legs although some gliding coordination may be present. The incidence of the large amplitude steps is a function of step frequency being much higher at low frequency. This is the only example of a consistent amplitude change in intact insects walking free on a horizontal surface.

In the late 1960s, important differences seemed to exist between the walking behavior of certain insects. The work to be described in following sections has shown that these differences arise from differences in experimental conditions rather than neural organization. Also the 1970s have been a period of consolidation in which it has been possible to show that the same fundamental pattern of behavior exists in all the insect species examined so far.

2.4 EVIDENCE FOR DIFFERENT GAITS IN THE SAME INSECT

One of the clues which suggested the involvement of only one basic mechanism was the discovery that a single animal could choose one of two gaits at a particular step frequency. Here gait is defined in its broadest context and implies not a particular temporal configuration of steps but the manner in which the pattern varies with step frequency as well. In the first instar stick insect two quite different gaits were found to occupy the same step-frequency range (Graham, 1972, 1977a). The predominant pattern in these animals is the simple tripod in which adjacent legs step with a relative phase of 0.5 and the same relationship is found for legs of the same segment. This step pattern is independent of the step frequency and may be found at the very lowest speeds. This gait is referred to as gait I and is directly comparable with the cockroach behavior reported by Delcomyn (1971a), although it differs from that reported by Hughes (1952) in which, at slow speeds, the step pattern becomes metachronal and the waves of protraction passing along each side become completely separated in time.

In the stick nymph a quite different walk behavior appeared under certain conditions. In this gait (II) the right and left sides of the body are not phased-locked and the lag between the protractions of adjacent legs on the same side is less than half the step period, giving an adult-type gait which only reaches the tripod condition at the highest step frequency. This type of behavior could be provoked by smearing leaf sap on the walk surface, and it sometimes appeared spontaneously toward the end of long walks using gait I. These results indicated a close affinity between gait I and gait II and that the basic mechanism could be quickly switched from one to the other (Fig. 10).

Is this gait duality unique to the stick nymph or is it a more general property of insect walking behavior? Such subtle changes are only likely to be detected where a considerable body of experimental data is available. Only the cockroach species *Blatta orientalis* and *Periplaneta americana* have been as thoroughly investigated as the stick insect. Pringle (1940) was the first to suggest a distinction between running and walking behavior in this animal, based upon the use of either fast or slow fibers in the leg muscles. Delcomyn and Usherwood (1973) and Delcomyn (1973) found significant differences in the number of impulses and their mean interval for the tibial extensors, and suggested that below 5 Hz the dependence of these parameters on step period differed from that above this step frequency. The recent work of Kozacik (1981) shows that the frequency of occurrence of a given step period is

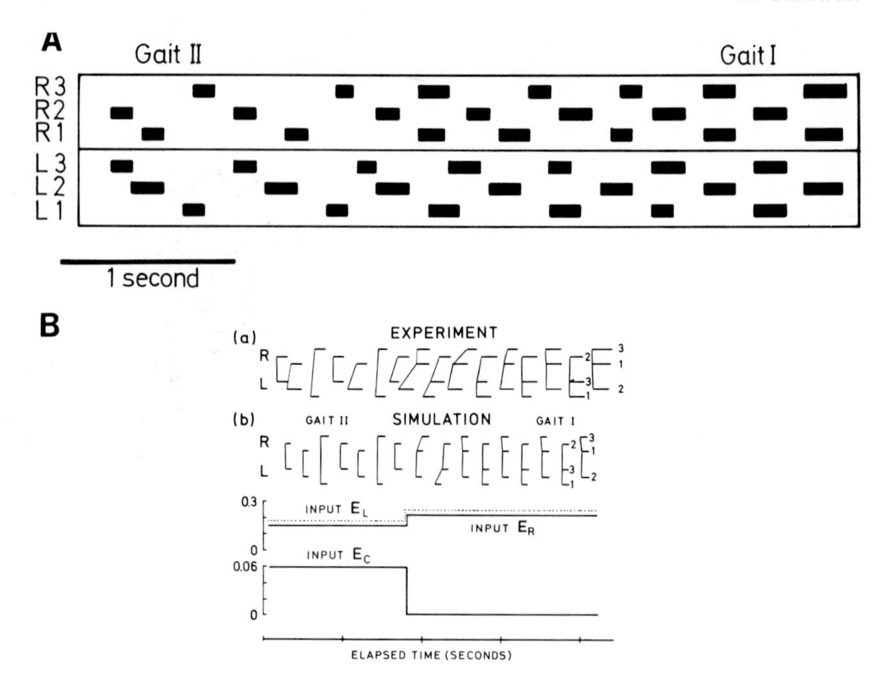

Fig. 10. Transition from gait II to gait I in a stick insect nymph. (A) shows an experimentally observed transition from a step pattern in which legs are recovered in pairs to the normal tripod pattern in which three legs are recovered together. (B) shows the same walk (a) with associated pairs and triplets marked; (b) a simulation of this transition using the model of Graham (1977a). Power inputs to right (E_R) and left sides (E_L) are increased together to accelerate the walk and input E_C determines the nature of the gait.

bimodal with a minimum at a step period of 200 ms, which corresponds to the transition reported by Delcomyn. Kozacik suggests that the two kinds of behavior may correspond to escape and nonescape behavior and reflect a difference in intent rather than a direct consequence of step frequency changes. The above evidence suggests that two separate domains exist but there is also some evidence that there may be significant overlap in these walking modes similar to that found in the stick nymph.

Dr. Delcomyn generously made his original thesis data available to the author during the preparation of a modeling paper (Graham, 1977a) and Fig. 11 shows a lag versus period plot for his cockroach data. While most of the data falls on a smooth curve, several data points lie well below the curve in the region comparable to the stick nymph using gait II. These data points are in the region below 5 Hz and although alone they were not of sufficient number to justify the proposal of an alternative gait, the quantitative measurements in the stick nymph indicate that two gaits may be possible, in

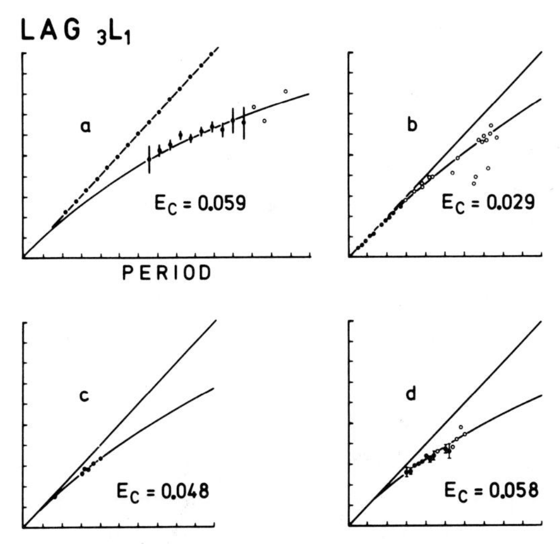

Fig. 11. Lag $_3L_1$ versus period simulations for different insects compared with the experimental data. Time unit 100 ms. Filled circles represent the mean of two or more measurements; open circles are single measurements. The error bars correspond to the error of the mean when this is larger than the filled circle. (a) First instar stick insect (Graham, 1972); (b) adult cockroach (Delcomyn, 1969); (c) adult locust (Burns, 1973); (d) adult grasshopper (Graham, 1978a). (From Graham, 1977b.)

this step-frequency range, for the cockroach. With the results of Kozacik (1981), who concentrated his attention on the slow unstartled behavior, this suggests that the cockroach may use a gait similar to that of gait II when browsing.

Thus two insects which appeared to have quite different walk characteristics show a fundamental equivalence in the patterning of their leg movements.

2.5 TURNING BEHAVIOR

Arthropods use several different methods for changing the direction of a walk (heading). One involves changing the step frequency on right and left sides. In the adult stick insect this produces a clearly observable change in coordination across the body. During such a turn, legs of the same segment may momentarily step together, giving a galloping appearance to the step pattern. Such turns are accompanied by a negligible change in amplitude of the leg movements and a relatively small change of heading. During such a turn only one step is lost on the inside of the turn and the heading changes by 20° (Graham, 1972).

In the first instar stick nymph when it uses gait I, the coordination of the legs need not alter during a turn. In gait I the body is supported by a tripod of legs (e.g., R1-L2-R3) at the midpoint of their range of movement, while the other set of legs (L1-R2-L3) swing forward in their recovery stroke. A turn is performed by reducing the propulsion produced by the middle leg. The body rotates about this leg and the turn is completed in one stance phase of the middle leg on the inside of the turn (Graham, 1972). The change in heading is similar to that observed in the adult but is achieved in half a leg cycle rather than over several steps. In gait II, the step frequency on right and left sides is rarely equal and the turns are similar to those found in the adult but there appears to be some amplitude variation as well. In very tight turns H. Cruse (personal communication) has observed backward walking on the inside of the turn in the stick nymph.

Turning behavior can be induced toward vertical black stripes in stick insects (Jander and Volk-Heinrichs, 1970) and Jander and Wendler (1978) used a rotating striped drum to control the rate of turning in a third instar stick insect walking on a driven sphere. They found similar turns to those of the first instar using gait I but were able to detect significant changes in the mean amplitude of the legs producing incremental turning behavior over many steps. The results showed an increase of the amplitude on the outside of the turn in addition to the shortening for the inside legs reported for the first instar nymph (Graham, 1972).

In the bee, Zolotov *et al.* (1975), using a similar stimulus, reported a hierarchy of turning methods related to the rate of turn required. At low rates of turning the legs change their stroke amplitude. As the rate of turn increases, the animal changes frequency on right and left sides. In faster turns the hind leg on the inside of the turn halts and anterior legs walk laterally. Finally, for a maximum rate of turn, the legs on the inside walk backward. In arthropods research has concentrated particularly on tight turning behavior initiated by visual input (Land, 1972) or courtship behavior (Bell and Schal, 1980; Franklin *et al.*, 1981). Extremely tight turning is often accompanied by a reversal of walking movements on one side of the body. In spiders this is accomplished while maintaining precise alternation of adjacent legs. The cockroach also uses almost pure rotation turns during courtship, which require that the legs on one side step backward. The timing of recovery strokes is disordered compared with spider rotations, although the frequency of occurrence of the tripod configuration when turning was significantly greater than that expected by chance. This pattern of leg coordination also forms an important component in other behavioral sequences requiring the legs, such as struggling which occurs when legs have no tarsal contact and recovery from being tipped over (Camhi, 1977); Sherman *et al.*, 1977).

The varying contributions of rotational and translational components during tight turning make analysis difficult even if one assumes a fixed distance between the body and the tarsus. However, this distance can be readily altered in most arthropods and such changes are an important feature of tight turning.

2.6 STARTING AND STOPPING

Obviously walks must begin and end and it is somewhat surprising that information on the initiation and termination of walks has only recently begun to appear in the insect literature. In arachnids, Land (1972) reported that walks could be momentarily stopped at any point in the cycle before reversing the direction of walking. This does not appear to be true of stick insects, which usually stop at a time when all legs are in contact with the ground and which corresponds to a moment of minimum velocity (personal observation). A permanent stop may be accompanied sometime later by a rearrangement of the legs into a symmetrical configuration (Dean and Wendler, 1984). As a result of this symmetrical arrangement the stick insect sometimes requires several steps before normal coordination is achieved in a new walk (see Fig. 12) (Graham, 1972). At such times an expected protraction may be missing or legs in the same segment may step at approximately the same time for several cycles.

The scorpion tends to walk in short bursts and is particularly suitable for such studies (Bowerman, 1981b). This animal walks with an overlapping metachronal step sequence with the short-latency wave progressing from front to rear. In starts this animal often requires several steps before the normal walking rhythm is fully established.

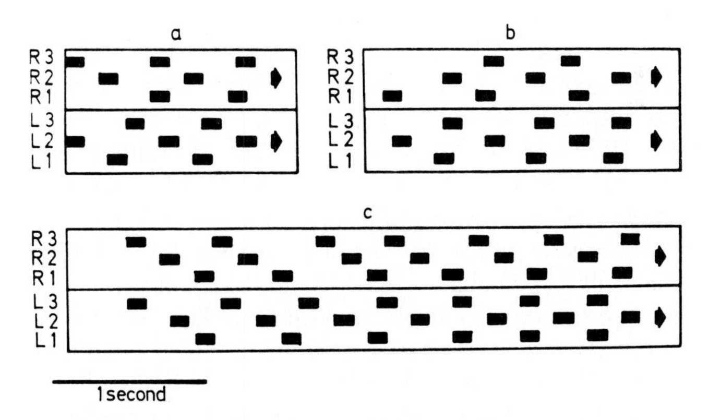

Fig. 12. Starting step patterns for an adult insect. (a) Typical start step sequence for an adult stick insect. (b) Rare start sequence. This may be considered as a reversal of the normal sequence on the right or the absence of an R3 protraction. (c) Rare "gallop" start sequence in which the legs of the same segment step in synchrony followed by a gentle turn to the right. (From Graham, 1972.)

These observations emphasize the flexible nature of the timing of leg movements and suggest that interactions across the body can be varied in strength. In scorpions, spiders, and insects there is no indication of any preliminary movements of the legs as a prelude to the onset of walking as has been observed in myriapods, for which Franklin (1984) reports that legs lift in groups in preparation for the start of a walk and may remain elevated for a short time after the walk stops.

2.7 FORCE MEASUREMENTS IN FREE WALKING

So far we have only considered the temporal patterning of the recovery movements in walking, but it is the propulsion stroke during the stance phase of the leg that generates the forward movement. While considerable attention has been devoted to force measurements and propulsion coordination in mammals, this has only recently been examined in arthropods (see review by Clarac, 1982). The small size of most insects (the stick insect *Carausius morosus* weighs less than 1 g) makes it difficult to attach force transducers directly to leg joints without significantly disturbing the inertia or flexibility of the joint. This would not necessarily apply to some of the larger stick insects such as *Acrophylla titans* (20 cm long) or *Extatosoma tiaratum* (14 cm) which weigh 30 g or more.

A novel approach is to use a birefringent plastic walking surface and measure the change in light transmission produced by localized stress. This technique has been used by Harris and Ghiradella (1980) to explore the vertical forces produced by standing and walking crickets. The results suggest that coordinated downward forces similar to those found during walking can be detected in standing animals just before and after a walk.

In order to measure forces in the stick insect the free-walking animal was encouraged to step onto a small force transducer which formed part of a much larger walking surface (Cruse, 1976a). The results for walks on a horizontal surface are shown in Fig. 13. The magnitude of the forces generated are of the same order as the body weight. When walking upright, the legs always produce downward forces when in contact with the ground and middle and hind legs provide most of the support. In the horizontal plane the component perpendicular to the body axis acts outward in the middle and hind legs and is approximately zero for front legs. The forces along the body axis sometimes show a negative or forward directed component. In the front legs the forces are positive toward the rear but in the hind and middle legs, during the first half of the stance phase, the forces applied oppose the direction of motion.

Cruse (1976a) suggested that these forces result from the static weight of the meso- and metathoracic segments forming a self-supporting bridge from front to rear legs via the thorax. Thus, the reaction to these forces is what

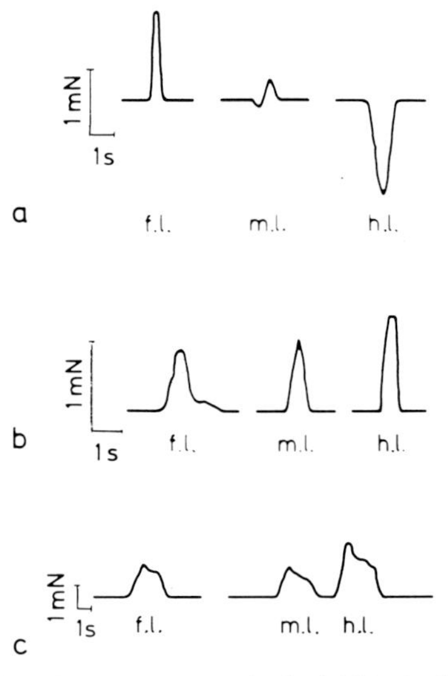

Fig. 13. The time course of the force components in the (a) longitudinal (backward = positive), (b) transverse (against the body = positive), and (c) vertical axis (downward = positive) when the animal is walking on a flat surface. The direction and intensity of the force exerted by each leg on the substrate can be read from the curves for a particular point in time. If the position of the individual leg joints is known for this time, the torque can be calculated for each joint. f.l., Fore leg; m.l., middle leg; h.l., hind leg. (From Cruse, 1976a.)

keeps the body aloft. This function is also performed by the outward-directed transverse components which form a similar self-supporting arch from leg to leg in the same body segment and appear to be sufficient to provide the necessary body support. An alternative or additional possibility is that the forces along the body axis are a direct consequence of the animal actively decelerating the body at regular intervals during a walk.

Measurements of the horizontal body velocity for an adult stick insect walking free, and the right and left wheel velocities for a similar walk on a two-wheel treadmill, can be compared in Fig. 14 (Graham, 1981). The body or the set of legs on each side makes large velocity changes in a very regular pattern during each leg cycle of gait I. Typically, the maximum velocity is achieved not at the end of the power stroke, as in most higher animals (Alexander, 1976; Cavagna *et al.*, 1977), but close to the midpoint of the stance phase. At the end of stance the body or leg movement is almost stopped. The body then accelerates again due to the action of the other leg tripod, while the first tripod is being recovered. When the recovered tripod touches down the legs again decelerate the body almost to a standstill. These large accelerations and decelerations which occur twice in each leg cycle are

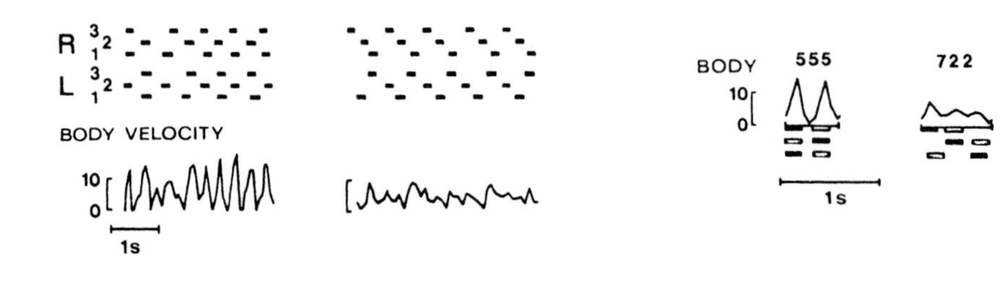

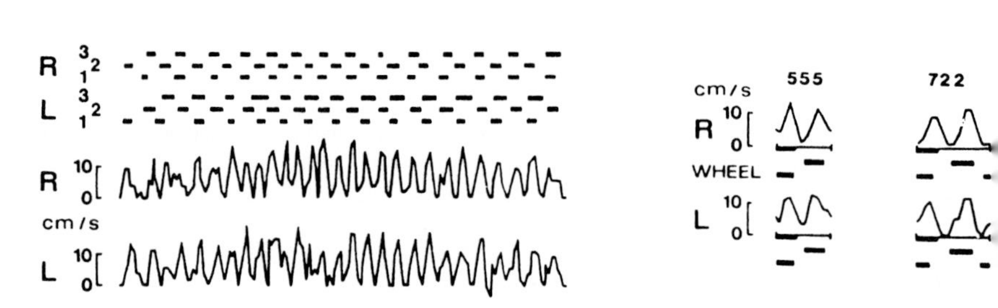

Fig. 14. Step patterns of protractions and velocity, (a) for a free adult stick insect with the velocity of the forward movement of the body at two different speeds; (b) for a walk on two light independently rotating wheels. On the right averages of six or more step cycles are shown at mean step periods of 555 and 722 ms. Filled and open blocks are for the legs on opposite sides of the body. (From Graham, 1981.)

clearly visible in the jerky or lurching motion which is characteristic of gait I in the stick insect nymph and high-speed walking in the adult.

These large decelerations are costly in the sense that they waste the kinetic energy acquired by the system and indicate a fundamental difference between the walking behavior of the stick insect and higher animals which only walk on horizontal or sloping surfaces and tend to conserve their kinetic energy. This particular feature may be a special characteristic of climbing animals in which maintenance of contact with the substrate is an essential part of the walk process.

2.8 WALKS ON INCLINED SURFACES

Several workers have examined the changes in insect walking behavior produced by climbing vertically or walking up or down a slope. von Holst

(1943) observed that the incidence of simultaneous recovery in legs of the same segment was high for *Locusta viridissima* when climbing vertically. Cruse (1976a) examined the amplitude, AEP, and PEP in climbing stick insects and reported a forward shift in both AEP and PEP and an increase in the amplitude of the leg stride. He also noticed behavior in the stick insect similar to that reported by E. von Holst regarding the timing of protractions (personal communication).

More recently, Spirito and Mushrush (1979) examined the changes in coordination for the cockroach moving up and down a plane inclined at 45° to the horizontal. In both situations the protraction duration is decreased relative to the step period when compared with that found in horizontal walks. For legs of the same segment these walks show a phase asymmetry across the body similar to that found in adult stick insects using gait II, with the exception of the pair L3-R3, which shows the opposite asymmetry to that found in the anterior leg pairs. The phase relations for legs on the same side of the body are similar to those found in horizontal walking with the exception of the pair R2-R3. The leg R3 is common to both exceptions, a finding which suggests that this leg plays a special role in uphill walking, and it is proposed that this leg is used as a stabilizing strut. Similar changes in coordination are present for the leg L1 in downhill walking and an analogous role is proposed for this leg. Again, these changes emphasize the flexibility of the control system coordinating the movements of the individual legs.

2.9 WALKING OVER OBSTACLES

One of the most important requirements of a general purpose walking system is its ability to deal with obstacles of various kinds. These obstacles can be divided into three overlapping classes. The first class of small-scale obstacles are those of a size range up to one-tenth of the height of the body above the surface in a normal standing position. Such obstacles may interfere with establishing foot placement and grip on the surface. The best procedure for dealing with them would appear to be one of placing the foot and waiting briefly for the foot to slide or roll into a position of support and take up a traction position. For the next class of obstacles ranging in size from one-tenth body height up to those approaching body height, a major modification in the position of the leg relative to the body is required, and when such a position of support is found it is advantageous in a multilegged animal to place more posterior legs in approximately the same position. Wendler (1964) found that the footprints of a stick insect walking on a flat surface tended to be grouped into sets of three prints on each side of the body, corresponding to the successive steps of front, middle, and hind legs on each side. Cruse (1979b) has shown, quantitatively, that the stick insect uses the behavioral trait found in donkeys, horses, and many quadrupeds of stepping close to their own forefoot prints. Such a strategy is particularly appropriate for an insect operating in a walking environment in which points of support are sparse and randomly placed such as in the twiggy interior of bushes. The only quantitative study of walking over precisely defined obstacles is that of Cruse (1976b). In this work the body height was examined for stick insects walking across a surface containing square section ditches and dykes 1 cm high. Negligible bending of the body occurred in dealing with such obstacles and the behavior can be represented by a simple model using passive springs for support. Thus the body adopts a position determined by the mean slope of the surface in the region of the obstacle.

For larger obstacles a major change in the orientation of the whole body would be necessary and in this case, unless the obstacle is avoided altogether, the walk system must attempt to clamber over or around the obstacle. Cruse also examined the reaction of the insect to large obstacles requiring a complete transition from horizontal to vertical walking and found that the animal avoided bending the body or losing contact with the middle legs by turning right or left during the approach to a wall. The final approach is made diagonally and is followed by a turn upwards when all legs have contacted the vertical wall. A similar approach is used for negotiating a cliff obstacle.

2.10 BODY STABILIZATION

Negotiation of large obstacles places heavy demands on the stabilizing system of a walking machine but a similar requirement exists for standing or walking on a horizontal surface. On vertical surfaces even when standing motionless some kind of control system is essential to prevent sideways or longitudinal folding of the legs. The stick insect appeared to be singularly prone to instability in this system as seen in the characteristic sideways rocking behavior often observed in these animals. Rocking sometimes follows immediately after walking or takes place while the animal is walking. Bouts of rocking can often be terminated by touching the animal or suddenly increasing the light intensity.

Pflueger (1977) has shown that this oscillatory behavior is central in origin rather than a relaxation of the stabilizing control system. The set point of the reflex control system is rhythmically changing. The significance of this behavior is not known in the stick insect but the behavior pattern appears to be similar to that found in standing locusts. Rocking movements were described by Wallace (1959), and Collet (1978) established that such movements are made by the locust to measure parallax and estimate the distance to nearby objects. This behavior may play a similar role in stick insects.

The existence of a simple servo-loop controlling femur–tibia angle has been conclusively demonstrated in a number of papers (Baessler, 1967, 1972b, 1973; Cruse and Storrer, 1977; Storrer and Cruse, 1977) for the standing animal. Similar reflex control is also present in the supporting coxal muscles that depress the femur (Wendler, 1964, 1972) and the protractor and retractor muscles of the coxa (Wendler, 1964; Graham and Wendler, 1981b). These control systems all act to maintain a constant joint angle against externally imposed forces primarily arising from gravitational forces acting on the body mass. Cruse and Pflueger (1981) and Cruse and Schmitz (1983) have shown that the femur–tibia stabilizing reflexes are also present in walking insects and act to maintain the required joint angles during a step.

3 Walks under controlled conditions

One of the major changes in the experimental approach to the study of walking behavior has been the increased use of sophisticated self-propelled or moving walking surfaces. Such devices are useful for several reasons. First, they simplify the gathering and analysis of data, and often allow complex temporal and spatial patterns to be statistically analyzed in minutes rather than months. Second, they allow a precisely defined modification of the walking environment of the animal. Parameters can be varied independently for either the whole system, selected groups of legs, or individual legs. These experimental perturbations permit manipulation of the system to expose the underlying logical structure. Third, they make possible direct electrophysiological recording and stimulation of the thoracic components of the walking system.

Such advantages, however, entail the risk that behavior may alter significantly on such devices. The only way that such risks can be minimized is by close comparison with free-walking behavior, when possible, and evidence of highly motivated, normally coordinated walks on the device when the animal is not subjected to a specific experimental stress.

3.1 ARTIFICIAL SUBSTRATES

A number of different methods have been devised to simplify data analysis by stabilizing the mean position of the animal or holding the body in a fixed position relative to a moving substrate. The Kramer–Heinecke sphere (Kramer, 1975, 1976) is an example of the first type. In this device a 50-cm-diameter sphere, weighing several kilograms, is rotated by two motors to maintain a target marker on the back of the animal at the vertical pole. The driven sphere is particularly valuable in orientation studies, as the animal is walking free on the sphere and the displacement produced by the motors produces a continuous map of the walk. The stability of the preparation is sufficient for trailing-lead electrode recording but the rapid cyclic changes in velocity which are characteristic of insect walks (Graham, 1972; Burns, 1973) cannot be followed by the high-inertia sphere and make it impossible to use hook or glass electrodes directly on the preparation (see Weber *et al.*, 1981, for a discussion of the sphere kinetics).

If the body is to be held rigidly for muscle, nerve, and neuron recording, then a self-propelled or driven substrate must be provided for the animal to walk on. Three possibilities have been explored in depth. The earliest technique described in detail by Wendler (1964, 1966) is the use of a self-propelled treadwheel. The animal is fixed above the wheel and drives it to the rear with the legs. This device has been modified (Graham, 1981) so that it is now possible to provide the animal with two light wheels, on a single axle, with approximately the same inertia as the insect body. The wheels can be counterbalanced to provide an upthrust against the legs equivalent to the supported weight. The rotational resistance to movement is less than 20 mg of force and, in the latest system, the rotational torque can be varied to alter the horizontal loading on the legs and simulate up- and downhill walking (Graham and Godden, 1985). The two-wheel configuration also permits turning behavior to be expressed, and has shown that coordination across the body is independent of mechanical coupling through the substrate and has its origin within the nervous system.

A similar approach to the treadwheel uses a light foam sphere as the self-propelled walking surface. In the earliest experiments the animal holds up the unsupported ball itself (Decomyn, 1973). In later work the sphere is supported on an air cushion (Carrel, 1972; Spirito and Mushrush, 1979) or water bath (Macmillan and Kien, 1983). These approaches have the advantage that the legs can maintain their normal relationship during turning in a manner closely analogous to free walking and the support provides a compliant upthrust on the legs.

The second approach is to use a motor-driven surface or belt. This approach has been particularly successful in crustacean preparations. It was first used to simplify data collection in lobsters by adjusting the belt speed to keep the free-walking animal under a camera (Macmillan, 1975). The same principle has been developed over the period of this review in a succession of reports (Evoy and Fourtner, 1973; Barnes, 1975a, 1977; Ayers and Davis, 1977a–c; Clarac, 1977, 1978, 1981). In the experiments of Ayers and Davis and those of Clarac, the body of the crustacean is clamped while the legs are forced to walk at a wide range of different speeds by a motor-driven belt. The latest reports of Chasserat and Clarac (1980, 1983) contain the most detailed comparison yet available between this driven-walking and the free-walking behavior and show that walking locomotion is more precisely coordinated under the driven condition (see also Clarac and Cruse, 1982, for a comparison of propulsive forces under these conditions).

Early attempts to produce a similar response in insects were unsuccessful. However, with a combination of walk-wheel and a driven belt, Foth (in preparation) has succeeded in inducing walks on the wheel which can be greatly prolonged if at least one leg walks on a driven belt. It is also possible to induce driven walking in all the legs of an insect with simultaneous stimulation and slow initial belt movement.

The third approach follows the discovery by Sherman *et al.* (1977) that insects can walk on a slippery glass surface. The features of such walks have been compared with free-walking behavior in stick insects using a mercury surface (Graham and Cruse, 1981) and a silicon oiled glass surface (Cruse and Epstein, 1982; Epstein and Graham, 1983). By careful adjustment of the body relative to the substrate long, normally coordinated, sequences of steps can be obtained. The importance of these experiments is that on such a surface all mechanical interactions via the substrate are abolished. The legs function independently of each other and the loading or unloading of individual legs can have no mechanical influence on the others. In this situation, the observation that legs on the same side often move to the rear at different velocities (Epstein and Graham, 1983) suggests that the substrate integrates the propulsion forces of the individual legs rather than the nervous system. Coordination between the legs becomes more sharply defined when mechanical interactions between the legs are removed.

3.2 EFFECTS OF VERTICAL AND HORIZONTAL LOADING

A considerable volume of research has been carried out on the influence of load on the crustacean walking system. Evoy and Fourtner (1973) examined the influence of loading on the crab when walking sideways up a 45° slope and on a braked treadwheel. Their results showed a broadening of the phase relation between adjacent legs on the same side. This effect is of sufficient magnitude under the heaviest loads to produce uncoordinated locomotion. The frequency of stepping was reduced from 2 to 0.2 steps per second. This was mainly caused by increases in the duration of the retraction stroke. The duration of the protraction stroke in loaded animals remained independent of step period and was equal to the shortest duration observed in unloaded animals. The slope of the protraction duration versus step duration plot for unloaded and loaded animals was 0.31 and 0.02, respectively.

Macmillan (1975) added weights to the body of a lobster or persuaded it to pull a load when walking free or on a moving belt. He reported no significant change in the mean phase relations between legs but a general broadening of the distribution of phase about the mean value and an overall reduction in mean velocity. The ratio of t_p/t_r was independent of loading for pulled loads up to 20% of body weight and carried loads up to 6% of body weight (not immersed). Higher loading produced serious perturbations in the phasing of leg movements similar to those reported in the crab. Barnes (1977) applied vertical loading to crayfish walking on a self-propelled treadwheel. This animal retained normal coordination under loads of up to 10% of the body weight and showed little change in step frequency. Grote (1981) also loaded the body vertically and compared locomotion under water and on land (carrying an additional load). The range of loading was from 1 g force (under water) through 15 g (the normal body weight on land) to an added weight of 27 g. This large load differential produced a change in the duration of the retraction stroke from 0.7 s to 2.5 s but no significant change in protraction duration.

In insects Pearson (1972) examined the changes in motor output to the tibial extensor muscles when a cockroach dragged a weight on a thread. He observed an increase in motor output but did not examine walking coordination (Fig. 15). Spirito and Mushrush (1979) investigated the changes in the pattern of stepping of the cockroach when walking at low speeds (two to seven steps per second) on an air-supported ball and free on a plane surface inclined at 0 and 45° to the horizontal. On the horizontal surface free-walking animals displayed coordination similar to that reported by Delcomyn (1971a). The predominant step pattern is the tripod with all adjacent legs

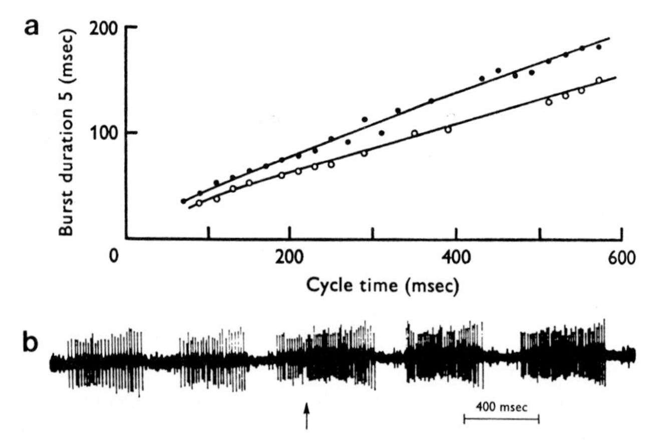

Fig. 15. (a) Decrease in levator burst duration with an increase in resistance to leg retraction. (●) No load; (○) dragging weight of 1.5. Each point is the mean duration of bursts in axon D_5 for cycle times between t and $t + 20$ ms. (b) Abrupt increase in discharge rate of axon D_6 (arrow) with a sudden increase in resistance to retraction of the metathoracic leg from which recordings were being taken. (From Pearson, 1972.)

having a relative phase of 0.5. Uphill and downhill walks showed significant changes in the phase relations between the legs. The step pattern became metachronal and asymmetric and the *p*/*r* ratio showed a lesser dependence on step frequency (Fig. 16). The new gait was qualitatively similar to that described as gait II in the stick insect. Ball walking showed a similar coordination to that found in walking up and down the sloping surface, but the dependence of *p*/*r* ratio on step frequency was similar to that found in the free-walking insect.

Walking up a slope involves two complementary changes in loading of the walk system. The component parallel to the surface is increased while that perpendicular to the surface is decreased. Therefore, the animal finds it easier to hold itself away from the surface but more difficult to move itself forward. It is possible that these changes may compete with one another and thereby weaken the observable effects of each influence on the walking system. In an attempt to isolate the two components and study them separately two light treadwheels separated by 28 mm were placed on a single fixed axle. The loading component parallel to the walk surface could be altered by applying friction at the wheel hub (Foth and Graham, 1983a,b) or applying a

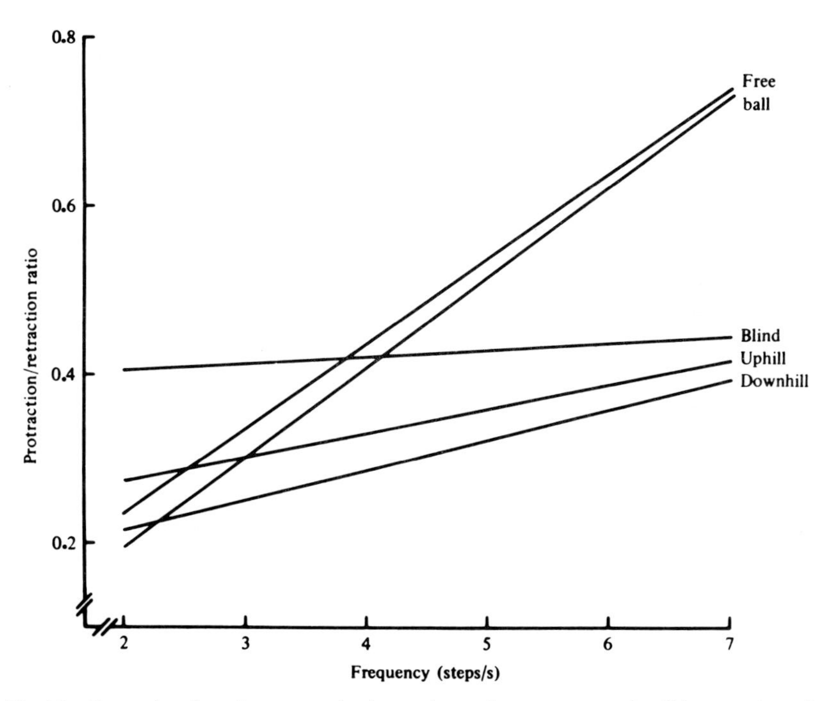

Fig. 16. Regression lines for protraction/retraction ratios at a range of walking speeds under five conditions. (From Spirito and Mushrush, 1979.)

controlled torque to each wheel using a low-inertia electric motor (Graham and Godden, 1985).

The behavior on this system is shown in Fig. 17 for an adult walking on the compensated treadwheels with a vertical component corresponding to a free walk (0.5 × body weight, because the abdomen is usually dragged and not supported by the legs). The wheel friction at the rim is less than 20 mg force and the gait is almost identical to that found in the first instar nymph using

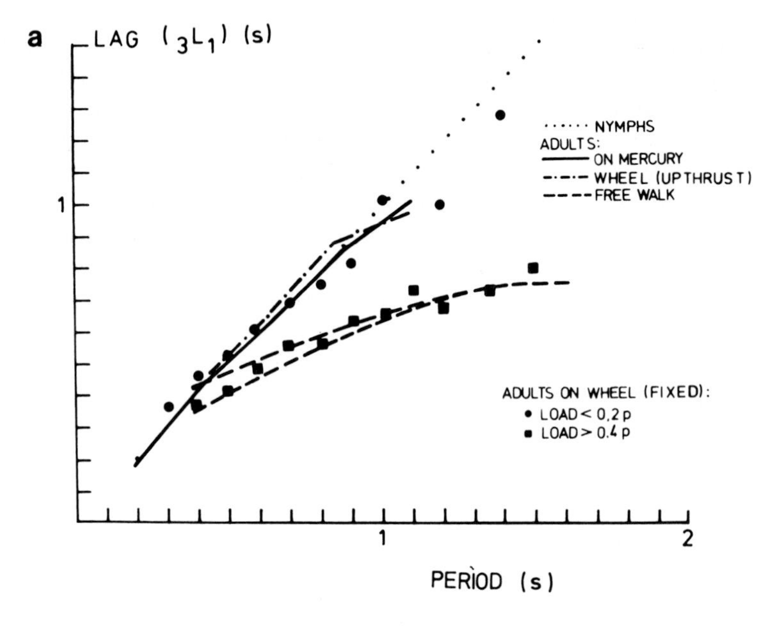

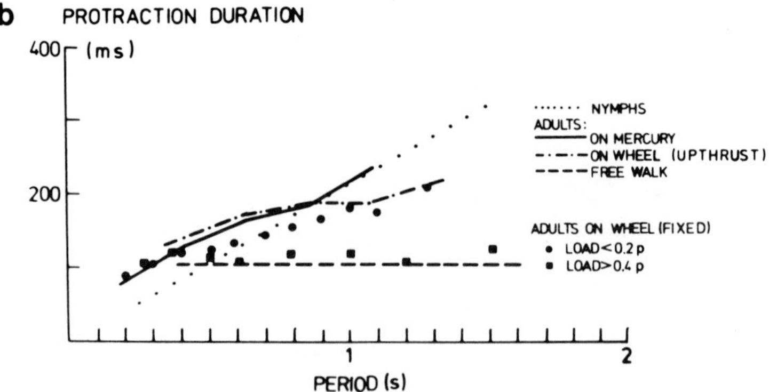

Fig. 17. Lag $_3L_1$ versus period curves for walks under different conditions (lines join mean value points). (b) Protraction duration versus period curves for the walks in (a). (From Foth and Graham, 1983a.)

gait I. With the axle fixed the results are similar to those found for the wheel with upthrust when the horizontal friction is low. As friction is increased the gait remains type I up to 20% of the body weight. Above this critical value the effect of increased horizontal loading is to progressively increase the duration of the stance or retraction stroke while the protraction duration remains short and is independent of load and step period. The duration of the protraction stroke is equal to the minimum value used by the free-walking or wheel-walking animal ($t_p = 100$ ms).

These experiments show that changes in horizontal load (load component parallel to the walk surface in this case) can explain the appearance of the type I gait in adults walking on light wheels, the difference between first instar gait (type I) and adult (type II), and the appearance of gait II in cockroaches walking uphill. They also explain the gait II behavior of the heavier free-walking insects which tend to "drag their abdomens behind them."

The response to increased load in insects consists of at least five separate components. There is an increase in motor output which attempts to maintain a short retraction duration (Pearson, 1972; also found in stick insects, Graham, unpublished). At higher loads the retraction duration progressively increases with load when no further increase in motor output is possible. In this region the protraction duration is minimized by using maximum leg-recovery motor output at all loads and speeds and t_p is independent of load and step period (Foth and Graham, 1983a). The relative independence of the lag (between adjacent legs) in response to load causes a change from the tripod step pattern to a separated metachronal pattern which places more legs on the ground at the same time. The observed changes of these components all act to improve power output and duty cycle and adjust appropriately to the externally imposed load. Finally, the mechanical advantage of the leg system is altered by moving both AEP and PEP forward relative to the unloaded condition (Cruse, 1976a; Baessler, 1977a). Similar results have been reported for some crustacea and these were reviewed and used for a model of load dependence by Cruse (1983).

3.3 TRANSFER OF LOADING INFORMATION

Baessler (1977b) and Cruse and Saxler (1980a) in insects, and Evoy and Fourtner (1973) in crustacea, all present evidence that the response to increasing load in one leg produces an increase in other legs which are not directly influenced by the load. These effects are most readily demonstrated in the wheel-loading experiments of Foth and Graham (1983b). In this preparation an increase in tangential loading on one side produces a dramatic increase in the rate of stepping on the unloaded wheel, showing that loading information is directly transferred to the unloaded side. These influences must be transferred via the nervous system and it has been suggested that stress-detecting sense organs produce this response. Such a conclusion represents only one possibility and it need not be inferred that the sense organs involved directly measure the stress created by the load. As an example, position-detecting organs could compare the current position of the leg with that expected for the motor output used. Such a comparison would make it unnecessary to know the orientation and magnitude of the gravity vector relative to the body and would, therefore, be independent of the exact configuration of the walk system in space. This would appear to be a particularly useful control concept for a climbing animal.

The use of position detectors to measure load indirectly could also explain how the stick insect nymph can change from gait I to gait II without any apparent change in external load. If this hypothesis were correct, it would merely require that the animal lower its motor output for retraction relative to the power stroke motor output normally required for that performance. This reduction in motor output, without a corresponding change in the expected performance, would then be interpreted by the walk system as an apparent increase in load and would produce the observed change from gait I to gait II. If load detection depended only on stress sense organs, such a change would require a complex calculation of the way in which the measured stress related to the current orientation of the body to the gravity vector.

3.4 UNIQUE PROBLEM OF DOWNHILL WALKING

Walking up a slope is similar, in some respects, to walking on a horizontal substrate against an increased resistance to motion or a positive load applied against the direction of motion of the substrate. The strategy for dealing with this problem has been described in the work of Pearson and Iles (1973), Cruse (1976a, 1983), Baessler (1977b), Spirito and Mushrush (1979) and Foth and Graham (1983a,b). It only requires the further assumption that stabilizing forces are present, including the possible use of a posterior leg as a strut or prop, in order to maintain the standing or walking animal in an appropriate orientation to the sloping surface.

Most insects, including adult stick insects, can also walk downhill. These insects have a preference for upward walking under certain conditions (Wendler, 1965, 1966; Baessler, 1965) but they will readily walk down steep slopes to avoid bright lighting or when disturbed during daylight. A downhill walk presents a special problem because the muscles controlling the leg movements must either reverse their role during the stance phase or include a braking stage in the cycle of leg movement, to avoid a precipitous descent in which the legs can no longer keep up with the acceleration of the body produced by the component of gravity acting down the plane.

In mammals this problem is, to some extent, avoided by holding the body vertically above the feet in bipeds or sitting back on the hind legs in quadrupeds. In this manner a long descent can be negotiated in an equilibrium state that only requires the maintenance of activity in the standing musculature and a hopping or sliding movement. The steepness of the slope that can be descended is determined by the animal's ability to maintain this standing equilibrium. In insects and legged vehicles that have long bodies relative to their ground clearance, such an equilibrium stance is not possible and such systems must adopt an alternative strategy.

One possibility is to go down the slope backward using insufficient retractor output and reversing the direction of the movement cycle, speed being controlled by the amount of force generated in the retractor muscles. This is the method used by bipedal climbers for very steep slopes. The author is not aware of any insects that use this strategy.

A second alternative would be to descend head first and use the muscles normally associated with forward movement of the leg to permit slow movement of the legs to the rear, controlling the rate of descent by the magnitude of the force generated in these muscles. Thus the protractor muscles would be continuously active throughout the leg cycle, with normal bursts of activity during protraction and a controlled level of activity during the retraction stroke to determine the speed down the slope.

Another alternative is to use protractor and retractor muscles in the normal way for the propulsion and recovery of the legs and insert a braking stage at the beginning or end of the stance phase by activating the protraction musculature while the legs are still in contact with the ground. Using this method the system moves down the slope in a series of lurches. Similar behavior has already been reported in horizontal free walking where the forward velocity of the body is reduced almost to zero twice in each leg cycle (Graham, 1972, 1983). Experimental measurements of protractor and retractor activity in middle legs on treadwheels of high inertia (Graham and Wendler, 1981a) show that bursts of protractor activity appear at the start of the stance phase and may continue for up to 60 % of the stance duration (Fig. 18).

In current studies on simulated downhill walking using torque, in the direction of substrate movement, applied to light treadwheels (Graham and Godden, in preparation) stick insects can use either the "braking" strategy or a continuously controlled descent. These results indicate that the system has been designed as a general purpose system and has not been optimized for horizontal walking alone, for which efficiency criteria would dictate that momentum be conserved from step to step.

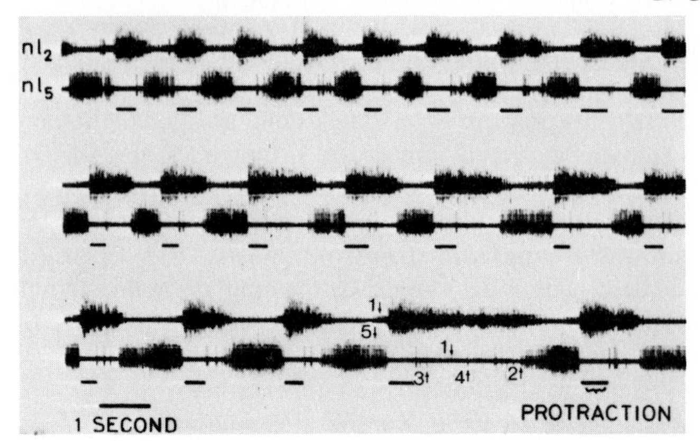

Fig. 18. Motor output from the mesothoracic protractor (nl_2) and retractor (nl_5) nerves at different walking speeds. Black bar shows the duration of protractions (± 20 ms). (From Graham and Wendler, 1981b.)

3.5 BACKWARD WALKING

Reversal of the normal forward walks is common in crustacea (Ayers and Davis, 1977a–c; Ayers and Clarac, 1978; Clarac, 1982) and has also been reported in arachnids (Land, 1972). In the hunting spider such reversals are primarily used to rapidly rotate the animal by walking forward on one side and backward on the other. The ipsilateral coordination is similar in both forward and backward walks and such reversals in the step sequence can take place at any point in the leg cycle. A considerable time interval can sometimes exist between the termination of a forward walk and the initiation of a backward walk on the same side, indicating that the system possesses a memory of the precise position in which each leg stopped. It is difficult to see how such a memory feature can be readily incorporated into an endogenous oscillator, and Land has suggested that the leg position itself must be a component of the oscillatory mechanism.

In insects spontaneous backward walks have not been reported in the literature, except on the inside of tight turns in the bee (Zolotov *et al.*, 1975) and the cockroach (Franklin *et al.*, 1981). However, it is possible to stimulate bilateral backward walks by slight traction on the antennae (Graham and Epstein, 1984). Coordination is irregular, and both anteriorly and posteriorly directed metachronal sequences of protractions can be found (Fig. 19). Often the front legs step with long strides at half the frequency of the middle and hind legs and show a coordination similar to that reported for forward walking in the long-legged grasshoppers (Graham, 1978a). The walking system is reversible but coordination is less precisely controlled. If only one

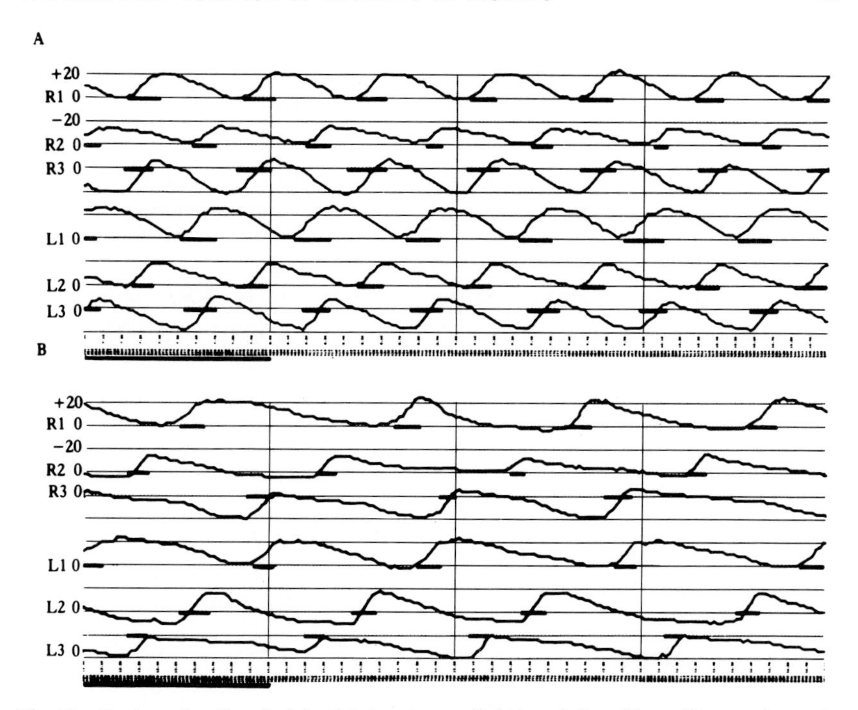

Fig. 19. Backward walks of adult stick insects on oiled glass, induced by pulling gently on the antennae. Protractions are shown by black bars along the line defining the position of the coxa–thorax articulation. The records show the movement of the distal tip of the femur in the direction parallel to the body axis. The time bar represents 1 s. Note that in this figure and in subsequent figures showing leg displacement the order of legs is from front to rear. This is essential for the correct presentation of the relative displacements of the legs. In this particular figure the vertical distance between traces does not represent the distance between the legs to avoid excessive overlap in the leg traces. (From Graham and Epstein, 1985.)

antenna is held, then backward walking is faster on that side, emphasizing the independence of the two sides.

This independence is most dramatically demonstrated when the animal walks on two separate treadwheels or on a slippery glass surface. Under this experimental condition one side may walk faster or slower than the other, and in the antenna holding experiments one can observe a fast backward walk on one side and slow backward or forward walking on the other.

3.6 INTERRUPTION OF LEG MOVEMENT

The use of treadwheels or a slippery surface and the consequent immobilization of the body make new kinds of experimentation possible. Interruption of the leg movement is a particularly good example. The protraction stroke

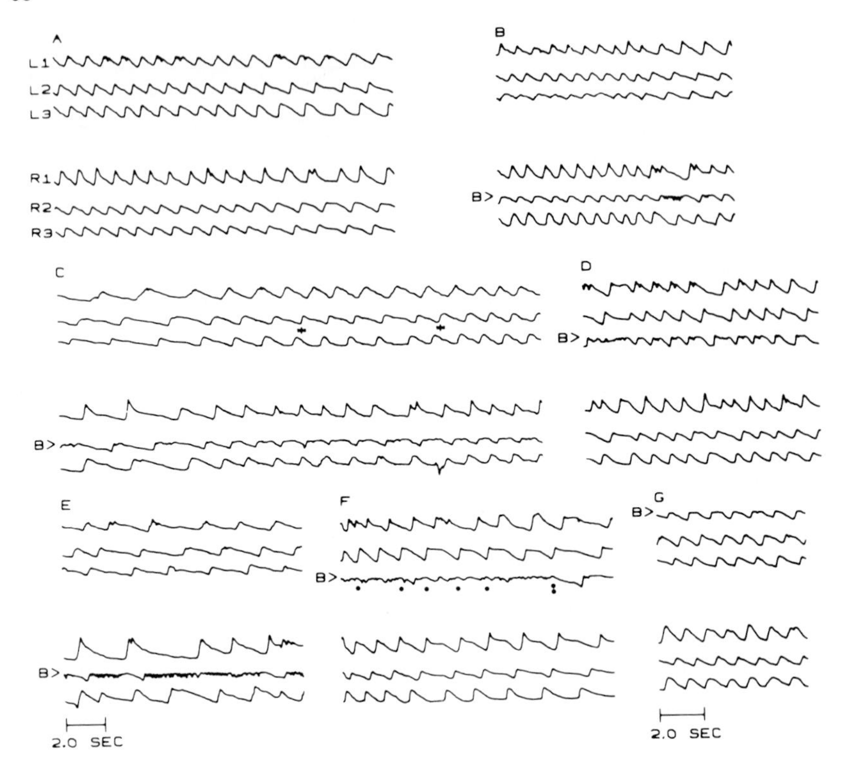

Fig. 20. (A–G) Stick insect locomotion with different barrier positions. All seven parts present step sequences in the format registered by the walking wheel (Wendler, 1978). Beginning at the top, the six traces represent the femur positions of the six legs in the following order: L1, L2, L3, R1, R2, R3. Movement of the leg in the anterior direction corresponds to displacement of the trace upward. The time scale for (A–F) is at the lower left; that for (G) is at the lower right. (A) Unobstructed walking. For all six traces, the most anterior position is a well-defined peak as each leg switches smoothly from protraction to retraction. For (B–G) the approximate position of the barrier present throughout each step sequence is indicated by a B; its true position is better expressed by the restriction it imposes on forward protraction of the obstructed leg. When this leg strikes the barrier, its rapid forward motion ends. During the rest of protraction, the femur remains near the barrier while it attempts to pass by it. Large up and down movements, rotations of the femur, and brief withdrawals from the barrier cause the small fluctuations in apparent anterior–posterior position particularly evident in (E) or in the second and third to last steps by R2 in (B). The most significant effect of the prolonged protraction during 1:1 walking (B–D) is to delay the protraction of the adjacent, rostral leg. (B) Middle leg (R2) obstruction; 1:1 stepping. (C) As (B); here the contralateral middle leg (L2) steps either just ahead of the blocked leg—between the two stars—or at the end of the blocked protraction. (D) Rear leg (L3) obstruction; 1:1 stepping. (E) More severe middle leg obstruction. Note that the prolonged protraction of R2 causes R1 to make fewer steps than the other four legs. (F) More severe rear leg obstruction. Preceding the final full retraction (colon) of the blocked leg (L3), small retraction movements (dots) accompany L2 protraction. (G) Front leg (L1) obstruction; 1:1 stepping. (From Dean and Wendler, 1982.)

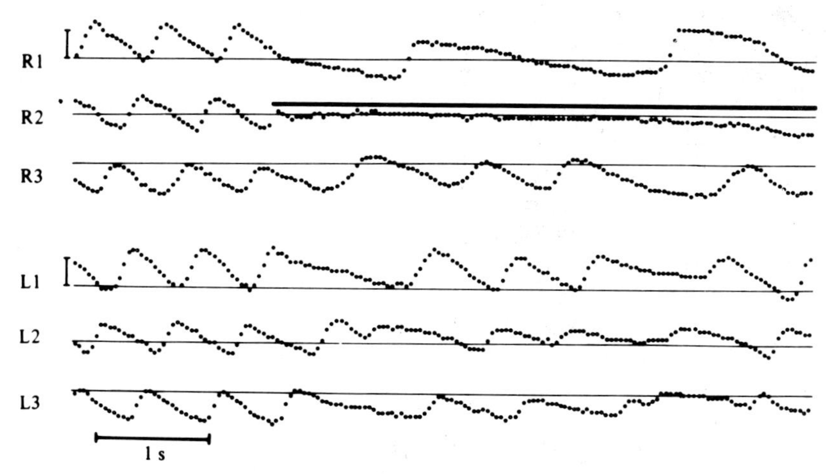

Fig. 21. Movement of the legs of an animal walking on a glass plate covered with silicon oil. The bar shows interruption of protraction of right middle leg. Position values are given for each leg separately and are related to position of the corresponding coxa (horizontal line). Vertical bar represents 20 mm. Upward deflection corresponds to protraction movement. (From Cruse and Epstein, 1982.)

can be prolonged by simply blocking forward movement by a wire or pin held vertically in front of the leg (Cruse and Saxler, 1980a). Experiments by Dean and Wendler (1982) and Cruse and Epstein (1982) show that the primary effect of the block is to reset the timing of the leg in front on the same side and alter its PEP, respectively (Figs. 20, 21, and 22). No demonstrable effect was found on the leg posterior to the blocked leg. Across the body no effects could be detected with the exception of blocking the hind leg, which tends to delay the protraction of the contralateral hind leg. These results indicate that the

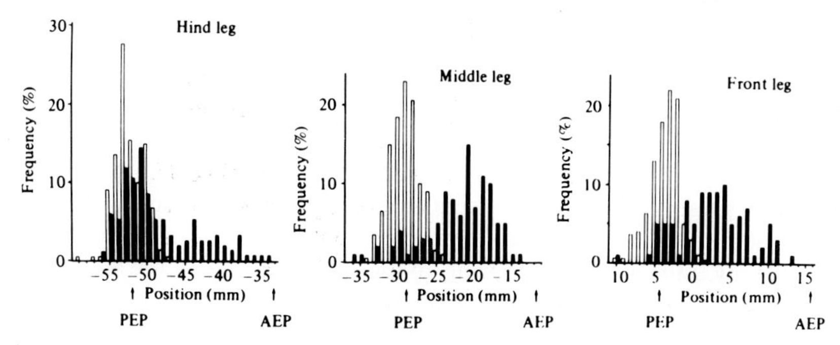

Fig. 22. Distribution of position of PEP when the leg is standing on the slowly moved platform (black columns) and of control (white columns). Mean values of PEP and AEP of the control are shown by arrows. Sample sizes: front leg, 98 (control, 297), middle leg, 102 (control, 200), hind leg, 152 (control, 220). (From Cruse and Epstein, 1982.)

anterior leg and, in hind legs, the contralateral leg are discouraged from protracting when the blocked leg fails to complete its own protraction stroke.

Blocking of the retraction stroke was first reported by Wendler (1964). The insect was persuaded to step onto a fixed stick close to the rim of the treadwheel. Further experimentation has been carried out by Baessler (1967, 1979b), Cruse and Saxler (1980a,b), Cruse and Epstein (1982), and Epstein and Graham (1983). When the leg steps onto the stick it is effectively removed from the ongoing walk behavior. This leg no longer makes protractions unless it is moved toward the normal PEP position. Thus the leg is not incorporated into the walk unless it is allowed to move into the PEP region. The leg actively produces a rearward force while standing, and this force is weakly modulated with the rhythm of the other walking legs. This form of immobilization is reversible and the leg will return to normal walk behavior whenever the stick, which it is grasping, is moved sufficiently toward the rear.

Precise comparison of the timing of the force modulation in the standing leg with the walking legs is difficult as it is not possible to define the force level that corresponds to the onset of the protraction stroke in a walking leg, if indeed the leg can be considered to be walking when in this state. If it is assumed that the force maximum is equivalent to the onset of retraction, the phase relations between the walking and standing legs differ markedly from those found in normal walking coordination. As an example, if two neighboring legs stand on sticks the force oscillations can be in phase. This is quite different from the behavior of neighboring legs when walking and suggests that standing legs are in a special state which may be only weakly related to normal walking.

If the standing leg is moved slowly to the rear the walking condition is more closely approximated. The PEP of the slowly moving leg is considerably further forward than in the controls (see Fig. 22), and the leg lifts in protraction at the time predicted by the other walking legs. This suggests that, as a walking leg moves toward the rear, it becomes steadily more disposed to protract, and the observation of coordinated lift-off before the normal PEP is reached indicates that delaying influences are followed by an "excitatory influence." This may correspond to the "inhibitory rebound" proposed by Pringle (1940) and Pearson and Iles (1973). The decision to protract is also facilitated by a decrease in the propulsive loading of the leg when close to the PEP. The leg may be prevented from protracting close to the PEP by using a weak leaf spring, to maintain the load on the leg, instead of a rigid stick (Cruse, personal communication).

3.7 TARGETING OF RECOVERY MOVEMENTS

Another new discovery resulting from treadwheel use is the ability of a leg to home in on the tarsus of the leg in front. Cruse (1979b) found that a

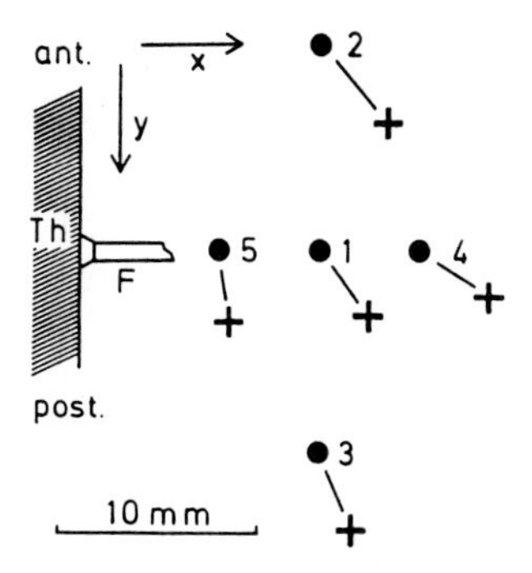

Fig. 23. The five test positions of the middle leg tarsus (dots) relative to the insect's thorax (Th) and the mean points for each of these test positions where the hind leg tarsus reached the platform (crosses), viewed from above. The proximal part of the femur of the middle leg is shown schematically (F). (From Cruse, 1979b.)

targeting interaction existed between adjacent legs on the same side. In the normal sequence of stepping a posterior leg places itself just behind or alongside the leg in front and, when it is in position, the leg in front protracts. The posterior leg does not place itself at the same position relative to the body, as might have been expected, but actually follows the position of the leg in front (Fig. 23).

It must be assumed that the hind leg has information on the whereabouts of the tarsus of the leg in front. This comes, in part, from the chordotonal organ of the anterior leg, because reversal of the afference from this organ (Graham and Baessler, 1981) causes incorrect lateral placement of the posterior leg, a finding consistent with the inverted signal from the operated sense organ. If incorrect placement leads to the posterior leg actually stepping onto the tarsus of the leg in front, then the posterior leg lifts again and repositions itself. Such errors appear most frequently in decerebrated animals (Graham, 1979b), but the response can be easily elicited in slow-walking intact animals by touching the middle leg tarsus with a fine brush as the hind leg completes its protraction. Stimulation at other parts of the leg cycle produce no lifting movement in the posterior leg, suggesting that the response depends upon the relative phase of the two legs (Fig. 24).

These recent discoveries emphasize the complex nature of the walking process, but they also indicate an elegant structure capable of dealing with a wide variety of external conditions using a minimal control system. The relative ease with which a particular subsystem can sometimes be separated for analysis by careful experimentation is most encouraging. As in this example many of the interactions between legs depend upon the controlling leg having a particular temporal or spatial relationship with the controlled leg. These phase-dependent responses have been investigated and cataloged by Baessler (1976, 1977a, 1979b, 1983), and indicate that the reflexes

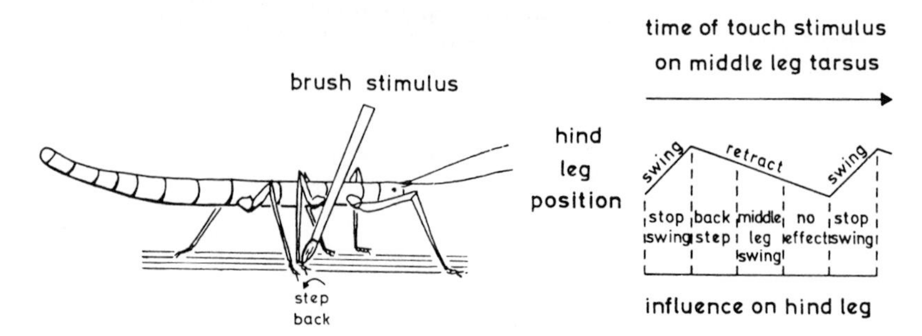

Fig. 24. The treading-on-tarsus (TOT) reflex. Left, the experimental setup; right, the response to middle leg stimulation during different phases of the hind leg step. (From Graham, see Baessler, 1983.)

controlling the cycle of leg movement in slow walking depend heavily upon where the leg is in its step cycle and the temporal or spatial phase of the other legs. Similar phase-dependent reflex properties have been reported in crustacea (Vedel, 1980; DiCaprio and Clarac, 1981) and vertebrates (Grillner and Wallen, 1982).

4 Motor output patterns during walking

So far our examination of insect walking has been restricted to the movements of whole legs and their interactions with each other. In the stick insect, individual leg movements are based upon a complex interaction of some 18 muscles in each leg (not including several muscles moving the tarsus). In the stick insect these can be combined loosely into four functional groups responsible for levation, depression, recovery, and propulsion. At different parts of the leg cycle most of these muscles must also be expected to make some contribution to stabilizing the system as a whole. A precise understanding of what happens during walking at the muscular level is difficult, but it may well be that in insects a simple and crude system has perforce been developed which is amenable to our limited analytical capabilities. It is hoped that either the above functional requirements may be separable at different times in the cycle or perhaps individual legs or joints may perform these functions almost exclusively. The most exciting new developments suggest that this may indeed be the case.

4.1 STRUCTURE AND FUNCTION OF THE LEG

It has already been mentioned that the structure of a leg and even the position of the leg on the body vary with the insect. Front legs appear to be modified for their forward arc of movement, while hind legs are often adapted to a jumping role and are usually oriented toward the rear. The middle leg is often something of a compromise.

In the stick insect the middle leg may operate either forward of a plane perpendicular to the body axis containing the coxa–thorax joint, like a front leg, or behind this plane, like a rear leg. Most often the leg steps across this plane with its AEP in front and PEP to the rear (Fig. 7). The configuration of the leg in this plane is shown in Fig. 25. If one constructs a line joining the subcoxal joint and the tarsus, then the leg as a whole is used rather like a rowing oar. The levator and protractor lift and swing the tarsus forward, and the depressor and retractor provide the support and propulsion for the body. The muscles moving the tibia extend the leg during swing and at the end of the propulsion stroke and may actively flex it during the early part of the stance phase (Cruse, 1976a; Epstein and Graham, 1983). Joint angles may be measured by chordotonal organs, strand receptors, multipolar tendon receptors, hair rows, and small hair fields situated close to the joints (see Section 6). There are also numerous companiform sensilla in these regions which respond to stress generated by the muscles and gravity acting on the body.

The individual legs of the stick insect are not as highly specialized as those of some insects and are similar in structure to the front and middle legs of the

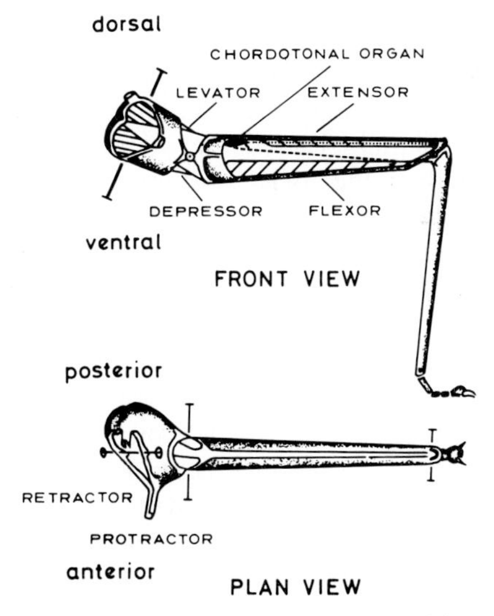

Fig. 25. Schematic drawing showing the articulation and some of the musculature that controls joint movement in the left leg of a stick insect. The tendons of the protractor and retractor coxae muscles are only partially shown. The dashed line joining the chordotonal organ to the flexor tendon represents the crossed apodeme operation. Normally, the chorodotonal organ apodeme runs parallel to the extensor tendon and attaches to the tibia dorsal to the femur–tibia articulation axis. (From Graham and Baessler, 1981.)

cockroach, locust, and grasshopper. The orthopterans often have large rear legs two to three times the length of the anterior legs, and the plane of action is perpendicular to the walk surface and almost parallel to the body axis. The cockroach also has larger hind legs but in this insect, as in many beetles, the plane of action is parallel to the walk surface. Thus in these insects the coxa is oriented to the rear and the levator and depressor of the femur and the flexor and extensor of the tibia flex and extend the leg. Here the hind leg acts like a piston while middle and front leg action is more oarlike.

Many insect species are winged and the motor apparatus for flight takes up a major part of the thoracic space. In the stick insect *Carausius morosus* the mature females are not winged and the thorax contains only the gut, respiratory tracts, nervous system, and musculature for driving the legs. The general arrangement of the thoracic muscles for the mesothoracic segment is shown in Fig. 26. In this scale diagram the body wall has been stretched transversely to form an almost planar surface. The protractor and retractor muscles lie parallel to each other along the anterior body wall, and respectively they pull the lateral and medial parts of the coxal rim forward. The axis of articulation of the coxa–thorax joint lies between these tendons and the muscle forces produce the recovery and propulsion movements of the whole leg. The upper end of this axis is a rolling and tilting joint and may be considered as a close approximation to a ball joint. The lower end is attached to the body wall by a short strut (trochantin), which lies approximately parallel to the body axis, pointing forward, and is articulated at both ends. This strut allows the lower end of the coxa–thorax joint to move toward or away from the body to a limited extent and permits the leg to lie parallel and in contact with the body in the "twig" posture typical of mimesis. The lower end of this joint is pulled toward the body by the sternal depressor muscles, and levation of the coxa is produce by a pair of muscles attached to the ventral surface of the body close to the sternal depressors, which pull down on the inboard part of the coxal rim and rotate the subcoxal axis outward about the upper ball joint.

The only other important thoracic muscles contributing to locomotion are the tergo- and pleuro-trochanteral muscles which span two segments. These two large muscles lie between the protractor and retractor and pass down into the coxa, where both are attached to the tendon of the depressor of the femur which is located in the coxa. The action of these muscles is complex and depends upon the orientation of the leg relative to the body. They can contribute to the later stages of protraction or retraction and probably provide an important contribution to depression of the leg when they are actively contracting. In the stick insect the description of muscle action given here is also true for the hind leg, with only minor modification for the specific orientation of this leg, as the musculature and nerve organization are similar.

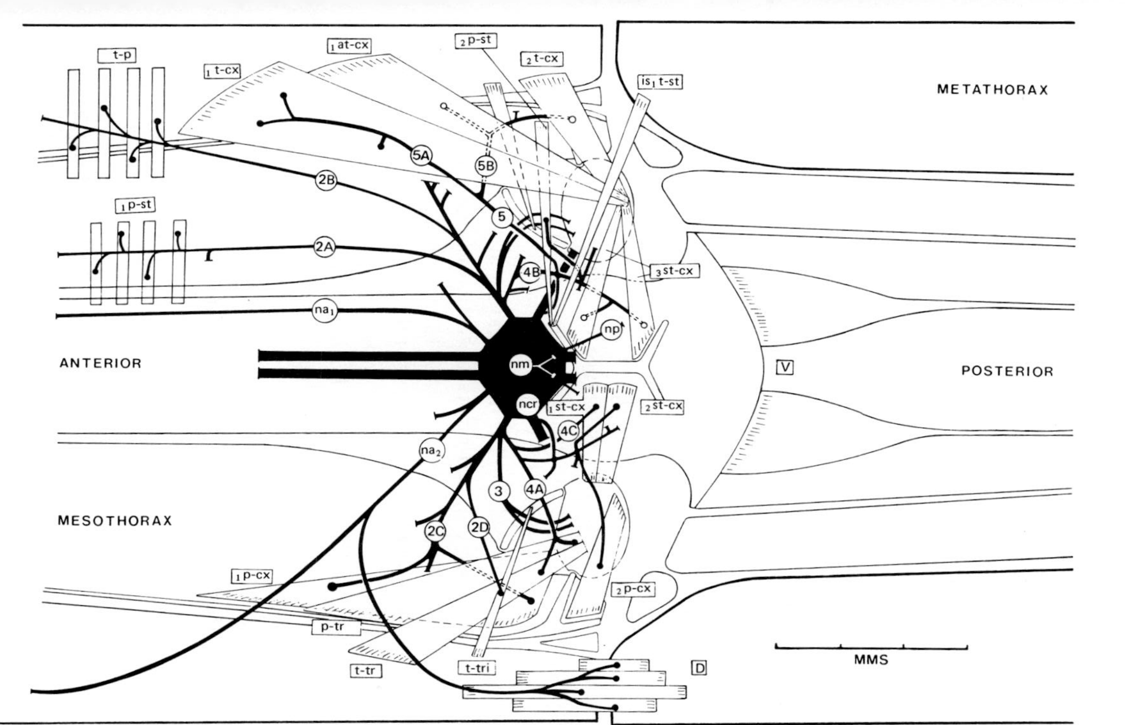

Fig. 26. Scale diagram of the thoracic neuromuscular anatomy of the mesothoracic leg of the adult stick insect *Carausius morosus*. Muscle designations are similar to those of Marquardt (1943) with some abbreviation and there are some differences in muscle attachment points and nerve branching. The final branch patterns are variable and the most divided state is shown. For example, in several preparations nerve 4C is not separated from 4B until it reaches muscle 1 st-cx. Levators: 3 st-cx. Depressors: 1 st-cx, 2 st-cx, p-tr, t-tr. Retractors: 1 t-cx, 1a t-cx. Protractors: 1 p-cx. Ventral muscles, v; dorsal muscles, D. Tergo-pleural and pleuro-sternal ventilation muscles, t-p, p-st. Tergo-trochantinal muscle, t-tri. Intersegmental muscles, t-st. Medial nerve, nm. Nervus crurus (main leg nerve), ncr. Anterior nerve, na; posterior nerve, np.

The front leg articulation is more complex with considerable longitudinal compression of the muscle system and considerably more freedom of movement.

4.2 MOTOR PATTERNS

The earliest recordings of the motor activity in walking insects are those of Hoyle (1964) and Ewing and Manning (1966). They examined the motor output to the extensor and flexor muscles of the tibia in the hind legs of locust and cockroach, respectively, using extracellular wire electrodes in a free-walking animal. The activity in these muscles showed considerable variability and suggested that movements could be achieved either by alternating bursts in the antagonists or more commonly by the modulation of activity in one muscle acting against a continuous level of activity in the other. The variability from one step to the next suggested that a wide variety of different motor patterns could produce the same externally observed movement. They proposed that such a system was best controlled by a sensory template which established the required movement, and actual motor output could attempt to produce this by a variety of different patterns. Later experimental work on these same muscles (Runion and Usherwood, 1966; Usherwood and Runion, 1970), paying careful attention to the avoidance of cross-talk between the recording electrodes, showed a consistent reciprocity between antagonists and no indications of the earlier diversity of motor behavior. Burns (1973) recorded the activity of these muscles in the front, middle, and hind legs of the locust and reported similar stereotyped behavior to that of Usherwood and Runion. Figure 27 shows phase–frequency data for the three axons of the extensor tibiae muscle and the more prominent spikes from a simultaneous myogram recording of the flexor tibiae muscle which is innervated by at least eight axons (Theophilidis and Burns, 1983). Both of these walking studies improved on the simple myogram recording techniques, the former using a biopotential analyzer to interprete recordings from main nerve branches supplying several muscles, and the latter recording directly from the motor nerve supplying the extensor tibia muscle by careful positioning of the recording electrode. These improvements in recording techniques made it possible to identify the patterns of activity in both the excitatory and inhibitory axons innervating this muscle.

An important feature of this work was an attempt to reconstruct the motor pattern to the muscle while recording the tension. This was first performed using a donor and recipient locust, after storing the nerve activity on magnetic tape (Usherwood and Runion, 1970). In subsequent experiments Burns and Usherwood (1978, 1979) used simulated patterns of activity to measure the mechanical responses of the muscle that might be expected during walking from the fast, slow, and inhibitory motor neurons.

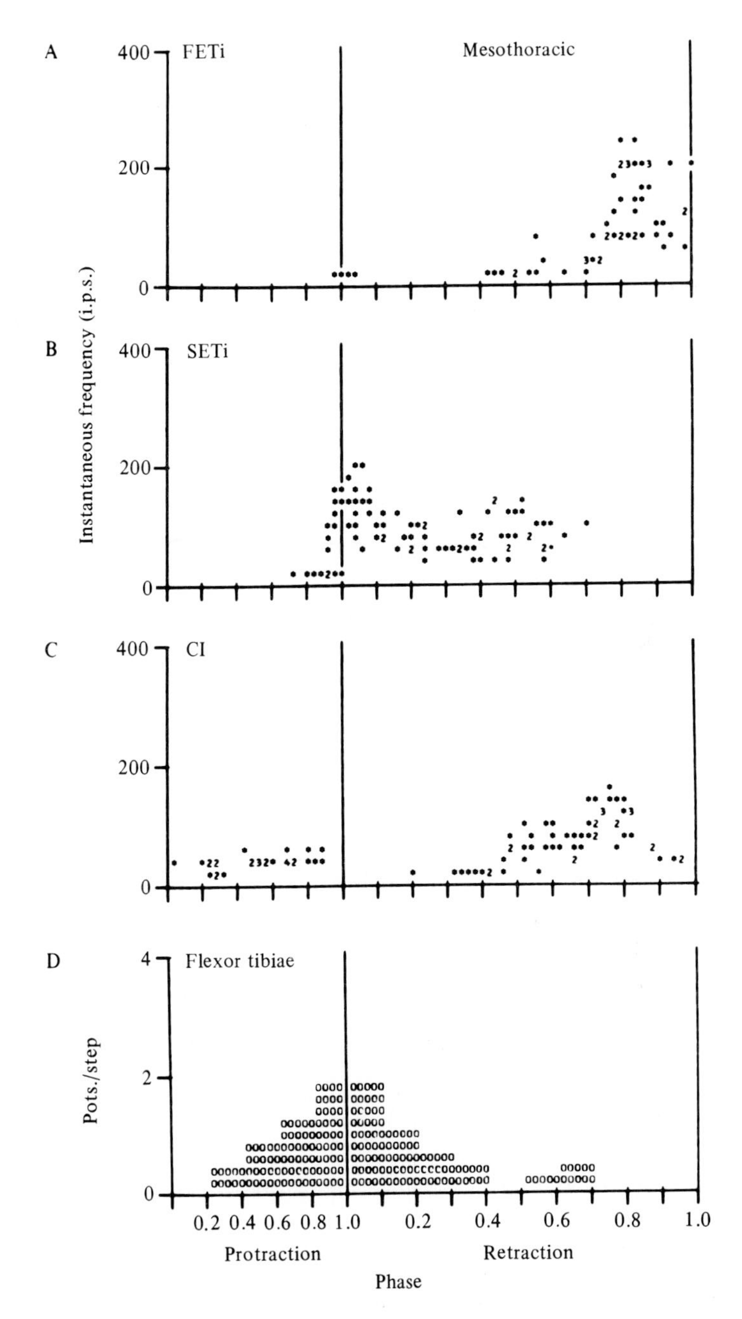

Fig. 27. Locust mesothoracic leg. Typical patterns of spike activity in the three extensor tibiae motor neurons (A–C) and in the flexor tibiae muscle (D) during nine steps of straight-line walking at 3.2 steps/s. (D) is from a different animal giving clearer flexor potentials. (From Burns, 1973.)

The muscle used in these experiments was that which lies in the femur and extends the tibia. This muscle is innervated by only three axons (excluding DUM neurons; Hoyle, 1977). For the middle leg of the locust which is usually oriented toward the rear, there is a fast neuron which is active toward the end of the stance phase in fast walking, a slow neuron which is active throughout the stance phase and is the only excitatory unit active during slow walks, and in addition a single common inhibitor neuron which produces a burst of activity toward the end of the power stroke and is weakly active during the swing phase. Under artificial stimulation the inhibitor produced a 10–20% reduction in tension for slow activity and slight improvement in the rate of

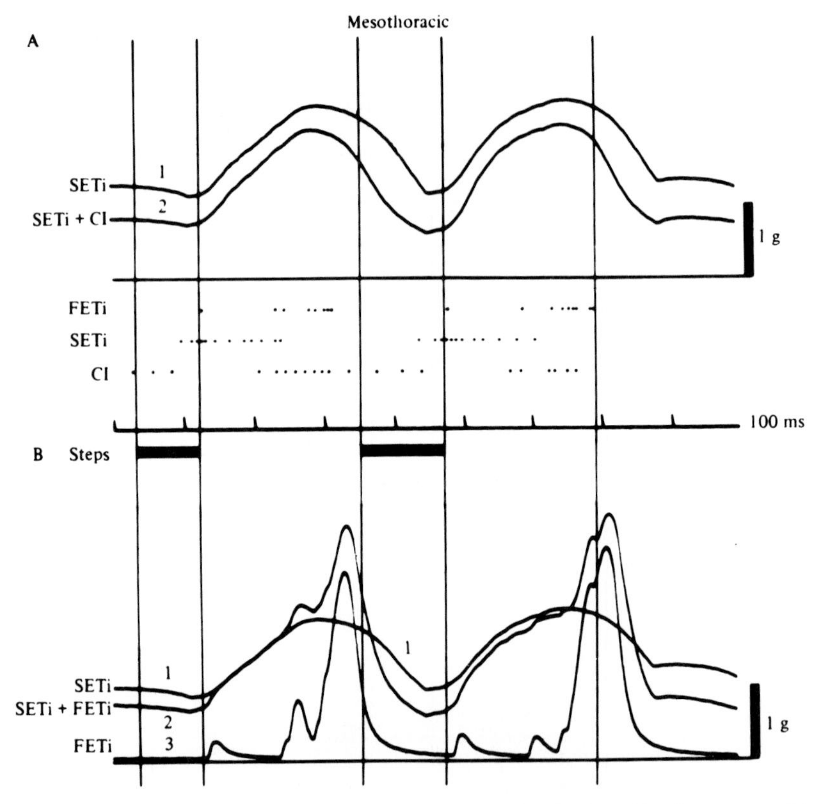

Fig. 28. Locust mesothoracic leg. Records of isometric tension developed by the extensor tibiae muscle in response to stimulation of the motor axons with patterns derived from a walking locust. Stimulus patterns are shown as dots for each axon and are those shown in Fig. 27 (3.2 steps/s). Protractions are marked as heavy bars in the steps line. SETi stimuli to the right of the last retraction in these plots are artificial and were added to maintain tension until the stimulation restarted. Each curve is marked with the stimulation program used. Rest tension = 1.16 g. (From Burns and Usherwood, 1979.)

relaxation (Fig. 28). At high walking speeds the influence of the inhibitor was negligible, and the major contribution to a rapid fall in the tension of the extensor during the protraction (flexion) of the leg was provided by late activation of the fast neuron. This fired several times near the end of the stance phase, producing a large momentary increase in the tension of the extensor followed by a very rapid and significant reduction in the overall tension of the muscle.

In the cockroach, Pearson (1972), Delcomyn and Usherwood (1973), and Krauthamer and Fourtner (1977) showed that muscle activity was reciprocal and highly stereotyped. The coxal levator and depressor muscles, which in this insect are equivalent in action to flexors and extensors and provide the propulsion stroke, showed precisely defined alternation between antagonists during free-walking behavior (Fig. 29). At the highest speeds of walking slight overlap appeared but the bursts were clearly defined and levators were assumed to be active only during leg recovery and depressors only during the propulsion phase.

These muscles are innervated by 12 and 5 axons, respectively. In static headless preparations it is possible to elicit bursts of activity in nerves 6B and 5, which innervate these muscles, similar to the myogram activity found during walking (Fig. 30). However, it is unlikely that these bursts actually correspond to walking because headless insects are generally considered to be

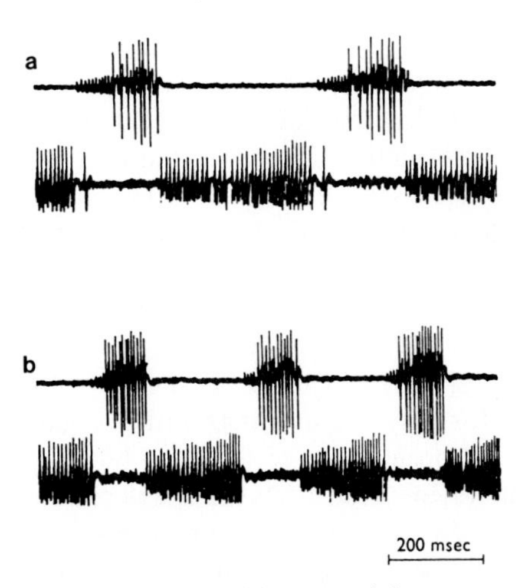

Fig. 29. Reciprocal activity recorded in coxal levator and depressor muscles during walking at two different speeds. (a) Records from levator muscle; (b) records from depressor muscle. The small and large junctional potentials recorded in the levator muscle correspond to activity in axons 5 and 6, respectively. The single junctional potential recorded in the depressor muscles corresponds to activity in axon D_1. Note that with an increase in walking speed there is an increase in the rate of discharge in all motor axons. (From Pearson, 1972.)

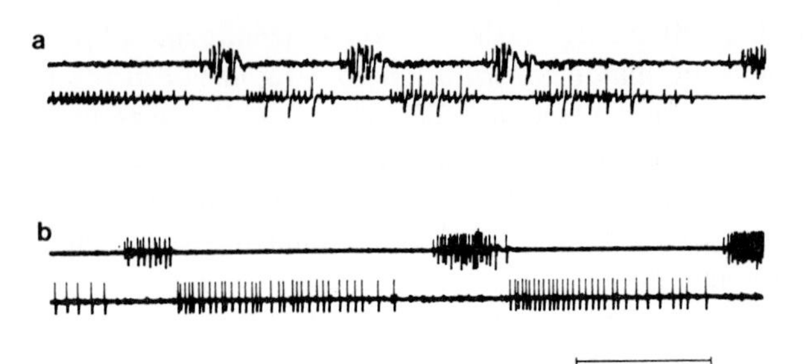

Fig. 30. Reciprocal activity in coxal levator and depressor motor axons after removal of all sensory input from leg receptors. Within each set, top traces are records from levator nerve; bottom traces are records from depressor nerve. (a) Reciprocal activity elicited by stimulation of the ipsilateral cercus. Interaction of the action potentials during levator bursts does not allow identification of the action potentials although the first spike in each burst is from axon 5. The small and large spikes seen during the depressor bursts arise from activity in axons D_s and D_f, respectively. (b) Spontaneously generated reciprocal activity. The two spikes in the first levator burst are from axons 5 and 6 (the spike from axon 6 being larger). Only axon D_s discharges during the depressor bursts. Note that the maximum discharge rate of axon D_s. occurred at the beginning of the burst. Time scale: (a) 80 ins; (b) 200 ms. (From Pearson, 1972.)

incapable of showing walking behavior (Roeder, 1937; Reingold and Camhi, 1977; Graham, 1979a). The single facilitating myogram potential in the depressor record corresponds to axon D_s. This slow axon is the primary generator of tension in the power stroke which may sometimes be accompanied by a fast unit axon, D_f, if the intact animal is startled.

Three more axons innervate this muscle and have been identified as separate "common" inhibitors. All become active during levation, which corresponds to the protraction or swing phase of walking behavior. The innervation of the levator is more complex, with some 12 axons of which one is an inhibitor which also innervates the depressor. At least two and possibly more excitatory units contribute to the protraction stroke (levator) and correspond to the firing of axons 5 and 6 in nerve 6B. Usually, in these headless preparations either depressors or levators are continuously active. Sometimes, alternating bursts of the kind shown in Fig. 30 are present. At other times the levator may fire in bursts without activity in the depressor. The depressor has not been observed to fire in bursts alone, suggesting that no spontaneous burst generator exists for the depressor neurons.

In the stick insect the pattern of motor output to some of the thoracic muscles which move the whole leg in retraction and recovery strokes has been examined by direct recording from the motor nerves close to their respective muscles during walking on a treadwheel (Graham and Wendler, 1981a; Graham and Baessler, 1981). In middle legs the results are similar to those found in locust and cockroach insofar as alternation of antagonists is concerned (Figs. 18 and 31).

At the highest speeds of walking the bursts of these antagonists show simple alternation of the kind reported in the cockroach, but at lower speeds the retractor bursts are divided into three distinct regions. At the start of the stance phase, as the tarsus first touches the ground, a distinctive short burst of spikes is observed in almost every step. After this burst relatively little activity is present in the retractor and this is usually associated with strong activity in the protractor motor nerve. This period of low activity in the retractor muscle may last from 50 to 600 ms and is followed by a conventional strong burst in the retractor nerve producing the normal power stroke. A study of the reflex properties and innervation of this muscle (Graham and Wendler, 1981b; Igelmund, unpublished) has shown that the main components of these bursts are all excitatory units (one slow, two semifast, and possibly two fast axons in the retractor and at least two semifast axons in the protractor).

The direct recording of activity from the motor nerves in a walking preparation and the stereotyped nature of the steps permit the average activity of these individual axons to be examined during a typical cycle of the leg. Figure 31 shows spike interval phase histograms for the motor axons of the retractor and the largest nerve potential of the protractor. This confirms that in fast-stepping simple alternation is present but at slower speeds the protractor is strongly active during the early part of the stance phase. The

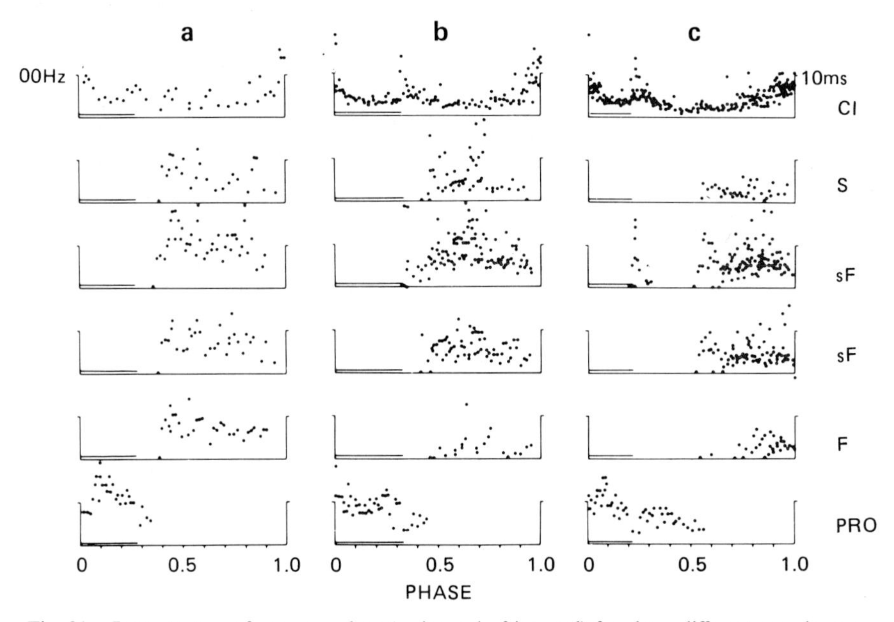

Fig. 31. Instantaneous frequency plots (reciprocal of interval) for three different speeds as a function of phase. The bar shows the protraction stroke. (a) Period 800 ms, one step; (b) period 1200 ms, two steps; (c) period 1600 ms, four steps. (From Graham and Wendler, 1981a.)

brief burst of activity at leg touchdown is generated by one of the semifast axons, and the activity of all excitatory axons is relatively constant throughout the power stroke, with a tendency for the fast unit to be concentrated toward the end of the stance phase in slow walking. The slow unit is sometimes silent during a walking step.

The inhibitory axon always shows a peak in activity before and during the transition from stance to swing phase and sometimes a weaker peak at the transition from swing to stance. In the cockroach the activity of the three common inhibitors of this animal shows a similar maximum distributed over the whole recovery stroke and suggests that these axons play a role in reducing tension in the antagonist (retractor) during the rapid swing phase. It is probable that a secondary peak in the locust extensor tibiae is present but is obscured by the flexor activity occurring at this time. However, Burns argues that the fast activity at the end of the power stroke is the primary cause of the rapid relaxation in the extensor. This is probably also a mechanism used by the stick insect, as it has not been possible to show any significant mechanical influence of the common inhibitor on the retractor while this same axon produces a reduction in the tension of the extensor tibiae muscle (Baessler and Storrer, 1980) similar to that found in the locust.

One of the difficulties in comparing motor output in different insects is the problem of the differing geometry in each instance and the varying role of the motor activity as the leg changes its orientation relative to the body. Taking the middle leg of the stick insect as an example of a generalized leg, in the early part of the stance phase, the flexor tibia probably contributes primarily to propulsion; in the middle of the stroke both the flexor and extensor muscles contribute more to lateral body stability; and, toward the end of the stroke, the extensor must assist in transferring propulsion forces to the body. In front and hind legs, with their more longitudinal orientation, these muscles become more important in the propulsion and recovery strokes. Throughout the leg movement these muscles must also be active in stabilizing the leg in its role of a supporting structure. To some extent the role of thoracic depressors and levators and the protractors and retractors is less complex but they must also assist in stabilizing the system.

An interesting feature of stick insect walking in this respect is the appearance of a static or "standing" phase in which the body velocity approaches zero. This happens twice in each cycle in the tripod gait (Graham, 1983). The body of this insect is supported well clear of the ground in most instars, unlike the cockroach which to some extent may toboggan over the surface, particularly at high speeds on smooth walking substrates. The "standing" phase in the stick insect occurs at the beginning of the stance phase for a middle leg and suggests that walking behavior is a process of throwing the body forward at regular intervals onto a tripod of legs which, initially, resist movement to the rear as they make contact with the ground. After a short interval, these legs throw the body forward onto the next support tripod. In the slower walks of the adult this brief "standing" state in each leg is distributed over the total cycle of all the legs and produces a smoother progression of the body because more legs are retracting at any instant than are attempting to stand still. This view of a walk seems somewhat unsatisfactory for high-speed walking, from a design and efficiency viewpoint, but it may be the simplest way to stabilize a walking machine with articulated legs and may be particularly appropriate for a climbing animal. Perhaps the inconvenience in the high-speed walks of this insect is the price that must be paid for a stable and fully supported general-purpose walking capability.

A reexamination of slow-walking motor output in the cockroach by Krauthamer and Fourtner (1978) suggested that these insects show a similar invasion of the early part of the stance phase by protractor (flexor) activity. At this time in the leg cycle there is a well-defined plateau in the leg displacement, which corresponds to the braking effect observed in the stick insect. One must be cautious with this comparison, however, as the cockroach hind leg extensor behavior on a supported ball is being compared with the middle leg retractor of the stick insect walking on a treadwheel. Furthermore, recent work on the stick insect shows that the motor output to the hind leg retractor and protractor muscles is much more variable than that for the middle leg. Figure 32 shows a simultaneous recording from the meso- and metathoracic retractor nerves. Certain features of the nerve activity make identification of the protraction easy in the middle leg record, but the broken firing pattern of the hind leg makes a movement record essential for the correct analysis of the record. These variable patterns are typical of hind leg recordings in the stick insect and suggest that this leg may be used to stabilize lateral movement of the body during the thrust of the depressor femur and extensor tibia muscles (Graham and Godden, 1985).

4.3 EFFECTS OF LOAD

Increasing the resistance to retraction of the leg was examined in the cockroach by adding a small weight (1.5 g) to the back of the animal or causing the animal to drag a similar weight attached by a thread. This horizontal loading produced a marked increase in the frequency of D_s firing and a shortening of the burst duration, producing faster protraction strokes (Pearson, 1972). A similar result is found for stick insects and in this insect a relatively light loading produces a constant protraction duration equivalent to that used at the highest walking speed.

Walking coordination changes because the step period increases without a

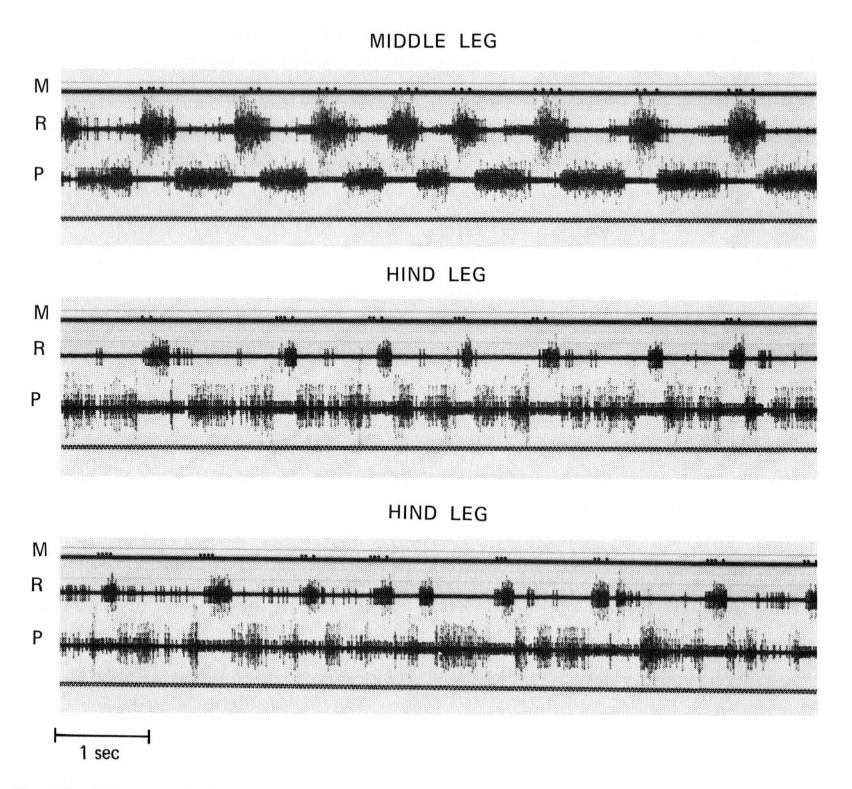

Fig. 32. The records show the activity in the nerves supplying the protractor coxa muscles P (1 p-cx: nerve 2C) and retractor coxa muscles R (1 t-cx, 1a t-cx: nerve 5A) during a walk on two light wheels for middle and hind legs of an adult stick insect. M is a marker showing the time at which the leg protracts. The middle leg record shows simple, well-defined, alternating bursts in the antagonists. The two examples of hind leg activity show much more variable behavior with a tendency for the protractors to fire intermittently during the retraction stroke.

corresponding change in the lag between adjacent legs. A similar effect was reported by Spirito and Mushrush (1979) for free cockroaches walking up a plane inclined 45° to the horizontal. In the cockroach similar changes in the firing rate of the power stroke muscle can be produced by squeezing the trochanter (Pearson, 1972), suggesting that the campaniform organs produce this effect by feedback to the depressor motor neurons. It seems likely that this is a distributed effect over all the legs, because Baessler (1979b) and Cruse and Saxler (1980a) have shown that insects with one leg on a force transducer show an increase in traction force when the other legs walking on a wheel are subjected to an increase in loading. This leads us to consider the extent to which the legs cooperate in producing propulsive force.

4.4 COORDINATION OF MOTOR OUTPUT

In a standing or walking leg the muscles of several segments must cooperate to transfer forces generated by the thoracic muscles to the tarsus or foot in contact with the ground. This is most simply illustrated by the transverse arch of support in a standing animal which consists of the body, coxa, femur, and tibial segments. Depressor muscles in the body force down the coxa and coxal depressors bear down on the femur, which in turn flexes the tibia if the leg is somewhat extended. Activity in various components of this chain of muscles is essential if the body is to be supported by the leg. For an upright animal these muscles must all be coordinated during the support phase and during recovery the leg must be lifted by concerted action in the levators of the coxa, femur, and tibia.

Intraleg reflexes distributed over several segments have been demonstrated in crustacea (Clarac, 1977, 1981) and in arachnids (Seyfarth, 1978). In insects, the work of Wilson (1965) suggested that intersegmental reflexes were weak and labile during imposed leg movements. More recent work with walking animals suggests that under these conditions inter- and intraleg reflexes are much more stable (Baessler, 1983). The only published attempt to explore such reflex pathways by direct stimulation in walking insects is that of Macmillan and Kien (1983). They used shocks of 30–40 V to stimulate campaniform sensilla in the legs of a locust walking on a water-supported sphere. Such stimulus trains near the end of stance phase increased the probability of a protraction, rather than inhibiting it as reported by Pearson (1972) and Baessler (1977b), who used loading and direct pressure as a stimulus. Several experiments (Baessler, 1983) have shown that movement, loading, or touch of a segment in a leg can produce complex reflex responses in several joints of the same leg or in other legs. These responses are used for step modifications such as obstacle avoidance, targeting, and correction of accidental stepping of one leg onto another.

It does not seem to be essential that propulsive forces arising from the different legs be balanced by a neural mechanism, and the results of running animals on a slippery surface suggest that this is indeed not the case in stick insects. In such an experiment, Epstein and Graham (1983) found that legs could move at different speeds over the surface despite the presence of normal phase-locked coordination (Fig. 33). Thus each leg provides thrust independently of the others but is constrained to lift at a time appropriate to the movements of the other legs. The ground surface integrates these forces and the body moves under the resultant of these combined forces.

The coordination between the legs for support purposes, however, is much more important as the precise timing of recovery strokes has shown. Coordinated stabilization of the body above the legs may also be a vital

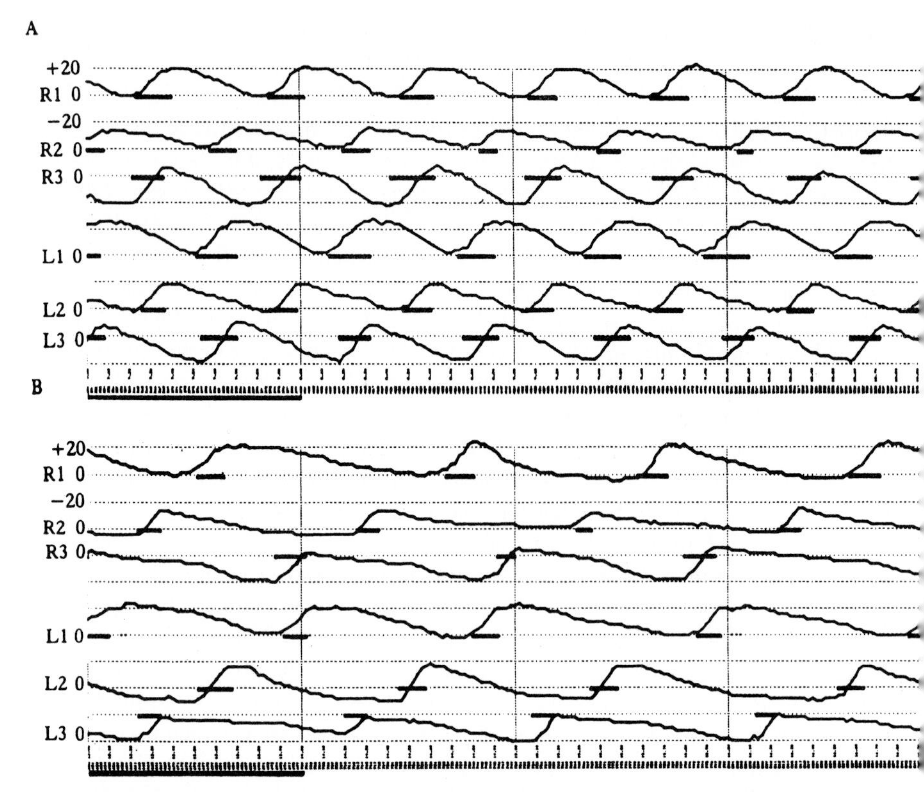

Fig. 33. Forward walks using (A) low viscosity (10^{-4} m^2 s^{-1}) oil and (B) high viscosity (10^{-3} m^2s^{-1}) oil. The displacement of the tarsus in the x direction (mm) relative to the subcoxal joint is projected onto the body axis. Bars superimposed on each trace indicate swing phase. Time bar is 1 s. (From Epstein and Graham, 1983.)

consideration but this area remains largely unexplored. Let us consider two extreme hypotheses for the actual support of the body and the lateral stabilization necessary to prevent it folding and collapsing to right or left in a standing animal:

Hypothesis I. It could be assumed that the extensors and flexors controlling the femur–tibia joint of the right leg are strongly active and lock this joint rigidly at an angle of approximately 90°. This creates a strut or rigid structure which props the body against a corresponding strut on the other side of the body (left leg). In the upright configuration this simple assumption would support the body above the ground with little or no activity in the depressors of the coxa and femur and would only require activity in the extensors.

Hypothesis II. Alternatively, we could assume that the depressors of the coxa and femur provide all the supporting forces and hold the femur parallel to the ground surface. Then the flexors and extensors of the right and left tibiae need to provide only weak forces to keep the body balanced over the almost vertical tibiae and prevent any tendency to collapse to right or left.

These extreme examples have been chosen to illustrate two alternative strategies but in practice they could be mixed to minimize the work being done by the muscles concerned or, alternatively, hypothesis I might be particularly appropriate for standing while hypothesis II might be more advantageous during walking.

Cruse (1976a) was able to show that a walking insect exerted forces on the ground directed away from the body, a finding which would suggest that the legs act as struts and prop up the body during the power stroke. Such forces would automatically provide lateral stabilization. These results favor hypothesis I and indicate that large forces should be generated in the tibial extensors. Recordings from crabs, cockroach, and scorpions suggest that the relevant motor output is very low in standing animals. Unfortunately, none of this work has been formerly published but is referred to by Yox *et al.* (1982) in the form of private communications by F. Clarac, C. R. Fourtner, and R. F. Bowerman, respectively. Similar observations have been made in stick insects (Graham, unpublished). These observations would tend to support hypothesis II.

Here we have the opportunity to consider the distinction between the active and passive forces generated by muscles (Fourtner, 1982). Friedrich (1933) noticed that the amputated legs of the stick insect tended to maintain a constant angle of 90° between femur and tibia. Chesler and Fourtner (1981) reported similar observations in amputated cockroach legs and noticed that they were quite resistant to bending. Measurements of the resting tension in these denervated legs produced surprisingly high forces when bent passively. Values for a femoral extensor muscle (177d; hind leg) were 0.25 g at minimum physiological length to 2.6 g at maximum physiological length. Such forces would be able to provide a significant contribution to the stabilization of the body above the standing legs and would require no nerve activity to maintain the system in standing equilibrium. In the case of the cockroach the passive forces may even be sufficient to provide the prop-stabilizing forces required by hypothesis I. The weaker passive forces in the stick insect are probably insufficient for hypothesis I, but are more than adequate for hypothesis II in a standing animal and are probably sufficient for a walking animal in the upright configuration. At other orientations to gravity some form of active servo-assistance would be required.

Newly emerged adults and the younger instars of the stick insect walk with a well-supported posture in which the femur–tibia angle is close to 90° at all times and the femur operates in a plane parallel to the walk surface. Gravid adults on the other hand tend to sag toward the ground and frequently drag their abdomens behind the walking legs.

In the cockroach, Delcomyn (1971c, 1973), Delcomyn and Usherwood (1973), Pearson (1972), and Pearson and Iles (1973) concentrated their attention on the levators and depressors of the femur, while Krauthamer and Fourtner (1977, 1978) recorded the activity in the extensor and flexor of the tibia. All experiments were performed on the larger hind legs, and due to the piston-like action of these legs both sets of muscles can be considered to act primarily as propulsive and recovery muscles in walking. Depressors and extensors contract synchronously to produce the power stroke and levators and flexors fold the leg during the recovery stroke.

The only simultaneous recordings of four muscles during walking are those of Graham and Baessler (1981) and Epstein and Graham (1983). Figure 34 shows myogram records from all four muscles for a stick insect walking on light treadwheels and additional records for three stick insects are shown in Fig. 35 for walking with the body supported above a slippery glass surface.

The typical action of the levators during a walk is to lift the leg as a prelude to protracting the leg in swing phase, but they are sometimes active in unsupported walking toward the end of the protraction stroke and possibly

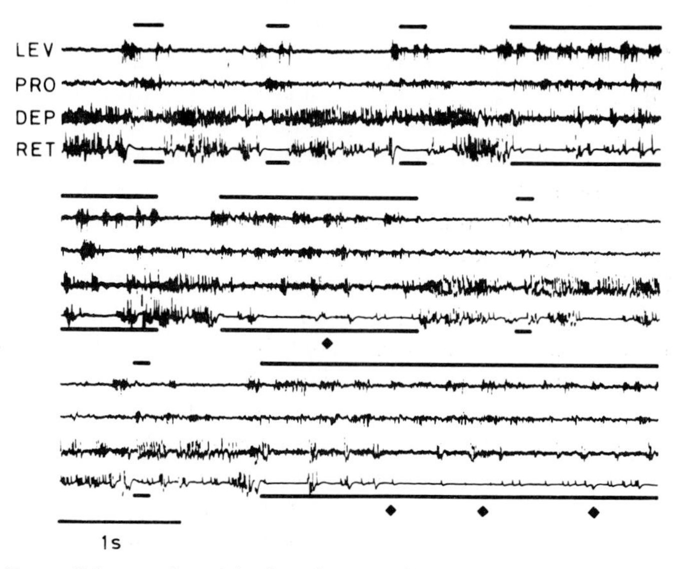

Fig. 34. Extracellular muscle activity from four muscles in the right middle leg in which the receptor apodeme is crossed (continuous record). The black bars are derived from a simultaneous film record and show that the leg is not in contact with the wheel. Short bars correspond to a normal protraction and the long bars represent a salute. The ◆ below two of the salutes denotes the beginning of a quiet period following a pronounced twitch observed in the film record. (From Graham and Baessler, 1981.)

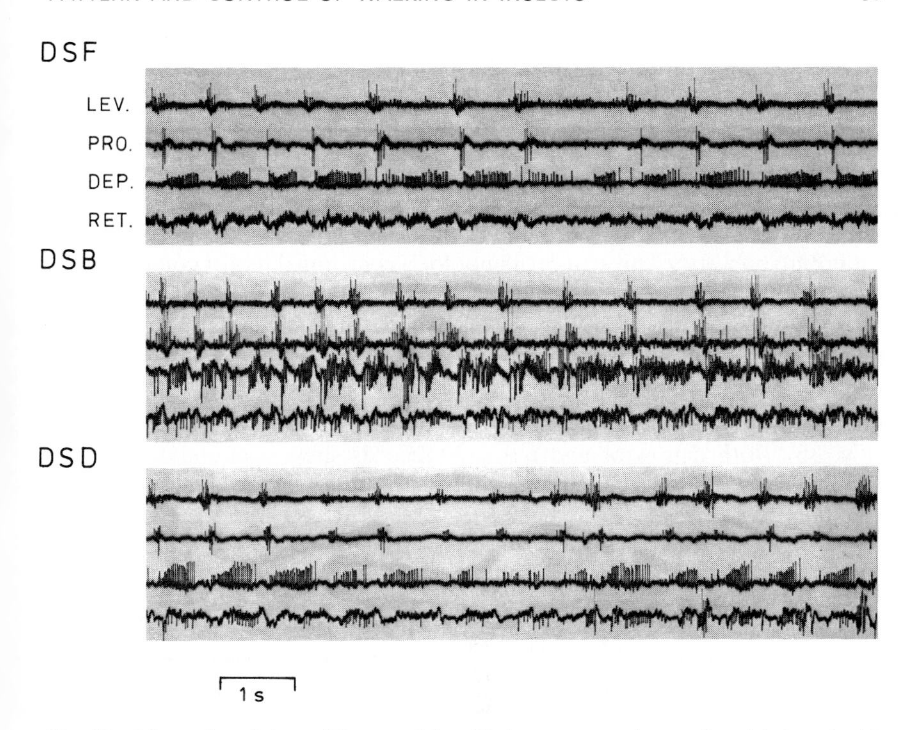

Fig. 35. The walks of three different adult stick insects on a glass surface lubricated with silicone oil. The protraction stroke of the middle leg is shown by a horizontal bar (from film record) above the myogram records for the LEVators of the femur, the PROtractors of the coxa, the DEPressors of the femur, and the RETractors of the coxa. Time bar, 1 s.

provide a "grappling" of the tarsal hooks to the walk surface. These late bursts do not appear in supported walks. This muscle is usually inactive throughout the retraction stroke in upright walking. No records are available for hanging walks but presumably this muscle would provide support forces to hold the body up close to the surface in such walks.

In unsupported, upright walking the depressors are strongly active during the whole retraction stroke with their maximum rate of firing in the early part of the stroke. This is in contrast with the retractor muscle, which tends to be more active towards the end of the stroke. The depressors may also contract briefly during the later part of the protraction stroke to overcome any residual tension in the levators and force the leg down onto the walk surface. The strong depressor activity during upright walking supports hypothesis II.

During supported walks on slippery glass the retractors show much weaker activity than on treadwheels, suggesting that load-dependent reflexes control the propulsion motor output either directly via stress-sensitive organs or by position-feedback control of the kind already proposed by Wendler (1964) and Baessler (1967) from static loading experiments, the walking studies of Cruse and Pflueger (1981), and Cruse and Schmitz (1983). A similar reduction in motor output can also appear in depressor recordings in the supported preparation and is particularly noticeable in some steps of animal DSD in Fig. 35.

The behavior of some muscles during "saluting" (see Section 6.4) and "standing on a stick" presents evidence for a centrifugal motor component in walking, arising either endogenously or from the walking behavior of the other legs. Figure 34 shows high frequency bursts of activity in the depressors, and to a much lesser extent in the retractors, which are modulated with the walking rhythm of the other legs. This is more clearly shown in the depressor records of a leg standing on a stick while the other legs walk on a treadwheel (Fig. 36). Both depressor and retractor records show a modulation which appears to be exactly coordinated with the hind leg on the same side.

When the standing leg is moved to a forward or rearward position the

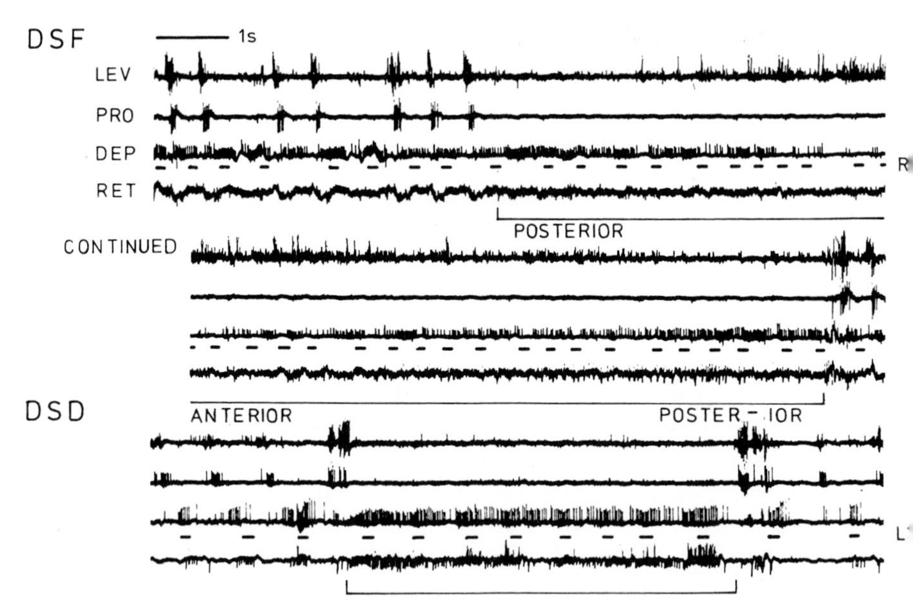

Fig. 36. A similar recording to that of Fig. 35 for animals DSF and DSD with the middle leg standing on a movable stick while the other legs walk. In the DSF record the left middle leg is caught on the stick and quickly moved to the rear (posterior). The stick is then moved forward to the AEP (anterior) and then moved slowly to the rear where the leg steps onto the walk surface. During this time the other legs continue to walk as shown by the marked protraction of R2. The DSD record shows the left middle leg standing on a fixed stick in the middle of its normal stroke. The protractions of L3 show that the other legs are stepping and both depressor and retractor activity are modulated with the rhythm of the walking legs.

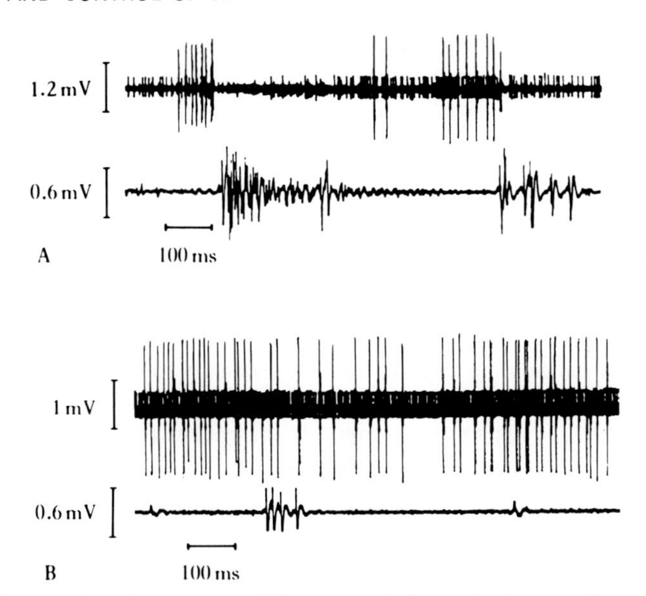

Fig. 37. (A) Metathoracic extensor and flexor records for a stick insect walking on a light wheel. The records were made, using the neuromyogram technique of Burns (1973), by Cruse and Schmitz (1983). The upper record shows the fast (large), slow (medium), and common inhibitor (small) nerve potentials of the extensor. The lower record is of the flexor muscle. The protraction stroke begins at the start of the flexor burst. (B) is a recording with the leg standing on a platform while the other legs walk on a wheel. (From Cruse and Schmitz, 1983.)

levator and depressor muscles, in addition to modulation, show that activation is transferred from the levator to the depressor. In the forward position the levator is strongly active while the depressor is weak. In the rearward position, the levator is inactive with a high frequency of spikes in the depressor. The muscle which is active in the extreme position would be expected to contribute to retraction of the limb. This suggests that in a standing leg the muscle which is best able to provide the force required is activated by a set of intersegmental reflexes within each leg of the kind found in crustacea.

Records of the activity of the extensor and flexor of the tibia in a supporting leg are available from the locust and can be examined in Fig. 27. The fast unit of the extensor and the flexor in this middle leg record show good alternation but the flexor is not restricted to the swing stroke. This is consistent with the observed movement of both locust and stick insect middle legs, which show a slight flexion at the end of the recovery stroke and an extension toward the end of the stance phase that continues during the early part of the swing. The slow activity in the first half of the stance phase agrees with the observed outward forces generated by walking stick insects as reported by Cruse (1976a), and supports the suggestion by Burns (1973) that this activity is involved in the stabilization of the body. The only other record available for these muscles in the stick insect is that of Cruse and Schmitz (1983), shown in Fig. 37. The activity is similar to that reported by Burns.

5 Nervous control of walking

Patterns of stepping, changes in external loading, blocking experiments, and targeting observations can suggest functional relationships between various parts of the walking system, but it is necessary to test these hypothetical communication pathways and determine the exact role of the muscles and sense organs. One approach is to modify the system and attempt to disprove some essential feature of the proposed mechanism. In attempting such studies, the experimenter runs a variety of risks. In operating on the animal the internal environment may be significantly altered. Even if this not the case, behavior may be seriously disturbed by damage or the stimulation of inappropriate receptors. The more extreme the operation, the more difficult becomes the task of interpretation, unless it can be clearly established that the behavior is essentially normal, in the operated condition, before the modification or stimulation is applied.

The behavioral evidence presented so far suggests that the most acceptable model for walking behavior in insects consists of independent leg movements coupled flexibly to permit a wide range of different gait patterns. This system is able to react to changes in loading, body orientation, obstacles, and the velocity and directional requirements of the animal. The overall problem of walking control is complex but can be separated into three functional levels, moving inward from the leg. First there is the detailed control of the muscles driving the individual standing and walking leg. The second is the level at which the individual legs are coupled to coordinate their actions and provide an efficient and reliable walking system that includes stabilization, support, and propulsion. Third, there is the highest level which examines input from the eyes and antennae and makes decisions about whole body movements. This hierarchical view comes from a number of experiments designed to isolate parts of the nervous system with minimal disruption to the internal environment.

An important advantage of insects in analyzing the patterning of behavior is the packaged nature of their central nervous system. Individual ganglia are associated with each pair of legs and these are connected by a bilateral pair of connectives. At the front end of the animal the supraesophageal ganglion in the head capsule processes incoming information from the eyes and antenna, and communicates via a pair of circumesophageal connectives with the subesophageal ganglion, which in turn interacts with the thoracic ganglia controlling the segmental leg pairs.

5.1 CONNECTIVE CUTTING

Roeder (1937) in his work on the mantis showed that the influence of the supraesophageal ganglion on motor behavior was primarily inhibitory. This was demonstrated by cutting the circumesophageal connective on one side. Operated animals showed continuous circling walks away from the operated side, indicating that the cutting of the connective released a spontaneous walking output on the operated side which was normally inhibited by the supraesophageal ganglion.

If both connectives were cut he reported that animals walked continuously! These bizarre results were partly confirmed in a later series of experiments on the stick insect (Graham, 1979a,b). Decerebrated insects in which both circumesophageal connectives were cut internally with minimal loss of hemolymph showed a hyperactivity of the kind described by Roeder. These animals did not walk continuously but in bursts of 30 steps at a time separated by brief intervals of rest, followed by whole body twitching which steadily increased in intensity until the animal started the next walk with a step period which began at 600–800 ms and ended at 2–3 s (Fig. 38). The overall activity decayed with time over several successive walks until the animal remained at rest. In this state a slight vibration or gentle touch was sufficient to start a new sequence of walks. Severing these connectives also abolished the cataleptic state usually found in these animals during daylight. Coordination varied between gait I and gait II and the only other unusual feature of the walks was a tendency for a leg to step onto the tarsus of the leg in front and the separation of the recovery stroke into separate lift, forward swing, and depression when the animal walked very slowly.

One can conclude from these observations that the supraesophageal ganglion acts as an inhibitor of spontaneous walk behavior, and to some extent stabilizes gait I or gait II and may determine the range of leg movement but is not involved in the overall organization and coordination of walking behavior.

The role of the subesophageal ganglion is uncertain. In experiments in which the connectives between the subesophageal and prothoracic ganglion were cut, corresponding to decapitation, the mantis (Roeder, 1937), cockroach (Reingold and Camhi, 1977; F. Delcomyn, personal communication), and stick insect (Graham, 1979a) showed no organized walking movements. This ganglion appears to be an essential component of the walking system. Whether it acts only as an excitatory input to thoracic ganglia or is actively involved in the coordination and the detailed control of leg movements is not

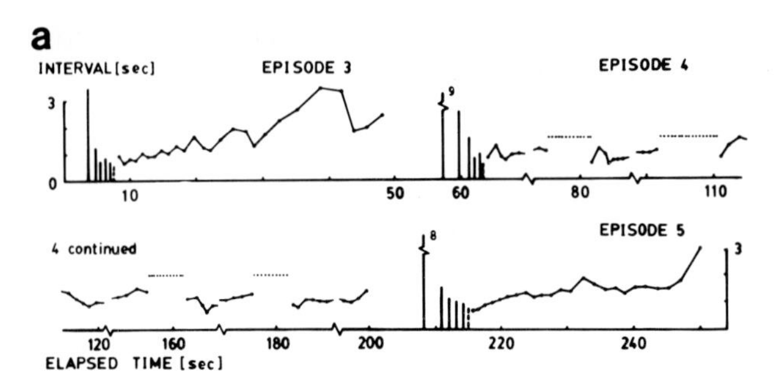

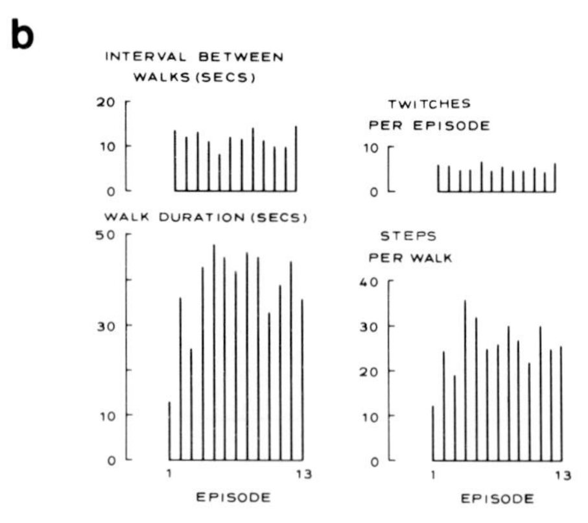

Fig. 38. (a) Twitch and walking behavior for an adult stick insect with cut circumesophageal connectives showing the intervals between twitches (vertical solid line), a rearward jerk (dashed vertical line), and the protraction of the left middle leg (connected dots). The vertical axis shows the interval in seconds. The horizontal axis is the elapsed time in seconds. A horizontal dotted line implies that the animal is removed from the walking surface. The animal walked continuously during the deleted parts of the record. (b) Parameters of the 13 consecutive walks of one decerebrate. The animal was briefly picked up at the beginning of several of the walks. (From Graham, 1979a.)

known. Decapitated stick insects adopt a head-down standing posture but can still show weak movements of individual legs. The only strong leg movements appear when the decapitated animal rights itself after being tipped over onto its back (Graham, 1979a). Several workers have reported the ability to initiate walking behavior, or at least regular alternating bursts from antagonist muscles in the leg, by stimulating neurons in the cervical connectives (Kien, 1981), thoracic connectives (Pearson, 1977), and a non-spiking interneuron (Fourtner, 1976). These observations suggest that the subesophageal ganglion may exert an excitatory or permissive influence on the thoracic nervous system.

Hughes (1957) and Greene and Spirito (1979) selectively cut connectives between the thoracic ganglia and reported that the presence of at least one connective is necessary between each ganglion for coordinated walking. The order of successive leg steps on the unoperated side was normal but changes in the order were present on the operated side. Lesion of the pro–meso connective caused the front leg to sometimes protract before or during the protraction of the middle leg rather than after it. Lesion of the meso–meta connective sometimes caused the hind leg to step earlier than the front leg on the same side. More often the leg was much less active and stepped much more slowly than the legs in front, which maintained their normal coordination with each other.

Unilateral connective lesions disturb the coordination of the legs concerned and indicate that precise timing information is weakened by the lesion. In static preparations, Pearson and Iles (1973) were able to show activity in connectives related to the cyclic motor output of a leg passing forward from the metathorax and to the rear from the mesothorax. The absence of walking behavior in the middle and hind legs when both anterior connectives are cut suggests that the ganglia depend upon input from the connectives to maintain their normal level of activity and do not receive sufficient input from the lateral and posterior periphery alone.

In all except the last experiment the animal remained intact with only small incisions to sever the connectives. In most cases similar experiments (in which the connective accidentally remained intact) acted as controls which showed that behavioral changes were produced only by the lesion

Another class of experiments that have helped to clarify the interleg organization of walking behavior with minimal damage to the system are those involving the removal of a whole leg.

5.2 AMPUTATIONS AND PROSTHESES

At first sight amputation might appear to be a particularly gross and disturbing modification. However, in several insects and crustaceans (Findlay, 1978), this is the normal strategy for dealing with a damaged leg. A small muscle in the trochanter shears a linear weakness in the cuticle causing autotomy or "breaking away" of the leg with minimal loss of body fluid. The changes in the timing of the remaining legs show that the walking system acts as a coupled system of individual legs. This plasticity of the nervous system (Bethe, 1930) is most evident in the middle leg amputee, as already described for both the cockroach and stick insect. However, subtle but quantifiable changes in the temporal and spatial patterning of movements of the remaining legs occur when any leg is removed (Hughes, 1957; Baessler, 1972a; Graham, 1977b). The amputee does not appear to require any learning period to adopt the new coordination.

If a middle leg is not removed but is tied down or immobilized, in some way other than standing on a separate support, the temporal relationship between front and hind legs becomes extremely disorganized and coordination is destroyed (Pearson, 1979; Graham, 1977b). This suggests that inappropriate movements of the "tied" middle leg disorganize front and hind leg activity when the middle leg is present. Only when it is autotomized is a stable coordinated gait possible. This appears to be an example of a disrupting afference which cannot be ignored because coordinated movements can only reappear when the "tied" leg is autotomized.

Restraint of a front or hind leg, however, causes only minor changes in coordination and is so effectively ignored that the animal quite often attempts to use a tripod of support in which one leg is tied up and falls over at the beginning of each new walk (Graham, 1977b). This unawareness of the state of front or hind legs seems to suggest that the middle leg plays a special role in coordinating the other legs.

An interesting feature of the effect of autotomy is that it appears to be reversible. Von Buddenbrock (1921) and Wendler (1964, 1966) mention that if a prosthesis (peg-leg) is glued to the stump, then normal coordination returns. A similar return to normal behavior has been reported in cockroach (Pearson, 1972), if the stumps of the amputated legs touch the ground, and in locust (Macmillan and Kien, 1983). Grote (1981) examined the behavior of crayfish after amputation of the hindmost leg and found that the leg stump moved irregularly compared with the leg in front on the same side. When this stump was allowed to contact the ground via a peg-leg, normal regular stepping and coordination were resumed. These results show that sense organs at the coxa–thorax joint (most proximal) determine the support state of the leg as a function of either the coxa position relative to the body or the stress acting on the joint. The system presumably uses this information to decide whether or not to incorporate a leg into the walking rhythm. Neither tarsal contact, grasping, or even the femur or tibia segments appear to be necessary for coordinated movements of the leg.

5.3 DEAFFERENTATION

The independence of action in individual legs, stereotyped nature of motor activity, and even normal intact coordination at high speeds of walking in some amputee insects (cockroach middle-leg amputees sometimes use the tripod step pattern; Delcomyn, 1971b) suggested that basic walking movements within a leg and between legs might be possible in the absence of peripheral feedback. A second alternative is that each leg is driven by an independent oscillator and these are coupled by the peripheral sense organs of one leg influencing other leg oscillators, or possibly by a reafference or corollary discharge associated with the protraction stroke of the controlling leg. A third possibility is that even the individual legs are controlled peripherally by a system of complex phase-dependent chain reflexes in which the performance of each part of a step depends upon the successful completion of the previous part. Needless to say the situation may be even more complex. Perhaps, at high speeds of stepping, peripheral mechanisms are too slow and the walking system changes over to a simplified endogenous mechanism using autonomous fixed activity patterns such as those found in grooming, righting, and struggling movements.

In other systems the approach to resolving these questions has been to isolate the central nervous system (CNS) from the lower levels of sensory input which are directly influenced by the output. Unfortunately, this entails considerable damage to the preparation which may make interpretation of the results difficult. Pearson and Iles (1973) applied this technique to the cockroach and their experiments have been repeated by Reingold and Camhi (1977). More recently a similar series of experiments has been performed on the stick insect (Baessler and Wegner, 1983). In this last preparation selective denervation was possible while the animal walked on a light wheel.

In the Pearson and Iles preparation the animal was opened ventrally, with the insect lying on its back, and all but one or two selected lateral motor nerves were severed from the thoracic ganglia. The thoracic connectives were left intact but the head (containing the subesophageal ganglion) was removed to give a more active preparation. The only uncut nerves were then severed from the muscles they applied and placed on recording electrodes. The animal was stimulated by touching the rearmost part of the body, which was left intact. In the cockroach, Pearson and Iles (1973) (Fig. 30) reported occasional bouts of reciprocal activation of the flexor and extensor tibiae similar to that found during walking behavior. In addition, a phase analysis of the motor activity for meso- and metathoracic ganglia in selected alternating bursts showed a tendency for a short latency between bursts in corresponding levator muscles. This indicated a central mutual inhibition minimizing the probability of meso and meta legs on the same side flexing together (Fig. 39). Reingold and Camhi (1977) carried out a similar series of experiments but reported that "even within a single leg, bursting activity was too variable to associate it clearly with the walking or grooming rhythms seen in intact behaving animals." They also showed experimentally that decapitation effectively reduced the incidence of walking to zero and concluded that the oscillatory activity and central coordination found in the headless preparations of Pearson and Iles (1973) probably arise from righting or grooming rather than walking behavior.

In the stick insect, in which only the circumesophageal connectives and lateral nerves on both sides were cut, alternating behavior is sometimes found

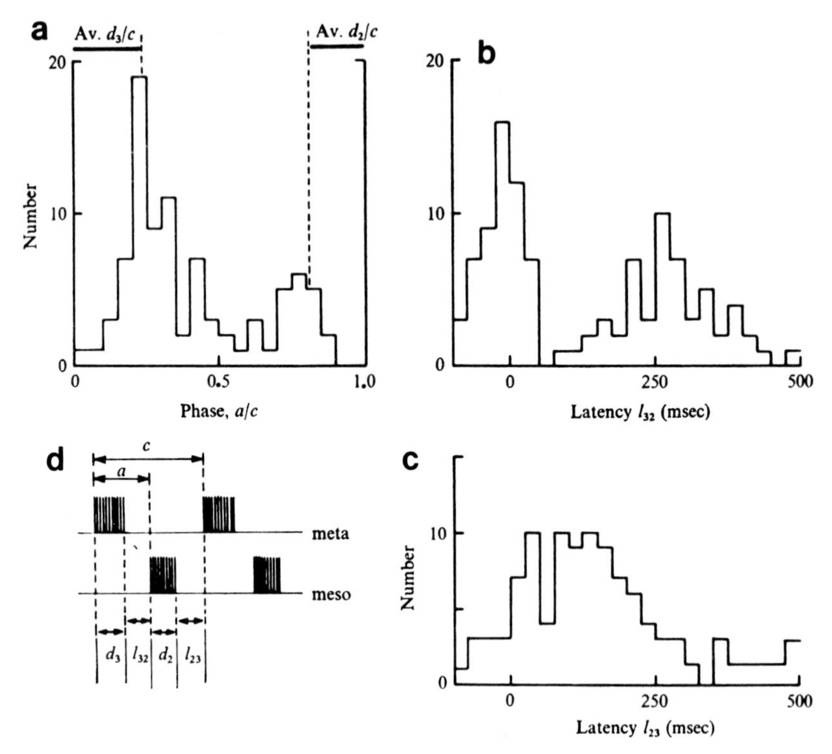

Fig. 39. (a) Histogram of the phases of mesothoracic bursts in the ipsilateral metathoracic cycle, a/c, observed in a typical preparation after removal of all sensory input from leg receptors. The two peaks in the histogram occur close to the average value of d_3/c and to one minus the average value of d_2/c, indicating the tendency for mesothoracic bursts to begin near the end of the metathoracic bursts or vice versa. This tendency is also seen in the latency histograms, which show (b) the duration of time between the end of the metathoracic bursts and the beginning of the next mesothoracic burst, l_{32}, and (c) the reverse, l_{23}. Each of these histograms has a prominent peak near zero. (d) illustrates the parameters measured. The histograms were constructed from data obtained from a headless preparation in which the cycle time varied from 100 to 1000 ms. (From Pearson and Iles, 1973.)

for the protractor and retractor muscles of the coxa, but is exceptional, and in all of the preparations either the protractor or retractor typically fired in long continuous bursts (Fig. 40). Even during the rare alternations, several features of the walking motor bursts reported by Graham and Wendler (1981a) are absent. In addition recordings were made from a nerve supplying the extensor tibia. This nerve shows continuous activity and on the rare occasions when it is modulated it is not coordinated with the coxal muscles as would be expected in a walking movement. In the deafferented stick insect no obvious coordination has been reported between ganglia.

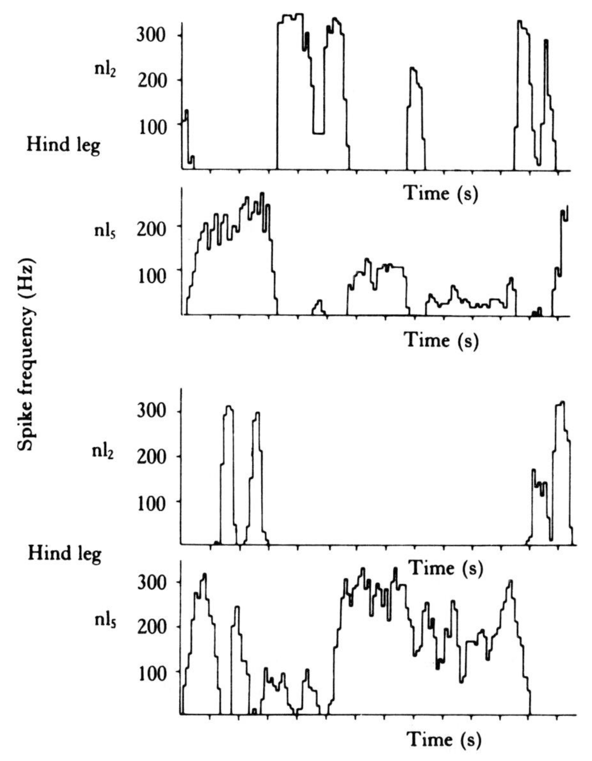

Fig. 40. The activity of the protractor (nl_2) and retractor (nl_5) motor neurons recorded from preparations in which the thoracic ventral nerve cord was totally denervated except for the connections to subesophageal and abdominal nervous systems. The animals were stimulated mechanically on the abdomen. (From Baessler and Wegner, 1983.)

One advantage in the stick insect experiments was the use of a walking wheel. Partial deafferentation showed that as long as one leg (usually a front leg) remained intact and walked on the wheel the motor output of the remaining deafferented hemiganglia showed regular alternation and a phase-locked relationship to the walking leg (Fig. 41). When the last leg was deafferented motor output rarely alternated and long bursts of continuous activity appeared in either protractors or retractors.

6 Sense organs of the leg

The role of individual sense organs in insect walking is varied and complex. The function of these organs ranges from the detection of the extremes of segment movement to the position-feedback control of joint angles in

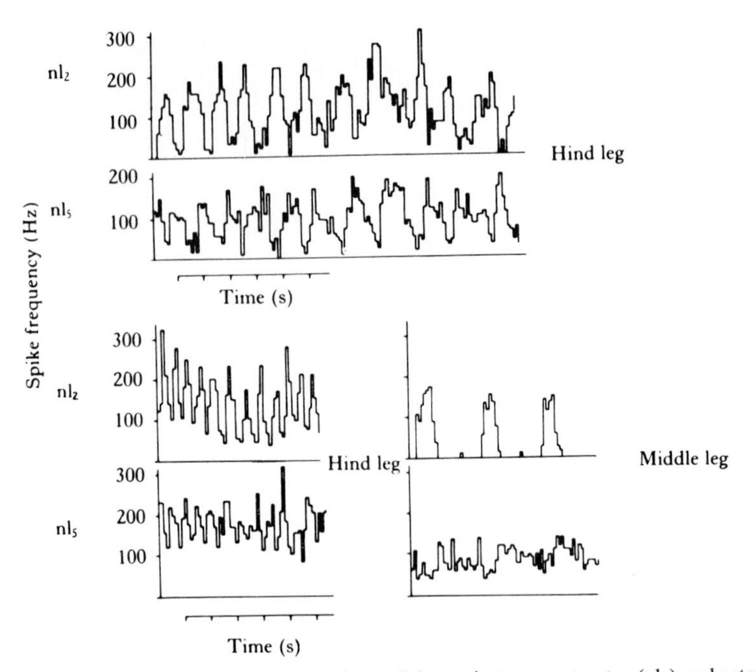

Fig. 41. Three typical examples of the activity of the excitatory protractor (nl_2) and retractor (nl_5) motor neurons recorded from a denervated ganglion half while the other legs are walking on a treadwheel. The modulation frequencies correspond to the step frequencies of the walking legs. (From Baessler and Wegner, 1983.)

stabilization, support, and probably propulsion, and they may even be essential to the actual generation of the motor output which creates the fundamental cycle of leg movement, as suggested by Land (1972). These organs are almost certainly responsible for the interleg coordination in slow-walking and climbing insects and even in faster moving ones, such as cockroach. However, at present, there is no evidence that any one sense organ fulfills this role. Research has tended to concentrate on the more peripheral organs but the "peg-leg" experiments indicate that the coxa–thorax joint is the place where such controlling sense organs might be found. Furthermore, coordinating responses may only appear when several sensory modalities are appropriately stimulated.

Sensory input from the legs is produced by three main classes of sense organ. The most obvious are the hairs distributed randomly or in spatially organized groups such as hair rows or fields. Hair rows are usually arranged along the major axis of a segment and hair fields are close to the joints. They respond to touch stimuli and inform the CNS that a specific part of the leg has contacted another object. If the hair field is close to a joint it can act as a proprioceptor providing information about the angular position of the joint (Wong and Pearson, 1976). In arthropods, segments are linked by flexible membranes to permit joint movement without loss of body fluids and these containment membranes roll and fold during movement of the joint. Strategically placed hairs depressed by these membranes can measure the joint angles over large ranges of movement. These organs are primarily position receptors and it seems unlikely that they would provide information on cuticular stress although it is not impossible to imagine hair configurations that could give information on cuticle deformation.

Another class of sense organs, the campaniform sensilla, are also located on the cuticle surface and measure local stress. Different surface modifications and orientations provide for measurements in any direction and these organs are usually situated close to the joints in regions of high stress such as the articulation axis (Zill *et al.*, 1977; Zill and Moran, 1981a,b). The sensilla are 2 μm in diameter and are usually arranged in small groups with a common orientation at leg joints.

Inside the femur of a leg the largest sense organ is the femur–tibia chordotonal organ. This consists of 200–300 sense cells in a nonisotropic matrix either attached directly to the cuticle wall or supported by several filaments. In the stick insect the matrix is attached to the anterior wall of the femur and extends into a single inelastic apodeme which is fixed to the tibia, dorsal to the joint axis. Relative movement of the femur and tibia produces an axial movement of up to 300 μm in the apodeme. This axial movement distorts the supporting matrix and stimulates the dendrites of the sense cells. The output from these organs can be directly recorded in the nerve, and phasic and tonic units responding specifically to extension and flexion have been found (T. Hofmann, private communication). The output from this organ is primarily concerned with position, but again the attachment of the cell matrix to the femur wall could provide information on local bending strains in the cuticle.

Tension receptors in the leg muscles of insects have been reported by Theophilidis and Burns (1979). There is evidence for tendon sense cells (Baessler, 1977a), multipolar sense cells (Hustert *et al.*, 1981), and strand receptors previously thought to be chordotonal organs (Braeunig and Hustert, 1980; Braeunig, 1982), all of which could provide information on the relative movement of leg segments. Figure 42 shows a schematic map of the sense organs associated with the more proximal leg joints of the locust, and Fig. 43 shows the location of some of the more distal sense organs in the stick insect. The role of some of these sensory structures has been examined in considerable detail and tendon receptors and multipolar cells appear to contribute little to the control of the femur–tibia joint when compared with the chordotonal organ (Baessler, 1977a).

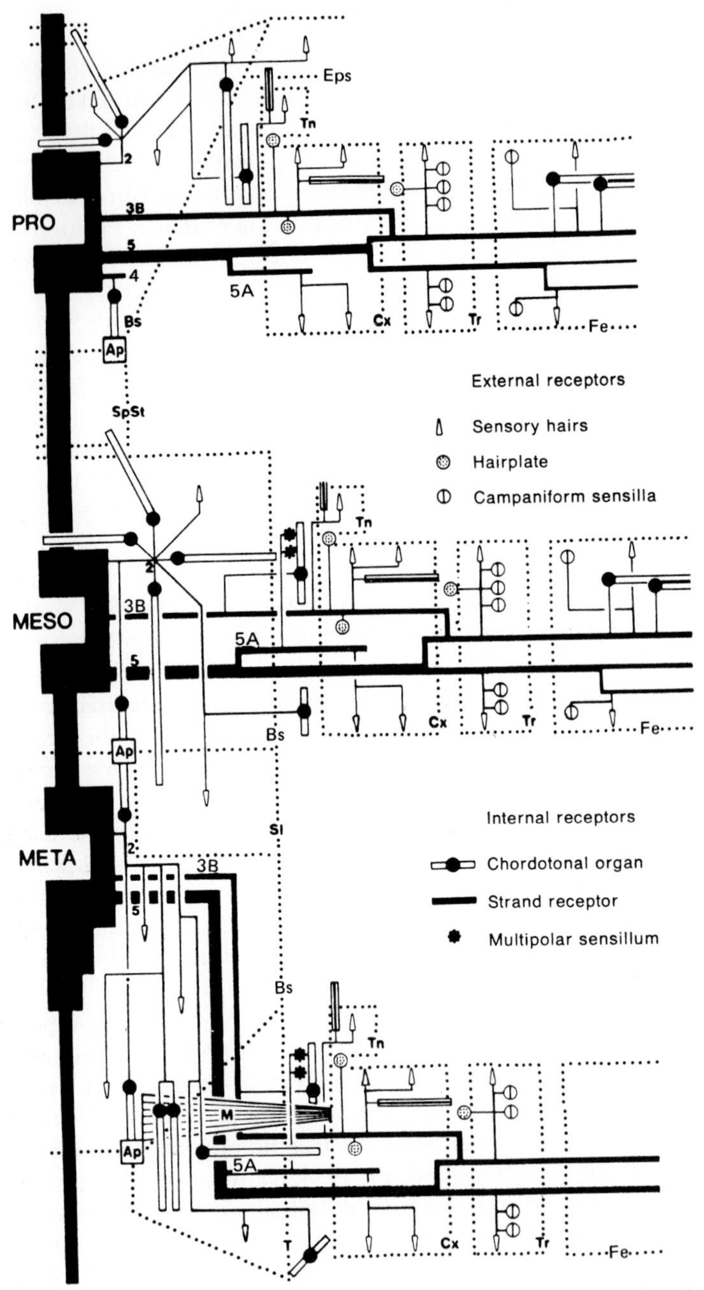

Fig. 42. Schematic representation of peripheral distribution and innervation of mechanoreceptors of locust thorax. Sclerites outlined by dotted lines; arabic numerals refer to nerves (solid lines). Ap, sternal apodeme; Bs, basisternum; Cx, coxa; Eps, episternum; Fe, femur; M, anterior coxal rotator muscle; Sl, sternellum; SpSt, spinasternite; T, triangulum; Tn, trochantin; Tr, trochanter. (From Hustert *et al.*, 1981.)

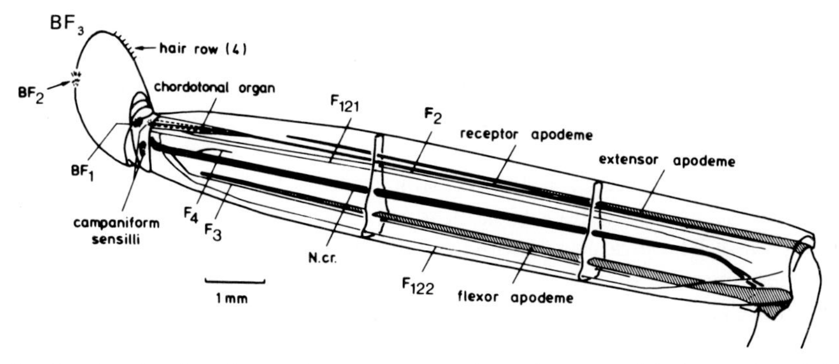

Fig. 43. Anterior view of the main sense organs of the stick insect and the innervation of the femur from a middle or hind leg. The transverse sections in the drawing show the three-dimensional nature of the reconstruction. (From Baessler, 1983.)

6.1 HAIR FIELDS

The earliest attempts to examine the role of specific receptors were the experiments of Wendler (1964). He removed mechanical input to the hair fields BF_2 (cxHPv) and BF_3 (cxHPd) by cutting away the hairs that were normally depressed by the forward movement of the coxa–thorax joint and showed that BF_2 played an important role in controlling the position of the free-hanging legs (Wendler, 1964). BF_3 produced no significant change in leg position. A similar experiment on BF_1 (trHP) at the trochanter–femur joint caused the leg to be lifted higher than normal. The removal of all these organs on both middle legs produced no significant change in interleg coordination but did cause a slight increase in the amplitude of the middle leg steps. More recently Wong and Pearson (1976) examined the influence of the equivalent hair field to BP_1 in the control of the coxa–trochanter joint of the cockroach. Two types of hair are present. The 30 longer hairs touch a fold of the intersegmental membrane and are arranged in rows, so that flexion of the joint causes successive rows to be depressed as the membrane rolls over them. The smaller hairs do not touch the membrane. Mechanical stimulation and recordings from the hair field nerve and levator and depressor muscles show that the hairs are excited by phasic flexions of the leg and inhibit the levator and excite the depressor. Removal of the hair field produces an influence on the walk behavior similar to that found by Wendler (1964). The leg protraction is exaggerated and the leg tends to collide with the leg in front.

The experimental study of the role of the trochanteral hair fields in controlling body height has been continued by Wendler (1972), whose recordings from the levator and depressor trochanteris show that this sense organ is part of a servo-control system which maintains the body height in the standing or hanging animal. If this organ is destroyed the animal is unable to support its own weight, while the intact animal is able to support up to four times its own body weight (Fig. 44). In this reflex, where stress might be expected to form part of the reflex loop, the hair field receptors (BF_1) appear to be a sufficient control element. Such a position-control system is particularly appropriate where the reflex loop is required to operate at any orientation of the body relative to the gravity vector.

The control of the coxa–thorax joint appears to be more complex and involves several receptors in addition to the BF_2 and BF_3 hair fields. Removal of BF_3 produces no obvious effects in standing or walking animals and demonstrates that it is not an important element of the reflex controlling the muscles retracting the coxa (Graham and Wendler, 1981b). BF_2 is important in limiting forward movement but its removal does not influence the retractor reflex either, and it is probable that internal joint receptors are responsible for this reflex. In *Extatasoma* (a large stick insect) and the locust, a multistrand chordotonal organ and several other sense organs span this joint and may be responsible for this reflex (Hustert *et al.*, 1983).

Baessler (1965) removed the hair rows along the posterior surface of the coxa and found a weak influence on the position of the leg under loading parallel to the body axis in standing animals. Removal of these hairs produced no detectable change in the retractor reflex mentioned above. Recent work on the targeting response mentioned in Section 3.7 shows that hair fields can modify this behavior (J. Dean, and J. Schmitz, personal communication).

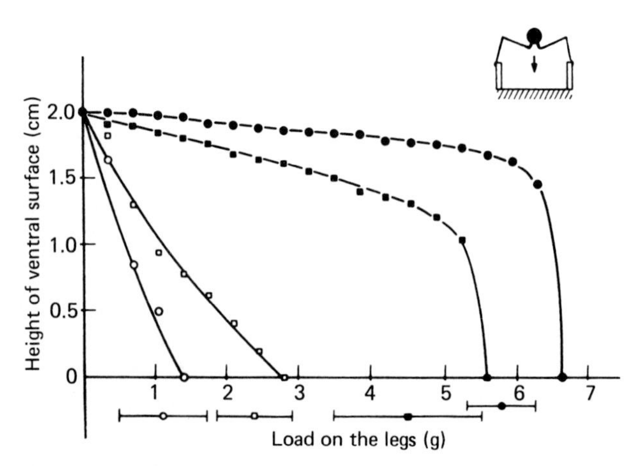

Fig. 44. The height of the ventral surface as a function of the load on the legs. (●) Intact adult. (□) An animal with all six trochanter hair fields (BF_1) covered with a water-soluble glue. (■) The loading curve after the glue is washed away. (○) show the effect of shaving the hair fields. (From Wendler, 1964.)

6.2 CHORDOTONAL ORGANS

Chordotonal organs, by virtue of the ease with which they can be mechanically stimulated, have received by far the most detailed examination of all leg sensory receptors. The first study of this sense organ in the hind leg of the locust (Usherwood *et al.*, 1968) showed both phasic and tonic responses in the chordotonal organ (CO) nerve to changes in the angle between femur and tibia at oscillatory frequencies in the range 0.1–30 Hz. These movements, in static preparations, produced a resistance reflex in the extensor and flexor muscles moving the tibia. When the apodemes of these organs were cut in the metathoracic legs of the walking locust, the activity in the hind leg extensors was reduced and the rate of stepping was halved when compared with that of the anterior legs. Usherwood *et al.* suggested that this implied the loss of coordination with the other legs. However, this behavior can appear spontaneously in grasshoppers (Graham, 1978a,b) and is not necessarily associated with a total loss of coordination with other legs.

In the locust the CO apodeme is only 3 mm in length but in some insects it may be 20 mm long, permitting reliable manipulation of the tendon and its associated organ. This organ in the stick insect has been the subject of extensive study as it is the major sense element in the servo-loop controlling the position of the tibia. The loop has been opened by cutting and manipulating the apodeme and its influence on the neural activity and forces generated by the extensor and flexor muscles has been investigated in different behavioral states (Baessler, 1967, 1972b,c, 1973, 1974). Cruse and Storrer (1977) have stimulated the behavior of this servo-system. More recently Cruse and Pflueger (1981) and Cruse and Schmitz (1983) have examined its role in walking and shown that the tibia–femur joint is under reflex control during walking movements if the leg is disturbed from its expected position by externally imposed forces.

The role of this organ in controlling motor output is well established for the tibia, but it also appears to extend its influence to other muscles of the same leg and even to other legs. This was shown by changing the attachment point of the apodeme to the other side of the femur–tibia joint. The animal walked quite normally at slow speeds after this operation, but measurement of wheel velocity on the two sides showed that the force developed by the operated middle leg was less than on the intact side. This effect was unlikely to be due to surgical damage as it is the thoracic muscles that generate all the propulsive force when the leg is perpendicular to the body. In this case a reversed or wrong afference reduced the force developed by muscles unrelated to its own joint. A second observation in these experiments was that hind legs on the operated side often missed contact with the walking wheels. They stepped outside the rim if the operated middle leg was strongly flexed and inside the rim if it was more extended than normal. In conjunction with the experiments of Cruse (1979b) on targeting of leg movements during walking, these results indicate that the afference from the middle leg CO is used by the hind leg (inappropriately in this instance) to decide where to place the tarsus at the end of the protraction stroke.

The importance of this organ in the coordination of walking movements is still uncertain. The experiments of Usherwood *et al.* (1968) should be repeated with quantitative measurement of the coordination changes as it is possible that in hind legs, with their plane of action parallel to the body axis, the CO is the major proprioceptor for stroke amplitude and tarsus position. If temporal coordination were still present after cutting the tendons of these organs, it would suggest that the information transferred to other legs may not be in the form of position information but more in the nature of "This leg is now in a protracting state." This information could be transmitted by either other sense organs at the base of the leg or motor reafference or possibly both.

6.3 CAMPANIFORM SENSILLA

The early experiments of Pringle (1940, 1961) and the more recent work of Pearson (1972) and Baessler (1977a) showed that stress applied to the trochanter, in the area where the campaniform sensilla organs are found, produces changes in propulsive muscle activity and inhibition of the protraction stroke in both the cockroach and the stick insect. Their influence on interleg coordination and the motor activity of other legs is not known.

Zill and Moran (1981a) have examined the responses to bending of the femur–tibia joint in the cockroach. Direct recording from the associated nerves has shown that they are stimulated by the normal movements of the joint. They respond primarily to bending of the joint in the plane perpendicular to the axis of joint rotation and bending perpendicular to the long axis of their segment and are, therefore, stimulated by gravitational and other leg forces as well as the muscles acting on their own joint. These organs do not appear to be important in fast walking but mediate load compensation during slow walking (Zill and Moran, 1981b). Similar responses are reported for the coxa-femur joint in the stick insect (Hofmann and Baessler, 1982), for which they probably serve a similar function.

6.4 AFFERENCE MODIFICATION

All ablation experiments in which sense organs are destroyed suffer from the important limitation that they remove an input to the system and their role may then be taken over by alternative sensory systems. This has been clearly demonstrated in the experimental work of Wendler (1965), in which the contributions of the antennae and leg receptors to gravity perception can only be assessed by counterbalancing the body. Under this condition the antennae are used to detect the gravity vector, although their removal from a free animal might have suggested that they played no role in gravity perception.

An alternative approach to the problem of accurately defining the behavioral function of a sensory receptor is to change the input–output relation for the sense organ. The advantage of this approach is that such changes are much more likely to produce a quantifiable response than simple ablation. The extent to which this modifies the behavior can be directly estimated and represents a quantitative measure of the importance of the organ in the behavior under investigation.

The "tied" leg experiments of Pearson (1972) and Graham (1977b) are examples of this approach but the stimulus modification was ill-defined. In the first of a series of such experiments on the stick insect, Baessler (1977b) used a clearly defined multiple stimulus to modify the walking behavior of a leg. He used a low-melting wax to depress the hair field BF_2 on the coxa–thorax joint, which signals the extreme forward position of the leg, and cut away the hair rows depressed during rearward movement of the leg. In the operated condition these organs signal that the leg is in the extreme forward position at all parts of the leg cycle. In this experiment the PEP of the leg is displaced conspicuously to the rear and the leg often fails to release itself from the ground. When it is finally dragged free it is recovered normally and the leg cycle is repeated. This behavior is only observed in front and middle legs while hind legs with this operation always step normally.

In a second experiment, the trochanteral joint of the leg was squeezed by a small wire clamp. This resulted in walks in which the leg moved to the rear in the first step but then remained in the posterior extreme position for the remainder of the walk and made no recovery strokes. If the clamp fell off or was removed normal walking was resumed. Similar behavior was reported for front, middle, and hind legs.

In the third series of experiments (Baessler, 1977b; Graham and Baessler, 1981), the output from the femoral chordotonal organ, which signals the angular position of the tibia relative to the femur, was inverted in the right middle leg by detaching the apodeme of this organ from its normal position close to the tendon of the extensor tibia and fixing it into a notch cut in the flexor tendon. This operation caused the CO to signal extension when the leg was flexed and flexion when it was extended. The walking behavior of the animal was normally coordinated at slow walking speeds but if it was stimulated to walk more quickly the operated leg lifted in a recovery stroke, was fully extended, and remained above the ground in a saluting posture. It stayed in this position for several steps of the other legs. Similar behavior is found in front, middle, and hind legs, but only the hind leg sometimes failed to release the ground in a manner similar to that described for the anterior hair plate depression experiments.

The conclusions drawn from these experiments were as follows. (1) The stimuli which cause a leg to remain in a particular phase of the cycle indicate that the phase has not been completed. (2) The cycle of leg movement can be interrupted by a suitable afference in at least three places: before the release of the tarsus, after the release but before recovery begins, and during the recovery stroke. In addition the "leg standing on a stick" experiments show that the cycle can also be interrupted in the early part of the stance phase.

These experiments establish that leg movements in the stick insect can only be accomplished when specific sensory requirements are met. Thus the leg cycle is a discrete sequence of movements in which each part must be successfully completed if the next part is to take place. Coordination has been examined quantitatively only in the last experiment of the series (Graham and Baessler, 1981). In this case, when the middle leg adopts the salute posture two changes in coordination appear. The first is an occasional absence of the front leg protraction, which would be expected to follow the protraction of the middle leg that begins the salute. The second is a change in the coordination of front and hind leg to that typically found in the middle-leg amputee. This change is maintained for the duration of the salute (Fig. 45). Thus, when a protraction stroke is unduly prolonged the opportunity for

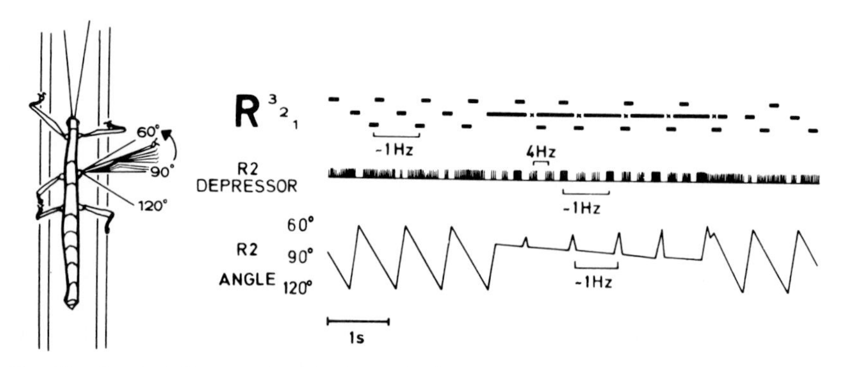

Fig. 45. On the left, the animal is shown fixed over two independent wheels. The receptor apodeme of the right middle leg is crossed. Top right, the step pattern of the legs on the right side during normal walking and saluting. Short black bars represent a swing phase; long bars, a salute. The crosses during saluting symbolize the occurrence of a visible movement of this leg (see bottom trace). Middle right, schematic representation of the electrical activity of the depressor muscle of the right middle leg during the walk shown in the top trace. Lower right, the angle formed by the femur of the middle leg and the longitudinal body axis during the same walk. The time scale is from left to right. (From Baessler, 1983.)

an anterior leg to protract may be lost, leading to a double-length stride. If the protraction is maintained for too long the new amputee gait is adopted.

These results and the observation, during temporary blocking of protraction movements, that the rhythm of the legs in front is reset (Dean and Wendler, 1982; Cruse and Epstein, 1982) suggest a sophisticated chain-reflex control system for individual leg cycles and the dominance of the periphery in the control of interleg coordination. Also, the use of position information rather than stress receptors to control height and possibly lateral stability probably reflects the preferred versatility of such a scheme in dealing with vertical and upside-down walking as well as the more conventional upright walks. In this respect the control of insect walking may be expected to differ markedly from the crustacea and vertebrates, which are primarily engaged in upright walking.

6.5 IMPOSED LEG MOVEMENT

The inaccessibility of the internal sense organs and the possibility of multiple-mode stimulation which might be associated with the control of the coxa–thorax joint suggested a direct approach toward assessing the role of sensory input by moving the whole leg stump during walks on a treadwheel.

The animal was supported above a counterbalanced double treadwheel, one leg was cut away at the middle of the coxa, and the open end was sealed, leaving only the most proximal leg joint. A drawing pen, driven by a pen motor, was clipped to the cut rim of the coxa, allowing the joint to be rotated backward and forward in an arc similar to that found in a normal walking movement. Wires inserted into the thorax recorded the activity of the retractor coxa muscle for the operated leg and the other legs on the same side (Graham, 1985).

When the insect stands still and the stump is moved, a normal resistance reflex is observed. When the stump is held still and the animal walks, weak bursts are observed coordinated with the walking legs. This could be produced directly by the movements of the other walking legs or as a modulation of standing leg motor output.

When the middle leg stump is moved with a period of 700 ms and the step period of the other legs is 1 s the activity in the retractor is modulated by the walking legs and waxes and wanes depending upon the phase relation with the walking legs (Fig. 46). In the hind leg the movements of the stump always produce strong bursts and show no modulation.

These preliminary results are interesting for two reasons. First they suggest how one leg may control the behavior of another at the motor output level. Second, they suggest a difference in the way in which motor output is generated in middle and hind legs. This difference in the role of the hind legs is

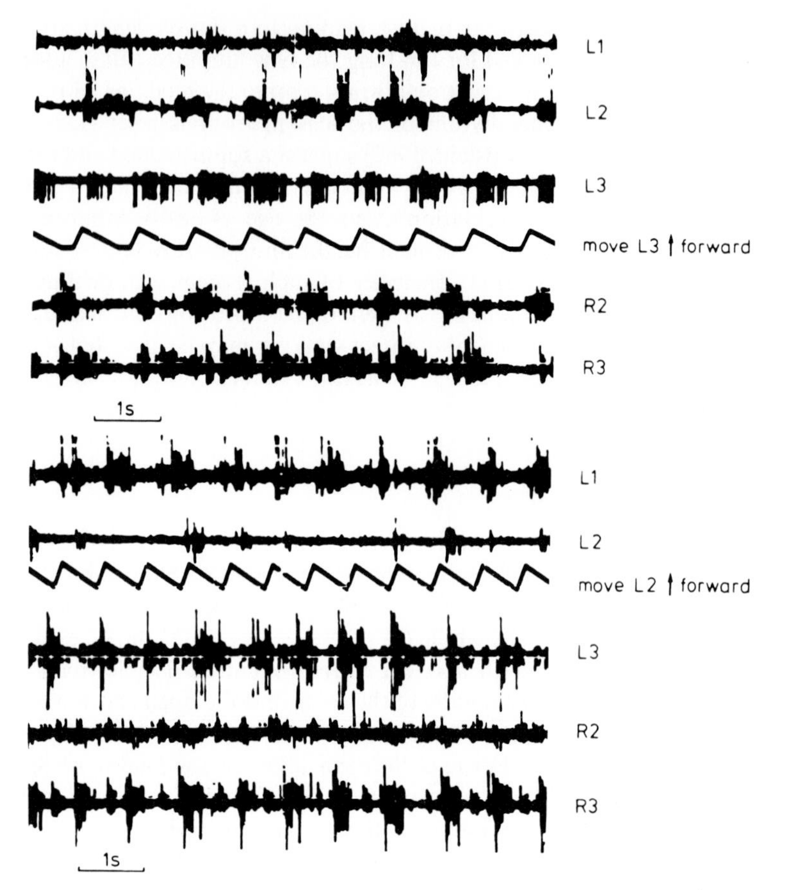

Fig. 46. Myograms for the retractor coxae muscles of several legs. In the upper picture the fourth trace shows the movement of the stump of leg L3. The animal is walking on a wheel. In the lower picture the stump of leg L2 is moved (third trace). (From Graham, 1985.)

also supported by the sensory modification experiments and more recently by observations on slippery glass walks, in which hind legs often fail to step while active stepping is maintained in the front and middle legs.

A summary of these results suggests that front legs may be used to some extent as antennae (Cruse, 1976a), but more commonly they are important as tractor elements (Graham, 1977b). This traction role is shared with the middle legs (Graham, 1983). Support is produced primarily by middle and hind legs (Cruse, 1976a). Lateral stability may be provided to a large extent by passive forces in the leg joints (Yox *et al.*, 1982). The hind legs do not show strong "braking" motor activity compared to middle legs (Graham and Godden, 1985); thus front legs must be expected to contribute strongly to "braking" during the other half of the middle leg cycle (as middle leg retractors are active until the leg lifts at the end of the power stroke; Graham and Wendler, 1981a). Hind legs also show much more variable motor activity during the stance phase than do middle legs, suggesting a stabilizing role for these legs (Graham and Godden, 1985). The most posterior legs appear to control the timing of those in front (Graham, 1978a,b; Cruse and Epstein, 1982; Dean and Wendler, 1982) during forward walking. However, in backward walking (Graham and Epstein, 1985) and when the hind leg is stepping slowly on a driven belt (Baessler, 1983), there is some evidence for a similar short-latency coordinating influence directed toward the rear which is not usually detected in normal forward walking. The precise characteristics of these interactions between the legs are difficult to determine during normal walking behavior and require a method for systematically perturbing the walk behavior in a reproducible manner.

6.6 THE PHASE RESPONSE CURVE

In order to examine the influence of one leg on another in the absence of a detailed understanding of the peripheral and central pathways involved, the periodicity of one leg can be modified and its influence on the period of the other leg measured. Stein (1974) has performed such an analysis on the swimmeret system of the crayfish and has reviewed the mathematics of the phase response curve (PRC). In his experiments Stein used a stimulus, equivalent to the movement of the controlling limb, at different times during a sequence of beats and compared an unstimulated period with the following one in which the stimulus appears. The result shows that the natural periodicity of the controlled limb either increased or decreased depending on the relative phase of stimulus and limb. The result is similar to that predicted by the "magnet effekt" of von Holst, and both Wendler and Heiligenberg (1969) and Ayers and Selverston (1979) have proposed neural models capable of generating such a PRC.

Two similar but less rigorous experiments have been carried out on walking arthropods. Such an experiment is possible in principle if one can assume that coupling across the body is negligible and that gliding coordination is present along the body, and if the control leg is stepping with a markedly different frequency compared to the controlled leg. This produces one or more natural or unstimulated steps of the controlled leg followed by a controlled step of that leg. If the difference in successive periods of the controlled leg is plotted as a function of the phase of the stimulus step then a PRC can be constructed. Such an experiment has been performed for the grasshopper (Graham, 1978b) and rock lobster (Chasserat and Clarac, 1983) and the results are shown in Fig. 47. In the lobster the stimulus step can increase or decrease the natural period of the leg immediately in front, and the result is similar to that for swimmeret movement and suggests a "magnet effekt" coupling. The result for the grasshopper is not a symmetrical result but shows that the natural period of the controlled leg tends to be increased if, and only if, the controlling leg steps just before the controlled leg. There appears to be no significant tendency for the controlling leg to decrease the natural period of the controlled leg. This type of interaction is most simply

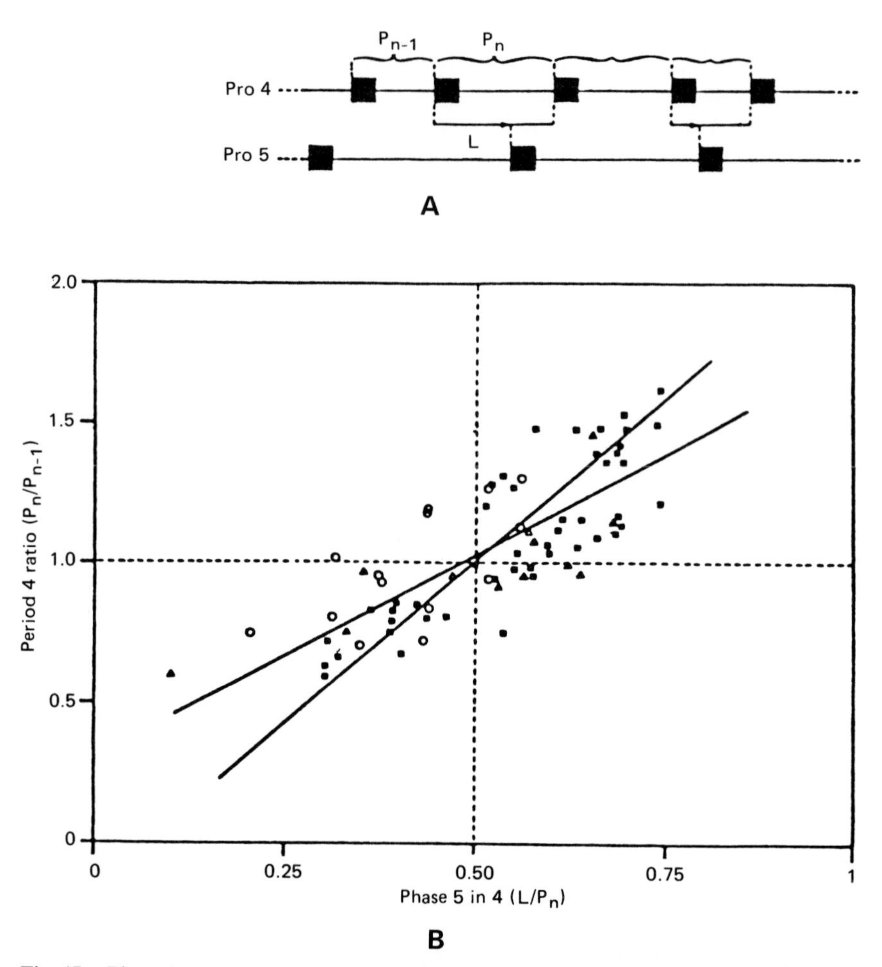

Fig. 47. Phase response curves are derived from the kind of step pattern shown in (A). The phase of the slow-stepping leg on the fast-stepping leg is plotted against the change in period for the nth and $(n - 1)$th step. In (B) for a rock lobster the ratio of these periods is plotted and shows that a protraction of the slow-stepping leg may delay or advance the protraction of the faster stepping leg depending on the phase. This is typical of a "magnet effekt." (*continued on p. 113*)

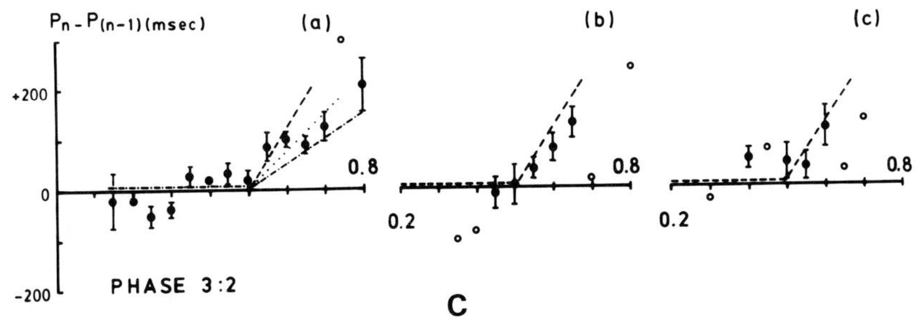

Fig. 47 (*continued*) (C) for a grasshopper the difference between successive periods of the controlled limb is shown as a function of the controlling limb phase. Three examples are shown for systematic absences (a), corresponding to the rock lobster condition, and for the second of a pair (b) and isolated absences (c). In the rock lobster condition the nth and $(n-1)$th steps are equal within experimental error for all phases except those in the range 0.6–0.8. This shows that the controlling leg only delays the controlled leg and is not able to advance the timing of the controlled leg. This is consistent with the forward-directed inhibition model, first proposed by Wilson (1966) and modified by Graham (1977a). For further details refer to Chasserat and Clarac (1983) and Graham (1978a).

represented by the forward-directed inhibition of protraction strokes proposed by Wilson (1966) and quantified in the model of Graham (1977a).

7 Models of walking behavior

Mathematical models can be constructed to represent the essential features of a system and the interactions between its parts. The simpler the model, the easier the task of predicting the future behavior of the system and comparing it with experiment. Models should not necessarily attempt a perfect description but rather should contain only those elements essential to function that make possible prediction of how the system will behave. Perhaps the most important feature of a quantitative model, from which accurate predictions can be derived, is that it forces the inventor to describe the system precisely and thereby show what it cannot predict or falsely predicts. This facilitates criticism and comparison and shows those areas in which further experimentation is required to resolve specific theoretical problems (see Cruse and Graham, 1984).

A number of models have been advanced to describe the essential features of coordinated locomotion in insects and these will be briefly described. The first major attempt to place the coordination of appendages on a firm theoretical base was the work of von Holst (1936, 1939a,b). In a series of papers he explored the temporal and spatial relationships between the beating of fish fins. He separated the spatial and temporal components and quantitatively defined phase-locked coordination and relative or gliding coordination, in which two oscillating systems influence each other's behavior when the inherent or natural frequency of the two systems differs. He did not propose a specific model for insect walking coordination but his concepts formed the basis of the first model of interleg coordination proposed by Wendler (1968). This model consisted of six oscillators driven by a common input which determined the same inherent frequency for all the oscillators. Each oscillator controlled the oscillator in front and on the opposite side by a generalized (and unspecified) "magnet effekt" interaction producing a phase-locked step pattern which changed with the level of the input (Fig. 48). At the highest step frequency a simulation of the model on an analog computer showed the tripod step pattern and at the lowest step frequency the legs on right and left sides separated into consecutive metachronal sequences as reported by Hughes (1952) for the cockroach.

Wilson (1966) had already proposed a model of similar structure in which the oscillator frequencies formed a hierarchy along the body axis with the lowest natural frequency at the rear, in which coupling along the body was produced by inhibiting or preventing the protraction of the leg in front. A similar mechanism was assumed to act mutually across the body. In order to explain Hughes' results, the inhibitory influence along the body axis was assumed to maintain an approximately constant latency and to produce a lag of fixed duration between the legs. Thus as the input to the oscillators decreased only the step period increased, producing the observed separation between the metachronal sequences on each side. Across-the-body coupling was rather more complex as it required the lag between right and left legs of the same segment to change at exactly the same rate as the step period. Later, following his work on tarantula walking (1967), Wilson (1968) suggested that one might also construct a model based on excitatory coupling with similar properties.

These inhibitory and excitatory concepts formed the basis for two quantitative models by Graham (1972, 1977a) and Cruse (1979a, 1980a,b), respectively. Graham had found that free-walking stick insects did not walk with a phase of 0.5 for legs in the same segment. Rather they displayed a pronounced asymmetry in their coordination across the body. When walking slowly, stick insects stepped with a phase of either 0.3 or 0.7 for right leg against left leg, rather than the symmetrical value of 0.5 found in high-speed walking. A corresponding difference was also observed in the lag versus period relation for legs on opposite sides, suggesting that the legs on one side determined the period of those on the other side but not their lag. This model was similar to that of the inhibitory model of Wilson but used an identical mechanism both along and across the body (Fig. 49). The duration of the lag between legs could be adjusted by a separate input, to produce the lag versus period characteristics for several different insects and the different gaits observed in

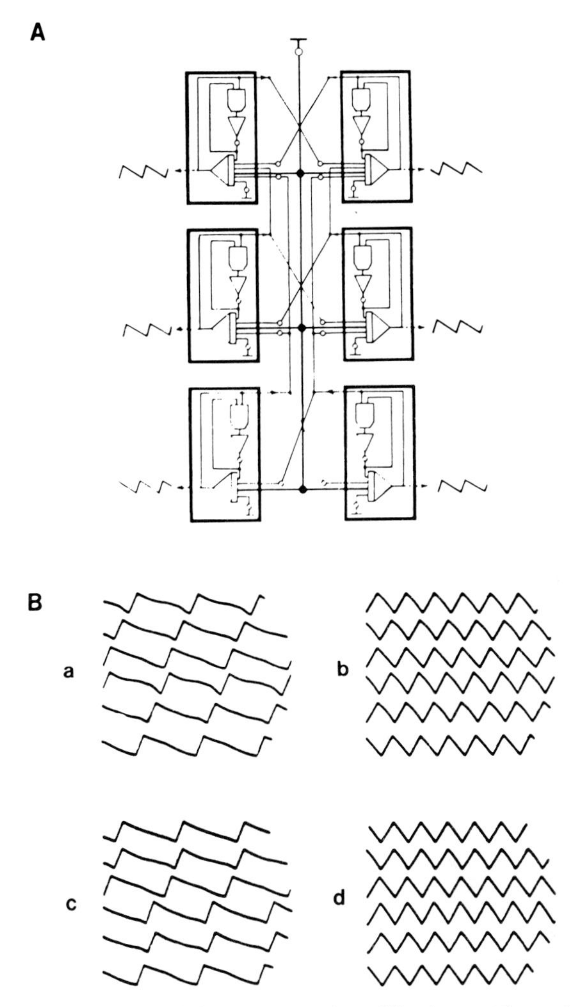

Fig. 48. (A) Model for the control of leg movement in walking insects. The model consists of six relaxation oscillators that can have different eigen (inherent or natural) frequencies. The direction and strength of coupling depend upon the relative phases of the oscillators. The overall step frequency is controlled by a nonrhythmic central input. The components can be interpreted biologically as follows: integrator—muscle and leg segments; feedback to comparator—hair fields; speed control—command fibers from the brain. When the model is stopped it begins again with the same phase relationship. (From Wendler, 1969.) (B) Analog computer simulation of the movements of the six legs in an intact slow (a) and fast (b) walking insect. In (c) and (d) the connectivity between the middle and front legs on each side is interrupted. In these figures the new relationships for the bilateral middle-leg amputee can be seen. Time runs from left to right. The plots represent from top to bottom R1, R2, R3, L1, L2, L3. Upward movement represents the protraction of the leg and downward movement shows the retraction stroke. (From Wendler, 1978b.)

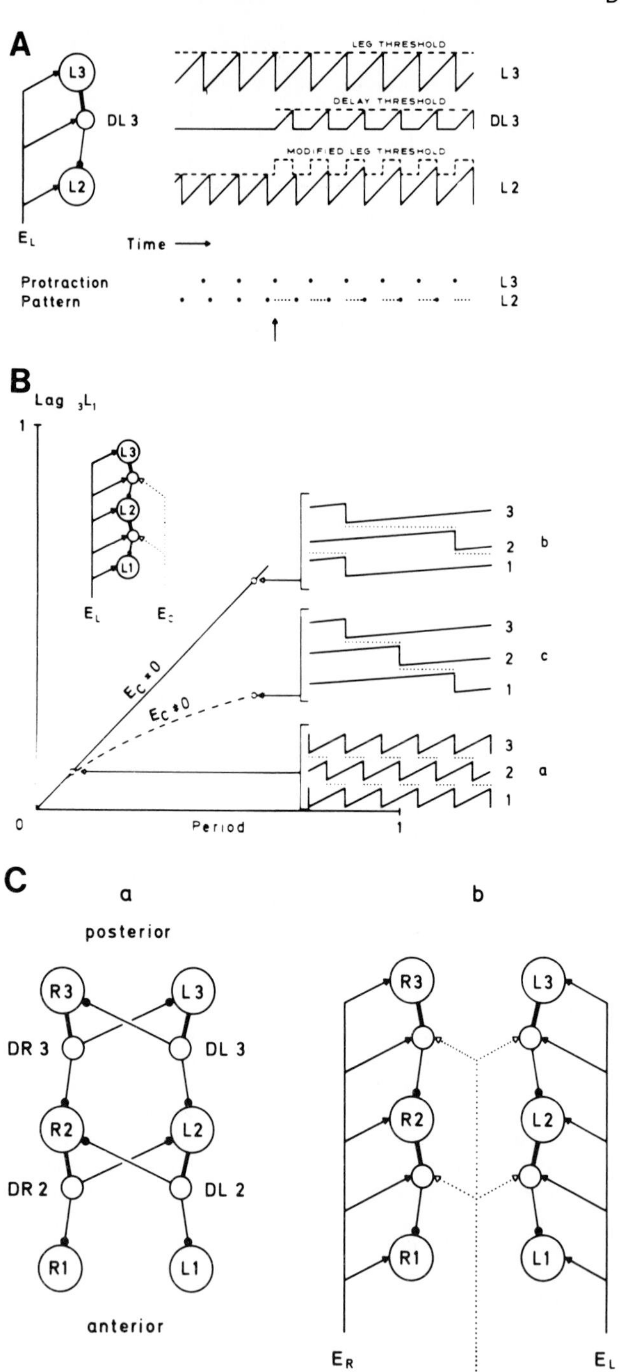
A
L3
DL 3
L2
E_L
LEG THRESHOLD
L 3
DELAY THRESHOLD
DL 3
MODIFIED LEG THRESHOLD
L 2
Time
Protraction
Pattern
L3
L2
B
Lag 3L1
1
0
Period
1
E_L
E_C
E_C = 0
E_C ≠ 0
3
2
1
b
c
a
C
a
b
posterior
anterior
R3
L3
DR 3
DL 3
R2
L2
DR 2
DL 2
R1
L1
E_R
E_L
E_C

adult and first instar stick insects. By changing the input level to right and left sides the model showed right and left asymmetry, predicted the coordination found in one form of turning, and predicted the possible step patterns in a wide range of amputations (Graham, 1977b).

This model requires a hierarchy of leg frequency along the body in order to maintain the normal coordination and couples the legs by raising the reset threshold of the controlled oscillator. One other difference is that the model uses different strength coupling across the body for the quantitative simulation of the adult stick insect behavior. Coupling is strongest between rear legs, weaker between middle legs, and negligible between front legs in order to explain lag asymmetry along the body in intact adults and the almost synchronous stepping of the front legs in single middle-leg amputees.

A new version of the Wendler analog model (Wendler, 1978b) uses the same connectivity as before and clarifies the control circuitry. The coupling mechanism in this model is a distributed function over the whole cycle, and the current state of an oscillator directly influences the current state of the controlled oscillator. Coupling influences are not restricted to the timing of the swing phase but take place continuously, and therefore involve relative movement of the legs contacting the ground at the same time. This model strategy differs significantly from that of the Wilson-type models in which the interaction is "phase dependent" and is restricted to prevention of a leg lifting if the controlling leg is in swing phase or is in the early part of its stance phase.

The model of Cruse (1979a, 1980a,b) has shown that an excitatory influence can be used to generate the step patterns of the intact stick insect if the metachronal rhythm is redefined to extend over twice the step period (Fig. 50) and the frequency hierarchy is such that the front legs tend to step with the highest frequency. The model requires different connections for the asymmetrical step patterns but generates the appropriate dependence of lag on period for the adult insect. One of the most valuable features of the model

---

Fig. 49. (A) The basic element of the model is shown on the left and consists of two leg oscillators L2 and L3 connected by a delay oscillator DL3. The internal state of each oscillator is shown as a function of time on the right with the dashed lines representing thresholds. The horizontal time arrow represents 100 ms. The vertical arrow shows the time at which coupling is introduced and the lower part of the figure shows the resulting step pattern. Here the forbidden times for protraction are shown as a dotted line (B) The lag ${}_3L_1$ versus period plot for a cascade of three leg oscillators. The inset shows the model configuration. Horizontal dotted lines show the duration of the delay producing the coordinated step pattern. The threshold of L3 is 100 units (L2 and L1 are at 80%). Input conditions are (a) $E_L = 1.0$ threshold unit per ms; $E_c = 0$ or 0.17. The larger value produces only a 20% decrease in the delay interval and is insignificant. (b) $E_L = 0.2$; $E_c = 0$. This represents a continuation of the normal tripod-type step pattern, gait I. (c) $E_L = 0.2$; $E_c = 0.17$. Now input to delay oscillator is greater by 2 and we have gait II leg timing. Time unit for lag and period is 1 s. (C). The organization of the complete model. (a) The interactions between the leg oscillators proposed for the intact stick insect; (b) the leg and delay oscillators and their respective inputs E_R, E_L, and E_C. (From Graham, 1977a.)

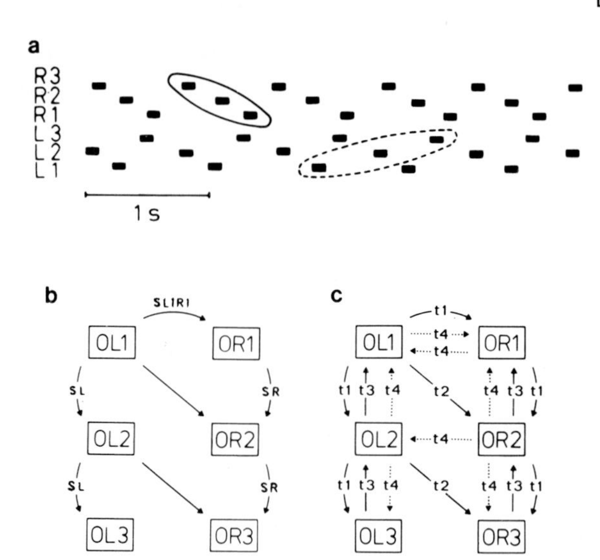

Fig. 50. (a) Typical coordination pattern of an adult stick insect. (From Graham, 1972.) (b) Cruse's "simple" model. OL1, Oscillator for the left foreleg; OR1, for the right foreleg; OL2, for the left middle leg; etc. Arrows indicate the direction of excitatory influences. (From Cruse, 1979a.) (c) Cruse's "extended" model. For details see original reference (Cruse, 1980b).

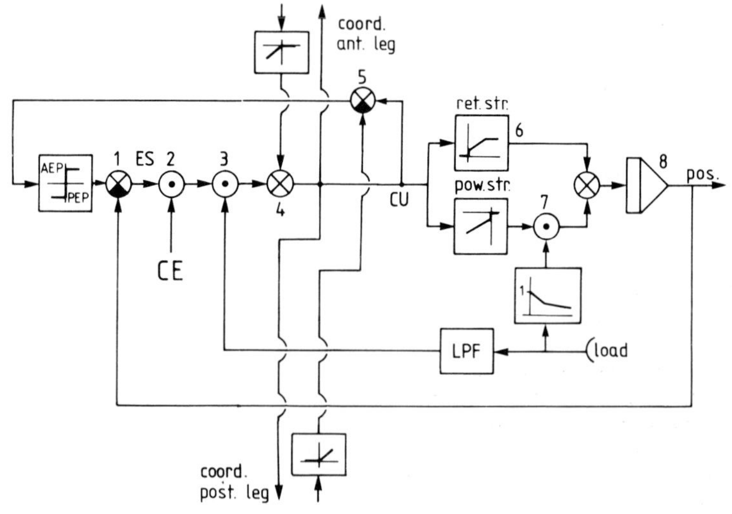

Fig. 51. Schematic of leg control system. (1) Comparison of peripheral position with leg state command. ES is the error signal. CU is the switch command for leg state. (2) Command input multiplier. (3) Load (stress) multiplier. (4) Position control input from anterior leg. (5) State control input from posterior leg. (6) Retraction stroke limiter. (7) Depressor multiplier for stress input from load. (8) Leg muscle system (integrator). CE is the tonic central excitation (Cruse, 1983.)

is a detailed hypothesis for the feedback control system which generates the leg propulsion forces (Fig. 51). This postulates a centrally generated reference position for each leg controlled by coordinating influences from other legs. If the leg falls behind the reference then motor output is increased to minimize the lag. Such a system describes the changes in motor output with increased load, and places on a quantitative footing several qualitative models proposed earlier to describe motor output in the cockroach, which included references to possible coordination inputs capable of producing the tripod gait.

In these descriptive single-leg models, Pearson and Iles (1970), Pearson (1972), and Delcomyn (1971a) considered that the activity of the antagonists in a leg was centrally programmed by a sinusoidal threshold. This produced bursts of motor activity at the required step frequency. These bursts activated the protractor musculature of the leg and at the same time inhibited an ongoing output to the retractor musculature. Combining these individual leg elements Pearson *et al.* (1973) proposed that coordination was produced by mutual inhibition between the protractor burst generators, and that campaniform sense organs were responsible for exciting the propulsion motor neurons and inhibiting the burst generator and slowing the rate of stepping (Fig. 52). The Cruse model includes these load-sensitive inputs but depends heavily upon the position of the leg when determining the level of motor output required in the propulsion stroke. He also addressed the problem of

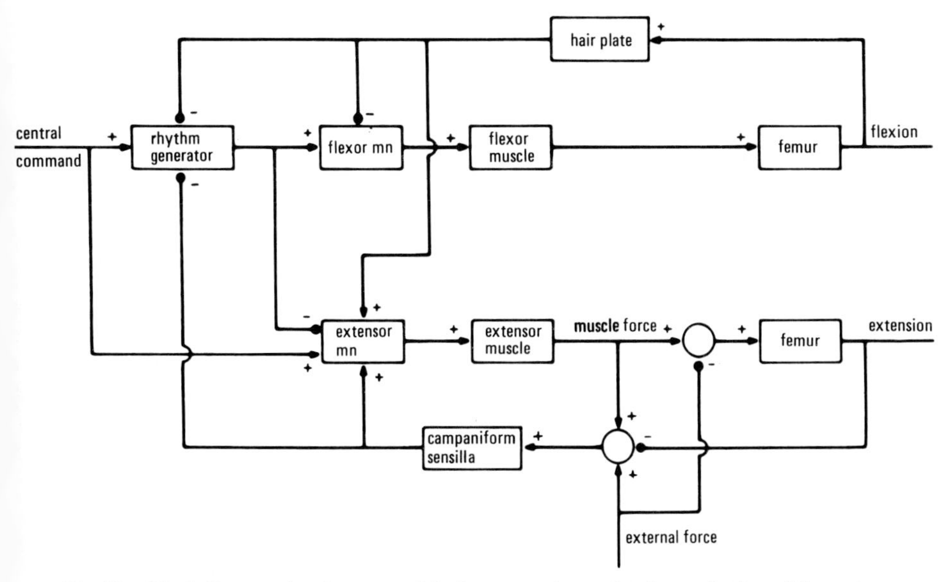

Fig. 52. Block diagram showing some of the known and postulated organization of the system controlling rhythmic movements of the femur in the hind legs of a walking cockroach. (From Pearson *et al.*, 1973.)

the transfer of loading information between legs and the changes in protraction duration with increased loading. In the qualitative neuronal models, coordination between legs is independent of loading.

Another model describing both motor output and load sensitivity is that of Land (1972). The model is complex because it attempts to describe both forward and backward walking in arachnids. The important feature of walking in the hunting spider is that the reversal of direction in a walk is accompanied by a memory of the precise position in which the leg stopped so that the new reversed step moves exactly back along the old trajectory. As long intervals may elapse between bouts of forward and backward walking, while the animal stands still, some form of long-term memory was required. Land proposed that the only satisfactory explanation for this and several other features of the walks was that the leg position must form a vital component of the leg oscillator. Land did not discuss interleg coordination in detail other than to point out that knowledge of the precise position of the other legs was essential for coordinated walking.

This exclusively peripheral model is shown in Fig. 53. Readers are advised to examine the original publication for a detailed description of its function. Essentially, the position of the leg as it approaches the end point of a stroke stimulates the antagonist which will reverse the direction of the limb. Depending upon the direction of walking either the levator or depressor is also activated. When the leg reaches the other end of the stroke the procedure is repeated. If neither of the two alternative "driving" feedbacks are facilitated by the command inputs for forward or backward walking then the system stops. When feedback is refacilitated the animal walks in the direction determined by the forward or backward command inputs. The important concept in this model is that the experimental evidence requires that the leg oscillators have a knowledge of leg position at all times and in particular when they are standing still. This is of importance in starting a walking system.

No model is yet able to explain all of the latest experimental results and a synthesis of certain aspects of the various models would be appropriate at this time (Cruse and Graham, 1984). The experiments of Graham (1978b), Cruse and Epstein (1982), Dean and Wendler (1982), and Foth and Graham (1983b) all indicate that the new model should be based upon a quantitative version of the Wilson hypothesis but must include load-induced effects such as the variation in protraction duration with step frequency. The original model (Graham, 1977a) is applicable to several insects and provides a description of the slow stepping in long hind legs as well as asymmetrical stepping and one form of turning. The peripheral version of this model is capable of incorporating the targeting observations of Cruse (1979b) and can be extended, by the addition of the servo-control of the leg muscles, to include a description of the changes in motor output and coordination consequent

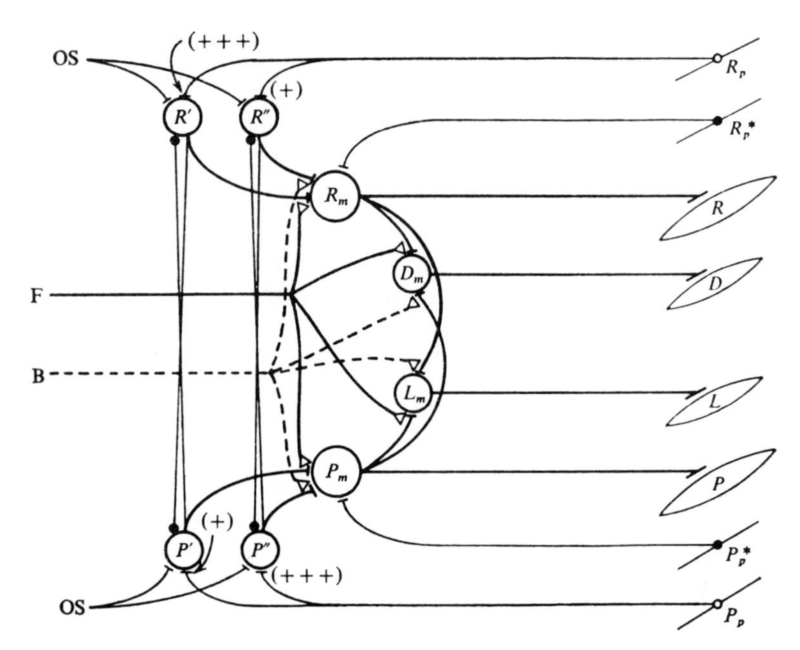

Fig. 53. Hypothetical neural model of the control system of one leg (see text). R, D, L, and P are the retractor, depressor, levator, and protractor muscles respectively, and R_m, D_m, L_m, and P_m are the motoneurons that drive them. R_p and P_p are length (or joint position) sensitive proprioceptors that are functionally in parallel with R and P, i.e., the discharge from R_p decreases as R shortens and similarly for P_p. R_p^* and P_p^* are proprioceptors which respond to tension (or cuticle deformation) and these are active when movements produced by R and P are opposed by a load. R′ and P′ and R″ are cells which constitute two oscillators; they are coupled by mutual inhibition and receive inputs of different strengths from the proprioceptors R_p and P_p, and from other segments (OS). R′ or R″ drives R_m, and P′ or P″ drives P_m, depending on the command fiber active. The command fibers (F, forward walking and B, backward walking) do not drive the motoneurons directly but facilitate (triangles) the synaptic endings of the oscillator cells. They also facilitate appropriate connections between R_m and P_m, and the levator and depressor motoneurons. Connections: all "synapses" are excitatory except closed circles (inhibitory) and triangles (facilitate exitatory synapses). In forward walking, the F command fibers are active on both sides, and the B fibers in backward walking. In turning, the F fibers are active on the side contralateral to the direction of turning, and the B fibers on the other. (From Land, 1972.)

upon an increase in loading in horizontal, uphill, and possibly downhill walking.

The preliminary results for downhill walking present an interesting test for such a model. At least 10% of downhill walks exhibit a rhythmicity which appears to be independent of the time at which a leg reaches its PEP. Forward and backward strokes are separated by prolonged "braking" periods whose duration appears to depend on an internally desired rate of stepping. In the remaining 90% these animals show a controlled-descent which is similar to horizontal walking behavior.

A model based upon a comparison of an internally desired state with that occurring in the periphery (a moving set point which is reset on each cycle by the periphery or a required velocity of movement) appears to be the only way in which certain aspects of downhill walking can be described. Such a model would be considerably more complex than any of those proposed so far and additional clarification of downhill walking and the fundamental coupling between various leg pairs is essential before attempting such a synthesis. At the present time the Wilson concept, based on a peripheral leg oscillator, appears to provide the simplest synthesis of the diverse behavior of walking insects and provides a quantitative description of most of the earlier observations.

8 Conclusions

The difficulties of quantitative comparison between walking systems have already been mentioned, but several classes of arthropods show very close parallels with the insects. The greatest volume of work is on the crustacea. Phase relations between walking legs are generally less precise, particularly in water-borne walkers and it would be difficult to detect the subtle changes found in cockroach and stick insect timing, which have given clues to the coordinating mechanisms in insects. The influence of loading on the crustacea ranges from no detectable effect on coordination to reports of a reduction in protraction duration and increase in step period similar to those reported for the stick insect. Motor output increases in response to load, and it seems likely that turning and uphill walking are similar, but again the greater variability of stepping has made analysis more difficult. However, the latest work on driven-walking behavior by Clarac and Chasserat (1983) and Chasserat and Clarac (1983) has shown a dramatic improvement in the reproducibility of timing, amplitude, and changes in imposed step frequency. Detailed analysis of spatial interactions between the legs has just begun and the results look most promising in evaluating the strategy used to deal with experimentally imposed changes in coordination.

The myriapods (Franklin, 1985), arachnids (Wilson, 1967; Seyfarth and Bohnenberger, 1980; Ferdinand, 1981), and scorpions (Bowerman, 1975, 1981) also show a close relationship to the insects, perhaps because they also have to fully support their body weight and show climbing ability.

A number of review articles have emphasized the similarities between legged locomotion in the arthropods and mammals. It is indeed true that all have legs which act in a repeatable cycle. The mammals use a piston-like action perpendicular to the walk surface, and support and propulsion are intimately related. Coordinated patterns of locomotion tend to switch from one gait to another with little overlap and up to four different gait patterns can usually be distinguished. Similarities appear to extend even to the motor patterns used in obstacle avoidance, but it is perhaps most valuable to mention some of the differences between these systems and the insects.

In mammals, the most obvious difference is the reduction in leg number to four, which makes a dynamic balancing mechanism essential. This complicates analysis, as most gaits in quadrupeds and bipeds are based on projecting the body forward over the substrate and correcting for incipient roll, pitch, or yaw by placing the legs in the best position to redirect the body at regular intervals. Even in slow-walking mammals the body mass has to be moved over whatever legs are available for support and is a dynamic and not a static process. This may account for the automated leg movements exhibited by "fictive locomotion" in cats. In such a system where dynamic control of balance is available, the weak peripheral control of a strong endogenous oscillator may possess significant advantages over the more peripheral type of oscillator found in insects.

Another distinctive difference is apparent in a comparison of the velocity profiles for insect and mammal walking. In fast-walking insects the legs in retraction accelerate the body to a maximum velocity at the midpoint of their stroke and decelerate the body almost to zero at the end of the stroke, while most quadrupeds maintain a more constant velocity with each leg contributing to the body velocity as it touches down. Such a strategy is only possible if an overall balancing mechanism is continually active to correct errors in the trajectory of the body.

The absence of grasping appendages in many quadrupeds limits them to upright locomotion. This appears to be a fundamental distinction between insects and upright-walking quadrupeds, and may have caused the evolution of quite different locomotor control systems. In insects this has produced a special emphasis on maintenance of the body in a suitable orientation to the walking surface for all positions of the body relative to the gravity vector.

In insects the investigation of the synaptic events which take place during walking behavior is just beginning with the advent of walking preparations in which such activity can be directly recorded (Godden and Graham, 1984) (Fig. 54). In the past our understanding of the origins of motor output has depended upon static inverted preparations only free to move parts of their legs. In this configuration experimenters have usually chosen to decapitate the insect, but several experiments indicate that walking behavior is not present in this state (see Section 5.3). Behaviors that are readily stimulated in this state are either grooming, rocking, or struggling and attempted righting movements. Such behavior may be similar in some respects to that found in high-speed walking, but significant differences in the timing and sequencing of motor output exist when comparing the deafferented behavior with both fast and slow walking. It has not been firmly established that walking

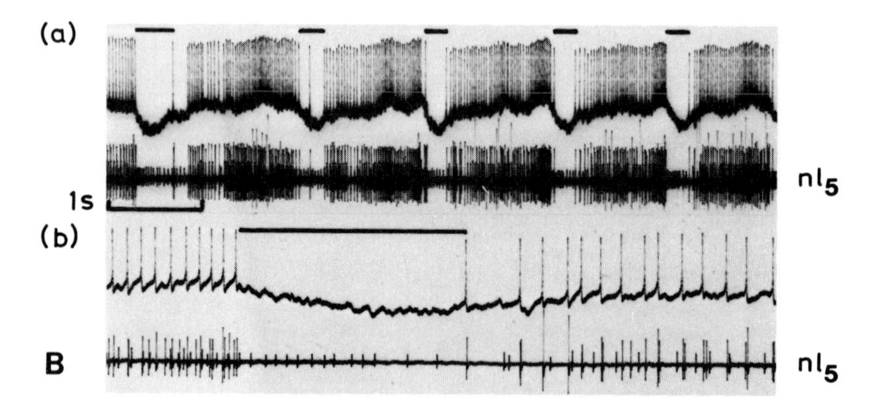

Fig. 54. (A) Setup for intracellular recording from treadwheel-walking animals. The thorax is opened dorsally and the gut removed to expose the ventral nerve cord for electrode insertion. (B) Intracellular recording from the neuropil of an sFRCx neuron (upper traces of a and b) and simultaneous recording from nl_5 (lower traces of a and b) from an animal on a treadwheel. The time scales of (a) and (b) differ. Protractions of the corresponding leg are designated by the black bars at the top. (From Godden and Graham, 1984.)

behavior is generated by endogenous oscillators, and most experiments on actual walking preparations suggest that events in the leg cycle depend upon the successful completion of earlier parts of the cycle so that the term *peripheral oscillator* might be more appropriate.

In flight, swimming, ventilation, and other rhythmic behaviors which can take place independently of the reaction of the surrounding medium, peripheral feedback is not essential but can confer advantages in terms of amplitude and frequency control. Numerous experiments have established that such systems can maintain a rhythm in the absence of peripheral feedback (Delcomyn, 1980). A self-supporting walking system might be expected to differ significantly from such homogeneous-environment locomotory systems, especially if the walking system is to respond to changes in substrate level, relative movement of legs, and the totally different motor output required for walks in which the animal walks downward or hanging below a substrate. Assuming that the system has been designed to depend heavily on feedback in order to accomplish these requirements, it may still be possible for it to ignore feedback under certain conditions of simplified locomotion and temporarily adopt an endogenous pattern normally used for some other purpose.

Recent work on the motor control capabilities of the locust metathoracic ganglion (Burrows and Siegler, 1976, 1978; Burrows, 1979, 1980; Siegler, 1981) shows that complex control can be initiated by specific interneurons. Spiking or nonspiking interneurons (Pearson and Fourtner, 1975; Pearson and Robertson, 1981) can provide a flexible and versatile coupling network capable of incorporating large numbers of motor neurons into a particular movement. Different movements can be produced by distinct sets of interneurons which interact with each other, descend onto the motor neuron pool of a leg, and are themselves influenced selectively by feedback from joint receptors.

Some interneurons have been shown to generate complete leg movements such as kicking and grooming and may be used as the basis for repetitive movements such as struggling or righting. It is possible that such action patterns may be utilized in high-speed walking behavior, although it has not been possible to show either interleg coordination or the patterned behavior of several muscles within one leg which would be expected in a walking step. Furthermore, in the flight system, proprioceptive control on a cycle to cycle basis has been demonstrated by Wendler (1973, 1978a), which indicates that such control is possible up to 20 Hz, which would include the fastest walking speed of the cockroach.

The control capabilities of interneurons appear to be almost without limitation and could cover co-contraction, antagonistic actions, differential contraction, and the time or proprioceptively influenced sequencing of muscles during a walking step of one or several legs. Precisely how this is achieved is an intriguing problem which will require multiple electrode recording from reliable walking preparations.

The complexity of the walking system and its distributed nature raises the question of whether it is possible to understand "walking behavior" in a complete sense at the neuronal level. While the individual neural building blocks are becoming better understood, the precise functional relationship between the pairs of ipsilateral legs in walking is still uncertain. The value of the simplistic functional approach to walking behavior is that it should help explain the overall control structure of the walking system and why parts of it are organized in the way they are. Furthermore, it can provide a guide to those parts of the system that are most likely to benefit from a detailed intracellular investigation. In conclusion, I would like to summarize our current understanding of walking in the insects in functional terms moving from neural components to walking behavior.

At the level of interneurons there is evidence for the endogenous generation of the motor activity appropriate for kicking, grooming, struggling, righting, and rocking or peering movements. However, it is not clearly established that such endogenous rhythms form the basis upon which the leg movements used in walking are built, except by inference from other systems, unless it is proposed that these endogenous oscillators are so strongly influenced by the periphery that they lose all claim to an independent rhythm.

Current work suggests that propulsion motor output is servo-controlled by a position or velocity reference. A number of experiments support the view that the transition from power stroke to recovery stroke is determined entirely by the periphery. Thus, the duration of the power stroke is determined by the motor output, the load the leg is experiencing, the relative timing of the other legs, and the information it receives about its position from its own sense organs. The AEP for front legs is probably determined by antennal information and the anterior legs act as targets for those behind. In the absence of peripheral sensory input, the system would be expected to remain in either power stroke or recovery stroke with the rate of firing proportional to the desired rate of walking. This is consistent with the most typical observations in deafferented preparations by Pearson and Iles (1973), Reingold and Camhi (1977), and Baessler and Wegner (1983).

At the motor neuron level no evidence has been found for direct inhibitory coupling between motor pools which function as antagonists during walking. This is not too surprising as such antagonists often show simultaneous activity in startle and twitch reactions in standing animals. The strong mutual inhibition observed in the walking state for power stroke and recovery muscles and support and levation muscles must arise either from premotor neurons or from collateral interneurons which become active in the walking state.

The control of joint angles involved in power stroke/recovery and depression/levation appears to be by simple activation of the muscle appropriate for the required direction of movement and simultaneous central inhibition of activity in the antagonist. This does not appear to be the case in all stabilizing muscles such as those controlling the extension and flexion of the tibia in the middle legs of the locust where co-contraction occurs. In walking, the role of inhibitors is obscure and as they appear to produce changes in tension of only 10–20% in the overall force generated, their effects can be ignored in low-speed walking (to a first approximation) but their effectiveness may determine the maximum possible speed of walking.

The sources of proprioceptive input in insects may be conveniently separated into three types of sense organ: (1) the hair fields which are capable of responding to position, velocity, and acceleration; (2) the chordotonal organs and strand receptors with similar properties; and (3) the campaniform sensilla, which are stimulated by cuticular deformation and indicate the level of stress applied to a joint. Hair fields control the overall range of movement of the joint and prevent damage by over extension or flexion. In joints which determine the body height they act as the primary sense element in a servo-controlled reflex. Individual hair fields at the coxa–thorax joint are involved in the position control of the leg during a walking step and their removal causes changes in AEP and PEP. In addition, when working in combination, they prevent the activation of the recovery musculature if they are modified to indicate that the leg is in the forward position. They do not appear to influence the coordination of the other legs following simple ablation, but selective stimulation of different fields could provide such control. These organs also provide the information used by posterior legs for targeting purposes.

The chordotonal organ is the major sensory element in the control of the femur–tibia joint which provides lateral stabilization in the middle leg of the stick insect, and may be important in the controlling the timing of power stroke and recovery for the front and hind legs as well as lateral targeting. Such organs have been found in the coxa–thorax joint, and are possibly responsible for the resistance reflexes of the power stroke and recovery muscles and may be important in the control of interleg coordination where the leg is oriented along the body axis.

The campaniform sensilla can be stimulated by stress to the legs either naturally or by local pressure applied to the cuticle. They enhance the power stroke motor output directly and increase the step period of the leg by suppressing activation of the recovery system.

The timing of the onset of the recovery stroke is influenced by several factors:

1. The position of the leg relative to the body, the probability of recovery activation increasing as the PEP is approached.
2. The stress on the leg joints, which tends to suppress the recovery stroke and possibly induces an increased likelihood of recovery when the stress is suddenly relieved.
3. Influences from the relative positions and/or stresses experienced by the other legs.

The strength of the recovery stroke may also be influenced by stress inputs, and a complex sharing or pooling of such inputs from legs in the stance phase increases the velocity of the recovery stroke under conditions of horizontal loading corresponding to that found in uphill walking. Vertical loading information is also transferred across the body axis and experiments on double wheels show that propulsive load inputs to only one side generate an increase in power stroke on both sides during walking and cause a higher frequency of stepping with the same amplitude on the unloaded side.

Placement of the leg tarsus is controlled by posteriorly directed information from the leg in front; this information is presumably generated by sense organs measuring position such as the chordotonal organs and hair fields. The timing of leg recovery on right and left sides depends upon the relative horizontal load experienced by the legs, and by inference can be presumed to be dependent upon the relative power output generated by the animal itself. The control mechanism appears to be a direct suppression of the tendency to recover the legs on the side that is under a lighter load and gives rise to those observed asymmetries in the timing of recovery strokes that appear under differential loading and to spontaneous differences in the power output to the two sides in free-walking animals. Turning is possible in such a system merely by a sufficient difference in output to the two sides, but experimental observations suggest that the interaction is weaker during this form of turning behavior.

This coordination hypothesis is sufficient to describe ipsilateral timing, and there is direct evidence to support it from slow-stepping behavior in the hind legs of grasshoppers and the blocking of recovery movements in stick insects. However, there is also some indication that posterior legs can excite or stimulate the recovery of the anterior leg. These experiments point the way toward future work on determining the interactions between ipsilateral legs by observing the changes in timing and position induced by the experimental manipulation of one rather than several legs.

One possible way to determine the exact nature of interleg coupling is to experimentally control the movement of a leg. This can be performed indirectly by providing a tank-track treadmill for front or hind leg. This method has been very successful in crustacea and has been used by Foth (private communication) to control the rate of stepping in the hind leg of the stick insect. These animals show similar behavior to that reported in the slow-stepping hind legs of the grasshopper.

Alternatively, the leg or leg stump may be directly driven backward and forward by a displacement device in a rhythm similar to that found in walking. The timing can be varied by the experimenter or computer controlled by the behavior of the other legs. An attempt at the former experiment has been made while recording from the coxal retractor muscles. In the hind leg motor output is directly related to leg stump movement and shows no obvious modulation related to the behavior of the other legs. These legs appear to be peripherally driven and they often stand still on a slippery surface while the more anterior legs walk, suggesting that they may require some movement or horizontal unloading in order to generate motor activity. The middle leg behaves rather differently, with a modulated motor output that may originate from the other walking legs. This is weakly expressed when the middle leg is held still. When the stump is moved during a walk this component is enhanced if the leg moves to the rear at an appropriate time (compared to the other legs) and is suppressed when it is out of step.

Such experiments show promise not only for an improved understanding of interleg coordination but also in establishing the interactions between the segments within one leg. The study of intersegmental reflexes, within one leg, has been particularly difficult to examine in the walking animal but such preparations may make it possible to record from a fixed leg that the animal believes to be moving.

Before 1966, descriptions of the mechanism of walking behavior consisted of sets of rules for the occurrence of protraction movements or recovery strokes. These rules were tested in the cockroach by Hughes (1957) and reviewed by Wilson (1966) and may be summarized as follows: (1) No leg protracts until the one behind is in position of support. (2) Legs of the same segment always alternate (phase 0.5). Wilson added two of his own to simplify modeling. (3) Protraction time is constant. (4) The intervals between the protractions of hind and middle legs and between middle and front legs are equal and constant.

The study of the cockroach in the early 1970s, with its emphasis on high-speed locomotion and with the prevailing climate of opinion which was eager to generalize the concept of central oscillators, has tended to obscure the steadily mounting evidence that hexapods which must carry their full weight on their legs and/or climb in a complex environment find it essential to control and coordinate their leg movements on a cycle to cycle basis via peripheral feedback. Under these conditions a simple centrally driven system is of limited value and is more likely to contribute to a rapid and uncontrolled descent than a steady climb.

The study of slow-walking insects has been instrumental in changing this view and there is now a large volume of evidence in conflict with the early centralist view of walking behavior in insects, which was based primarily upon the deafferentation experiments of Pearson and Iles (1973). It would be unwise to reject endogenous oscillators entirely in insect walking, as clear examples of their momentary presence is available from at least three deafferentation studies. The question is rather, "To what extent is the endogenous oscillator used in struggling (righting), grooming, or rocking, which probably underlies the deafferentation results, used by the walking system?" Current studies indicate that the central contribution is small during walking at step frequencies below 4 Hz but may be of more importance at high speeds on flat surfaces where support is not too important and a sledging or swimming type of locomotion is possible.

Recent studies have shown that most of the Wilson rules can be broken under certain experimental conditions and Wilson's extra rules depend upon the insect under consideration. However, these generalized rules have helped to develop the view that the insect walking system is best considered to be an assembly of relatively autonomous proprioceptively controlled relaxation oscillators (the individual legs), which may be strongly or weakly coupled together in various patterns for specific functions. The interaction along the body axis is particularly strong but can be disrupted by interrupting the recovery stroke of a leg. Across the body the interactions may be much weaker, to achieve turning behavior, for example, or during some start sequences. However, if the animal does decide to walk with phase-locked locomotion across the body, it is difficult to disrupt it by externally imposed forces. This is illustrated by differential loading of the two-wheel treadmill, in which right and left sides can be forced to operate at step frequencies in the ratio of 2 or 3 to 1 while maintaining phase-locked step patterns.

For this reason, across-the-body coupling has provided the major opportunity for the study of the interactions between legs in walking insects. Under the condition described above, coupling normally remains in effect and by suitable differences in frictional loading it has been possible to establish that slowing one side reduces the natural periodicity of the other side. As the loading is increased the step period increases until a second step of the unloaded side can be completed before the next step of the loaded side. In this configuration the unloaded side shows the short–long–short–long variation in period to be expected for the inhibitory model of Wilson. No tendency for the unloaded side to speed up the rate of stepping by shortening the stride length on the loaded side has been observed. This would be expected if short latency coupling influences were excitatory in nature.

Interactions between the legs along the body axis have presented greater experimental difficulties. In some insects with long hind legs 2:1 stepping may appear spontaneously, and gliding coordination has permitted a phase analysis of the interaction between middle and hind legs. These results indicate that the coupling interaction is forward directed and simply delays the protraction of a leg in front only when the hind leg has just begun its protraction stroke. Interruption of protraction shows a similar effect on the leg in front and suggests that the mechanism is very similar to that found in across-the-body interactions.

The models proposed for insect locomotion show that an assembly of three legs in which the controlling leg exerts an influence on the other leg of the segment and that immediately in front is sufficient to describe all the intact, forward-walking patterns observed so far in insects and most of the simple unilateral and bilateral amputations as well. Other more complex models are also possible in which the information flow is reversed, excitatory, and links diagonal leg pairs. However, models of this kind have not, so far, been successful in describing turning and amputee behavior, and are not compatible with the differential loading and PRC experiments.

Recent studies of changes in PEP following blocking, stepping of front legs in apodeme-switching experiments, stepping of middle legs when hind legs step slowly, and the motor output from driven leg stumps all tend to confirm the anterior flow in forward-walking insects but under some experimental conditions other pathways appear. If the pathways are multiple and sometimes redundant in certain experimental situations, then it should be possible to reveal these subtle influences by modifying the behavior of the first leg and observing its influence on the legs in front and behind it for a variety of experimental conditions. This approach is just beginning for the legs on the same side.

Current work on walking under nonhorizontal conditions (Pflueger and Hustert, 1985) is suggesting new concepts in walking control, and it is hoped that studies on the motor output in upside-down walking and walking backward will help to generalize our understanding of insect walking and its motor control and establish a detailed model capable of explaining the diverse walking behavior of insects. Quantitative studies of the control strategy for dealing with defined obstacles are now possible, and it is to be hoped that some field studies of the context of walking behavior will appear soon (Pearson and Franklin, 1984).

New, low cost, automated methods of measuring the movement of legs during walking behavior involving digital analysis of video pictures are now available (Godden and Graham, 1983) and should facilitate the analysis of subtle changes in walk behavior.

Experience with treadwheel devices and both passively driven and powered systems in which the body is rigidly fixed will permit the most advanced techniques available to the neurophysiologist to be applied to the problem of this neural control network. This is of particular importance as the walking behavior of insects appears to be one of the few areas in the study of neural control systems in which the periphery may play a vital role in the actual generation of motor output. The possibility that certain legs play a special role in stabilization, traction, or timing suggests the problem of walking in insects may be separable into distinct functional units. This prospect is an exciting one, and if it is found to be correct should greatly assist in the analysis of insect walking behavior. In this context the current increasing interest in robotics and the respectability of attempts to create hexapod vehicles (McGee, 1976; Sutherland and Ullner, 1984; Raibert and Sutherland, 1983) suggest that transfer of information between these disciplines may assist in the understanding of insect control systems. In return the insect provides us with a working example of one of natures most versatile walking machines which is particularly amenable to experimental analysis.

Acknowledgment

I wish to express my gratitude to Prof. D. Fleming for giving me my first foothold in biology; Prof. R. K. Josephson for stimulating my interest in insect behavior; Prof. P. N. R. Usherwood for his insight into the insect nerve and muscle system; Prof. G. Wendler for introducing me to the stick insect; and Prof. U. Baessler for his encouragement and the opportunity to develop my research. My grateful thanks are also given to all those scientists and students who have devoted their labor and thought to producing the work this review has attempted to describe. I wish to thank Prof. U. Baessler, Prof. H. Cruse, Dr. F. Delcomyn, Dr. D. Godden, and Dr. M. J. Berridge who have kindly commented on and corrected errors in the manuscript. Those that remain are my responsibility and I apologize for them in advance. I also wish to apologize to those whose work I may have overlooked or misunderstood. Finally, I wish to thank all those who have stimulated the ideas in this review which at first sight might appear to be my own.

References

Alexander, R. McN. (1976). Mechanics of bipedal locomotion. *In* "Perspectives in Animal Biology" (P. Spencer-Davis, ed.), pp. 493–504. Pergamon, Oxford.

Ayers, J. L., and Clarac, F. (1978). Neuromuscular strategies underlying different behavioural acts in a multifunctional crustacean leg joint. *J. Comp. Physiol.* **128**, 81–94.

Ayers, J. L., and Davis, W. J. (1977a). Neural control of locomotion in the lobster, *Homerus americanus*. I. Motor programs for forward and backward walking. *J. Comp. Physiol.* **115**, 1–27.

Ayers, J. L., and Davis, W. J. (1977b). Neural control of locomotion in the lobster, *Homerus americanus*. II. Types of walking leg reflexes. *J. Comp. Physiol.* **115**, 29–46.

Ayers, J. L., and Davis, W. J. (1977c). Neural control of locomotion in the lobster, *Homerus americanus*. III. Dynamic organization of walking leg reflexes. *J. Comp. Physiol.* **123**, 289–298.

Ayers, J. L., and Selverston, A. I. (1979). Monosynaptic entrainment of an endogeneous pacemaker network: A cellular mechanism for von Holst's magnet-effect. *J. Comp. Physiol.* **129**, 5–17.

Baessler, U. (1965). Propriorezeptoren am Subcoxal- und Femur-Tibia-Gelenk der Stabheuschrecke *Carausius morosus* und ihre Rolle bei der Wahrnehmung der Schwerkraftrichtung. *Kybernetik* **2**, 168–193.

Baessler, U. (1967). Zur Regelung der Stellung der Femur-Tibia-Gelenkes bei der Stabheuschrecke *Carausius morosus* in der Ruhe und im Lauf. *Kybernetik* **4**, 18–26.

Baessler, U. (1972a). Zur Beeinflussung der Bewegungsweise eines Beines von *Carausius morosus* durch Amputation anderer Beine. *Kybernetik* **10**, 110–114.

Baessler, U. (1972b). Der "Kniesehnenreflex" bei *Carausius morosus*: Ubergangsfunction und Frequenzgang. *Kybernetik* **11**, 32–50.

Baessler, U. (1976). Reversal of a reflex to a single motoneurone in the stick insect *Carausius morosus. Biol. Cybern.* **24**, 47–49.

Baessler, U. (1973). Zur Steuerung aktiver Bewegungen des Femur-Tibia-Gelenkes der Stabheuschrecke *Carausius morosus*: *Kybernetik* **13**, 38–53.

Baessler, U. (1974). Vom femoralen Chordotonalorgan gesteuerte Reaktionen bei der Stabheuschrecke *Carausius morosus*: Messung der von der Tibia erzeugten Kraft im aktiven und inaktiven Tier. *Kybernetik* **16**, 213–226.

Baessler, U. (1976). Reversal of a reflex to a single motoneurone in the stick insect *Carausius morosus. Biol. Cybern.* **24**, 47–49.

Baessler, U. (1977a). Sense organs in the femur of the stick insect and their relevance to the control of position of the femur-tibia joint. *J. Comp. Physiol.* **121**, 99–113.

Baessler, U. (1977b). Sensory control of leg movement in the stick insect *Carausius morosus. Biol. Cybern.* **25**, 61–72.

Baessler, U. (1979). Interaction of central and peripheral mechanisms during walking in first instar stick insects, *Extatosoma tiaratum. Physiol. Entomol.* **4**, 193–199.

Baessler, U. (1983). Neural basis of elementary behavior in insects. "Studies of Brain Function 10." Springer-Verlag, Berlin and New York.

Baessler, U., and Storrer, J. (1980). The neural basis of the femur-tibia-control-system in the stick insect. *Carausius morosus* I: Motoneurons of the extensor tibiae muscle. *Biol. Cybern.* **38**, 107–114.

Baessler, U., and Wegner, U. (1983). Motor output of the denervated thoracic ventral nerve cord in the stick insect *Carausius morosus. J. Exp. Biol.* **105**, 127–145.

Barnes, W. J. P. (1975a). Leg co-ordination during walking in the crab, *Uca pugnax. J. Comp. Physiol.* **96**, 237–256.

Barnes, W. J. P. (1975b). Nervous control of locomotion in crustacea. *In* "Simple Nervous Systems" (P. N. R. Usherwood and D. R. Newth, eds.), pp. 415–441. Arnold, London.

Barnes, W. J. P. (1977). Proprioceptive influence on motor output during walking in the crayfish. *J. Physiol. (Paris)* **73**, 543–564.

Barnes, W. J. P., Spirito, C. P., and Evoy, W. H. (1972). Nervous control of walking in the crab, *Cardisoma guanhumi*. II. Role of resistance reflexes in walking. *Z. Vergl. Physiol.* **76**, 16–31.

Bell, W. J., and Schal, C. (1980). Patterns of turning in courtship orientation of the male German cockroach. *Anim. Behav.* **28**, 86–94.

Bethe, A. (1930). Studien ueber die Plastizitaet des Nervensystems. I. Mitteilung. Arachnoideen und Crustaceen. *Pfluegers Arch. Ges. Physiol.* **224**, 793–820.

Bowerman, R. F. (1975). The control of walking in the scorpion. I. Leg movements during normal walking. *J. Comp. Physiol.* **100**, 183–196.

Bowerman, R. F. (1977). The control of arthropod walking. *Comp. Biochem. Physiol.* **56A**, 231–247.

Bowerman, R. F. (1981a). An electromyographic analysis of the elevator/depressor muscle motor programme in the freely-walking scorpion, *Paruroctonus mesaensis*. *J. Exp. Biol.* **91**, 165–177.

Bowerman, R. F. (1981b). Arachnid locomotion. *In* "Locomotion and Energetics in Arthropods" (C. F. Herried, II and C. R. Fourtner, eds.), pp. 73–102. Plenum, New York.

Brady, J. (1969). How are insect circadian rhythms controlled? *Nature (London)* **223**, 781–784.

Braeunig, P. (1982). Strand receptors with central cell bodies in the proximal leg joints of orthopterous insects. *Cell Tissue Res.* **222**.

Braeunig, P. and Hustert, R. (1980). Proprioceptors with central cell bodies in insects. *Nature (London)* **283**, 768–770.

Braeunig, P., Hustert, R., and Pflueger, J. H. (1981). Distribution and specific central projections of mechanoreceptors in the thorax and proximal leg joints of locusts. I. Morphology, location and innervation of internal proprioceptors of pro- and metathorax and their central projections. *Cell Tissue Res.* **216**, 57–77.

Buddenbrock, W., von (1921). Der Rhythmus der Schreitbewegungen an der Stabheuschrecke *Dixippus*. *Biol. Zentralbl.* **41**, 41–48.

Burns, M. D. (1973). The control of walking in orthoptera. *J. Exp. Biol.* **58**, 45–58.

Burns, M. D., and Usherwood, P. N. R. (1978). Mechanical properties of locust extensor tibiae muscles. *Comp. Biochem. Physiol.* **61A**, 85–95.

Burns, M. D., and Usherwood, P. N. R. (1979). The control of walking in orthoptera. II. Motor neurone activity in normal free-walking animals. *J. Exp. Biol.* **79**, 69–98.

Burrows, M. (1979). Graded synaptic interactions between local pre-motor interneurones of the locust. *J. Neurophysiol.* **42**, 1108–1123.

Burrows, M. (1980). The control of sets of motoneurones by local interneurones in the locust. *J. Physiol. (London)* **298**, 212–233.

Burrows, M., and Siegler, M. V. S. (1976). Transmission without spikes between locust interneurones and motor neurones in the metathoracic ganglion of the locust. *J. Physiol. (London)* **285**, 231–255.

Burrows, M., and Siegler, M. V. S. (1978). Graded synaptic transmission between local interneurones and motor neurones in the metathoracic ganglion of the locust. *J. Physiol. (London)* **285**, 231–255.

Camhi, J. M. (1977). Behavioural switching in cockroaches: Transformations of tactile reflexes during righting behaviour. *J. Comp. Physiol.* **113**, 283–301.

Carrel, J. S. (1972). An improved treading device for tethered insects. *Science* **175**, 1279.

Cate, ten, J. (1941). Quelques remarques à propos de l'innervation des mouvements locomoteurs de la blatte (*Periplaneta americana*). *Arch. Neerl. Physiol.* **25**, 401–409.

Cavagna, G. A., Heglund, N. C., and Taylor, R. C. (1977). Walking, running and galloping: mechanical similarities between different animals. *In* "Scale Effects in Locomotion" (T. J. Pedley, ed.), pp. 111–126. Academic Press, New York.

Chasserat, C., and Clarac, F. (1980). Interlimb coordinating factors during driven walking in crustacea. *J. Comp. Physiol.* **39**, 293–306.

Chasserat, C., and Clarac, F. (1983). Quantitative analysis of walking in a decapod crustacean, the rock-lobster *Jasus lalandii*. II. Spatial and temporal regulation of stepping in driven walking. *J. Exp. Biol.* **107**, 219–245.

Chesler, M., and Fourtner, C. R. (1981). The mechanical properties of a slow insect muscle in a cockroach. *J. Neurobiol.*

Clarac, F. (1977). Motor coordination in crustacean limbs. *In* "Identified Neurons and the Behavior of Arthropods" (G. M. Hoyle, ed.), pp. 167–187. Plenum, New York.

Clarac, F. (1978). Locomotory programs in basal leg muscles after limb autotomy in crustacea. *Brain Res.* **145**, 401–405.

Clarac, F. (1981). Postural reflexes coordinating walking legs in a rock lobster. *J. Exp. Biol.* **90**, 333–337.

Clarac, F. (1982). Decapod crustacean leg coordination during walking. *In* "Locomotion and Energetics in Arthropods" (C. F. Herried, II and C. R. Fourtner, eds.), pp. 31–71. Plenum, New York.

Clarac, F., and Chasserat, C. (1983). Quantitative analysis of walking in a decapod crustacean, the rock-lobster *Jasus lalandii*. I. Comparative study of free and driven walking. *J. Exp. Biol.* **107**, 189–219.

Clarac, F., and Cruse, H. (1982). Comparison of forces developed by the leg of the rock lobster when walking free or on a treadmill. *Biol. Cybern.* **43**, 109–114.

Collet, T. S. (1978). Peering—a locust behavior pattern for obtaining motion parallax information. *J. Exp. Biol.* **76**, 237–241.

Cruse, H. (1976a). The function of the legs in the free walking stick insect *Carausius morosus*. *J. Comp. Physiol.* **15**, 235–262.

Cruse, H. (1976b). The control of body position in the stick insect *Carausius morosus*, when walking over uneven surfaces. *Biol. Cybern.* **24**, 25–33.

Cruse, H. (1979a). A new model describing the coordination patterns of the legs of a walking stick insect. *Biol. Cybern.* **32**, 107–113.

Cruse, H. (1979b). The control of the anterior extreme position of the hindleg of a walking insect, *Carausius morosus*. *Physiol. Entomol.* **4**, 121–124.

Cruse, H. (1980a). A quantitative model of walking incorporating central and peripheral influences. I. The control of the individual leg. *Biol. Cybern.* **37**, 131–136.

Cruse, H. (1980b). A quantitative model of walking incorporating central and peripheral influences. II. The connections between the different legs. *Biol. Cybern.* **37**, 137–144.

Cruse, H. (1983). The influence of load and leg amputation upon coordination in walking crustaceans: A model calculation. *Biol. Cybern.* **49**, 119–125.

Cruse, H., and Epstein, S. (1982). Peripheral influences on the movement of the legs in a walking insect *Carausius morosus*. *J. Exp. Biol.* **101**, 161–170.

Cruse, H., and Graham, D. (1984). Models for the analysis of walking in arthropods. *Soc. Exp. Biol. Semin. Ser.* **24**.

Cruse, H., and Pflueger, H-J. (1981). Is the position of the femur-tibia-joint under feedback control in the walking stick insect? II. Electrophysiological recordings. *J. Exp. Biol.* **92**, 97–107.

Cruse, H., and Saxler, G. (1980a). Oscillations of force in the standing leg of a walking stick insect (*Carausius morosus*). *Biol. Cybern.* **36**, 159–163.

Cruse, H., and Saxler, G. (1980b). The coordination of force oscillations and of leg movement in a walking insect (*Carausius morosus*). *Biol. Cybern.* **36**, 165–171.

Cruse, H., and Schmitz, J. (1983). The control system of the femur-tibia joint in the standing leg of a walking stick insect *Carausius morosus*. *J. Exp. Biol.* **102**, 175–185.

Cruse, H., and Storrer, J. (1977). Open loop analysis of a feedback mechanism controlling the leg position in the stick insect *Carausius morosus*: Comparison between experiment and stimulation. *Biol. Cybern.* **25**, 143–153.

Dean, J., and Wendler, G. (1982). Stick insects walking on a wheel: Perturbations induced by obstruction of leg protraction. *J. Comp. Physiol.* **148**, 195–207.

Dean, J., and Wendler, G. (1984). Stick insects walking on a wheel: Patterns of stopping and starting. *J. Exp. Biol.* **110**, 203–216.

Delcomyn, F. (1971a). The locomotion of the cockroach *Periplaneta americana*. *J. Exp. Biol.* **54**, 443–452.

Delcomyn, F. (1971b). The effect of limb amputation on locomotion in the cockroach *Periplaneta americana*. *J. Exp. Biol.* **54**, 453–469.

Delcomyn, F. (1971c). Computer aided analysis of a locomotor leg reflex in the cockroach, *Periplaneta americana*. *Z. Vergl. Physiol.* **74**, 427–455.

Delcomyn, F. (1973). Motor activity during walking in the cockroach *Periplaneta americana*. II. Tethered walking. *J. Exp. Biol.* **59**, 643–654.

Delcomyn, F. (1980). Neural basis of rhythmic behavior in animals. *Science* **210**, 492–498.
Delcomyn, F. (1981). Insect locomotion on land. *In* "Locomotion and Energetics in Arthropods" (C. F. Herreid and C. R. Fourtner, eds.), pp. 103–125. Plenum, New York.
Delcomyn, F., and Usherwood, P. N. R. (1973). Motor activity during walking in the cockroach *Periplaneta americana*. I. Free walking. *J. Exp. Biol.* **59**, 629–642.
DiCaprio, R. A., and Clarac, F. (1981). Reversal of a walking leg reflex elicited by a muscle receptor. *J. Exp. Biol.* **90**, 197–203.
English, A. W. (1979). Interlimb coordination during stepping in the cat: An electromyographic analysis. *J. Neurophysiol.* **42**, 229–243.
Epstein, S., and Graham, D. (1983). Behavior and motor output for an insect walking on a slippery surface. I. Forward walking. *J. Exp. Biol.* **105**, 215–229.
Ewing, A., and Manning, A. (1966). Some aspects of the efferent control of walking in three cockroach species. *J. Insect Physiol.* **12**, 1115–1118.
Evoy, W. H., and Fourtner, C. R. (1973). Nervous control of walking in the crab, *Cardisoma guanhumi*. III. Proprioceptive influences on intra- and inter-segmental coordination. *J. Comp. Physiol.* **83**, 303–318.
Ferdinand, W. (1981). The locomotion of Thomisid spiders and Wilson's model of metachronal rhythms. *Zool. Jb. Physiol.* **85**, 46–65.
Findlay, I. (1978). The role of the cuticular stress detector, CSD1, in locomotion and limb autotomy in the crab *Carcinus maenas*. *J. Comp. Physiol.* **125**, 79–90.
Foth, E., and Graham, D. (1983a). Influence of loading parallel to the body axis on the walking coordination of an insect. I. Ipsilateral changes. *Biol. Cybern.* **47**, 17–23.
Foth, E., and Graham, D. (1983b). Influence of loading parallel to the body axis on the walking coordination of an insect. II. Contralateral changes. *Biol. Cybern.* **48**, 149–157.
Fourtner, C. R. (1976). Central nervous control of cockroach walking. *In* "Neural Control of locomotion" (R. Herman, S. Grillner, P. S. G. Stein, and D. G. Stuart, eds.), pp. 401–418. Plenum, New York.
Fourtner, C. R. (1982). Role of muscle in insect posture and locomotion. *In* "Locomotion and Energetics in Arthropods" (C. F. Herried and C. F. Fourtner, eds.), pp. 195–213. Plenum, New York.
Franklin, R. (1985). The generation of locomotion in the millipede *Oxidus gracilis*. (In preparation).
Franklin, R., Bell, W. J., and Jander, R. (1981). Rotational locomotion by the cockroach *Blatella germanica*. *J. Insect Physiol.* **27**, 249–255.
Friedrich, H. (1933). Nerven physiologische Studien an Insekten. I. Untersuchen ueber das reissphysiologische Verhalten der Extremitaeten von *Dixippus morosus*. *Z. Vergl. Physiol.* **18**, 536–561.
Godden, D. H. (1973). A re-examination of circadian rhythmicity in *Carausius morosus*. *J. Comp. Physiol.* **19**, 1377–1386.
Godden, D. H., and Graham, D. (1983). Instant analysis of movement. *J. Exp. Biol.* **107**, 505–509.
Godden, D. H., and Graham, D. (1984). A preparation of the stick insect *Carausius morosus* for recording intracellularly from identified neurons during walking. *Physiol. Entomol.*, **9**, 275–286.
Graham, D. (1972). A behavioural analysis of the temporal organization of walking movements in the 1st instar and adult stick insect *Carausius morosus*. *J. Comp. Physiol.* **81**, 23–52.
Graham, D. (1977a). Simulation of a model for the coordination of leg movement in free walking insects. *Biol. Cybern.* **26**, 187–198.
Graham, D. (1977b). The effect of amputation and leg restraint on the free walking coordination of the stick insect *Carausius morosus*. *J. Comp. Physiol.* **116**, 91–116.
Graham, D. (1978a). Unusual step patterns in the free walking grasshopper *Neoconocephalus robustus*. I. General features of the step pattern. *J. Exp. Biol.* **73**, 147–157.

Graham, D. (1978b). Unusual step patterns in the free walking grasshopper *Neoconocephalus robustus*. II. A critical test of the leg interactions underlying different models of hexapod coordination. *J. Exp. Biol.* **73**, 159–172.

Graham, D. (1979a). The effects of circumoesophageal lesion on the behavior of the stick insect *Carausius morosus*. I. Cyclic behaviour patterns. *Biol. Cybern.* **32**, 139–145.

Graham, D. (1979b). The effects of circumoesophageal lesion on the behaviour of the stick insect *Carausius morosus*. II. Changes in walking coordination. *Biol. Cybern.* **32**, 147–152.

Graham, D. (1981). Walking kinetics of the stick insect using a low inertia, counter-balanced, pair of independent treadwheels. *Biol. Cybern.* **40**, 49–57.

Graham, D. (1983). Insects are both impeded and propelled by their legs during walking. *J. Exp. Biol.* **104**, 129–137.

Graham, D. (1985). Influence of coxa-thorax joint receptors on retractor motor output during walking (*Carausius morosus*). *J. Exp. Biol.*, in press.

Graham, D., and Baessler, U. (1981). Effects of afference sign reversal on motor activity in walking stick insects (*Carausius morosus*). *J. Exp. Biol.* **91**, 179–193.

Graham, D., and Cruse, H. (1981). Coordinated walking of stick insects on a mercury surface. *J. Exp. Biol.* **92**, 229–241.

Graham, D., and Epstein, S. (1985). Behaviour and motor output for an insect walking on a slippery surface. II. Backward walking. *J. Exp. Biol.*, in press.

Graham, D., and Godden, D. H. (1985). In preparation.

Graham, D., and Wendler, G. (1981a). Motor output to the protractor and retractor coxae muscles in stick insects walking on a treadwheel. *Physiol. Entomol.* **6**, 161–174.

Graham, D., and Wendler, G. (1981b). The reflex behaviour and innervation of the tergo-coxal retractor muscles of the stick insect *Carausius morosus*. *J. Comp. Physiol.* **143**, 81–91.

Greene, S., and Spirito, C. (1979). Interlimb coordination during slow walking in the cockroach. II. The effects of cutting thoracic connectives. *J. Exp. Biol.* **78**, 245–253.

Grillner, S. (1975). Locomotion in vertebrates central mechanisms and reflex interaction. *Physiol. Rev.* **55**, 247–304.

Grillner, S. (1977). On the neural control of movement: A comparison of different basic rhythmic behaviours. *In* "Function and Formation of Neural Systems" (G. S. Stent, ed.), pp. 197–224. Dahlem Konferenzen, Berlin.

Grillner, S., and Wallen, P. (1982). On peripheral control mechanisms acting on the central pattern generators for swimming in the dogfish. *J. Exp. Biol.* **98**, 1–22.

Grote, J. R. (1981). The effect of load on locomotion in crayfish. *J. Exp. Biol.* **92**, 277–288.

Harris, J., and Ghiradella, H. (1980). The forces exerted on the substrate by walking and stationary crickets. *J. Exp. Biol.* **85**, 263–279.

Hofmann, T., and Baessler, U. (1982). Anatomy and physiology of trochanteral campaniform sensilla in the stick insect, *Cuniculina impigra*. *Physiol. Entomol.* **7**, 413–426.

Holst, E., von (1936). Ueber den "Magnet-Effekt" als koordinierendes Prinzip im Rueckenmark. *Pfluegers Arch. Ges. Physiol.* **241**, 655-682.

Holst, E., von (1939a). Ueber die nervoese Funktionsstruktur des rhythmisch taetigen Fischrueckenmarks. *Pfluegers Arch. Ges. Physiol.* **241**, 569–611.

Holst, E., von (1939b). Die relative Koordination als Phaenomen und als Methode zentralnervoeser Funktionsanalyse. *Ergebn. Physiol.* **42**, 288–306.

Holst, E., von (1943). Ueber relative Koordination bei Arthropoden. *Pfluegers Arch. Ges. physiol.* **246**, 847–865.

Hoyle, G. (1964). Exploration of neuronal mechanisms underlying behavior in insects. *In* "Neural Theory and Modelling" (R. F. Reiss, ed.), pp. 346–376. Stanford Univ. Press, Stanford, CA.

Hoyle, G. (1976). Arthropod walking. *In* "Neural Control of Locomotion" (R. M. Herman, S. Grillner, P. S. G. Stein, and D. G. Stuart, eds.), pp. 137–179. Plenum, New York.

Hoyle, G. (1977). The dorsal, unpaired, median neurons of the locust metathoracic ganglion. *J. Neurobiol.* **9**, 43–57.

Hughes, G. M. (1952). The coordination of insect movements. I. The walking movements of insects. *J. Exp. Biol.* **29**, 267–284.

Hughes, G. M. (1957). The coordination of insect movements. II. The effect of limb amputation and the cutting of commissures in the cockroach *Blatta orientalis*. *J. Exp. Biol.* **34**, 306–333.

Hustert, R., Pflueger, J. H., and Braeunig, P. (1981). Distribution and specific central projections of mechanoreceptors in the thorax and proximal leg joints of locusts. III. The external mechanoreceptors: The campaniform sensilla. *Cell Tissue Res.* **216**, 97–111.

Jander, R., and Volk-Heinrichs, I. (1970). Das strauch-spezifische visuelle Perceptor-System der Stabheuschrecke (*Carausius morosus*). *Z. Vergl. Physiol.* **70**, 425–447.

Jander, R., and Wendler, G. (1978). Zur Steuerung des Kurvenlaufs bei Stabheuschrecken *Carausius morosus*. *In* "Kybernetik 1977" (G. Hauske and E. Butenandt, eds.), pp. 388–392. Oldenbourg, Muenchen.

Kien, J. (1981). Motor control by plurisegmental interneurones in the locust. *Adv. physiol. Sci.* **23**, 515–535.

Kozacik, J. J. (1981). Stepping patterns in the cockroach, *Periplaneta americana*. *J. Exp. Biol.* **90**, 357–360.

Kramer, E. (1975). Orientation of the male silkmoth to the sex attractant Bombykol. *In* "Olfaction and Taste V" (D. A. Denton and J. P. Coghlan, eds.), pp. 1. 329–335.

Kramer, E. (1976). The orientation of walking honeybees in odour fields with small concentration gradients. *Physiol. Entomol.* **1**, 27–37.

Krauthamer, V., and Fourtner, C. R. (1977). Activity of the flexor and extensor tibiae in the cockroach. *Am. Zool.* **17**, 962.

Krauthamer, V., and Fourtner, C. R. (1978). Locomotory activity in the extensor and flexor tibiae of the cockroach, *Periplaneta americana*. *J. Insect Physiol.* **24**, 813–819.

Land, M. F. (1972). Stepping movements made by jumping spiders during turns mediated by the lateral eyes. *J. Exp. Biol.* **57**, 15–40.

McGee, R. B. (1976). Robot locomotion. *In* "Neural Control of Locomotion" (R. M. Herman *et al.*, eds.), pp. 237–264. Plenum, New York.

Macmillan, D. L. (1975). A physiological analysis of walking in the American lobster (*Homarus americanus*). *Philos. Trans. R. Soc. Ser. B* **270**, 1–51.

Macmillan, D. L., and Kien, J. (1983). Intra- and intersegmental pathways active during walking in the locust. *Proc. R. Soc. London Ser. B* **218**, 287–308.

Marquardt, F. (1943). Beiträge zur Anatomie der Muskulatur und der peripheren Nerven von *Carausius* (*Dixippus*) *morosus*. *Zool. Jb. Abt. Anat. Ontol.* **66**, 63–128.

Moffett, S. (1977). Neuronal events underlying rhythmic behaviours in invertebrates. *Comp. Biochem. Physiol.* **57A**, 187–195.

Moorhouse, J. E. (1971). Experimental analysis of the locomotor behaviour of *Schistocerca gregaria* induced by odour. *J. Insect Physiol.* **17**, 913–920.

Moorhouse, J. E., Fosbrooke, I. H. M., and Kennedy, J. S. (1978). 'Paradoxical driving' of walking activity in locusts. *J. Exp. Biol.* **72**, 1–16.

Pearson, K. G. (1972). Central programming and reflex control of walking in the cockroach. *J. Exp. Biol.* **56**, 173–193.

Pearson, K. G. (1977). Interneurons in the ventral nerve cord of insects. *In* "Identified neurons and behavior of anthropods" (G. Hoyle, ed.), pp. 329–338. Plenum, New York.

Pearson, K. G., and Fourtner, C. R. (1975). Non spiking neurones in the walking system of the cockroach. *J. Neurophysiol.* **38**, 33–52.

Pearson, K. G., and Franklin, R. (1984). Characteristics of leg movements and patterns of coordination in locusts walking on rough terrain. *Int. J. Robatics Res.* **3**, 101–112.

Pearson, K. G., and Iles, J. F. (1970). Discharge patterns of coxal levator and depressor motoneurones of the cockroach, *Periplaneta americana*. *J. Exp. Biol.* **52**, 139–165.

Pearson, K. G., and Iles, J. F. (1973). Nervous mechanisms underlying intersegmental coordination of leg movements during walking in the cockroach. *J. Exp. Biol.* **58**, 725–744.
Pearson, K. G., and Robertson, R. M. (1981). Interneurons coactivating hindleg flexor and extensor motoneurons in the locust. *J. Comp. Physiol.* **144**, 391–400.
Pearson, K. G., Fourtner, C. R., and Wong, R. K. (1973). Nervous control of walking in the cockroach. *In* "Control of Posture and Locomotion" (R. B. Stein, K. G. Pearson, R. S. Smith, and J. B. Redford, eds.), pp. 495–514. Plenum, New York.
Pflueger, J. H. (1977). The control of the rocking movements of the phasmid *Carausius morosus.* Br. *J. Comp. Physiol.* **120**, 181–202.
Pflueger, J. H., and Hustert, R. (1985). In preparation.
Pringle, J. W. S. (1940). The reflex mechanism of the insect leg. *J. Exp. Biol.* **17**, 8–17.
Pringle, J. W. S. (1961). Proprioception in arthropods. *In* "The Cell and the Organism" (J. A. Ramsey and V. B. Wigglesworth, eds.), pp. 256–282.
Raibert, M H., and Sutherland, I. E. (1983). Machines that walk. *Sci. Am.* **248**, 32–41.
Reingold, S. C., and Camhi, J. M. (1977). A quantitative analysis of rhythmic leg movements during three different behaviours in the cockroach, *Periplaneta americana. J. Insect Physiol.* **23**, 1407–1420.
Roeder, K. D. (1937). The control of tonus and locomotor activity in the praying mantis (*Mantis religiosa* L.). *J. Exp. Zool.* **76**, 353–374.
Runion, H. I., and Usherwood, P. N. R. (1966). A new approach to neuromuscular analysis in the intact free-walking preparation. *J. Insect Physiol.* **12**, 1255–1263.
Seyfarth, E.-A. (1978). Lyriform slit sense organs and muscle reflexes in the spider leg. *J. Comp. Physiol.* **125**, 45–57.
Seyfarth, E.-A., and Bohnenberger, J. (1980). Compensated walking of tarantula spiders and the effect of lyriform slit sense organ ablation. *Proc. Int. Congr. Arachnol.* **8**, 249–255.
Sherman, E., Novotny, M., and Camhi, J. M. (1977). A modified rhythm employed during righting behaviour in the cockroach, *Gromphadorhina portentosa. J. Comp. Physiol.* **113**, 303–316.
Siegler, M. V. S. (1981). Postural changes alter synaptic interactions between non-spiking interneurons and motor neurons of the locust. *J. Neurophysiol.* **46**, 310–323.
Spirito, C. P., and Mushrush, D. L. (1979). Interlimb coordination during slow walking in the cockroach. I. Effects of substrate alterations. *J. Exp. Biol.* **78**, 233–243.
Stein, P. S. G. (1974). Neural control of interappendage phase during locomotion. *Am. Zool.* **14**, 1003–1016.
Storrer, J., and Cruse, H. (1977). Systemanalytische Untersuchung eines aufgeschnittenen Regelkreises, der die Beinstellung der Staheuschrecke *Carausius morosus* kontrolliert: Kraftmessungen an dem Antagonisten Flexor und Extensor tibia. *Biol. Cybern.* **25**, 131–143.
Sutherland, I. E., and Ullner, M. K. (1984). Footprints in the asphalt. *Int. J. Robatics Res.* **3**, 29–36.
Theophilidis, G., and Burns, M. D. (1979). A muscle tension receptor in the locust leg. *J. Comp. Physiol.* **131**, 247–254.
Theophilidis, G., and Burns, M. D. (1983). The innervation of the mesothoracic flexor-tibiae of the locust. *J. Exp. Biol.* **105**, 373–388.
Thomson, I. (1985). Ph. D. thesis. Department of Zoology, Leeds University.
Usherwood, P. N. R., and Runion, H. I. (1970). Analysis of the mechanical responses of metathoracic extensor-tibiae muscles of free-walking locusts. *J. Exp. Biol.* **52**, 39–58.
Usherwood, P. N. R., Runion, H. I., and Campbell, J. I. (1968). Structure and physiology of a chordotonal organ in the locust leg. *J. Exp. Biol.* **48**, 305–324.
Vedel, J. P. (1980). The antennal motor system of the rock lobster: Competitive occurrence of resistance and assistance reflex patterns originating from the same proprioceptor. *J. Exp. Biol.* **87**, 1–22.
Wallace, G. K. (1959). Visual scanning in the desert locust *Schistocerca gregaria* Forskal. *J. Exp. Biol.* **36**, 512–525.

Weber, T., Thorson, J., and Huber, F. (1981). Auditory behaviour of the cricket. I. Dynamics of compensated walking and discrimination paradigms on the Kramer treadwheel. *J. Comp. Physiol.* **141**, 215–232.

Wendler, G. (1964). Laufen und Stephen der Stabheuschrecke *Carausius morosus*: Sinnesborstenfelder in den Beingelenken als Glieder von Regelkreisen. *Z. Vergl. Physiol.* **48**, 197–250.

Wendler, G. (1965). Ueber den Anteil der Antennen an der Schwererezeption der Stabheuschrecken. *Z. Vergl. Physiol.* **51**, 60–66.

Wendler, G. (1966). The coordination of walking movements in arthropods. *Symp. Soc. Exp. Biol.* **20**, 229–249.

Wendler, G. (1968). Ein Analogmodell der Beinbewegungen eines laufenden Insekts. *In* "Kybernetik 1968" (H. Marko and G. Faerber, eds.), pp. 68–74. Oldenbourg, Munich.

Wendler, G. (1972). Koerperhaltung bei der Stabheuschrecke: Ihre Beziehung zur Schwereorientierung und Mechanismen ihrer Reglung. *Verh. Dtsch. Zool. Ges.* **65**, 215–219.

Wendler, G. (1973). The influence of proprioceptive feedback on locust flight coordination. *J. Comp. Physiol.* **88**, 173–200.

Wendler, G. (1978a). The possible role of fast wing reflexes in locust flight. *Naturwissenschaften* **65**, 65.

Wendler, G. (1978b). Erzeugung und Kontrolle koordinierter Bewegungen bei Tieren—Beispiele an Insekten. *In* "Kybernetik 1977" (G. Hauske and E. Butenandt, eds.), pp. 11–34. Oldenbourg, Munich.

Wendler, G., and Heiligenberg, W. (1969). Relative Koordination bei gekoppelten, rhythmisch taetigen Modellneuronen. *Verh. Dtsch. Zool. Ges. Zool. Anz. Suppl.* **33**, 477–482.

Wilson, D. M. (1965). Proprioceptive leg reflexes in cockroaches. *J. Exp. Biol.* **43**, 397–409.

Wilson, D. M. (1966). Insect walking. *Annu. Rev. Entomol.* **11**, 103–122.

Wilson, D. M. (1967). Stepping patterns in tarantula spiders. *J. Exp. Biol.* **47**, 133–151.

Wilson, D. M. (1968). An approach to the problem of control of rhythmic behaviour. *In* "Invertebrate Nervous Systems" (C. A. G. Wiersma, ed.), pp. 219–229. Univ. of Chicago Press, Chicago.

Wong, R. K. S., and Pearson, K. G. (1976). Properties of the trochanteral hair plate and its function in the control of walking in the cockroach. *J. Exp. Biol.* **64**, 233–249.

Yox, D. P., Dicaprio, R. A., and Fourtner, C. R. (1982). Resting tension and posture in Arthropods. *J. Exp. Biol.* **96**, 421–425.

Zill, S. N., and Moran, D. T. (1981a). The exoskeleton and insect proprioception. I. Responses of tibial campaniform sensilla to external and muscle-generated forces in the American cockroach, *Periplaneta americana. J. Exp. Biol.* **91**, 1–24.

Zill, S. N., and Moran, D. T. (1981b). The exoskeleton and insect proprioception. III. Activity of tibial campaniform sensilla during walking in the American cockroach, *Periplaneta americana. J. Exp. Biol.* **94**, 57–76.

Zill, S. N., Varela, F. J., and Moran, D. T. (1977). Modulation of locomotor activity by adjacent campaniform sensilla varies with sensillum orientation. *Neurosci. Abstr.* **3**, 191.

Zolotov, V., Frantsevich, L., and Falk, E. (1975). Kinematik der phototaktischen Drehung bei der Honigbiene *Apis mellifera. J. Comp. Physiol.* **97**, 339–353.

Cyclic Nucleotide Metabolism and Physiology of the Fruit Fly *Drosophila melanogaster*

John A. Kiger, Jr.

Department of Genetics, University of California, Davis, California

Helen K. Salz

Department of Biology, Princeton University, Princeton, New Jersey

Discovery of the cyclic nucleotides has been followed by an expanding search for their physiological roles. Biochemical studies have elucidated the basic features of their synthesis, degradation, and mode of action through modulation of the phosphorylation state of target proteins (for recent reviews see Gilman, 1984; Ingebritsen and Cohen, 1983; Nestler and Greengard, 1983). Many studies designed to relate the biochemical roles of cyclic nucleotides to physiological roles have been carried out employing cultured

ADVANCES IN INSECT PHYSIOLOGY, VOL. 18

cells, dissected organs, or tissue homogenates under conditions established by the investigator to optimize the study of one aspect or another of the system. Other studies have employed drugs, whose biochemical actions have been studied *in vitro*, to alter the biochemistry and physiology of cells *in vivo*. From such studies one attempts by extrapolation to determine the physiological significance of the cyclic nucleotide system in the intact organism.

A complementary approach to determining the physiological significance of any biochemical system is to alter the system genetically and to observe in the intact organism the consequences of that alteration. In this approach an alteration in the genotype of the organism is identified, and the effect of that alteration is observed by noting the change in the phenotype of the organism. The key to this approach is knowing which biochemical element in the system is affected by the genetic alteration, and how it is affected.

Because of the accumulated genetic knowledge of the past 75 years, the insect most amenable to study by the genetic approach is the laboratory fruit fly, *Drosophila melanogaster*. Indeed, the ease of genetic manipulation of *D. melanogaster* makes it the organism of choice for studying basic physiological mechanisms that pertain to all higher organisms. The purpose of this article is to review the current state of knowledge of the cyclic nucleotide system in *D. melanogaster* and to explain how the genetic approach has begun to elucidate the physiological roles of this system in the intact organism. In these studies biochemical and genetic progress has been made hand-in-hand. The goal of this review is to point out the unique contributions that a combined biochemical analysis and genetic analysis can make to an understanding of the physiological roles of cyclic nucleotides. Our current understanding is quite incomplete; the many gaps in our knowledge that remain to be filled will be apparent.

1 Cyclic nucleotide phosphodiesterase activities

The phosphodiesterases are, at present, the best characterized elements of the cyclic nucleotide system in *Drosophila*. The hydrolysis of cAMP and of cGMP to 5′-AMP and 5′-GMP, respectively, is easily measured in crude homogenates of adult flies. Analysis of these activities by centrifugation (105,000 *g* for 1 h) demonstrates that 50–75 % of the cAMP phosphodiesterase activity and 80–90 % of the cGMP phosphodiesterase activity are soluble and that most of the remaining activities can be recovered in the particulate material of the crude homogenate (Davis and Kiger, 1980). Fractionation of the supernatants of crude homogenates demonstrates the presence of two activities that are easily separated by gel filtration, velocity sedimentation, or ion-exchange chromatography (Kiger and Golanty, 1979; Davis and Kiger, 1980).

The larger of the two activities, form I, is a cyclic nucleotide phosphodiesterase activity that hydrolyzes both cAMP ($K_m = 4.4 \times 10^{-6}$ M) and cGMP ($K_m = 4.2 \times 10^{-6}$ M). Each cyclic nucleotide is a competitive inhibitor of the hydrolysis of the other. Cyclic GMP inhibits cAMP hydrolysis with a $K_i = 3.8 \times 10^{-6}$ M, and cAMP inhibits cGMP hydrolysis with a $K_i = 3.4 \times 10^{-6}$ M. These observations may suggest that form I has a single active site for cyclic nucleotide hydrolysis (Davis and Kiger, 1980). However, the possibility that it has two active sites, each of which is allosterically regulated by the other, remains to be tested. The smaller of the two activities, form II, is specific for cAMP hydrolysis ($K_m = 2.0 \times 10^{-6}$ M) (Davis and Kiger, 1980).

Considerable caution is required in correlating activities with unique proteins. The biochemical relationship between these two forms of phosphodiesterase activity and between soluble and particulate activities has been clarified by genetic analysis. Indeed, genetic analysis of cAMP and cGMP hydrolysis in crude homogenates of flies actually preceded the biochemical analysis described above. The choice of genetics as the first approach to analysing phosphodiesterase activities in *Drosophila* was based upon consideration of numerous biochemical studies of phosphodiesterase activities in other, primarily mammalian, species that seemed to indicate an incredible complexity. The results of the genetic analysis have provided a solid foundation upon which to base a biochemical analysis.

1.1 CYTOGENETIC ANALYSIS OF cAMP AND cGMP HYDROLYSIS

Drosophila is a diploid organism and normally possesses two copies of every gene, one on each pair of homologous chromosomes. An exception is found for genes located on the X chromosome, where females have two X chromosomes but males have only one. However, almost without exception, genes on the X chromosome are expressed at comparable levels in both females and males: this sex-specific difference in the level of expression of X chromosome genes (measured on a per gene basis) is called "dosage compensation." A priori, the amount of a polypeptide synthesized in cells is expected to be proportional to the number of structural genes encoding that polypeptide. When cells possess more or less than the normal (euploid) number of copies of a gene encoding a polypeptide, they would be expected to synthesize proportionately more or less of that polypeptide. This expectation has been observed for a number of autosomal and X chromosome genes in both sexes of *Drosophila.*

Gene dosage can be manipulated experimentally in two different ways. First, this may be accomplished through the use of many different deficiency chromosomes (*Df*), lacking cytologically and genetically defined portions of the chromosome, and duplication chromosomes (*Dp*), carrying similarly defined extra portions of particular chromosomes. The banded nature of the *Drosophila* giant polytene chromosomes in larval salivary glands provides landmarks to identify the cytological extent of such deficiencies and duplications, while genetic analysis makes it possible to identify their genetic extent.

Second, duplications and deficiencies for virtually the entire genome of *D. melanogaster* can be systematically synthesized (when *Df* or *Dp* chromosomes are not available) by crossing strains heterozygous for different chromosome translocations, a technique known as "segmental aneuploidy." Numerous reciprocal translocation strains have been produced in which there is a break in one arm of the Y chromosome, whose arms are genetically marked by B^s and y^+, and another break in the X, or one of the autosomes (Fig. 1). In these strains the Y centromere carries the portion of the other chromosome distal (with respect to its centromere) to the break, and the other centromere carries the distal portion of the broken Y chromosome arm attached at the point of the break (Lindsley *et al.*, 1972; Stewart and Merriam, 1973, 1975). In these strains the X and autosome translocation breakpoints have been located cytologically, and the breakpoints in the Y have been determined by noting the segregation in crosses of the Y arms marked by B^s and y^+.

In a cross between two translocation strains, whose Y breakpoints are in different arms and whose other two breakpoints on the X or on an autosome are reasonably close to one another, segregation at meiosis in the two parents produces several identifiable types of siblings depending on the type of segregation that occurs (Fig. 1). First are euploid siblings that carry two doses of every gene. Second are duplication siblings that carry three doses of those genes located between the two translocation breakpoints. Third are deficiency siblings that carry only one dose of those genes located between the two breakpoints. By assaying the amount of the polypeptide of interest (e.g., by comparing V_{max} for an enzyme) in each class of sibling, a proportionality between gene dosage and polypeptide content may be noted if the structural gene is located between the two X or autosome breakpoints. Applying this approach systematically to survey the entire genome can lead to the identification of a chromosomal location for the gene of interest.

This approach was employed to survey the genome for chromosomal regions exhibiting a dosage effect on the rate of hydrolysis of cAMP and of cGMP (Kiger and Golanty, 1977). The distal quarter of the X chromosome, when duplicated, was found to increase the rate of cAMP hydrolysis but not of cGMP hydrolysis. Next, employing *Df* and *Dp* chromosomes, this region was narrowed to chromomeres 3D3–3D4. As indicated in Table 1, females with two normal X chromosomes and the duplication $Dp(1;2)w^{+51b7}$ (see Fig. 2) exhibit approximately 1.6 times the rate of cAMP hydrolysis of euploid females. Females with one normal X chromosome and any one of the deficiency chromosomes $Df(1)dm^{75e19}$, $Df(1)N^{64i16}$, and $Df(1)N^{71h24-5}$ exhibit between 0.6 and 0.7 times the normal euploid rate of cAMP hydrolysis.

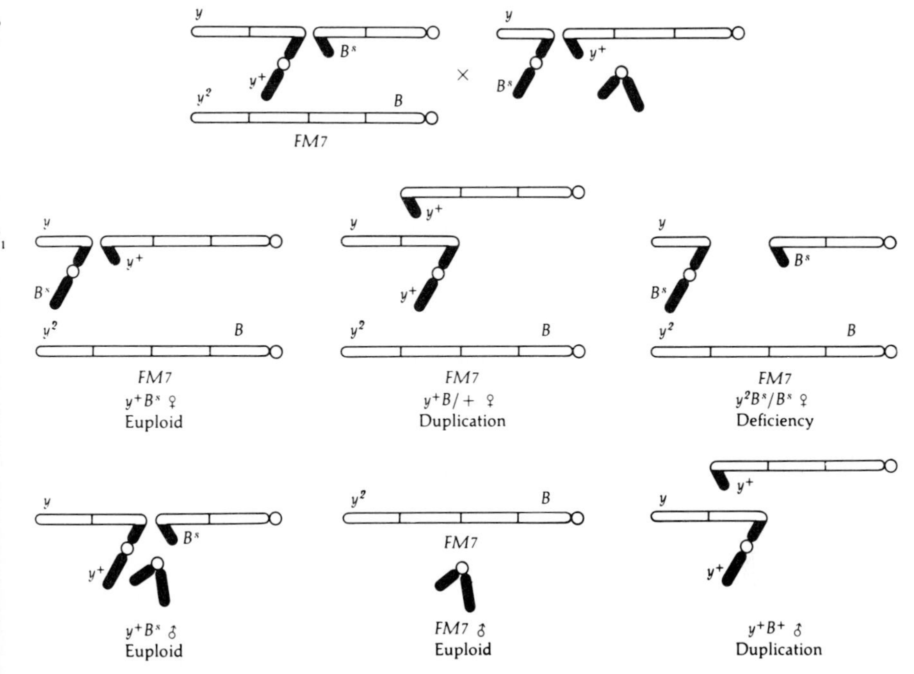

Fig. 1. Segmental aneuploidy. Crossing males and females of two different translocation strains permits the synthesis of duplication and deficiency aneuploids for the segment of the chromosome located between the two translocation breakpoints. The example illustrated here involves translocations between the X (open) and Y (black) chromosomes. All X chromosomes carry yellow mutants (y and y^2). *FM7* is a balancer X chromosome that is marked with the dominant Bar (B) mutation and possesses a number of inversions that prevent recombination with the translocation X chromosome in the mother. The translocation Y chromosomes have arms marked with y^+ and B^s (B^s is distinguishable from B). The father also carries a normal Y chromosome. The genetic markers permit identification of each of the F_1 progeny types indicated here. The deficiency aneuploid is male lethal and is not illustrated. Additional X; Y translocation strains (not illustrated) would be used to produce segmental aneuploids for the remaining portions of the X chromosome. The synthesis of segmental aneuploids involving the autosomes employs autosome; Y translocations and is identical, in principle, to the scheme illustrated here. (After Stewart and Merriam, 1975.) (From Ayala and Kiger, 1980, reproduced with permission.)

Females heterozygous for a normal X and the deficiency $Df(1)N^{64j15}$ exhibit an intermediate value of approximately 0.8. The duplication and the first three deficiency chromosomes all alter the dosage of chromomere 3D4. Chromosome $Df(1)N^{64j15}$ retains 3D4 but its right breakpoint is very close to or within 3D4 (Fig. 2).

These observations provided the first clues that a gene affecting the rate of cAMP hydrolysis is located in chromomeres 3D3–3D4. It should be noted that this study failed to locate a chromosomal region exhibiting similar

TABLE 1
Rates of cAMP and cGMP hydrolysis in homogenates of aneuploid females relative to euploid females[a]

Aneuploid	Aneuploid V_{max} / Euploid V_{max}	
	cAMP	cGMP
Dp(1; 2)w^{+51b7}	1.58 ± 0.26	
Df(1)dm^{75e19}	0.662 ± 0.028	1.20 ± 0.23
Df(1)$N^{71h24-5}$	0.638 ± 0.077	0.95 ± 0.08
Df(1)N^{64i16}	0.725 ± 0.093	1.10 ± 0.15
Df(1)N^{64j15}	0.815 ± 0.063	1.02 ± 0.13

[a] The 95% confidence limits of the *t* distribution are given. Data from Kiger and Golanty (1977).

properties upon cGMP hydrolysis. Possible pitfalls in the use of this approach have been discussed by Kiger and Golanty (1977).

Each of the deficiencies shown in Fig. 2 is lethal when homozygous in females or hemizygous in males, causing death of the organism during embryogenesis due to the absence of one or more essential genes (Campos-Ortega and Jiminez, 1980). However, *Dp*(1; 2)w^{+51b7} carries, on the second chromosome, all the essential genes missing from the *Df*(1)*N* chromosomes, permitting the survival of males with a *Df*(1)*N* chromosome when the duplication is present. Mating such males with females heterozygous for the *Df*(1)dm^{75e19} chromosome produces zygotes that carry both deficiency chromosomes; one-half of these zygotes lack the duplication on the second chromosome due to segregation. These latter zygotes lack all genes present in the interval where the two different deficiencies overlap (Fig. 2). Surprisingly these zygotes develop to give adult females, indicating that no genes essential for development are present in this interval of the X chromosome (Kiger, 1977).

Females of genotypes *Df*(1)$N^{71h24-5}$/*Df*(1)dm^{75e19} and *Df*(1)N^{64i16}/*Df*(1)dm^{75e19}, constructed by the above procedure, lack chromomere 3D4 and exhibit at least two obvious abnormalities: they lack detectable form II cAMP-specific phosphodiesterase activity (Kiger and Golanty, 1979; Kiger *et al.* 1981); and a large fraction of them are sterile, the remainder producing an abnormally small number of progeny. Of these few progeny about 15% are morphologically abnormal, lacking one or more legs, halters, wings, tergites, or genitalia. Appropriate crosses show that these morphological

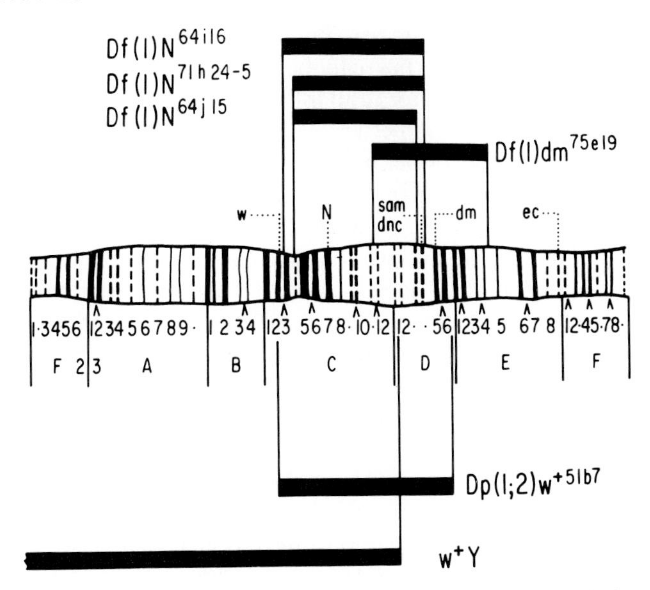

Fig. 2. A portion of the polytene X chromosome of *D. melanogaster* showing the locations of the following genes: *white* (*w*); *Notch* (*N*); *sperm-amotile* (*sam*); *dunce* (*dnc*); *diminutive* (*dm*); and *echinus* (*ec*). Bars indicate the cytological extent of the missing portions in the *Df* chromosomes and the extra portions in the *Dp* chromosomes described in the text.

defects are a consequence of the maternal genotype, i.e., the lack of chromomere 3D4 in the mother (Kiger, 1977). In contrast, females of genotype $Df(1)N^{64j15}/Df(1)dm^{75e19}$, which possess one dose of chromomere 3D4, exhibit a reduced form II cAMP-specific phosphodiesterase activity and a more normal degree of fertility, and produce no defective offspring (Kiger, 1977; Kiger *et al.*, 1981).

Males of genotype $Df(1)N^{71h24-5}/w^{+}Y$ and $Df(1)N^{64i16}/w^{+}Y$ lack chromomeres 3D3 and 3D4 and survive because essential genes lacking in these deficiencies are present in the duplication w^{+} carried on the Y chromosome (Fig. 2). These males lack detectable form II activity and are sterile, exhibiting nonmotile sperm upon dissection in *Drosophila* Ringer's solution. Males of genotype $Df(1)N^{64j15}/w^{+}Y$ exhibit reduced form II activity and are fertile, possessing motile sperm (Kiger, 1977; Kiger and Golanty, 1979; Kiger *et al.*, 1981).

The genetic removal of one form of phosphodiesterase activity from adult flies provides organisms whose residual phosphodiesterase activity can be characterized in crude homogenates and compared to the total phosphodiesterase activity in similar homogenates of the normal organisms. Such comparisons can complement the more traditional approach to biochemical analysis of enzyme activities by purification and subsequent characterization. Purification is usually a lengthy procedure during which changes in enzymatic properties can be induced by proteolysis and the removal of endogenous activators and inhibitors. This is particularly true for form II and form I phosphodiesterase activities. Limited proteolysis can increase the activity of the former by 150 % and the latter by 200–300 %, with loss of Ca^{2+} sensitivity (Kauvar, 1982; see Davis and Kauvar, 1984, for discussion). The effects of Mg^{2+} and Ca^{2+} on cAMP hydrolysis in crude homogenates of males with and without chromomere 3D4 are shown in Fig. 3. At concentrations of Mg^{2+} lower than 10^{-3} *M*, Ca^{2+} inhibits activities in homogenates of both genotypes whereas above 10^{-3} *M* Mg^{2+}, Ca^{2+} stimulates the residual activity present in the deficiency 3D4 homogenate. In addition, the form II activity present in the normal homogenate is markedly activated by Mg^{2+} whereas the residual activity present in the deficiency 3D4 homogenate is insensitive to Mg^{2+} concentration in the absence of Ca^{2+}. Moreover, the Ca^{2+}-chelator EGTA inhibits activity in homogenates of normal flies by only 10 %, whereas the activity in homogenates of deficiency 3D4 flies is inhibited by 50 %, supporting the idea that only the residual activity in deficiency 3D4 flies is Ca^{2+} sensitive (Kiger and Golanty, 1979).

Measurement of cAMP content in flies with 0, 1, 2, or 3 doses of chromomere 3D4 demonstrates that flies lacking 3D4 have elevated levels of cAMP compared to those with 1 or 2 doses and that flies with 3 doses have depressed levels of cAMP. Thus, the changes in form II phosphodiesterase activity associated with changes in gene dosage are physiologically significant (Davis and Kiger, 1978).

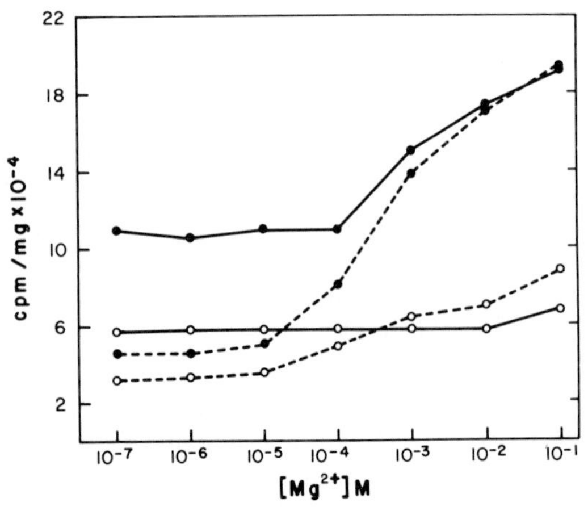

Fig. 3. Hydrolysis of [^{3}H]cAMP (10 μM) as a function of Mg^{2+} and Ca^{2+} in crude homogenates of normal, *Basc*/w^+ *Y* (●), and deficiency 3D4, *Df*(1)N^{64i16}/w^+ *Y* (○), male flies. Solid lines are activities in the absence of added Ca^{2+}, and dashed lines are activities in the presence of 1 m*M* Ca^{2+}. (From Kiger and Golanty, 1979, reproduced with permission.)

Thus cytogenetic analysis employing deficiencies and duplications permits several physiological functions to be assigned to chromomere 3D4: male fertility, normal female fertility, normal oogenesis, the presence of form II cAMP-specific phosphodiesterase activity, and normal cAMP level. This assignment raises the question of whether the multiple defects caused by lack of 3D4 are all a consequence of lack of a single gene. If so, is that gene a gene encoding form II cAMP phosphodiesterase? The first question is amenable to further genetic analysis. The second question requires both biochemical and genetic analyses.

1.2 GENETIC ANALYSIS OF THE PHYSIOLOGICAL EFFECTS OF CHROMOMERE 3D4

Genetic analysis requires selection of mutations induced in normal chromosomes that map to the region of interest, an examination of their mutant phenotypes, and complementation analysis of those phenotypes in combination with the *Df* chromosomes described above. A large collection of recessive X chromosome mutations affecting female fertility has been created by J. D. Mohler, many of which have maternal effects on development (Mohler, 1977). Some of these mutations, mapping to the distal portion of the X chromosome, were screened for failure to complement the female-sterility phenotype associated with $Df(1)N^{64i16}$, i.e., females with the deficiency chromosome and a recessive female-sterile mutation were produced and tested for fertility. Two female-sterile mutations were found in this way. These mutations, however, combined with $Df(1)N^{64j15}$, give females that are fertile (Salz *et al.*, 1982). Thus, these mutant genes must be located in chromomere 3D4. These two mutations fail to complement each other's female sterility phenotype and are, therefore, mutations of the same gene, i.e., alleles.

Complementation analysis soon led to the association of an additional mutant phenotype with the female sterility phenotype of Mohler's mutants. A number of mutations affecting learning ability had been selected by Seymour Benzer and colleagues (Dudai *et al.*, 1976). These mutations identify five different genes of the X chromosome, one of which is called *dunce* (*dnc*) (see Quinn and Greenspan, 1984, for review). Two mutants alleles of *dunce*, now designated dnc^1 and dnc^2, were known, and the latter also exhibited recessive female sterility. Nothing was known concerning any biochemical defect of these mutants, nor was the precise chromosomal location of the *dunce* gene known. However, Duncan Byers discovered that the two female-sterile alleles of Mohler failed to complement the female sterility of dnc^2, suggesting that dnc^2 is an allele of these mutants, and thus located in chromomere 3D4. Moreover, dnc^1 and dnc^2 were found to be deficient in form II cAMP phosphodiesterase activity and to exhibit elevated cAMP levels (Byers *et al.*, 1981), and Mohler's mutants were found to be defective in learning (Byers, 1979). Because of the precedence of the dunce designation, Mohler's mutants are now called dnc^{M11} and dnc^{M14}. The dnc^1 allele, however, appeared to be anomalous, since homozygous dnc^1 females were fertile. This anomaly was later shown to be due to the presence of a dominant suppressor of female sterility gene, *Su(fs)*, located elsewhere in the dnc^1 chromosome. When this suppressor is removed by recombination, the dnc^1 allele exhibits female sterility (Salz *et al.*, 1982).

An additional mutant allele of *dunce* was recovered by screening males carrying mutagenized X chromosomes for mutations causing a decreased cAMP phosphodiesterase activity. This mutant, dnc^{ML}, also exhibits recessive female sterility (Salz *et al.*, 1982). These different approaches to a mutational analysis of chromomere 3D4 lead to the conclusion that there is a single gene therein that is required for normal cAMP phosphodiesterase activity, normal female fertility, normal oogenesis, and normal learning ability.

Males carrying *dnc* mutations are fertile, suggesting that a second gene, affecting male fertility, may reside in 3D4. To resolve this question mutagenized X chromosomes were selected that gave fertile males in the presence of $Dp(1;2)w^{+51b7}$ but gave sterile males in its absence. These male-sterile mutations were then tested for fertility in the presence of the w^+Y duplication and found to be sterile. Thus they must map between the endpoints of the two duplications (Fig. 2) in the neighborhood of 3D4. These mutations, designated sam^1 and sam^2 (for *sperm-amotile*), have the same phenotype as the deficiency for 3D4. Moreover, females homozygous for sam^1 or sam^2 are fertile, indicating that these mutations affect a different gene than is affected by *dnc* mutations. This indication is supported by the observation that *sam* mutations complement *dnc* mutations in heterozygous females, i.e., females of genotype *sam*/*dnc* are fertile, indicating that the *sam* chromosome carries a normal *dnc* gene (Salz *et al.*, 1982; Salz and Kiger, 1984).

Recombinational analysis of *sam* and *dnc* alleles demonstrates that alleles of each gene are contiguous to one another on the chromosome and that the *sam* gene is to the left of the *dnc* gene (closer to the tip) on the chromosome (Salz *et al.*, 1982; Salz and Kiger, 1984). Thus *sam* must be located within chromomere 3D4 and not in chromomeres 3D5–3D6: a more precise location than could be assigned from the cytogenetic analysis.

1.3 BIOCHEMICAL ANALYSIS OF CYCLIC NUCLEOTIDE PHOSPHODIESTERASES

The genetic analysis described above establishes that form I and form II activities are genetically distinct forms of phosphodiesterase. The dosage relationship between chromomere 3D4 and the level of form II activity suggests, but does not prove, that *dunce* is the structural gene for this enzyme. It might, for example, be a gene for a regulatory protein required, in a dosage-dependent manner, for the synthesis or activity of this enzyme. Proof that *dunce* is the structural gene requires a demonstration that it encodes the amino acid sequence of the enzyme. Evidence bearing on this point has been provided by a biochemical comparison of form II activities in normal flies and in mutants of the *dunce* gene (Davis and Kiger, 1981; Kauvar, 1982) and will be discussed later.

Studies of the phosphodiesterase activities in crude homogenates of normal and mutant flies have been materially aided by Shotwell (1983) who has developed assays that allow form I and form II activities to be determined independently of each other without purification. Assay of form II activity in the presence of form I activity is possible by taking advantage of the fact that cGMP is a competitive inhibitor of form I cAMP hydrolysis. Thus, in the presence of excess unlabeled cGMP, hydrolysis of [^{3}H]cAMP becomes a measure of form II activity. This assay has been employed to confirm the relationship between the dosage of chromomere 3D4 and the level of form II phosphodiesterase activity and to investigate the anatomical distribution of the two activities (Shotwell, 1983). The data in Table 2 demonstrate that both forms of phosphodiesterase are present in all the body parts analyzed. Form I activity is particularly high in the head, and this is due in large part to its presence in the brain. Form II activity is also high in the head and brain but is also high in many of the other parts examined. The ovary is markedly low in

TABLE 2
Phosphodiesterase activities in body parts and clonal cell lines[a]

		Specific activity [substrate] (pmol min^{-1} μg protein^{-1})		
		Form II [cAMP]	Form I [cAMP]	[cGMP]
Female	Whole fly	1.800	0.700	1.200
	Head	2.900	2.500	6.600
	Thorax	1.200	0.110	0.420
	Abdomen	2.000	0.750	0.860
	Brain	4.700	3.700	9.300
	Gut	3.400	0.730	1.200
	Ovary	0.250	0.200	0.270
Male	Whole fly	1.900	0.560	1.400
	Head	3.200	2.800	6.900
	Thorax	2.000	0.130	0.510
	Abdomen	1.700	0.630	0.950
Clonal cell lines	CL 7	0.150	0.030	0.120
	CL 12	0.130	0.018	0.160
	CL 14	0.058	0.036	0.084

[a] From Shotwell (1983), reproduced by permission.

both forms of activity. Analysis of clonal cell lines demonstrates that both forms are present in the same cell type.

The form I and form II enzymes have been carefully characterized and purified to near homogeneity by Kauvar (1982), making it possible to estimate their molecular weights with some degree of ambiguity caused by proteolysis occurring during the procedure. Form I, which in impure preparations is activated by Ca^{2+}/calmodulin (Kiger and Golanty, 1979; Yamanaka and Kelly, 1981; Solti *et al.*, 1983) by binding of two molecules of calmodulin per molecule of enzyme (Walter and Kiger, 1984), is no longer stimulated by Ca^{2+} in the pure preparation. This loss of Ca^{2+} sensitivity was also observed by Davis and Kiger (1980). The pure preparations of form I and form II have molecular weights of 120,000 and 35,000 respectively, by SDS–polyacrylamide gel electrophoresis. Estimates of the molecular weights of the unpurified molecules determined by gel filtration (168,000 and 68,000) would suggest that form I is monomeric, but, since the sites of proteolysis of form II are unknown, it is possible that form II is either monomeric or dimeric.

Using the logic of the Shotwell assay and the different thermolabilities of form I and form II enzymes (Kiger and Golanty, 1979), Kauvar (1982) has distinguished a third phosphodiesterase activity in crude homogenates of normal flies by employing excess unlabeled cAMP to inhibit form I activity and assaying [^{3}H]cGMP hydrolysis (Fig. 4). This activity is extremely labile, requires Mg^{2+} (20 m*M*), and is not affected by Ca^{2+} (1 m*M*). The extreme lability of this activity has precluded its purification, and in view of the theoretical analyses of Vincent and Thellier (1983), its existence as an independent, genetically unique enzyme requires further demonstration. Furthermore, if form I contains two active sites, as discussed previously, it is possible that this activity may be due to a form I activity that is not allosterically inhibited by cAMP due to the presence of some other component of the crude homogenate. The properties of the *Drosophila* phosphodiesterases are compiled in Table 3.

Byers and Gustafsson (unpublished data) have carried out an extensive survey of potential inhibitors of form I and form II activities, in the presence of EGTA (2–4 m*M*), and they have identified a number of compounds effective at micromolar concentrations on one or both forms. The most potent inhibitors are 1-methyl-3-isobutylxanthine and some of its derivatives (1–22 μM). Caffeine and theophylline are relatively weak inhibitors (300–700 μM) while papaverine is intermediate (18–41 μM). Some inhibitors are relatively selective for one form over the other by ratios as high as 20:1 (Table 3).

The form II activities present in *dunce* mutant strains have also been characterized. Table 4 shows the activities of form II present in crude

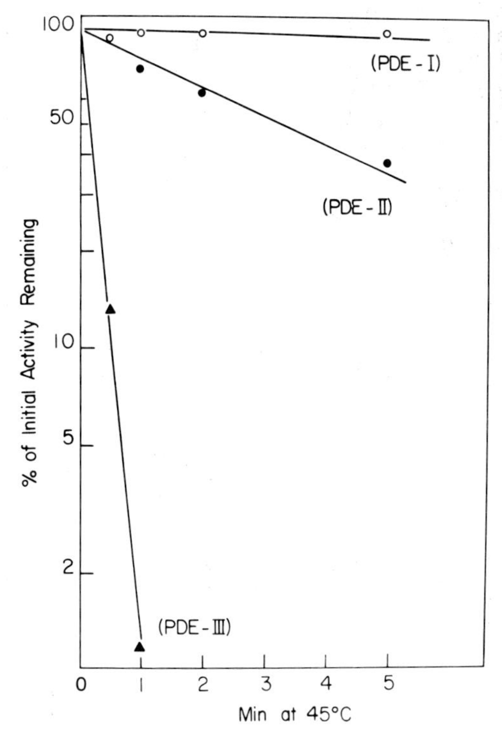

Fig. 4. Thermal inactivation at 45°C of three forms of phosphodiesterase (PDE) activities, assayed at 21°C, in crude homogenates of normal male flies. PDE III is assayed by hydrolysis of [^{3}H]cGMP (20 μM) in the presence of 50-fold excess of unlabeled cAMP. PDE II is assayed by hydrolysis of [^{3}H]cAMP (20 μM) in the presence of excess unlabeled cGMP. PDE I is the residual activity which is not specific for either [^{3}H]cAMP or [^{3}H]cGMP. (From Kauvar, 1982, reproduced with permission.)

homogenates of flies of different genotypes involving *dunce* employing the Shotwell assay. The mutants dnc^{M14}, dnc^{M11}, and dnc^{ML} exhibit an activity identical to that of the *dnc* deficiency $Df(1)N^{64i16}/w^+Y$ and appear to abolish detectable form II activity. The mutants dnc^1 and dnc^2 exhibit significant but decreased form II activity as does dnc^{CK}, an X-ray-induced mutation, selected as female sterile, which is associated with a translocation breakpoint near the *dnc* gene (Salz *et al.*, 1982; Salz and Kiger, 1984). The residual activities exhibited by the *dnc* null mutations (3–4 % of normal activity) are probably a consequence of form I activity that is not inhibited by excess cGMP.

The decreased form II activity exhibited by homozygous dnc^1 and dnc^2 flies pertains to both soluble and particulate activities, indicating that soluble and particulate activities are not genetically distinct components (Davis and Kiger, 1981).

TABLE 3
Properties of *Drosophila* phosphodiesterases

Phosphodiesterase	cAMP hydrolysis (% Total activity)	cGMP hydrolysis (% Total activity)	Effect of Ca^{2+}/calmodulin	Effect of Mg^{2+}	Strong selective inhibitors (K_iI:K_iII)
I	(~30) $K_m = 4–5 \times 10^{-6}$ *M*; completely inhibited by cGMP (2 m*M*); $K_i = 3.8 \times 10^{-6}$	(~70) $K_m = 4–25 \times 10^{-6}$ *M*; completely inhibited by cAMP (2 m*M*); $K_i = 3.4 \times 10^{-6}$ *M*	Stimulated (binds two molecules of Ca^{2+}/calmodulin)	Not stimulated	8-*t*-Butyl-IBMX[a] (1:17) 8-Methyl-IBMX (1:10)
II	(~70) $K_m = 2–8 \times 10^{-6}$ *M*; no inhibition by cGMP (3 m*M*)		Not stimulated (no binding of Ca^{2+}/calmodulin)	Stimulated	SQ-20009 (20:1) 2-Chloroadenosine (18:1)
III		(~30) $K_m = 30 \times 10^{-6}$ *M*; active in presence of cAMP (2 m*M*)	Ca^{2+} independent	Stimulated	

[a] IBMX is 1-methyl-3-isobutylxanthine.

TABLE 4
Form II phosphodiesterase activities in homogenates of males[a]

Genotype	Activity (pmol min^{-1} μg $protein^{-1}$) $\pm$ 2 SEM	Fraction of normal activity (%)
Canton S	1.66 ± 0.16	100
dnc^1	0.85 ± 0.08	51
dnc^{CK}	0.58 ± 0.04	35
dnc^2	0.25 ± 0.03	15
dnc^{M14}	0.07 ± 0.01	4
dnc^{M11}	0.07 ± 0.01	4
dnc^{ML}	0.05 ± 0.03	3
$Df(1)N^{64i16}/w^+Y$	0.07 ± 0.02	4

[a] From Salz and Kiger (1984), reproduced by permission. Copyright Genetics Society of America.

The residual form II activity in homozygous dnc^1 flies exhibits an abnormal thermolability compared to the normal form II activity, and flies of heterozygous genotype ($dnc^1/+$) exhibit activities with two distinct thermolability components, like those of dnc^1 and normal, in the proportion expected if each allele is expressed at the level observed in the homozygous flies (Fig. 5). The difference in thermolability of form II activity in dnc^1 flies and normal flies is also evident after extensive purification (1000- to 2000-fold), suggesting that the difference is an intrinsic property of the enzyme, as might be expected of a change in an amino acid (Kauvar, 1982).

The residual form II activity in homozygous dnc^2 flies is also abnormal, exhibiting a K_m of approximately 60×10^{-6} M, more than 10-fold higher than the normal enzyme, and being much more sensitive to inhibition by papaverine and theophylline. Flies of heterozygous genotype ($dnc^2/+$) exhibit two kinetic components as would be expected if the altered K_m is an intrinsic property of the dnc^2 enzyme (Fig. 6). The altered K_m of the dnc^2 enzyme persists after extensive purification (Kauvar, 1982).

The altered thermolability of the dnc^1 enzyme is unique to dnc^1 and not shared by the dnc^2 enzyme. Similarly, the altered enzymatic properties of the dnc^2 enzyme are not shared by the dnc^1 enzyme. The composite properties of normal and mutant enzyme activities exhibited by $dnc^1/+$ and $dnc^2/+$ heterozygotes are reproduced by mixing equal amounts of homogenates of normal and mutant flies (Kauvar, 1982). Together with the dosage relationship between chromomere 3D4 and the level of form II activity, these observations strongly suggest that *dunce* is the structural gene for form II cAMP-specific phosphodiesterase. This conclusion is substantiated by a

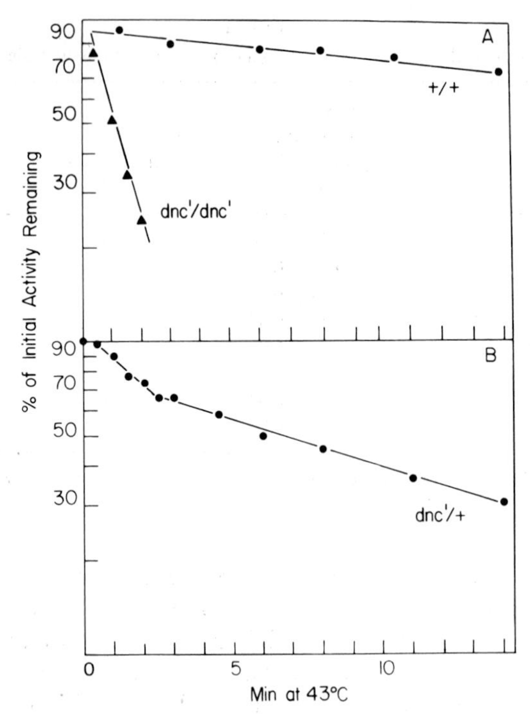

Fig. 5. Thermal inactivation at 43°C of form II phosphodiesterase activities, assayed at 21°C, in crude homogenates of (A) females of genotypes dnc^+/dnc^+ and dnc^1/dnc^1 and (B) females of genotype dnc^1/dnc^+. (From Kauvar, 1982, reproduced with permission.)

recombinational analysis of the enzymatic phenotypes of dnc^1 and dnc^2 that demonstrates their genetic map position on the X chromosome to be between the *white* and *echinus* genes as expected for alleles of the *dunce* gene (Fig. 2). Thus the enzymatic defects are intrinsic to the *dunce* gene, ruling out the possibility that other genes on the X chromosome might act in some other way to alter the form II enzyme present in *dunce* flies (Kauvar, 1982).

Attempts to isolate mutants affected in cAMP metabolism by screening mutagenized X chromosomes for temperature sensitivity to theophylline, propranolol, and dihydroergotoxin have been reported (Savvateeva and Kamyshev, 1981). One mutant sensitive to theophylline and a second sensitive to propranolol are reported to exhibit abnormally low levels of phosphodiesterase activity, but it is not clear which form(s) of activity are affected. A third mutant, while not sensitive to any of these compounds, is a temperature-sensitive lethal and exhibits abnormally high phosphodiesterase activity (Savvateeva *et al.*, 1982). These mutants require further characterization.

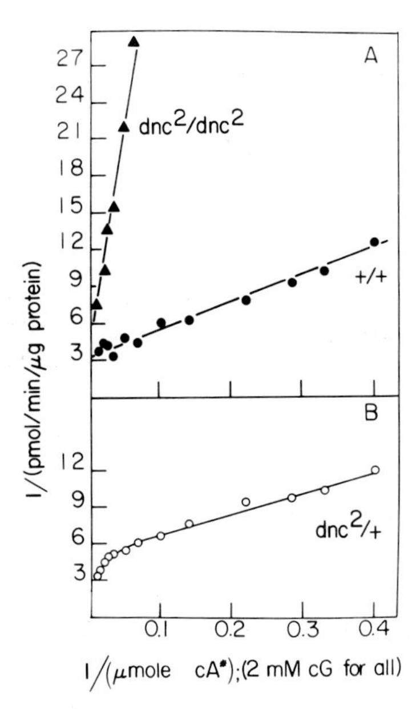

Fig. 6. Kinetics of [^{3}H]cAMP hydrolysis, presented in Lineweaver–Burke double reciprocal plots, by form II phosphodiesterase in crude homogenates of (A) females of genotypes dnc^+/dnc^+ and dnc^2/dnc^2 and (B) females of genotype dnc^2/dnc^+. (From Kauvar, 1982, reproduced with permission.)

1.4 FINE STRUCTURE ANALYSIS OF THE *dunce* GENE

The *dnc* point mutations have been subjected to recombinational analysis using restoration of female fertility to detect recombination between pairs of *dnc* mutations. Females of the general genotype $w\ dnc^X\ ec^+/w^+\ dnc^Y\ ec$; *Cy*, $Dp(1;2)w^{+51b7}/\pm$ were bred, and male progeny of genotypes *w ec* and w^+ ec^+, lacking *Cy*, $Dp(1;2)w^{+51b7}$, were selected and bred to establish females homozygous for each recombinant chromosome. Testing the fertility of each of these females permits detection of those recombinants, produced by crossing over between the two *dnc* alleles, that restore female fertility. The order of the two *dnc* mutations is determined by the flanking marker genotypes, *w ec* or $w^+\ ec^+$. Female-fertile recombinant chromosomes were then tested for restoration of normal form II enzyme activity to confirm that the recombinant chromosomes are dnc^+ (Salz and Kiger, 1984).

Using this procedure, a fine structure map has been constructed showing the relationship of the different *dnc* mutations to one another and to the adjacent mutations in the *sam* gene (Fig. 7). Four mutations form a closely linked cluster that has not been resolved and that contains both null enzyme alleles and the alleles encoding altered enzymes. Thus, this cluster must

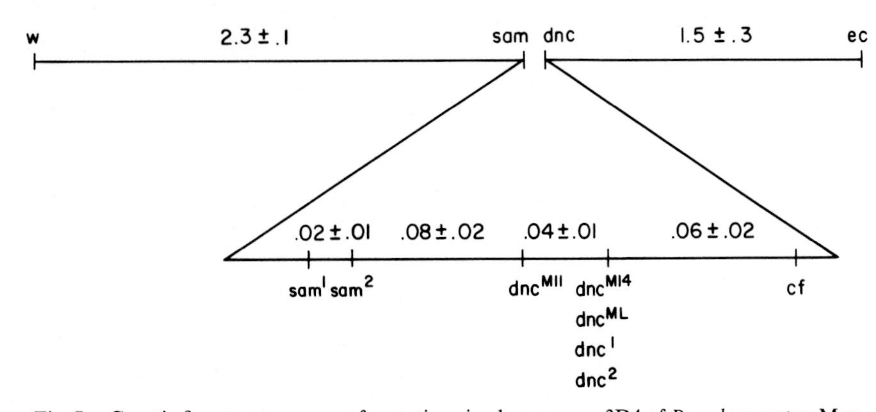

Fig. 7. Genetic fine structure map of mutations in chromomere 3D4 of *D. melanogaster*. Map distances are given in centimorgans (cM). (From Salz and Kiger, 1984, reproduced with permission.)

identify at least a portion of the coding region. The null enzyme allele dnc^{M11} maps 0.04 cM to the left of this cluster.

Female-fertile recombinants between *dnc* alleles are expected to exhibit normal form II enzyme activity, and this expectation is met by recombinants between dnc^{M11} and the dnc^{M14} and dnc^2 alleles in the cluster. However, all female-fertile recombinants between dnc^1 and the alleles dnc^{M14}, dnc^{M11}, and dnc^2 exhibit a reduced and thermolabile form II activity characteristic of the dnc^1 allele. To account for these observations, the dnc^1 chromosome is postulated to contain two closely linked mutations: dnc^1, responsible for the reduced and thermolabile activity of form II enzyme, and cf^1 (control of fertility), responsible for the female sterility of the dnc^1 cf^1 chromosome. The other chromosomes must contain a cf^+ allele to account for the female-fertile recombinants, which must be dnc^1 cf^+. The data indicate that *cf* maps 0.06 cM to the right of dnc^1 (Fig. 7). Because dnc^1 cf^1/dnc^{M14} cf^+ females are sterile, *cf* must act in cis to regulate the female sterility phenotype associated with dnc^1. This observation suggests that *cf* may identify a regulatory sequence for the *dnc* structural gene that acts in some way to regulate that *dnc* expression associated with female fertility (Salz and Kiger, 1984).

The fine structure analysis suggests that the dnc^1 mutation, unlike the other *dnc* alleles, is not female sterile. The thermolabile enzyme, therefore, must provide enough activity for female fertility. However, when cf^1 is adjacent to the dnc^1 locus, female fertility is not observed, unless another gene, *Su*(*fs*), located at the other end of the X chromosome, is present (Salz *et al*., 1982). Suppression of the female sterility phenotype by *Su*(*fs*) does not affect either the activity or the thermolability of the form II enzyme (Salz *et al*., 1982; Salz, 1983). The suppressor is allele specific; therefore it may be acting to suppress the effect of the mutant cf^1 locus present with dnc^1. If this is the case, then the wild-type product of the suppressor gene may normally interact with *cf* to allow female fertility. How these genetic components affect the learning phenotype is not known. The only strain to be tested in a learning paradigm is the original strain isolated by Dudai *et al.* (1976). This strain is the triple mutant—carrying both the mutant cf^1 locus and *Su(fs)*.

1.5 CLONING THE DNA OF CHROMOMERE 3D4

Recombinant DNA technology provides a high resolution approach to understanding how DNA structure can affect gene expression and ultimately affect the physiology of the organism. Application of recombinant DNA technology generally requires prior knowledge of the biochemical function of a gene and identification of its mRNA before its cloned DNA can be identified for study. Because of its advanced genetics and cytogenetics, *Drosophila* now provides the opportunity to investigate any gene without prior knowledge of biochemistry or the availability of mRNAs, which may be very rare for a gene of interest. This is possible due to the development with *Drosophila* of techniques termed "chromosome walking and jumping" (Bender *et al.*, 1983). These techniques permit the DNA of a gene that has been located cytogenetically to be identified in a library of cloned DNA sequences from the entire genome. Beginning with a previously identified cloned DNA sequence, one moves along the chromosome, selecting overlapping cloned sequences from the library, until the gene of interest is reached. A restriction endonuclease map of the chromosome is constructed from each cloned DNA sequence in the process, and the progress of the walk is monitored by comparing the restriction site map of the normal chromosome with those for chromosomes that have rearrangement breakpoints (deficiencies, inversions, or translocations) near or at the gene of interest. In this way the DNA of chromomere 3D4 has been identified and made available for study (Davis and Davidson, 1984).

By using restriction site polymorphisms as genetic markers, dnc^2 has been mapped to a 10- to 12-kb DNA segment. The right breakpoint of $Df(1)N^{71h24\text{–}5}$, which eliminates form II activity, is also located within this DNA segment. The right deficiency breakpoint of $Df(1)N^{64i16}$, which also eliminates form II activity, is located 10–20 kb to the right. Therefore, it appears that as much as 30 kb of DNA may be required for *dnc* gene expression (Davis and Davidson, 1984).

These results are in general agreement with the results of the genetic fine structure analysis (Salz and Kiger, 1984). The *dnc* point mutations all map within 0.04 cM of dnc^2. Using the metric of 300 kb per cM found for the nearby *Notch* region (Artavanis-Tsakonas *et al.*, 1983), *dnc* must be at least

12 kb long. Furthermore, *cf* maps 0.06 cM, or about 18 kb, to the right of *dnc*. The genetic analysis would therefore predict that 30 kb of DNA is required for proper *dnc* gene expression.

Evidence for an additional regulatory region to the left of the *dnc* gene is provided by chromosomal breakpoints in this region. The $Df(1)N^{64j15}$ chromosome exhibits reduced form II enzyme activity (Table 1) as does the dnc^{CK} translocation (Table 4). The dnc^{CK} breakpoint has been mapped cytogenetically to the left of the *dnc* gene (Salz and Kiger, 1984). The dnc^{CK} breakpoint has been located on the DNA restriction map of chromomere 3D4 within 5 kb of the right breakpoint of $Df(1)N^{64j15}$, about 35 kb to the left of the *dnc* region. However, there are two additional deletions and a small insertion in the $Df(1)N^{64j15}$ chromosome located closer to *dnc* (R. L. Davis, personal communication), and it is not known which alteration might be responsible for the effect on *dnc* expressions.

2 Other elements of the cyclic nucleotide system

Other elements of the cyclic nucleotide system in *Drosophila* will be discussed here. Additional elements of the system that have not yet been demonstrated directly might be inferred from their existence in other organisms, but the need is evident for their demonstration in *Drosophila.*

2.1 ADENYLATE CYCLASE

The defective learning phenotype of *dunce* mutants, together with cellular studies of conditioning and sensitization in molluscan neurons, suggests that cAMP plays an important role in these phenomena (Castellucci *et al.*, 1980; Kaczmarek *et al.*, 1980; Byers *et al.*, 1981; Kandel and Schwartz, 1982; Siegelbaum *et al.*, 1982; Hawkins *et al.*, 1983; Carew *et al.*, 1983; Alkon *et al.*, 1983). Therefore, other *Drosophila* mutants defective in learning may also exhibit defective cAMP metabolism. This reasoning has led to the examination of adenylate cyclase activities in wild-type and a number of behavioral and learning mutant strains (Uzzan and Dudai, 1982).

Adenylate cyclase activity in particulate matter from heads of *Drosophila* has been found to be very similar to adenylate cyclase activities in mammalian tissues (Ross and Gilman, 1980). This activity has a K_m of 50 μM for ATP, and basal activity is stimulated 1.4-fold by GTP with a maximum at 10 μM. The poorly hydrolyzed analog guanylyl imidodiphosphate (GppNHp) also is stimulatory and exhibits a maximum stimulation of 10-fold at 10 μM with a half-maximum of 0.5 μM. NaF exhibits a 9-fold stimulation at 10^{-2} M. Among putative neurotransmitters, octopamine exhibits of 5- to 6-fold stimulation and smaller amounts of stimulation (1.3- to 1.4-fold) are seen for serotonin and dopamine. Additivity in stimulation suggests that these three compounds act on different receptors (Uzzan and Dudai, 1982). Basal activities in 10 μM GTP in head particulate matter of the learning mutants *dunce* (*dnc*[1] and *dnc*[2]), *amnesiac*, *cabbage*, and *rutabaga* were compared with the activity in the parental wild-type strain. All the mutants are reported to exhibit a higher activity (1.3- to 1.4-fold) than wild type except for *rutabaga* which is the same as wild type. However, the activity in whole-head homogenates of *rutabaga* was observed to be variable compared to the others but generally lower, between 60–70% of normal (Uzzan and Dudai, 1982).

Studies of adenylate cyclase activities in normal and mutant *Drosophila* heads and abdomens by Livingstone *et al.* (1984) have confirmed and extended the studies of Uzzan and Dudai (1982) on the particulate activity in normal heads but have led to different conclusions concerning the mutants. Basal activities in crude homogenates, in the absence of GTP, in heads and abdomens of wild-type and mutant strains are shown in Table 5. These data show marked differences between *rutabaga* and the other strains, particularly in homogenates of abdomens. They do not, however, support the observation of Uzzan and Dudai (1982) that the other mutants have higher than normal activities. Keeping in mind that the two groups have not used identical conditions, the different observations might be explained if the basal activity

TABLE 5
Adenylate cyclase activities in learning mutants and wild type[a,b]

Genotype	Adenylate cyclase activity	
	Heads	Abdomens
Canton-S	3.86 ± 0.08 (100%)	0.40 ± 0.02 (100%)
rutabaga	2.67 ± 0.04 (69%)	0.15 ± 0.01 (38%)
dunce	3.26 ± 0.13 (84%)	0.42 ± 0.02 (105%)
amnesiac	4.30 ± 0.30 (111%)	0.51 ± 0.06 (128%)
cabbage	3.56 ± 0.11 (92%)	0.36 ± 0.05 (90%)
turnip	3.01 ± 0.05 (78%)	0.35 ± 0.02 (88%)

[a] From Livingstone *et al.* (1984), reproduced by permission. Copyright M.I.T.

[b] Adenylate cyclase activity was measured in crude homogenates of heads or abdomens from male flies in the presence of 0.2 mM ATP without added guanyl nucleotides or other ligands. Activity is expressed as pmol cyclic AMP formed min^{-1} mg protein^{-1} and as a percentage of the wild-type level.

of the wild-type strain measured by Uzzan and Dudai (1982) was abnormally low for some unaccounted reason.

The adenylate cyclase activities in crude homogenates are composed of soluble and particulate activities. The soluble activity differs from the particulate activity in that it is not stimulated by GppNHp (50 μM), forskolin (50 μM), or Mn^{2+} (5 mM). This activity has not been further characterized because the difference in activity between wild-type and *rutabaga* strains resides in the particulate fraction (Livingstone *et al.*, 1984).

The following observations are important in determining how the *rut* mutation affects adenylate cyclase activity. The particulate activity is stimulated by GppNHp in wild type and in *rutabaga* by 1.9- and 2.4-fold, respectively, and NaF stimulates by about the same amounts. Both wild type and *rutabaga* show the same concentration dependencies for GppNHp and NaF stimulation, suggesting that both possess the N_s (G_s) component of adenylate cyclase, upon which these compounds act (Gilman, 1984). Adenylate cyclase activities stimulated by Mn^{2+} (5 mM) are inhibited by GppNHp in both wild type and *rutabaga* by 73 and 65 %, respectively, suggesting that both possess the N_i (G_i) component of cyclase (Gilman, 1984). However, GppNHp does not inhibit *Drosophila* cyclase activated by forskolin as is seen for mammalian cyclases. Furthermore, activities from both strains are stimulated by octopamine, wild type by 2.8-fold and *rutabaga* by 3.2-fold, and both show the same concentration dependencies. Thus the different activities of adenylate cyclase in wild type and *rutabaga* do not appear to be due to an altered octapamine receptor or to altered N_s (G_s) or N_i (G_i) regulatory moieties, suggesting that the difference may reside in the catalytic moiety.

Forskolin (50 μM) stimulates activities of both strains by 5- to 6-fold, and Mn^{2+} (5 mM) stimulates both by 2- to 3-fold. Both forskolin and Mn^{2+} are believed to act on the catalytic moiety. Wild-type and *rutabaga* activities, however, differ in their response to Ca^{2+}. As shown in Fig. 8, Ca^{2+} has a stimulatory effect at low concentrations and an inhibitory effect at higher concentrations on wild-type activity in both heads and abdomens. Activities from similar preparations of *rutabaga* exhibit only the inhibitory effect, suggesting the presence of two different cyclase activities in wild type and the absence of one of these, the Ca^{2+}-stimulated activity, in *rutabaga.* Consistent with this interpretation, the Ca^{2+}-chelator EGTA (0.25–0.50 mM) decreases activity in wild type but increases activity in *rutabaga.*

The difference in wild-type and *rutabaga* activities (Fig. 8) is not due to a deficiency of endogenous calmodulin in *rutabaga* compared to wild type. When membrane preparations from wild type and *rutabaga* are stripped of endogenous calmodulin by washing in chlorpromazine, exogenous bovine calmodulin (50 μg/ml) activates wild-type abdominal cyclase 1.9-fold but slightly inhibits the *rutabaga* cyclase. Thus rutabaga appears to completely

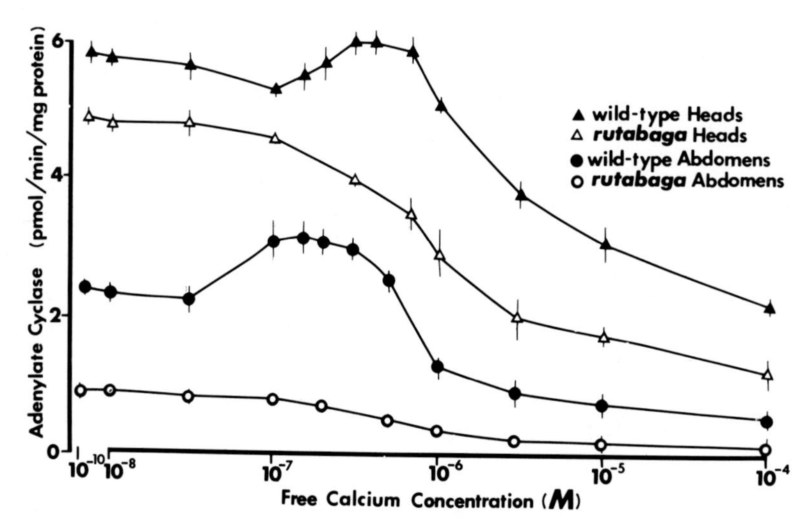

Fig. 8. Response to Ca^{2+} of adenylate cyclase activities in particulate matter prepared from normal and *rutabaga* heads and abdomens. (From Livingstone *et al.*, 1984, reproduced with permission. Copyright M.I.T.)

lack a calmodulin-activated component of adenylate cyclase activity. This component comprises the majority of cyclase activity in the abdomen and is a minority component in the head (Table 5).

The *rutabaga* mutant strain was selected for a defect in learning following mutagenesis and isolation of X chromosomes from a wild-type strain (Canton-S) (Dudai *et al.*, 1976). The mutagenesis procedure employed frequently leads to the isolation of chromosomes carrying more than one induced mutation. In order to determine if the learning defect and the defect in adenylate cyclase activity are caused by a single mutation, these two phenotypes have been subjected to genetic analysis (Livingstone *et al.*, 1984). Females of genotype *rut*/*y cv v f car* were bred, and males carrying recombinant X chromosomes, produced by single crossovers between each of the morphological markers (yellow, *y*; crossveinless, *cv*; vermilion, *v*; forked, *f*; and carnation, *car*) distributed along the length of the chromosome, were selected for analysis of the learning phenotype. This analysis places *rut* between *v* (map position 33.0) and *f* (map position 56.7). In order to determine whether the two defective phenotypes map to the same chromosomal region, 14 independent recombinants between *v* and *f* were assayed for both phenotypes. The results shown in Fig. 9 demonstrate complete cosegregation of the two phenotypes. If the two phenotypes are caused by two independent mutations then the two must be within 4 cM of one another at the 95% confidence limits (Livingstone *et al.*, 1984).

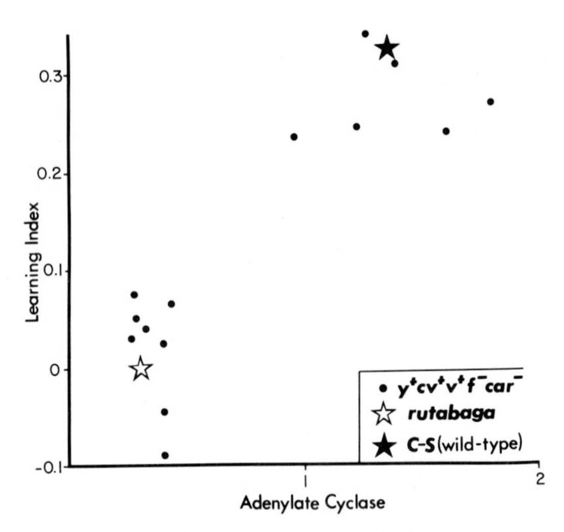

Fig. 9. Segregation of the learning phenotype and of the adenylate cyclase phenotype associated with *rutabaga* in recombinants produced by crossing over between vermilion and forked. These data indicate that each of the 14 crossovers recorded occurred either to the left or to the right of a locus responsible for each of these phenotypes and suggest that both phenotypes are controlled by a single gene, i.e., *rutabaga*. If any points were to fall off of the diagonal between the two stars, i.e., on the horizontal and vertical axes of the two stars, two genetic loci would have have to be postulated to account for the two phenotypes. (From Livingstone *et al.*, 1984, reproduced with permission. Copyright M.I.T.)

A more precise chromosomal location has been derived from complementation analysis with *Df* chromosomes in which the deficiencies are located between *v* and *f* (Campos-Ortega and Jiminez, 1980). The *Df*(1)*KA9* chromosome, a deletion of chromomeres 12E1 to 13A5, is the only *Df* chromosome tested that fails to complement both phenotypes exhibited by *rut*. These analyses suggest that both phenotypes are caused by mutation of the same gene.

The *Df*(1)*KA9* chromosome is lethal when homozygous or hemizygous, indicating that it lacks one or more essential genes. However, the fact that *rut*/*Df*(1)*KA9* adults are recovered indicates either that *rut* is nonessential or that the *rut* mutation retains sufficient gene activity to permit development. Data pertinent to the nature of the *rut* mutation are presented in Fig. 10. The particulate, calmodulin-stripped, abdominal adenylate cyclase activities, measured with different activators, are compared in the figure as a function of the *rut* genotype. The activities measured in the presence of Mn^{2+} are most informative since, in mammals, Mn^{2+} directly activates the catalytic moiety of adenylate cyclase (Neer, 1979). These data demonstrate that flies of genotype *rut*/wild type and *Df*(1)*KA9*/wild type exhibit virtually the same

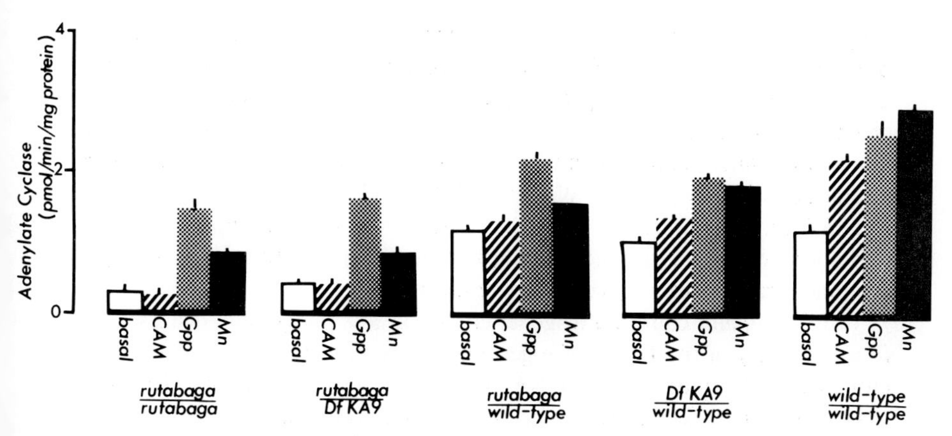

Fig. 10. Responses of calmodulin-stripped, abdominal, particulate adenylate cyclase activities from flies of different genotypes to 50 μg/ml bovine calmodulin (CAM), 50 μ*M* guanylyl imidodiphosphate (Gpp), and 5 m*M* Mn^{2+}. (From Livingstone *et al.*, 1984, reproduced with permission. Copyright M.I.T.)

level of Mn^{2+}-activated activity, a level almost exactly intermediate between that of homozygous wild type and that of *rut/rut* and *rut/Df*(1)*KA*9. These observations are consistent with the hypotheses that the *rut* mutation abolishes measurable gene activity and is equivalent to the deficiency for the gene, and that there is a dosage relationship between the rut^+ gene and the level of a catalytic moiety of adenylate cyclase. The fact that neither *rut/rut* or *rut/Df*(1)*KA*9 are completely lacking in catalytic activity implies that the residual cyclase activity is under the control of one or more genes located elsewhere. The effect of exogenous calmodulin on the activities presented in Fig. 10 indicates that the catalytic moiety conferred by rut^+ is specifically activated by calmodulin, while the residual cyclase activity is not. The effect of GppNHp on the activities observed for the different genotypes suggests that N_s (G_s) activates both types of catalytic moiety. These observations are consistent with the hypotheses that (1) rut^+ is the structural gene for a calmodulin-activated cyclase or that (2) rut^+ encodes a regulatory product that acts in a dosage-dependent manner, and is required for synthesis or activity of a calmodulin-activated cyclase. At this time evidence is not available to distinguish between these hypotheses.

2.2 CALMODULIN

The existence of calmodulin in *Drosophila* was first demonstrated by Yamanaka and Kelly (1981), who purified a heat-stable protein from *Drosophila* heads that activated calmodulin-deficient phosphodiesterase preparations from rat brain and *Drosophila.* This protein comigrates with porcine calmodulin upon SDS–polyacrylamide gel electrophoresis. In SDS gels it migrates faster in the presence of Ca^{2+} than in the presence of EGTA, a characteristic of calmodulin from a number of sources (Burgess *et al.*, 1980). These observations have been confirmed on calmodulin isolated from larvae and adults of *Drosophila* by Walter and Kiger (1984).

The level of calmodulin has been determined by radioimmunoassay in abdomens of wild type and *rutabaga* to be 2.1 ± 0.2 and 2.5 ± 0.3 ng/mg abdomen, respectively (Livingstone *et al.*, 1984).

The *Drosophila* calmodulin gene, cloned in a λ phage, has been selected from a library of the *Drosophila* genome by Yamanaka *et al.* (1983) using a cloned cDNA made from mRNA from the electroplax of the electric eel (Munjaal *et al.*, 1981). *In situ* hybridization of the cloned *Drosophila* DNA to salivary gland polytene chromosomes is reported to localize the calmodulin gene to region 49, in the right arm of the second chromosome (Yamanaka *et al.*, 1983). The effects of segmental aneuploidy for this region on the levels of calmodulin in flies have not yet been reported, nor have mutations that affect its activity been identified.

2.3 CYCLIC AMP-BINDING PROTEINS

The physiological effects of cAMP are mediated through its action as a ligand of cAMP-binding proteins. These proteins have, thus far, been well characterized only in early embryos of *Drosophila.* The soluble matter of early embryos contains two major protein components that bind cAMP: the regulatory moiety of cAMP-dependent protein kinase; and a much larger protein, designated H, whose function is unknown (Tsuzuki and Kiger, 1975).

The H protein has been extensively purified (Tsuzuki and Kiger, 1974, 1975). It has an apparent molecular weight of about 200,000, an isoelectric point of 5.10, and a dissociation constant for cAMP of 1.2×10^{-7} *M*. Binding of cAMP is inhibited by ATP, ADP, and AMP, but not by cGMP, cUMP, or cCMP. The H protein also binds adenosine with a dissociation constant of 1.6×10^{-7} *M* (Tsuzuki and Kiger, unpublished). Adenosine inhibits cAMP binding and cAMP inhibits adenosine binding, but the kinetics of inhibition are complex in both cases. Based upon these properties, the H protein appears to be similar to cAMP-binding proteins that have been observed in mammalian erythrocytes (Yuh and Tao, 1974) and in mammalian liver (Sugden and Corbin, 1976; Ueland and Døskeland, 1977, 1978). Based upon its ligand affinities, Sugden and Corbin (1976) have called it the adenine analog-binding protein (AABP). A similar protein has also been purified from human lymphoblasts and placenta and from mouse liver, and identified as *S*-adenosylhomocysteine hydrolase by Hershfield and Kredich (1978), who have speculated on its role in mediating adenosine toxicity.

The H protein does not exhibit cAMP phosphodiesterase activity, adenylate cyclase activity, or protein kinase activity, with or without cAMP, but it has not been assayed for *S*-adenosylhomocysteine hydrolase activity.

The other cAMP-binding protein in embryos is the regulatory component of cAMP-dependent protein kinase, which is present both free and complexed with the catalytic moiety of this enzyme (Tsuzuki and Kiger, 1975). The molecular weight of the heteromeric complex (RC) is 70,000 (Tsuzuki and Kiger, 1975). In the presence of cAMP the complex dissociates into a catalytic moiety (C) of molecular weight 49,000 and a regulatory moiety (R) of molecular weight 21,000 (Tsuzuki and Kiger, 1978). These properties make *Drosophila* embryo cAMP-dependent protein kinase quite different from the mammalian enzymes, whose structures are R_2C_2, in which the R moiety is larger than C, and remains in a dimer (R_2) upon dissociation (Taylor *et al.*, 1981).

The dissociation constants for cAMP of purified R and purified RC are 9.3×10^{-10} *M* and 2.7×10^{-8} *M*, respectively, indicating that C influences the binding of cAMP to R. This is confirmed by Scatchard plots in which pure R yields a straight line while that for RC is hyperbolic (Tsuzuki and Kiger, 1978).

A kinetic analysis of cAMP binding to R and to RC has been performed and rate constants evaluated by a curve-fitting technique. The results indicate a mechanism in which a ternary complex (LRC) of R, C, and cAMP (L) is an important intermediate in activation of protein kinase by cAMP.

$$\mathrm{L + R} \underset{k_{-1}}{\overset{k_1}{\rightleftarrows}} \mathrm{LR} \qquad (1)$$

$$\mathrm{L + RC} \underset{k_{-3}}{\overset{k_3}{\rightleftarrows}} \mathrm{LRC} \underset{k_{-4}}{\overset{k_4}{\rightleftarrows}} \mathrm{LR + C} \qquad (2)$$

The estimated rate constants at 0°C for reactions (1) and (2) are $k_1 = 2.3 \times 10^6\ M^{-1}\ \mathrm{s}^{-1}$; $k_{-1} = 1.1 \times 10^{-3}\ \mathrm{s}^{-1}$; $k_3 = 3.5 \times 10^6\ M^{-1}\ \mathrm{s}^{-1}$; $k_{-3} = 7.3 \times 10^{-1}\ \mathrm{s}^{-1}$; and $k_4 = 3.8 \times 10^{-2}\ \mathrm{s}^{-1}$. The LRC complex, once formed, does not necessarily dissociate to LR + C (active) but may dissociate to L + RC (inactive); indeed, under the experimental conditions employed about 95% of LRC dissociates without activation (Tsuzuki and Kiger, 1978). Thus, activation of the protein kinase may be buffered against small fluctuations in cAMP level under physiological conditions and may only be activated by a dramatic increase in cAMP level.

The genes encoding the H, R, and C proteins have not been identified in *D. melanogaster*, so mutants that would permit an assessment of the physiological roles of these proteins in the whole organism remain to be discovered and studied.

3 Physiological roles of cyclic nucleotides

Drosophila offers unique insights on the roles of cyclic nucleotides and of specific elements of the cyclic nucleotide system that may be of general importance to higher organisms.

The giant polytene chromosomes of the larval salivary gland have long been a model system for studying the organization and function of interphase chromosomes. The expansion and regression of puffs at specific chromosomal positions have been correlated with changes in gene expression in the region of the puffs during normal development and during experimental manipulation of hormone levels and environmental stimuli, such as heat shock. Indirect immunofluorescent techniques have been employed to visualize the distribution of cAMP and cGMP along the length of polytene chromosomes (Spruill *et al.*, 1978). Cyclic GMP, but not cAMP, is observed to be associated with specific regions of the chromosomes: puffs, diffuse bands, and interbands, but not with condensed bands or the nucleolus. The intensity of the cGMP-specific fluorescence varies greatly from location to location and shows no obvious correlation with interband or puff size. Heat shock dramatically alters the pattern of puffs and of RNA transcription along the chromosomes and is accompanied by the appearance of a number of heat-shock puffs whose appearance precedes the new synthesis of specific proteins. The most actively transcribed heat-shock puff is at 93D, and this region upon heat shock becomes intensely fluorescent when probed for cGMP (Spruill *et al.*, 1978). These observations suggest that cGMP may play a role in processes associated with transcriptional activity.

Evidence for more specific physiological roles for certain elements in the cyclic nucleotide system is provided by the phenotypes of mutants that have already been mentioned.

3.1 LEARNING

Drosophila larvae and adults are capable of modifying their behavior in response to natural or experimental stimuli. Learning can be demonstrated in a number of different laboratory paradigms designed to measure associative or nonassociative types of learning. Experience-dependent behavioral modifications that also reflect learning can be observed during the courtship routine.

Quinn *et al.* (1974) developed an associative learning paradigm to train populations of flies to associate, and then to subsequently avoid, electric shock coupled with one of two odors (A and B). The learning index, Λ, for the population is defined as the fraction of the population avoiding the shock-associated odor minus the fraction avoiding the other odor (averaged for different populations trained to associate shock with odor A or odor B). Thus, perfect learning would have a $\Lambda = 1$, and no learning would have a $\Lambda = 0$. In this particular paradigm only about 65% of normal flies actually learn: $\Lambda = (0.65 - 0.35) = 0.30$.

This paradigm was used to screen many populations of flies derived from single, mutagen-treated X chromosomes of the Canton-S wild-type strain. The first mutant strain identified in this way had a $\Lambda = 0.04 \pm 0.02$ (Canton-S: $\Lambda = 0.31 \pm 0.02$) and was named *dunce* (*dnc*^{1}) (Dudai *et al.*, 1976). Since then a number of other mutant strains have been identified in the same way (for recent reviews see Aceves-Pina *et al.*, 1983; Quinn and Greenspan, 1984). The rutabaga *mutant* was also discovered in this way: $\Lambda = 0.00 \pm 0.02$ (Livingstone *et al.*, 1984).

Learning mutants have been tested and compared with the parent Canton-S strain in a number of different paradigms. Aceves-Pina and Quinn (1979) showed that normal first and third instar larvae can learn to associate electric shock with one of two odors in a paradigm very similar to that described above, whereas *dnc*^{1} larvae fail to learn. Thus *dunce* affects the nervous systems of both larvae and adults.

Dudai (1979, 1983) has modified the paradigm discussed above to test for learning over much shorter time spans following the initial training. Normal flies were found to learn with high efficiency when tested only 30 s after training and to retain an appreciable memory of what was learned after 7 min. In contrast, *dunce* and *rutabaga* mutants were observed to have learned with a reduced efficiency when tested 30 s after training and to have forgotten completely when tested 7 min after training (Dudai, 1983). These results suggest that the learning defects of these mutants may be due to an inability to retain short-term memory.

Impaired learning in *dunce* and *rutabaga* has also been measured by Folkers (1982) in a paradigm associating a visual stimulus with severe shaking. The learning index in this paradigm for normal flies increases linearly with the number of training sessions, no plateau being attained even after 24 training sessions. Learning is detected in *dunce* and *rutabaga* mutant strains; however, the maximum learning index for rutabaga is attained after only 9 training sessions and for dunce after 16 sessions. At these points the mutants have learned only half as well as normal flies.

A memory defect in *dunce* and *rutabaga* mutants can also be demonstrated employing a positive reinforcement paradigm. Hungry flies can be trained to selectively choose an odor that has previously been associated with sucrose (Tempel *et al.*, 1983). In normal flies tested in this paradigm learning and memory consolidation requires about 100 min, and memory persists for as long as 24 h. The mutant *dunce* strain learns in this paradigm as well as normal flies when tested immediately after training but forgets 25 times as fast as normal. No memory of the training is evident after only 2 h. In contrast the mutant *rutabaga* strain performs, initially, only half as well as normal flies and shows no evidence of training after 1 h.

The above paradigms all measure learning as a parameter of a population of normal or mutant flies. Booker and Quinn (1981) have adopted the paradigm of Horridge (1962), developed with the cockroach, to study operant conditioning in individual flies. In this paradigm the leg of an immobilized fly serves as a switch, completing a circuit and causing an electric shock when extended and breaking the circuit when flexed (or vice versa). Normal flies learn to avoid shocks by breaking the circuit, but they learn rather inefficiently unless the head is removed. Headless normal flies learn very quickly to avoid shocks as do headless cockroaches. More than 90% of headless normal flies are observed to meet a fixed criterion for learning in this paradigm while only 25–45% of headless flies of genotypes dnc^1 and dnc^2 meet this criterion. Intact *dunce* flies also learn more poorly than intact normal flies. Interestingly, some individual *dunce* flies learn as well as do normal flies, but, as a group, *dunce* flies do less well than normal flies. The source of this variability is not known.

The previous sets of experiments demonstrate that flies can modify their behavior based on past experience and that mutations in the *dunce* and the *rutabaga* genes alter their ability to do so. It appears that the mutants can learn for short periods of time, but they are unable either to store or to retrieve this information for future use. These mutants also show abnormal learning patterns in nonassociative paradigms. Two types of nonassociative learning, habituation and sensitization, have been shown in *Drosophila* (Duerr and Quinn, 1982). When sugar water is applied to the foreleg, the fly extends its proboscis in search of food. Unless the fly receives food, the proboscis extension reflex will soon stop; this is habituation. Feeding sugar water to the fly upon stimulation will increase its responsiveness even when a neutral stimulus is applied to the foreleg; this is sensitization. Both *dunce* and *rutabaga* mutants habituate about half as well as wild-type flies. Sensitization, on the other hand, appears normal at 15 s, and after that time both mutants show a rapid decline in their responses compared to normal flies. By 60 s both mutants are significantly less sensitized than wild type. Thus *dunce* and *rutabaga* mutants are defective in both associative and nonassociative learning paradigms.

Mutants of *dunce* and *rutabaga* also interfere with experience-dependent behavioral modifications that normally occur during courtship. Courtship in *Drosophila* consists of a series of olfactory, visual, and auditory signals exchanged between males and females (for a detailed review see Quinn and

Greenspan, 1984). Naive mature males cannot discriminate between immature males and females, possibly because the immature male produces a pheromone similar to that of females. The naive mature male will attempt to court immature males, but after being rejected several times the mature male will no longer court an immature male. Both *dunce* and *rutabaga* males, on the other hand, will persist in courting immature males even after repeated rejections (Gailey *et al.*, 1982). Similarly, normal males, after having been rejected by fertilized females, will not court any female for 2–3 h. Mutant males, on the other hand, will continue to court fertilized females (Gailey *et al.*, 1984).

Courting males produce a characteristic song by vibrating their wings. Normal virgin females stimulated with an electronic version of this song are more receptive to mating than females that are not prestimulated. Homozygous *dunce* and *rutabaga* females are oblivious to the effects of such prestimulation. Heterozygous females, on the other hand, show some effect of prestimulation if they are tested immediately, but the effect is gone by 1 min (Kyriacou and Hall, 1984).

In all of the behavioral trails described here, the phenotypes of *dunce* and of *rutabaga* mutants are very similar, in spite of the fact that they affect the metabolism of cAMP in fundamentally different ways.

The effects of each of these mutations and of the double combination of the two mutations on enzyme activities and on cAMP level are compared with the normal in Fig. 11 (Livingstone *et al.*, 1984). In abdomens, the *dunce* mutant exhibits a markedly elevated cAMP level, but the *rutabaga* mutant does not exhibit a cAMP level significantly different than normal. The double mutant combination, however, exhibits an intermediate level of cAMP, suggesting that both mutants affect the cAMP level in at least one intracellular pool. However, the double mutant genotype exhibits virtually the same learning-defective phenotype as each of the mutants separately (Fig. 11). Moreover, flies of genotype dnc^{M11} rut/dnc^{+} rut also have a learning index near zero (Livingstone *et al.*, 1984). These observations suggest that, while both mutants are defective in learning due to a failure to modulate cAMP level, their effects on learning may not be mediated by a common mechanism.

The possibility that the learning defects of the two mutants do not share a common mechanism is also suggested by biochemical studies. Walter and Kiger (1984) have shown that the cAMP-specific form II phosphodiesterase coded by the *dunce* gene does not interact physically with, nor is its activity affected by, Ca^{2+}/calmodulin. On the other hand, the *rutabaga* gene specifically affects a Ca^{2+}/calmodulin-dependent adenylate cyclase activity. The primary physiological defect in *dunce* mutants, leading to their neurological phenotype, may well be due to a direct failure to regulate cAMP level in

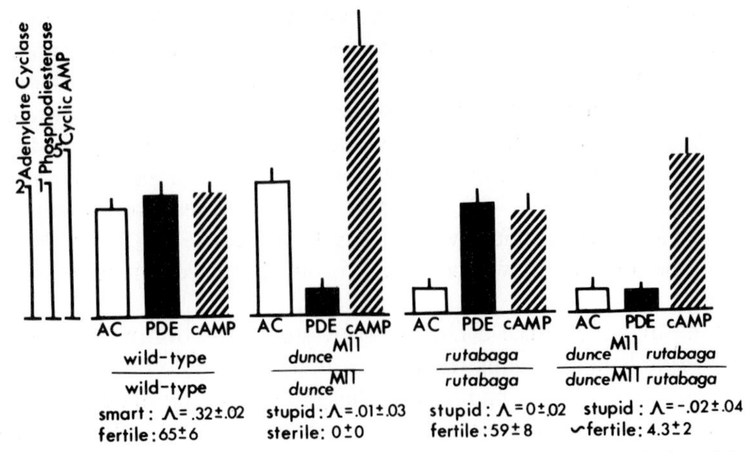

Fig. 11. Effects of different genotypes at the *dunce* and *rutabaga* loci on particulate abdominal adenylate cyclase (AC) activities (pmol cAMP formed min^{-1} mg $protein^{-1}$), soluble abdominal cAMP phosphodiesterase (PDE) activities (nmol cAMP hydrolyzed min^{-1} mg $protein^{-1}$), and endogenous abdominal cAMP content (pmol mg $protein^{-1}$). Λ is the learning index. Fertility is the number of progeny per female per 10 days. (From Livingstone *et al.*, 1984, reproduced with permission. Copyright M.I.T.)

neurons. The biochemical defect of *rutabaga* suggests that its neurological phenotype may be due to a failure to mediate a Ca^{2+} signal and to increase the synthesis of cAMP. Presynaptic facilitation, mediated by cAMP, in neurons is accompanied by Ca^{2+} influx (Kandel and Schwartz, 1982). The two mutants may affect different pools of cAMP in neurons or even affect entirely different neurons.

3.2 FEMALE FERTILITY

Mutations of the dunce gene (with the exception of dnc^1 cf^+) severely reduce female fertility. Superficially, oogenesis appears normal. Females produce mature oocytes but lay fewer than normal eggs. Those eggs that are laid appear normal although they seldom hatch. The *rutabaga* mutant, on the other hand, is fully fertile. The sterility of *dunce* females is seldom complete (Salz *et al.*, 1982). The majority of females homozygous for the different *dunce* alleles produce no progeny, and those that do produce only a few (Table 6). The variability in female fertility seems to be dependent on genetic background.

The dnc^1 allele is an exception. The sterility of the dnc^1 chromosome is due to a closely linked mutation, cf^1, as discussed in a previous section (Salz and Kiger, 1984). The sterility of dnc^1 cf^1 females can be suppressed by a distantly

TABLE 6

Effects of *Su(fs)* on the female fertility of different *dunce* alleles[a,b]

*dnc**	*dnc*/dnc**	*dnc* Su(fs)/dnc* Su(fs)*	*dnc¹cf¹ Su(fs)/dnc**	*dnc* Su(fs)/dnc¹cf¹*	*dnc*/dnc¹cf¹*
dnc^1cf^1	6/30 (5 ± 2)	48/48 (59 ± 3)	29/30 (74 ± 3)	29/30 (74 ± 3)	6/30 (5 ± 2)
$dnc^{M14}cf^+$	0/36	12/39 (5 ± 2)	62/62 (35 ± 1)	32/51 (31 ± 3)	2/19 (3 ± 2)
$dnc^{M11}cf^+$	0/33	11/24 (8 ± 4)	44/51 (39 ± 2)	58/91 (29 ± 2)	8/23 (10 ± 3)
dnc^2cf^+	2/44 (11 ± 6)	19/36 (26 ± 4)	50/50 (48 ± 2)	70/97 (40 ± 2)	9/25 (19 ± 5)
dnc^{CK}	1/21 (2)	31/55 (9 ± 1)	49/49 (39 ± 1)	66/99 (10 ± 1)	5/26 (7 ± 2)

[a] Modified from Salz *et al.* (1982).

[b] The data recorded are the number of females producing progeny over the total number of females tested. The numbers in parentheses are the average number of progeny produced by the nonsterile females in 6 days at 25°C. Confidence intervals are 1 SEM.

linked gene, *Su(fs)* (*Salz et al.*, 1982). This suppressor restores female fertility without affecting either the level of activity or the thermolability of the form II enzyme (Salz *et al.*, 1982; Salz, 1983). The effect of *Su(fs)* on the expression of the other *dunce* alleles is shown in Table 6. Only dnc^1 cf^1 is fully fertile in combination with *Su(fs)*, and from this observation it thus appears that the suppressor is specific for dnc^1 cf^1. Increases in female fertility observed for the other *dunce* alleles might well be attributed to changes in genetic background introduced in the crosses required to construct the new combinations of alleles shown in Table 6.

Suppression of the female sterility of the null-enzyme allele, dnc^{M11}, by *rutabaga* has been reported by Livingstone *et al.* (1984). The double mutant produces 4.3 ± 2 progeny per female per 10 days (Fig. 11). The degree of suppression reported is weak since normal females produce 65 ± 6 progeny per female per 10 days. This weak suppression could be due to the fact that the cAMP level in the double mutant combination is reduced compared to the cAMP level in dnc^{M11} (Fig. 11). However, using the same criterion for suppression as that employed in analyzing the data for *Su(fs)* (Table 6), the increase in fertility seen in the *dunce–rutabaga* double mutant would be attributed to a change in genetic background introduced during the crosses required to construct the combination. Clearly, the distinction between specific suppression by a single gene and phenotypic variation, due to nonspecific changes in genetic background, is a difficult distinction to make.

The question of whether *rutabaga* mutations interact with *dunce* mutations to affect the sterility of the latter females will have to be answered by different genetic techniques, e.g., the induction in *dunce* chromosomes of new suppressor mutations that can be shown to be new *rutabaga* alleles.

Severe female sterility is exhibited by those *dunce* mutations that severely reduce or eliminate form II phosphodiesterase activity (compare Tables 4 and 6). The exceptional dnc^1 mutation exhibits the mildest effect on form II activity of all the mutations and only causes female sterility when coupled with cf^1. The site identified by *cf* acts in cis to regulate in some way the expression of dnc^1 as it is involved in oogenesis. It is possible that the high cAMP level caused by deficiency in form II phosphodiesterase activity is a direct cause of female sterility. Elevated cAMP levels are associated with meiotic arrest in yeast (Matsumoto *et al.*, 1983), amphibians (Schorderet-Slatkine *et al.*, 1982), and mammals (Olsiewski and Beers, 1983). It is possible that meiotic arrest is the direct cause of female sterility in *dunce* mutants. On the other hand, the female sterility may also be a consequence of some defect in neural activity required for normal oogenesis. It is also possible that a common defect in membrane current regulation may underlie both the defects in learning and in oogenesis.

4 Conclusion

The goal of this review has been to point out the unique contributions that a combined biochemical and genetic analysis can make to an understanding of the physiological roles of cyclic nucleotides in higher organisms and to describe the results of such an analysis in *Drosophila*. *Drosophila* is particularly suitable for such an analysis, but many of the techniques applicable to *Drosophila* may soon be applicable to other organisms due to advances in recombinant DNA techniques, monoclonal antibodies, and the genetics of somatic cells. As should be evident from this review, much remains to be accomplished to complete the analysis in *Drosophila*. The mutants already available, along with the techniques discussed here, provide the tools for future progress.

Acknowledgments

We thank our collaborators over the years, and we thank our other colleagues who have communicated their results to us and given their permission to reproduce their data here. Hugo J. Bellen has made valuable comments on the manuscript. The clarity of the manuscript has been improved by the careful reading of Heidi M. Biedebach. The research of the authors has been supported by U.S. Public Health Service Grants GM21137 and GM07467.

References

Aceves-Pina, E. O., and Quinn, W. G. (1979). Learning in normal and mutant *Drosophila* larvae. *Science* **206**, 93–96.

Aceves-Pina, E. O., Booker, R., Duerr, J. S., Livingstone, M. S., Quinn, W. G., Smith, R. F., Sziber, P. P., Tempel, B. L., and Tully, T. P. (1983). Learning and memory in *Drosophila* studied with mutants. *Cold Spring Harbor Symp. Quant. Biol.* **48**, 831–839.

Alkon, D. L., Acosta-Urquidi, J., Olds, J., Kuzma, G., and Neary, J. T. (1983). Protein kinase injection reduces voltage-dependent potassium currents. *Science* **219**, 303–306.

Artavanis-Tsakonas, S., Muskavitch, M. A. T., and Yedvobnick, B. (1983). Molecular cloning of *Notch*, a locus affecting neurogenesis in *Drosophila melanogaster*. *Proc. Natl. Acad. Sci. U.S.A.* **80**, 1977–1981.

Ayala, F. J., and Kiger, J. A., Jr. (1980). "Modern Genetics," 1st Ed. Benjamin-Cummings, Menlo Park, California.

Bender, W., Spierer, P., and Hogness, D. S. (1983). Chromosomal walking and jumping to isolate DNA from the *Ace* and *rosy* loci and the bithorax complex in *Drosophila melanogaster*. *J. Mol. Biol.* **168**, 17–33.

Booker, R., and Quinn, W. G. (1981). Conditioning of leg position in normal and mutant *Drosophila*. *Proc. Natl. Acad. Sci. U.S.A.* **78**, 3940–3944.

Burgess, W. H., Jemiolo, D. K., and Kretsinger, R. H. (1980). Interaction of calcium and calmodulin in the presence of sodium dodecyl sulfate. *Biochim. Biophys. Acta* **623**, 257–270.

Byers, D. (1979). Ph.D. thesis. California Institute of Technology, Pasadena.

Byers, D., Davis, R. L., and Kiger, J. A., Jr. (1981). Defect in cyclic AMP phosphodiesterase due to the *dunce* mutation of learning in *Drosophila melanogaster*. *Nature* (*London*) **289**, 79–81.

Campos-Ortega, J. A., and Jimenez, F. (1980). The effect of X-chromosome deficiencies on neurogenesis in *Drosophila*. *In* "Development and Neurobiology of *Drosophila*" (O. Siddiqi, P. Babu, L. M. Hall, and J. C. Hall, eds.), pp. 201–222. Plenum, New York.

Carew, T. J., Hawkins, R. D., and Kandel, E. R. (1983). Differential classical conditioning of a defensive withdrawal reflex in *Aplysia californica*. *Science* **219**, 397–400.

Castellucci, V. F., Kandel, E. R., Schwartz, J. H., Wilson, F. D., Nairn, A. C., and Greengard, P. (1980). Intracellular injection of the catalytic subunit of cyclic AMP-dependent protein kinase simulates facilitation of transmitter release underlying behavioral sensitization in *Aplysia*. *Proc. Natl. Acad. Sci. U.S.A.* **77**, 7492–7496.

Davis, R. L., and Davidson, N. (1984). Isolation of the *Drosophila melanogaster dunce* chromosomal region and recombinational mapping of *dunce* sequences with restriction site polymorphisms as genetic markers. *Mol. Cell. Biol.* **4**, 358–367.

Davis, R. L., and Kauvar, L. M. (1984). *Drosophila* cyclic nucleotide phosphodiesterases. *In* "Advances in Cyclic Nucleotide and Protein Phosphorylation Research" (S. J. Strada and W. J. Thompson, eds.), Vol. 16, pp. 393–402. Raven, New York.

Davis, R. L., and Kiger, J. A., Jr. (1978). Genetic manipulation of cyclic AMP levels in *Drosophila melanogaster*. *Biochem. Biophys. Res. Commun.* **81**, 1180–1186.

Davis, R. L., and Kiger, J. A., Jr. (1980). A partial characterization of the cyclic nucleotide phosphodiesterase of *Drosophila melanogaster*. *Arch. Biochem. Biophys.* **203**, 412–421.

Davis, R. L., and Kiger, J. A., Jr. (1981). *Dunce* mutants of *Drosophila melanogaster*: Mutants defective in the cyclic AMP phosphodiesterase enzyme system. *J. Cell Biol.* **90**, 101–107.

Dudai, Y. (1979). Behavioral plasticity in a *Drosophila* mutant $dunce^{DB276}$ (dnc^2). *J. Comp. Physiol.* **130**, 271–275.

Dudai, Y. (1983). Mutations affect storage and use of memory differentially in *Drosophila*. *Proc. Natl. Acad. Sci. U.S.A.* **80**, 5445–5448.

Dudai, Y., Jan, Y., Byers, D., Quinn, W. G., and Benzer, S. (1976). *Dunce*, a mutant of *Drosophila* deficient in learning. *Proc. Natl. Acad. Sci. U.S.A.* **73**, 1684–1688.

Duerr, J. S., and Quinn, W. G. (1982). Three *Drosophila* mutations that block associative learning also affect habituation and sensitization. *Proc. Natl. Acad. Sci. U.S.A.* **79**, 3646–3650.

Folkers, E. (1982). Visual learning and memory of *Drosophila melanogaster* wild type C-S and the mutants $dunce^1$, *amnesiac, turnip*, and *rutabaga*. *J. Insect Physiol.* **28**, 535–539.

Gailey, D. A., Jackson, R., and Siegel, R. W. (1982). Male courtship in *Drosophila*: The conditioned response to immature males and its genetic control. *Genetics* **102**, 771–782.

Gailey, D. A., Jackson, R., and Siegel, R. W. (1984). Conditioning mutations in *Drosophila melanogaster* affect an experience-dependent behavioral modification in courting males. *Genetics* **106**, 613–623.

Gilman, A. G. (1984). G proteins and dual control of adenylate cyclase. *Cell* **36**, 577–579.

Hawkins, R. D., Abrams, T. W., Carew, T. J., and Kandel, E. R. (1983). A cellular mechanism of classical conditioning in *Aplysia*: Activity-dependent amplification of presynaptic facilitation. *Science* **219**, 400–405.

Hershfield, M. S., and Kredich, N. M. (1978). *S*-Adenosylhomocysteine hydrolase is an adenosine-binding protein: A target for adenosine toxicity. *Science* **202**, 757–760.

Horridge, G. A. (1962). Learning of leg position by the ventral nerve cord in headless insects. *Proc. R. Soc. Ser. B* **157**, 33–52.

Ingebritsen, T. S., and Cohen, P. (1983). Protein phosphatases: Properties and role in cellular regulation. *Science* **221**, 331–338.

Kaczmarek, L. K., Jennings, K. R., Strumwasser, F., Nairn, A. C., Walter, J., Wilson, F. D., and Greengard, P. (1980). Microinjection of catalytic subunit of cyclic AMP-dependent protein kinase enhances calcium action potentials of bag cell neurons in cell culture. *Proc. Natl. Acad. Sci. U.S.A.* **77**, 7487–7491.

Kandel, E. R., and Schwartz, J. H. (1982). Molecular biology of learning: Modulation of transmitter release. *Science* **218**, 433–443.

Kauvar, L. M. (1982). Defective cyclic adenosine 3′: 5′-monophosphate phosphodiesterase in the *Drosophila* memory mutant *dunce*. *J. Neurosci.* **2**, 1347–1358.

Kiger, J. A., Jr. (1977). The consequences of nullosomy for a chromosomal region affecting cyclic AMP phosphodiesterase activity in *Drosophila*. *Genetics* **85**, 623–628.

Kiger, J. A., Jr., and Golanty, E. (1977). A cytogenetic analysis of cyclic nucleotide phosphodiesterase activities in *Drosophila*. *Genetics* **85**, 609–622.

Kiger, J. A., Jr., and Golanty, E. (1979). A genetically distinct form of cyclic AMP phosphodiesterase associated with chromomere 3D4 in *Drosophila melanogaster*. *Genetics* **91**, 521–535.

Kiger, J. A., Jr., Davis, R. L., Salz, H., Fletcher, T., and Bowling, M. (1981). Genetic analysis of cyclic nucleotide phosphodiesterases in *Drosophila melanogaster*. *Adv. Cyclic Nucleotide Res.* **14**, 273–288.

Kyriacou, C. P., and Hall, J. C. (1984). Learning and memory mutations impair acoustic priming of mating behavior in *Drosophila*. *Nature* (*London*) **308**, 62–64.

Lindsley, D. L., Sandler, L., Baker, B. S., Carpenter, A. T. C., Denell, R. E., Hall, J. C., Jacobs, P. A., Miklos, G. L. G., Davis, B. K., Gethmann, B. C., Hardy, R. W., Hessler, A., Miller, S. M., Nozawa, H., Parry, D. M., and Gould-Somero, M. (1972). Segmental aneuploidy and the genetic gross structure of the *Drosophila* genome. *Genetics* **71**, 157–184.

Livingstone, M. S., Sziber, P. P., and Quinn, W. G. (1984). Loss of calcium/calmodulin responsiveness in adenylate cyclase of *rutabaga*, a *Drosophila* learning mutant. *Cell* **37**, 205–215.

Matsumoto, K., Uno, I., and Ishikawa, T. (1983). Initiation of meiosis in yeast mutants defective in adenylate cyclase and cyclic AMP-dependent protein kinase. *Cell* **32**, 417–423.

Mohler, J. D. (1977). Developmental genetics of the *Drosophila* egg. Identification of 59 sex-linked cistrons with maternal effects on embryonic development. *Genetics* **85**, 259–272.

Munjaal, R. P., Chandra, T., Woo, S. L. C., Dedman, J. R., and Means, A. R. (1981). A cloned calmodulin structural gene probe is complementary to DNA sequences from diverse species. *Proc. Natl. Acad. Sci. U.S.A.* **78**, 2330–2334.

Neer, E. J. (1979). Interaction of soluble brain adenylate cyclase with manganese. *J. Biol. Chem.* **254**, 2089–2096.

Nestler, E. J., and Greengard, P. (1983). Protein phosphorylation in the brain. *Nature* (*London*) **305**, 583–588.

Olsiewski, P. J., and Beers, W. H. (1983). cAMP synthesis in the rat oocyte. *Dev. Biol.* **100**, 287–293.

Quinn, W. G., and Greenspan, R. J. (1984). Learning and courtship in *Drosophila*: Two stories with mutants. *Annu. Rev. Neurosci.* **7**, 67–93.

Quinn, W. G., Harris, W. A., and Benzer, S. (1974). Conditioned behavior in *Drosophila melanogaster*. *Proc. Natl. Acad. Sci. U.S.A.* **71**, 708–712.

Ross, E. M., and Gilman, A. G. (1980). Biochemical properties of hormone-sensitive adenylate cyclase. *Annu. Rev. Biochem.* **49**, 533–564.

Salz, H. K. (1983). Ph.D. thesis. University of California, Davis.

Salz, H. K., and Kiger, J. A., Jr. (1984). Genetic analysis of chromomere 3D4 in *Drosophila melanogaster*. II. Regulatory sites for the dunce gene. *Genetics* **108**, 377–392.

Salz, H. K., Davis, R. L., and Kiger, J. A., Jr. (1982). Genetic analysis of chromomere 3D4 in *Drosophila melanogaster*: The *dunce* and *sperm-amotile* genes. *Genetics* **100**, 587–596.

Savvateeva, E. V., and Kamyshev, N. G. (1981). Behavioral effects of temperature sensitive mutations affecting metabolism of cAMP in *Drosophila melanogaster*. *Pharmacol. Biochem. Behav.* **14**, 603–611.

Savvateeva, E., Labazova, I., and Korochkin, L. (1982). Isozymes of cyclic nucleotide phosphodiesterase in *Drosophila* temperature-sensitive mutants with impaired metabolism of cAMP. *Dros. Inf. Serv.* **58**, 133.

Schorderet-Slatkine, S., Schorderet, M., and Bauleu, E. E. (1982). Cyclic AMP mediated control of meiosis: Effects of progesterone, cholera toxin and membrane-active drugs in *Xenopus laevis* oocytes. *Proc. Natl. Acad. Sci. U.S.A.* **79**, 850–854.

Shotwell, S. L. (1983). Cyclic adenosine 3′: 5′-monophosphate phosphodiesterase and its role in learning in *Drosophila*. *J. Neurosci.* **3**, 739–747.

Siegelbaum, S. A., Camardo, J. S., and Kandel, E. R. (1982). Serotonin and cyclic AMP close single K^+ channels in *Aplysia* sensory neurones. *Nature (London)* **299**, 413–417.

Solti, M., Dévay, P., Kiss, I., Londesborough, J., and Friedrich, P. (1983). Cyclic nucleotide phosphodiesterases in larval brain of wild type and *dunce* mutant strains of *Drosophila melanogaster*: Isoenzyme pattern and activation by Ca^{2+}/calmodulin. *Biochem. Biophys. Res. Commun.* **111**, 652–658.

Spruill, W. A., Hurwitz, D. R., Lucchesi, J. C., and Steiner, A. L. (1978). Association of cyclic GMP with gene expression of polytene chromosomes of *Drosophila melanogaster*. *Proc. Natl. Acad. Sci. U.S.A.* **75**, 1480–1484.

Stewart, B., and Merriam, J. R. (1973). Segmental aneuploidy of the X-chromosome. *Dros. Inf. Serv.* **50**, 167–170.

Stewart, B., and Merriam, J. R. (1975). Regulation of gene activity by dosage compensation at the chromosomal level in *Drosophila*. *Genetics* **79**, 635–647.

Sugden, P. H., and Corbin, J. D. (1976). Adenosine 3′: 5′-cyclic monophosphate-binding proteins in bovine and rat tissues. *Biochem. J.* **159**, 423–437.

Taylor, S. S., Kerlavage, A. R., Zoller, M. J., Nelson, N. C., and Potter, R. L. (1981). Nucleotide-binding sites and structural domains of cAMP-dependent protein kinases. *In* "Protein Phosphorylation" (O. M. Rosen and E. G. Krebs, eds.), pp. 3–18. Cold Spring Harbor Laboratory, Cold Spring Harbor, New York.

Tempel, B. L., Bonini, N., Dawson, D. R., and Quinn, W. G. (1983). Reward learning in normal and mutant *Drosophila*. *Proc. Natl. Acad. Sci. U.S.A.* **80**, 1482–1486.

Tsuzuki, J., and Kiger, J. A., Jr. (1974). A simple method for density gradient zonal electrophoresis with imidazole-glycine buffer and its application to a study of cyclic AMP-binding protein. *Prep. Biochem.* **4**, 283–294.

Tsuzuki, J., and Kiger, J. A., Jr. (1975). Cyclic AMP-binding proteins in early embryos of *Drosophila melanogaster*. *Biochim. Biophys. Acta* **393**, 225–235.

Tsuzuki, J., and Kiger, J. A., Jr. (1978). A kinetic study of cyclic adenosine 3′: 5′-monophosphate binding and mode of activation of protein kinase from *Drosophila melanogaster* embryos. *Biochemistry* **17**, 2961–2970.

Ueland, P. M., and Døskeland, S. O. (1977). An adenosine 3′: 5′-monophosphate-adenosine binding protein from mouse liver. *J. Biol. Chem.* **252**, 677–686.

Ueland, P. M., and Døskeland, S. O. (1978). An adenosine 3′: 5′-monophosphate-adenosine binding protein from mouse liver. *J. Biol. Chem.* **253**, 1667–1676.

Uzzan, A., and Dudai, Y. (1982). Aminergic receptors in *Drosophila melanogaster*: Responsiveness of adenylate cyclase to putative neurotransmitters. *J. Neurochem.* **38**, 1542–1550.

Vincent, J. C., and Thellier, M. (1983). Theoretical analysis of the significance of whether or not enzymes or transport systems in structured media follow Michaelis-Menten kinetics. *Biophys. J.* **41**, 23-28.

Walter, M. F., and Kiger, J. A., Jr. (1984). The *dunce* gene of *Drosophila*: Roles of Ca^{2+} and calmodulin in adenosine 3′:5′-cyclic monophosphate-specific phosphodiesterase activity. *J. Neurosci.* **4**, 495–501.

Yamanaka, M. K., and Kelly, L. E. (1981). A calcium-calmodulin-dependent cyclic adenosine monophosphate phosphodiesterase from *Drosophila* heads. *Biochim. Biophys. Acta* **674**, 277–286.

Yamanaka, M., Tobin, S. L., Saugstad, J., and McCarthy, B. J. (1983). Transcriptional modulation during development of the *Drosophila* calmodulin gene. *J. Cell Biol.* **97**, 145a.

Yuh, K-C. M., and Tao, M. (1974). Purification and characterization of adenosine-adenosine cyclic 3′: 5′-monophosphate binding protein factors from rabbit erythrocytes. *Biochemistry* **13**, 5220–5226.

Note added in proof

An important erratum to the paper by Livingstone *et al.* (1984) appears in *Cell* **38**, 342 (1984).

We overlooked two papers on adenylate cyclase activity and one on learning performance:

Dudai, Y., Uzzan, A., and Zvi, S. (1983). Abnormal activity of adenylate cyclase in the *Drosophila* memory mutant rutabaga. *Neurosci. Lett.* 42, 207–212.

Dudai, Y., and Zvi, S. (1984). Adenylate cyclase in the *Drosophila* memory mutant rutabaga displays an altered Ca^{2+} sensitivity. *Neurosci. Lett.* **47**, 119–124.

Folkers, E., and Spatz. H.-Ch. (1984). Visual learning performance of *Drosophila melanogaster* is altered by neuropharmaca affecting phosphodiesterase activity and acetylcholine transmission. *J. Insect Physiol.* **30**, 957–965.

One paper, which appeared after the manuscript was completed, should also be noted:

Savvateeva, E. V., Peresleny, I. V., Ivanushina, V., and Korochkin, L. I. (1985). Expression of adenylate cyclase and phosphodiesterase in development of temperature-sensitive mutants with impaired metabolism of cAMP in *Drosophila melanogaster. Develop. Gen.* **5**, 157–172.

The Developmental Physiology of Color Patterns in Lepidoptera

H. Frederik Nijhout

Department of Zoology, Duke University, Durham, North Carolina

ADVANCES IN INSECT PHYSIOLOGY, VOL. 18

1 Introduction

Interest in biological pattern formation has experienced somewhat of a revival in the past 15–20 years, both with experimental biologists who have begun to obtain novel insights into many old and difficult problems, and with theoreticians who have been exploring the broader implications of Turing's ideas on the origins of pattern. Next to evolution, pattern formation is undoubtedly one of the most complex and difficult phenomena in biology, and model systems that have a simple morphology and that are easily manipulated are desirable. The color patterns of Lepidoptera are of interest in this regard because they are both structurally simple and exceedingly diverse in their morphology. The attractive feature of lepidopteran color patterns is that they develop in a tissue that is essentially a two-dimensional monolayer of cells in which there is no significant growth nor cell movement. Furthermore, the wing is a tissue that is nonessential for the proper development and survival of the individual so that it can be manipulated with a considerable amount of freedom. The majority of color patterns are patterns of melanin deposition. To make a melanin one needs but a single enzyme, presumably a single gene product, so that many color patterns may simply represent two-dimensional patterns of gene activation and enzyme synthesis. Spatially restricted biomolecular events are of considerable interest in developmental physiology. Yet in spite of their structural simplicity, color patterns are exceedingly diverse. The Lepidoptera are a fairly recent and monophyletic taxon and, next to the Coleoptera, perhaps the largest such taxon among living things. Their color patterns, therefore, present us with an unparalleled richness of material for comparative studies that can be used as a framework for the elucidation of the development and the evolution of pattern.

2 Wing Development

2.1 IMAGINAL DISKS AND VENATION

In view of the fact that the color pattern bears a very intimate relationship to the pattern of wing veins it will be worthwhile to consider in some detail how the wing and its venation pattern develop. Depending on the species the development of the wing imaginal disk begins either shortly before or sometime after the young larva hatches from the egg (Comstock, 1918; Köhler, 1932; Kuntze, 1935). The wing imaginal disk originates as a slight thickening followed by an invagination of the epidermis on either side of the meso- and metathoracic segments at a location that corresponds to the position of the spiracles on the other body segments (Kuntze, 1935). A tracheal loop that branches off the main lateral trachea becomes associated with the tip of the invagination (Fig. 1a) and, in later stages, this tracheal loop will give rise to the tracheal system of the wing. In the course of larval life the wing imaginal disks gradually enlarge by periods of cell division associated with each larval molt and by the enlargement of the cells of the inner (proximal) face of the invaginated pocket (Fig. 1b). Throughout their development, the walls of the imaginal disk and the future wing, like the epidermis with which they are always continuous, remain only one cell thick.

During or shortly after the molt to the penultimate instar (in *Samia cynthia*) or antepenultimate instar (in *Ephestia kühniella*), an imaginal disk

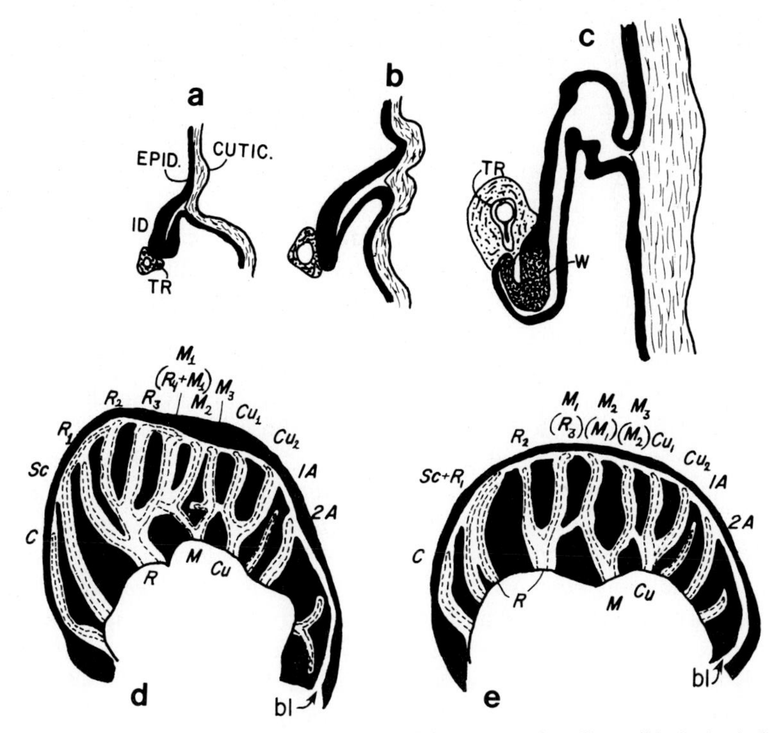

Fig. 1. Early stages of wing development in *Philosamia cynthia* (Saturniidae). (a–c) Cross sections of successive stages of wing imaginal disk development during early larval life. The portion that will form the wing proper is indicated in (c) by heavy stippling. Tr, Trachea from which wing vein tracheae will develop. (d and e) Wing imaginal disks of the forewing and hindwing, respectively, of the prepupa showing the primary venation system (see text). Lacunae are white; tracheae are outlined by dashed lines. Labels indicate the conventional nomenclature of the veins as follows: C, costa; Sc, subcosta; R, radius; M, media; Cu, cubitus; A anal veins; bl, bordering lacuna. Venation in parentheses refers to the name of the trachea of the vein in question when this differs from the conventional nomenclature for that vein. (Redrawn from Kuntze, 1935.)

undergoes a process of growth and folding which establish that Köhler (1932) has called the wing anlage. As Fig. 1c shows, the wing anlage arises as an invagination of the thicker proximal face of the wing imaginal disk in the vicinity of the tracheal attachment site. The wing anlage undergoes considerable growth during the penultimate instar. Late during that instar strands of tracheole-building cells grow out from the tracheal loop at the base of the wing anlage and begin to invade its lumen.

The pattern of venation and the shape of the adult wing become established during the last larval instar. During the growth phase of the final instar larva the wing anlage enlarges considerably and flattens into a crescent-shaped lamina (Fig. 1d,e). The basement membranes of the two cell layers that now make up the wing anlage become fused over the entire surface of the anlage and remain fused throughout the further development of the wing except in the regions where the future wing veins will develop. The sequence of events in the development of the incipient venation differs somewhat from species to species. In *S. cynthia* a branching pattern of tubular lacunae forms first, due to local separation of the basement membranes and filling of the resulting spaces with hemolymph. Soon after the lacunae are formed, and while their tips are still "growing" out toward the margin of the wing anlage, they are invaded by tracheal branches growing out from the main trachea at the base of the anlage (Kuntze, 1935). In *E. kühniella* this sequence is reversed (Köhler, 1932) in that tracheal branches that grow out from the basal trachea appear to penetrate between the basement membranes before any lacunae are formed. In pushing the basement membranes apart the tracheae thus appear to be instrumental in forming the lacunae whose branches subsequently continue to dilate somewhat as they fill with hemolymph. The manner in which initial pathfinding is accomplished is not known for either of the above cases. The initial phase of lacuna and trachea formation end when a bordering lacuna is formed (Fig. 1d,e). The bordering lacuna establishes the outline of the adult wing. All cells distal to this lacuna will degenerate during the pupal stage (Süffert, 1929a). The path of the bordering lacuna defines not only the overall shape and proportions of the adult wing but also the finer details of its morphology. The tails on the hindwings of swallowtails and tailed moths, for example, are laid out at this stage as an outward loop of the bordering lacuna.

The lacuna and tracheal systems that develop at this time, about the middle of the final instar, are referred to as the primary lacuna and the primary tracheal systems, respectively (Fig. 1d,e). The primary tracheal system is, at this time, still fluid filled and will not fill with air and become functional until the larva pupates. The term primary tracheal system is therefore synonymous with pupal tracheal system.

The elements of the Comstock–Needham system for nomenclature and homology in the wing venation system of insects are clearly discernible, even at this very early stage of development (Fig. 1d,e). In the forewing the costa, subcosta, radius (five-branched), media (two-branched), cubitus (two-branched), and anal veins each arise from a single stem and each receives an independent trachea (the exception being R_4 which also receives the anterior branch of the media trachea and is, by convention, referred to as M_1, not R_4). The situation in the hindwing is similar except that the tracheae of the subcosta and R_1 lie in a common lacuna and the radius lacuna has two branches instead of five. While the homologies between the tracheal and lacunal systems of fore and hindwing appear to be straightforward there is nevertheless an unsatisfactorily resolved convention that designates the posterior branch of the radius (R_3) as M_1 and the only two branches of the media as M_2 and M_3, respectively (see Enderlein, 1902; Comstock, 1918, for discussions). Toward the end of the growth phase of the final instar larva, the base of the media and the 1A lacunae constrict considerably so that their diameter becomes resticted to that of the tracheae that they enclose. These portions of the lacunae in effect disappear (Fig. 5). The bordering lacuna also disappears and its former course is difficult to find except after pupation when the portion of the pupal wing that will degenerate can be recognized by a clear line of demarcation on the pupal cuticle (Süffer, 1929a; Fig. 9a).

The growth phase of the final instar larva ends shortly after ecdysone (the molting hormone) secretion begins. In Lepidoptera there are two temporally and functionally distinct periods of ecdysone secretion prior to pupation. The first of these provokes a number of behavioral changes in the larva, which now ceases to feed, voids the contents of its gut, begins to wander about for a more or less prolonged period of time in "search" of a proper site for pupation, and in species that do so, spins a cocoon (Nijhout and Williams, 1974; Lounibos, 1974). This first period of ecdysone secretion also appears to be instrumental in the switchover of the developmental "program" from that of a larva to that of a pupa, in that certain larva-specific genes are turned off while certain pupa-specific ones are turned on (Riddiford and Truman, 1978; Willis, 1974; Nijhout and Wheeler, 1982). The second period of ecdysone secretion generally occurs sometime after the larva has established itself in a suitable site for pupation. This second period of ecdysone secretion provokes the actual molt to the pupal stage (for a description of physiology of molting and metamorphosis in Lepidoptera, see Riddiford and Truman, 1978). Soon after ecdysone secretion begins the epidermis separates from the larval cuticle (*apolysis*). Depending on the species the prepupal stage may last 1–8 days. In most species color pattern *determination* begins at about this time.

Shortly after apolysis the wing anlage and remainder of the imaginal discs are evaginated by simply being "pushed out" of the slit-like opening remaining after invagination of the imaginal disk earlier in larval life. The tissues of the imaginal disk that are not part of the wing anlage now become part of the thoracic integument and wing hinge, while the wing anlage itself slips as a flat flap between the epidermis and the old larval cuticle. In this confined space the wings now undergo a period of extremely rapid growth during which they become severely crumpled. Upon pupation the wing expands, becomes smoothed, and the primary tracheae fill with air.

About the middle of the pupal stage a new set of tracheae branch from the main tracheal loop at the base of the wing and begin to penetrate the lacunae. These tracheae are the secondary tracheal system which will provide the air supply for the adult wing. The secondary tracheae fill with air about the time that pigment begins to develop in the wing, usually a few days before adult eclosion (Kuntze, 1935). The primary tracheae remain intact until adult eclosion when they are simply torn and fragmented as the adult wing expands. Pieces of primary trachea remain as debris in the adult wing veins (Kuntze, 1935). The venation system of the adult wing thus corresponds to the pattern of secondary tracheation and of the lacunae through which these tracheae run. It is an abbreviated venation system relative to that of the larval wing anlage because it lacks the base of the media and first anal vein. The large cell that arises between the bases of the radius and cubitus when the media lacuna disappears is called the *discal wing cell.* In many species one or more crossveins develop that close off the discal wing cell distally. These discal crossveins as a rule arise as secondary lacunae without tracheae (Köhler, 1932; Behrends, 1935).

After pupation, depending on the species, the pupa may enter a more or less prolonged period of diapause or may immediately continue development to the adult. Endocrinologically, the difference between these alternatives is the timing of a brief period of ecdysone secretion. In pupae that do not diapause, ecdysone secretion begins a few hours to several days after pupation, depending on the species. In species that diapause as pupae, by contrast, ecdysone secretion is delayed, usually until the animal has experienced a prolonged period of cold followed by a return to higher temperatures (Williams, 1947). In some species photoperiodic cues control the onset and termination of pupal diapause (Beck, 1968). Ecdysone triggers the continuation of development.

2.2 DEVELOPMENT OF THE SCALES

Ecdysone secretion is followed almost immediately by apolysis and by a period of mitosis in the wing epidermis. Epidermal mitoses serve two distinct purposes. First, they serve to increase the cell number and hence the potential surface area of the wing. Köhler (1932) has presented a detailed account of the timing and pattern of these mitoses in the wing epidermis of *Ephestia.* Second, a small subpopulation of epidermal cells undergoes two successive "differentiation divisions" that yield scale cells and scale socket cells. There appears to be, however, a considerable amount of species specificity in the pattern and timing of this second type of mitosis. Köhler (1932), Braun (1936), and Stossberg (1938) have shown that in *Ephestia* both types of mitoses occur simultaneously but that they can be easily distinguished from one another by the orientation of the spindle axes. The spindle axis of a normal epidermal cell always lies parallel to the surface of the wing, while the spindle axis of scale mother cells lies perpendicular to the wing surface. The two cell types are otherwise indistinguishable. Soon after division the daughter cell that is produced toward the core of the wing degenerates. The daughter cell nearest the surface of the wing undergoes another mitotic division. This time the spindle axis is inclined about 45° to the surface of the wing and parallel to the long axis of the wing (Stossberg, 1938). The proximal daughter cell of this division becomes the scale cell while the distal daughter cell wraps around the former to form a socket. Approximately one out of every 4 cells in the pupal wing of *Ephestia* is a scale mother cell (Pohley, 1959). In *Precis* about one of every 15 cells of the wing epidermis becomes a scale cell (Nijhout, 1980b).

In butterflies an additional complication is added to this scheme in that the scale-building cells become arranged in perfect equidistant parallel rows running perpendicular to the long axis of the wing (Süffert, 1937; Nijhout, 1980b; Fig. 2a). In *Araschnia levana* and *Aglais urticae* the scale mother cells can be recognized early on by the fact that they are significantly larger than normal epidermal cells. They are, in turn, divided into subpopulations of two slightly different sizes. The larger ones become recognizable first, about 18–24 h after pupation. They are found to lie in rows, and within the row they alternate with the cells of the smaller size class which become recognizable about 6 h later (Süffert, 1937). The larger cells will subsequently give rise to the cover-scales which are about twice as long as, and as their name implies completely cover, the under-scales produced by the smaller cells. After the scale mother cells become enlarged each cell undergoes two cell divisions which appear to be equivalent to the two differentiation divisions described above for *Ephestia* (Süffert, 1937; Köhler and Feldotto, 1937). In *Precis coenia* no such mitotic period follows the appearance of enlarged scale-building cells (Nijhout, 1980b), so that it appears that in this species the differentiation divisions may precede the enlargement of the scale-building cells, as is the case in *Ephestia* (Stossberg, 1937, 1938). The enlargement of the scale-building cells appears to be associated with a polyploidization of their nuclei, and Henke and Pohley (1952) have reported that the size of the scale that will eventually be produced is proportional to the ploidy level of the scale-building cell. This would mean that the regular alternation of cover- and under-scales on butterfly wings arises from rows of cells whose ploidy

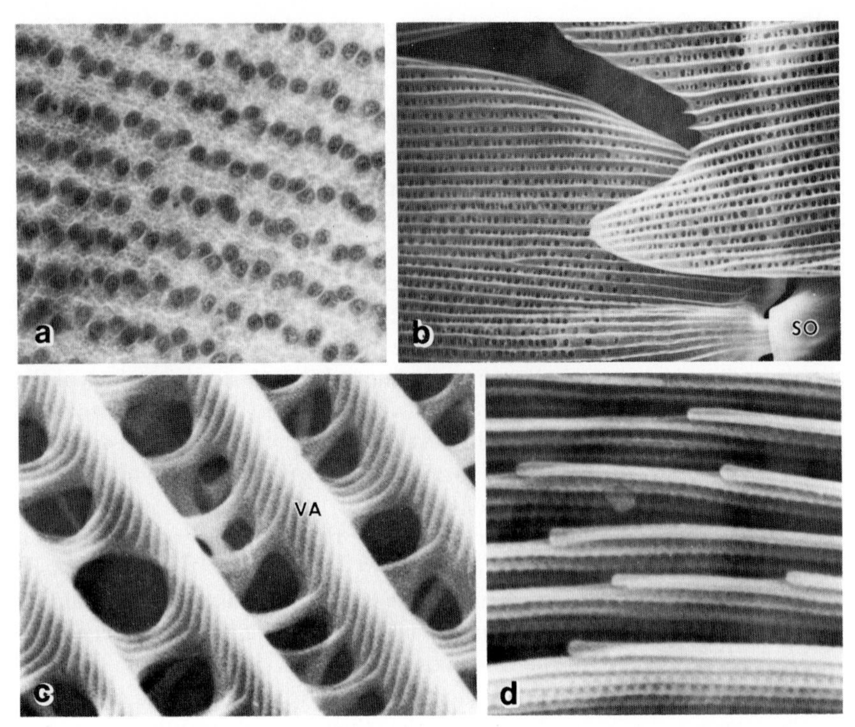

Fig. 2. (a) Light micrograph of rows of scale-forming cells in the early pupil wing of *Precia coenia* (from Nijhout, 1980b). (b) SEM of two scales of *Morpho rhetenor*, showing the longitudinal vanes and a socket (SO). (c) SEM of the vanes (VA) of a brown scale of *M. rhetenor*, also showing the supporting strutwork. (d) SEM of the vanes on an iridescent blue scale of *M. rhetenor*, showing the nearly horizontal striations that make up the interference reflector.

level alternates. Divisions of the unspecialized epidermal cells increases the surface area of the wing two- to fourfold. This increase in surface area is accommodated by a fine corrugation of the wing surface (Nijhout, 1980b). Wing expansion upon emergence of the adult is accomplished by smoothing out this wrinkled surface.

About the middle of adult development each scale-building cell begins to send out a finger-like process directed toward the distal margin of the wing. This process gradually lengthens and then flattens into a spatulate lamina, connected to the scale-building cell by a narrow neck (Stossberg, 1937; Köhler and Feldotto, 1937). This cytoplasmic lamina then acquires a very complex extracellular cuticle. The cuticle of the outer surface develops a large number of evenly spaced parallel ridges or vanes (Fig. 2b,c), which in iridescent scales may have an exceedingly complex structure (Nijhout, 1981; Fig. 2d). Within the lamina there develops a system of evenly spaced struts that are eventually connected to the outer and inner cuticles of the lamina and provide a trusslike support. Very little is known about the cell biology of scale differentiation. A number of descriptive studies have been published by Köhler and Feldotto (1937), Stossberg (1938) Süffert (1938), Kühn (1946), and others but these address only the gross morphology, developmental timing, and comparative structure of the scales. Little is known about how the structure of the scale cuticle is controlled in such a delicate manner as to yield the dramatic structural colors of many butterflies though Ghiradella and Radigan (1976) have presented a quasi-theoretical model that may explain their formation. Paweletz and Schlote (1964), Overton (1966), Ghiradella (1974), and Greenstein (1971) have described the distribution of microfilaments and microtubules in developing scales. The distribution of these cytoskeletal elements coincides with the pattern of vanes on the scale and it has been suggested that development of these vanes may simply be a response to mechanical stress. Little is known, however, about how these cytoskeletal elements could set up the appropriate stresses (Locke, 1967; Ghiradella, 1974), and whether or not they also affect other aspects of scale micromorphology. Nothing appears to be known about the way in which the dense internal strutwork of the scale—which is topologically extracellular and must therefore involve the development of holes through the cell—becomes established. Shortly prior to eclosion of the adult, about the time that pigment synthesis begins in the scales, the scale-building cells and the epidermal cells of the wing degenerate.

3 Sources of color

The color pattern on the wings of Lepidoptera resides exclusively in the scales. The cuticle of the wing is colorless or brownish and transparent. Epidermal cells are absent in the adult wing. As a rule the color pattern resides in a subpopulation of scales, the cover-scales. Although the underscales bear pigment in most cases, they are usually much less intensely colored than the cover-scales. The color of a scale may arise either from chemical pigments or from extremely fine physical structures in the scales whose dimensions are in the neighborhood of the wavelengths of visible light.

3.1 STRUCTURAL COLORS

Most whites, almost all blues, and all iridescent (metallic) colors on lepidopteran wings are structural. The greens of most butterflies are also structural but the greens of several species of moths have been shown to be due to (as yet unidentified) pigments (Ford, 1945). The work of Mason (1926,

1927a,b) remains one of the most thorough accounts of the origin of structural colors in insets.

Whites. Mason (1926) points out that white is observed whenever a colorless cuticle presents many small and irregular reflecting surfaces. Hence, unless special provisions are made, lepidopteran scales will appear white even in the absence of white (colorless) chemical pigments (technically, of course, all white pigments are "structural whites" as well). Whether a white is chalky, pearly, or metallic depends on the intensity and the amount of scattering of the reflected light (Mason, 1926).

Scales become "colored" either by the incorporation of a chemical pigment that absorbs specific wavelengths of visible light, or by developing nonrandom reflecting surfaces so that reflected light experiences destructive interference at certain wavelengths. Occasionally both a chemical pigment and a structural color are found in the same scale. Metallic violets, for instance, appear always to be due to iridescent blue over a red pigment in the same scale. Also, since iridescent scales owe their appearance to reflected light they are always darkly melanized. The melanin absorbs most of the transmitted light, preventing backscatter, and making the structural color appear very brilliant.

Iridescent colors. Two structurally different types of iridescent scales have been described in Lepidoptera. They are referred to as the *Urania* type and the *Morpho* type, respectively, after the genera in which they were first studied. No species is known to possess both types of iridescent scales.

The Urania type. In these scales the upper surface of the lamina is composed of multiple thin films so that the whole surface of the lamina serves as an interference reflector. This type of scale has been found in *Urania* and *Chrysiridia* (Uraniidae), in *Ornithoptera* and some *Papilio* (Papilionidae), in *Aenea* and *Limenitis* (Nymphalidae), and in some Lycaenidae (Süffert, 1924; Mason, 1927a).

The Morpho type. In this type of scale the source of structural coloration does not reside in the lamina but only in the vanes that run along the upper surface of the lamina (Fig. 2d). These vanes bear fine striations, inclined slightly toward the base of the scale, and light reflecting from these striations experiences destructive interference at certain wavelengths. Scales of this type are found in *Morpho* (Morphidae), *Apatura* and *Callicore* (Nymphalidae), and *Ancyluris* (Nemeobiidae) (Mason, 1927a; Anderson and Richards, 1942; Lippert and Gentil, 1959; Nijhout, 1981). The UV-reflecting scales of *Eurema* (Pieridae) have been shown to be of the *Morpho* type but with vannal striations so fine as to produce constructive interference at wavelengths between 330 and 350 nm (Ghiradella *et al.*, 1972). The spectrum of reflected light is quite broad in both the *Morpho* and the *Urania* type scales, even though the reflected colors may appear quite pure to the unaided eye.

Morpho scales, for instance, which appear to be pure blue from a distance, can be shown upon microscopical examination to reflect flecks of yellowish-green through purple from various portions of the vanes. Also, *Morpho* scales appear to be even more brilliant in the UV than they are in the blue region of the spectrum (R. E. Silberglied, personal communication).

The standard method for demonstrating the presence of a structural color is to replace the air in and around the scales by a liquid with an index of refraction that approximates that of the chitin in the scales (~ 1.55). Structural colors then vanish but reappear, completely unaffected, when the liquid is allowed to evaporate.

3.2 CHEMICAL COLORS

The majority of chemical pigments (zoochromes) that have been found in the wings of Lepidoptera belong to four categories: melanins, pterins, flavonoids, and ommochromes (Fig. 3). Several bile pigments have also been found. In addition there are a number of unidentified zoochromes that have been detected in singular instances and which may not belong to any of the above categories. In the sections below I shall give a brief account of the origin, distribution, and methods for characterization of each of the major pigment categories.

Melanins. All black pigments and the vast majority of brown pigments in the wings of Lepidoptera are melanins. Melanins are formed from orthodiphenols and indoles by the action of enzymes known as tyrosinases (properly: *o*-diphenol:O_2 oxidoreductases). The amino acid tyrosine is the most common precursor for melanin synthesis (Nicolaus, 1968; Needham, 1974; Richards, 1951; Blois, 1978), although the wing scale tyrosinases of the Buckeye butterfly, *Precis coenia*, cannot use tyrosine as a substrate for melanogenesis (Nijhout, 1980b). Melanins occur in a variety of colors from black and brown (eumelanins) to red and yellow (erythro- and phaeomelanins). In *P. coenia*, for instance, three colors of melanin occur side by side on a single wing (Nijhout, 1980b). The chemical or structural differences that are responsible for the various colors of melanin are not well understood. Pure tyrosine is converted into a black melanin by tyrosinase but a mixture of tyrosine and certain sulfhydryl-containing compounds such as cysteine is converted to a red melanin (Riley, 1977). Prota and Thomson (1976) have suggested that the control over eumelanin vs phaeomelanin synthesis in mouse hair resides in the control over the cysteine content of the melanocytes. Inagami (1954) has shown that when tyrosinase acts on mixtures of dopa and 3-OH-kynurenine (3-OH-K) it is possible to obtain "melanin" that ranges in color from black through brown to red, depending on the amount of 3-OH-K in the mixture. Whether or not this latter mechanism is responsible for producing true red melanins is somewhat unclear because, in the presence of

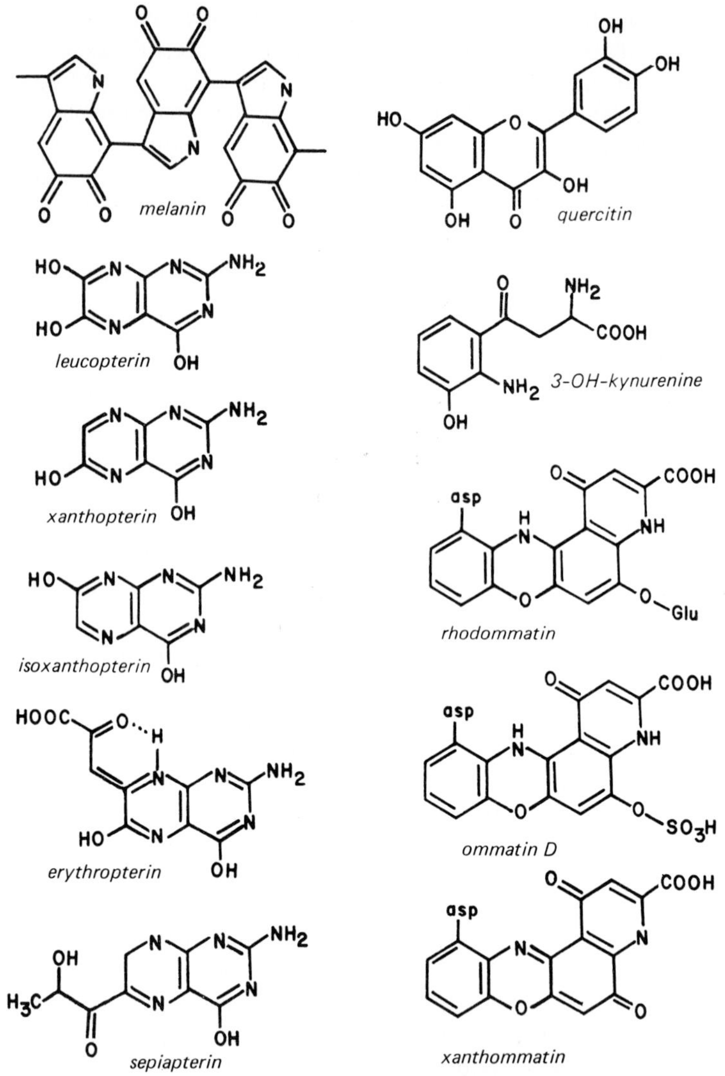

Fig. 3. Structural formulas for representatives of the major categories of wing pigments in Lepidoptera. Melanin is represented by a fragment of an indolic eumelanin. The identification of quercitin as the flavonoid in *Melanargia* (see text) should be considered tentative.

dopa and tyrosinase, 3-OH-K can be oxidized to the reddish-brown xanthommatin (Riley, 1977). Tyrosinases in the scales of *P. coenia* convert dopamine into a black melanin and *N*-actetyldopamine into a red melanin (Nijhout, 1980b), but there is no persuasive evidence that the red pigments produced in either of the above-mentioned cases in any way resemble real erythromelanins.

The problem in melanin identification lies in the fact that these compounds are virtually insoluble, indeterminately large, and, *in situ*, often complexed with proteins (Needham, 1974). Melanins are usually identified by their resistance to extraction and solubilization. Pure eumelanin is insoluble in dilute acids and alkalis but dissolves readily in hot alkali, in hot sulfuric acid, and in some organic solvents such as diethylamine and ethylene chlorhydrin (Lea, 1945; Needham, 1974). Erythromelanins and phaeomelanins are soluble in dilute HCl (Needham, 1974). The harsh extraction regimes required often destroy the melanin, which makes it difficult to accurately deduce the structure of natural melanins.

Pterins. Pterins are derived from guanosine triphosphate by a complex multibranched pathway shown in Fig. 4 (Needham, 1974; Descimon, 1977). Of the pterins that are found in lepidopteran wings leucopterin is colorless and acts as a structural white. The others range from pale yellow to red (Fig. 4). Although pterins are common in Lepidoptera, as wing pigments they

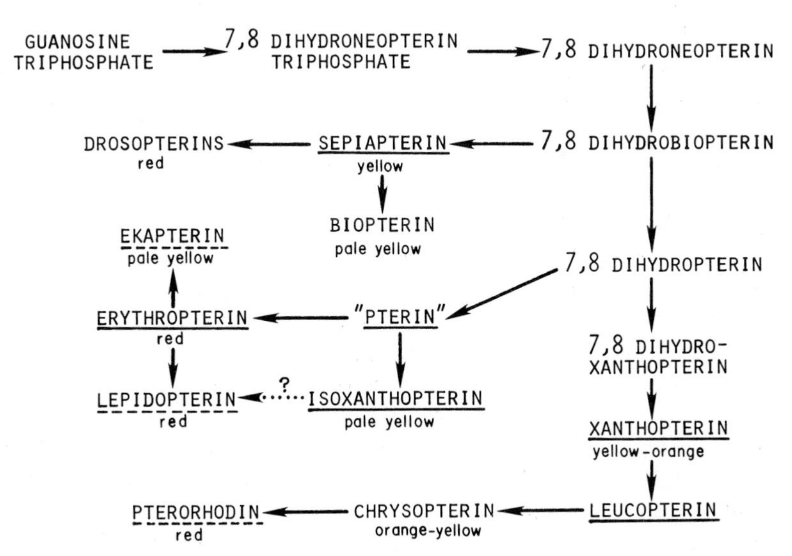

Fig. 4. Biosynthetic pathway for the pterins. Unless indicated, the compounds are colorless (structural whites). Compounds that are underlined are known to occur in lepidopteran wings. Dashed underlining denotes compounds that have been found in Lepidoptera but not as wing pigments. (After Needham, 1974; Descimon, 1977.)

have so far been found only in the Pieridae (Watt, 1964, 1967, 1974; Descimon, 1977) and Heliconiidae (Baust, 1967). Pterins are slightly soluble in water but are quite soluble in dilute acids and alkalis and can be readily extracted from wings in hot carbonate/bicarbonate buffer at pH 10 (Watt, 1964, 1967; Needham, 1974). Ford (1947) has described a simple modification of the murexide test whereby pterins may be demonstrated *in situ* in intact butterflies and moths. This test will detect pterins in single scales and makes it possible to distinguish the localization of pterins in different components of the wing pattern. To perform this test one simply exposes a specimen to molecular chlorine for 20 min followed by fuming with ammonia for a few minutes. Shortly after this treatment pterins begin to change to a purple color which intensifies over a period of hours. A more brilliant purple is obtained if the ammonia step is omitted but color development may take several days.

Flavonoids. Flavonoids are primarily plant pigments. They cannot be synthesized by animals which must obtain them from plant food (Ford, 1941; Needham, 1974). The flavonoids that occur in the wings of Lepidoptera range from white to yellow but their chemical structure has not yet been elucidated, the exception being the flavonoid in *Melanargia* (Satyridae), which appears to be quercitin (Fig. 3) (Thomson, 1926a,b). Ford (1941) has surveyed the occurrence of flavonoids in Lepidoptera and has shown that, while uncommon, they are widespread. Where they occur they are not alone responsible for the white or yellow colors. Flavonoids are common pigments in the subfamily Dismorphiinae of the Pieridae but are uncommon elsewhere in this family. They also occur in the Papilionidae, where they are characteristic of the Leptorcini (*Graphium, Eurytides, Lamproptera*), the genus *Parnassius*, and the New World *Parides* (Ford 1941, 1944a,b). Flavonoids have also been found in a few species of Satyridae and Riodinidae. Among the moths flavonoids have been found in the Sphingidae, Agaristidae, Agrotidae, and Castniidae (Ford, 1941) but this survey is far from exhaustive. The presence of flavonoids can usually be demonstrated in intact specimens by a nondestructive test described by Ford (1941). When wings are fumed with strong ammonia a yellow color is produced. This color change is unstable and the specimen rapidly returns to its original color. This reaction is highly sensitive and diagnostic and can be used to detect minute quantities of flavonoids in the wing scales. As a demonstration it works particularly dramatically on species of *Melanargia*. If the suspected flavonoid is itself yellow or when it is masked by other pigments it may be extracted with ethyl acetate (which dissolves flavonoids but not pterins) followed by reaction with sodium carbonate which yields a deep yellow color. The latter method is, however, much less sensitive than the former (Ford, 1941).

Ommochromes. Ommochromes are brown, red, or yellow pigments (Fig.

3) derived from tryptophan via the kynurenine pathway (Linzen, 1974; Needham, 1974). The ommochromes are by far the best studied of the insect pigments and, with exception of melanins, the most widespread. Their occurrence as wing pigments in Lepidoptera appears to be limited to the Nymphalidae (Linzen, 1974). Rhodommatin and ommatin D have been found in the wings of the European nymphalids: *Vanessa*, *Aglais*, *Inachis*, and *Argynnis*. In *Heliconius*, some reds have been shown to be di-H-xanthommatin, browns xanthommatin, and yellows 3-OH-kynurenine (Brown, 1981; L. E. Gilbert, personal communication). The occurrence and diversity of ommochromes in lepidopteran wings have not yet been systematically investigated, and the distribution mentioned above reflects in large part the limited attentions of a few investigators (but see Linzen, 1974, for documented absences of ommochromes in many species). Ommochromes are best characterized by their solubility properties. They are insoluble in water, organic solvents, and dilute acids, but are readily soluble in alkaline solutions and strong acids. They are most readily extracted by methanol containing 1–5% HCl (xanthommatin), by 0.1 *M* Na_2HPO_4, or, for wings in particular, by dilute ammonia (rhodommatin and ommatin D). After extraction they may be separated and characterized by circular paper chromatography, though with some difficulty (Linzen, 1974).

Bile pigments. Several true blue wing pigments are known. These are collectively referred to as "pterobilins" and two have been identified by Choussy and Barbier (1973) as phorcabilin 1 and sarpedobilin. These pigments have been demonstrated in the wings of Attacidae, Nymphalidae, Papilionidae, and Pieridae. Their most spectacular occurrence is in the genus *Nessaea* (Nymphalidae) where they occur in large sky-blue patches on the dorsal wing surfaces. The ventral wing surfaces of *Nessaea* are bright green, and this color has been shown to be due to a mixture of a blue bile pigment with an unidentified yellow pigment (Vane-Wright, 1979).

Other pigments. Ford (1942) has described five different red pigments from butterflies belonging to the Papilionidae (two pigments), Nymphalidae (one pigment), and Pieridae (two pigments). The two pierid pigments show a positive murexide test, suggesting that they are pterins, but one can be easily changed to a yellow pigment by HCl (and returned to red by NH_3) while the other cannot. Descimon (1975, 1977) has suggested that these two pigments may simply be erythropterin conjugated with proteins in different ways. The other three red pigments have not yet been identified. The yellow and white papiliochromes found in the wings of some Papilionidae by Umebachi and Takahashi (1956) are kynurenine derivatives complexed with a quinone and therefore could be consideed to belong to a collateral branch of the ommochrome pathway. The green pigments of Geometridae and Sphingidae (Ford, 1945) have not yet been identified but may provide to be bilins.

Cockayne (1924) has described the distribution among Lepidoptera of a number of pigments that show fluorescence *in situ.* No attempt has yet been made to identify or characterize these fluorescent compounds.

4 Morphology of the color pattern

4.1 DISTRIBUTION OF PIGMENTS AMONG SCALES

The color pattern on lepidopteran wings is a finely tiled mosaic of monochrome scales. Each scale is generally homogeneous in color and may contain but a single pigment. Because each scale is the product (appendage) of a single cell, this implies that during development, each scale-building cell becomes determined to deposit only one out of a wide array of possible colors. In reality, the repertoire of colors in any one species is fairly limited. The vast majority of Lepidoptera have only two or three colors of scales and even the most colorful of the butterflies seldom have more than five colors of scales. In *Precis coenia* (Nymphalidae), for instance, the colorful pattern on the dorsal forewing is composed exclusively of melanins (Nijhout, 1980b). There are three colors of melanin in this species: red, brown, and black. Judging from the dynamics of synthesis there is reason to believe that there may be two molecularly different browns (Nijhout, 1980b). The absolute amount of pigment in each scale may vary, so that the red melanin gives color not only to the red scales but also to the buff-colored scales, where it occurs in very low concentration. The brown melanin likewise occurs in at least three different "intensities" in different scales.

In the wing of *P. coenia* as well as in those of virtually every other species of lepidopteran, great variations in hues, and smooth gradients of transition from one color to another, are accomplished by variations in the ratios of the few scale colors characteristic for the species. Mixing of various colors of scales in precisely controlled proportions can lead to unexpected visual effects. The purple on the wing tips of *Hyalophora cecropia* (Saturniidae), for instance, is actually a mixture of roughly equal proportions of white, red, and black scales. Similarly, the "green" on the ventral hindwing of many Pieridae (e.g., *Colias, Euchloe*) contains no structural green but is simply a mixture of black and yellow scales.

While a given scale appears to be of a single homogeneous color it is not clear whether it likewise contains a single pigment, so that the color pattern is a mosaic of "primary" colors, or whether the color of a scale is itself achieved by a mixture of pigments. The melanins of *P. coenia* appear to be pure, in the sense that each scale synthesizes only a single species of melanin (Nijhout, 1980b). The yellow scales of *Colias eurytheme*, on the other hand, probably bear a mixture of pterins. While the yellow color in this species is due to xanthopterin, a significant amount of leucopterin and small quantities of erythropterin and sepiapterin are present as well (Watt, 1964). It is not at all clear, however, whether the minority pigments in this case are present as an accident of the biochemical pathway of pterin synthesis (Watt, 1972), or whether they represent a balanced (and presumably adaptive) mix designed to achieve a particular yellow chroma. The latter seems not an unlikely possibility considering the great range of species-specific tones of yellow and orange one finds in the Pieridae. The white scales of *Pieris* likewise contain sepiapterin, xanthopterin, and isoxanthopterin as minority pigments (Watt and Bowden, 1966) and the same considerations apply.

The fact that a given scale contains one or perhaps two major pigments, and the observation that adjacent scales often differ dramatically in the type of pigment they contain, raise questions as to how exactly a specific pigment becomes incorporated in a particular scale. Several possibilities exist. (1) A pigment could circulate freely in the hemolymph, having been synthesized in the fat body or absorbed from the food, and could be incorporated only by certain scale-building cells and not by others. This is likely to be the case for the flavonoids, for instance, which must be obtained from plants. In order to get a specific distribution of pigments it would then be necessary that only some scale-building cells, and not others possess an active transport mechanism for the accumulation of these pigments. (2) Scale-building cells could differ in the enzymes they synthesize so that only certain pigment synthetic pathways, or portions thereof, are enabled to operate. This appears to be the mechanism in *P. coenia.* In this species the enzymes for melanin synthesis (tyrosinases) can be demonstrated to be present in an active form in the scales long before melanin synthesis begins (Nijhout, 1980b). Pigment synthesis begins when substrates for melanogenesis are supplied via the circulatory system. Each of the melanins is synthesized in a highly stereotyped temporal pattern. Red melanin synthesis occurs first in the presumptive red scales followed some 6 h later by black melanin synthesis in the presumptive black scales. Brown melanin synthesis begins some 4–6 h after the initiation of black melanin synthesis. Each scale synthesizes only one of the melanins. Thus during the early stages of black melanin synthesis the red scales continue to manufacture only red pigment while the presumptive brown scales remain colorless without a trace of pigment synthesis. Evidently the enzymes contained in each scale type are different and have nonoverlapping affinities for the substrates employed to produce each of the melanins. Since melanins require but a single enzyme for their synthesis, it may well be that in this case the essential difference between the different colors of scales resides in the choice of one of four alternate enzymes. (3) Scale-building cells could differ in the type or the amount of substrate that they take up from the hemolymph. The work of Descimon (1966) and Watt (1974) on the control over color polymorphism in *Colias* suggests that this may be a mechanism for controlling alternate color morphs. Watt (1974) has shown that the outcome of the reactions of the pterin pathway (Fig. 4) depends strongly on the supply of precursors to the overall pathway. Changes in the supply of precursors alters the kinetic balance among the various reactions in this pathway because the enzyme, xanthine dehydrogenase, which oxidizes xanthopterin to leucopterin, is strongly inhibited by the precursor of xanthopterin, 7,8-dihydroxyxanthopterin. This "feed-forward" inhibition is believed to account for the fact that in the presence of large amounts of precursor one finds an accumulation of xanthopterin (yellow), while under low levels of precursor the major endproduct is leucopterin (colorless) because its synthesis is no longer inhibited. Although this mechanism has been proposed only to account for the overall color of the wing in the yellow/white polymorphism of *Colias*, it seems reasonable to suppose that control over precursor availability could be exercised at the level of the individual scale. Thus individual scale-building cells could control the outcome of their pterin pathway by the rate at which they take up precursors from the hemolymph. The possibility that at least some of the color specificity of the scales in *P. coenia* is likewise due to selective uptake of different substrates cannot be excluded at present.

4.2 DISTRIBUTION OF PIGMENTS IN THE WINGS

The color pattern we observe on the wing of a butterfly or moth comes about through the nonrandom distribution of pigments among the population of scales on the wing. The problem of pattern determination in this system is to discover how a given scale cell "decides" what color to make. As shown above, such a decision is manifested by either the synthesis or the uptake of a few specific molecules.

The problem can be approached at three different levels. First, we may attempt to elucidate the process or mechanism whereby determination occurs in an individual scale cell. This involves discovering the nature of the inducing signals from the environment and the manner in which these are transduced into specific pigment synthesis. Second, we may attempt to discover the basis for the local pattern within a single developmental field (which, as will be seen below, usually corresponds to a wing cell, the area bounded by two wing veins). This involves elucidating the nature of the organizing centers for the local pattern, their number and origin, and the manner in which they interact with each other and with the population of scale cells they influence. Third, we may attempt to develop an understanding of how the overall pattern is organized, of how development in the various fields is coordinated to form an integrated design. These three levels of resolution are similar, in fact, to the levels at which we attempt to study and explain any other developmental system, the principal difference being that the present system is morphologically extremely simple. The pattern we are studying here is one of chemicals (the pigments) distributed in a two-dimensional monolayer of cells. In the sections that follow I shall discuss the three points mentioned above in roughly reversed order, reflecting the sequence in which our understanding of color patterns and their development has evolved.

4.3 SCHWANWITSCH/SÜFFERT "NYMPHALID GROUNDPLAN"

The most comprehensive understanding of the gross morphology of lepidopteran color patterns was developed independently by Schwanwitsch (1924) and Süffert (1927) in the form of the "Nymphalid Groundplan" (Fig. 5). This groundplan represents a hypothetical pattern, from which the color patterns of extant Lepidoptera may be derived. We owe the term "Nymphalid Groundplan" to the work of Schwanwitsch (1924), who developed it to explain homologies in the wing patterns of the butterfly family Nymphalidae (*sensu lato*; i.e., all Rhopalocera except Pieridae, Papilionidae, Danaidae, and Hesperiidae). Süffert (1927, 1929b) and others (Henke, 1936; Sokolow, 1936; Henke and Kruse, 1941) subsequently showed that this groundplan is equally useful for the interpretation of the color patterns of virtually all the major families of butterflies and moths. Many families have unique pattern idiosyncrasies, to be sure, which are not easily accounted for by the elements of the Nymphalid Groundplan; nevertheless, the Nymphalid Groundplan has proven to be a robust model for understanding the gross morphology of color patterns in the majority of Lepidoptera. The elements of the Schwanwitsch/Süffert Nymphalid Groundplan are briefly described below. As before

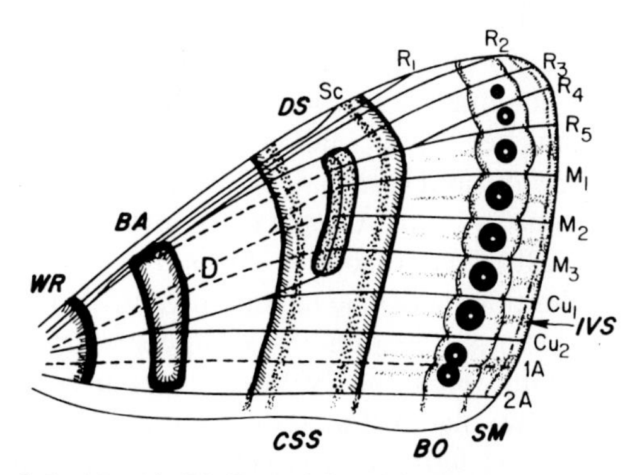

Fig. 5. Elements of the Nymphalid Groundplan. BA, Basal symmetry system; BO, border ocelli; CSS, central symmetry system; D, discal wing cell; DS, discal spot; IVS, intervenous stripe; SM, submarginal bands; WR, wing root band. Venation pattern shown is the secondary pattern (see text). Names of veins as in Fig. 1. (After Schwanwitsch, 1924; Süffert, 1927.)

(Nijhout, 1978, 1981), I have adopted a (free) translation of Süffert's terminology for the elements of the groundplan, because his terminology has always struck me as more functional that the one used by Schwanwitsch. The synonyms of Schwanwitsch are indicated for each pattern element.

The Discal Spot (*Discalis I* of Schwanwitsch, 1924). This is a pigmented spot or stripe that lies on the crossveins that close off the discal cell (Fig. 5). It is present in the vast majority of species and serves as a convenient landmark for finding the other elements of the color pattern. In some species the discal spot is highly elaborated into an eyespot.

The Central Symmetry System (*Media I and II*, and *Granulata I and II* of Schwanwitsch, 1924). The central symmetry system consists of a set of pigmented bands, one on each side of the discal spot, that run from the anterior to the posterior margin of the wing. The central symmetry system derives its name from the fact that the pigment composition of these two bands is symmetric along a line that passes through the discal spot.The idea of symmetry here applies only to the sequence or distribution of pigments in the bands, not to their path across the wing. In fact, the "shape" of the bands is as a rule highly asymmetric. Süffert (1929) has given a detailed account of the comparative morphology of the central symmetry system.

The Basal Symmetry System. This is the *Hohlbinde* of Süffert (1927) and the *Discals II* of Schwanwitsch (1924). This band has the characteristics of a symmetry system in that left and right portions are mirror images of one another and it can undergo modifications similar to those of the central symmetry system (Nijhout, 1978).

The Wing Root Band (*Basalis* of Schwanwitsch, 1924). This is a single band at the base of the wing in some species of Nymphalidae. This band, like the basal symmetry system, is lacking in most species.

The Border Ocelli. In its basic form this is a system of small circular patterns near the margin of the wing and characterized by two important features: (1) there is potentially one border ocellus per wing cell; and (2) the midpoint of an ocellus lies on the midline of a wing cell. The border ocelli correspond to the *Ocellata* and *Circuli* of Schwanwitsch (1924). Note that the wing cell between veins Cu_2 and 2A appears to possess two ocelli rather than one (Figs. 5b,c, 10c, 22b). The reason for this is that this region was, in fact, two cells earlier in development. The lacuna of vein 1A, which is part of the primary venation system (see Section 2.1), disappears prior to pupation and is not part of the adult venation. Evidently pattern determination in the border ocelli system takes place before the secondary (adult) venation system becomes established. The term *border ocellus* is somewhat of a misnomer because these pattern elements usually do not resemble eyespots. In fact, as will be seen below (Section 5.5.1), they frequently deviate so drastically from circularity that homology of this pattern element in different species is easily subject to confusion and misinterpretation. The outermost circles of pigment in the ocelli of adjacent cell often appear to fuse into a symmetry system that flanks the row of border ocelli.

Submarginal Bands. These are thin dark-colored bands that run close to and along the distal margin of the wing. Schwanwitsch (1924) recognized three such bands (*Externa I, II, and III*). Süffert (1927) recognized only the distalmost two of these (Fig. 5). It appears that Schwanwitsch's Externa I is considered to be part of the border ocelli system by Suffert.

Dependent Patterns. These are color patterns whose disposition depends on structural features of the wing, usually the wing veins. Strictly speaking, the discal spot and the distalmost submarginal band are dependent patterns in that they coincide with the discal crossveins and wing border, respectively. As a rule, however, the term *dependent pattern* is reserved for color patterns that coincide with the wing veins or with the midline of the wing cells (Figs. 5, 22a). These patterns are called *venous* and *intervenous strips*, respectively (Schwanwitsch's *Venosa* and *Intervenosa*). Süffert did not consider these pattern elements as formal parts of the Nymphalid Groundplan. Figure 22a shows that venous stripes occur within the discal cell and in the cell between veins Cu_2 and 2A, where no veins exist in the adult. These patterns correspond to the locations of the base of the media and the first anal vein of the primary (pupal) venation pattern (Section 2.1). Evidently, the development of dependent patterns, like the border ocelli, depends on the primary, not the secondary, venation system.

4.4 BACKGROUND COLORATION AND PATTERNS

The patterns listed above are referred to as the *elements* of the Nymphalid Groundplan. They are in essence the spots' stripes and bands of pigment that are responsible for the structure and detail of the color pattern. The background upon which these pattern elements develop is sometimes of a single homogeneous color. Often, however, the background itself is made up of a coarse pattern of several *color fields* (Süffert, 1929b; Nijhout, 1978), which contribute significantly, sometimes overwhelmingly, to the "appearance" of the wing. Color fields as a rule occupy substantial portions of the area of a wing, they are not restricted by the wing veins, but they are frequently restricted by one or more of the elements of the Nymphalid Groundplan. No systematic study has yet been done to describe or catalog the diversity of color fields or to establish their developmental/physiological origin. The color patterns of *Heliconius* (Heliconiidae) have been classified as color fields (Nijhout, 1978). Nevertheless, a number of *Heliconius* patterns, particularly on the hindwings, are clearly dependent patterns, and the fore- and hindwing patterns of heliconiid genera like *Dione*, *Eueides*, and *Agraulis* show strong affinities to patterns that may be generated from the Nymphalid Groundplan. The frequent interaction of color fields with elements of the Nymphalid Groundplan and the apparent continuity between color fields and the Nymphalid Groundplan in Heliconiidae suggest a developmental link that deserves attention. Perhaps the genetic studies of the color patterns of *Heliconius* done by Turner (1971a,b) and more recently by L. E. Gilbert (University of Texas, unpublished) will shed some light on the origin and significance of color fields.

Iridescent scales, and the structural blues and greens of butterflies, also appear to be part of the background coloration of the wing. There is no case, to my knowledge, in which elements of the Nymphalid Groundplan are expressed in iridescent scales alone. These structural colors, however, appear to follow the same general rules of patterning as the color fields described above. That is, while their distribution is sometimes *limited* by an element of the Nymphalid Groundplan, they usually appear as patches or broad bands of indefinite form and outline. Perhaps the best example of this can be found in the South American nymphalids of the genera *Callicore* (*Catagramma*), *Diaethria*, and *Perisama*, in which, depending on the species, apparently homologous color fields may be expressed in pigment or as a structural color.

A final feature of the color pattern in Lepidoptera are the *ripple patterns* (Süffert, 1929b). These consist of irregular rhythmical patterns that resemble the ripples on sand dunes and are arranged perpendicular to the course of the longitudinal wing veins (Fig. 6a). Ripple patterns also appear to be part of the

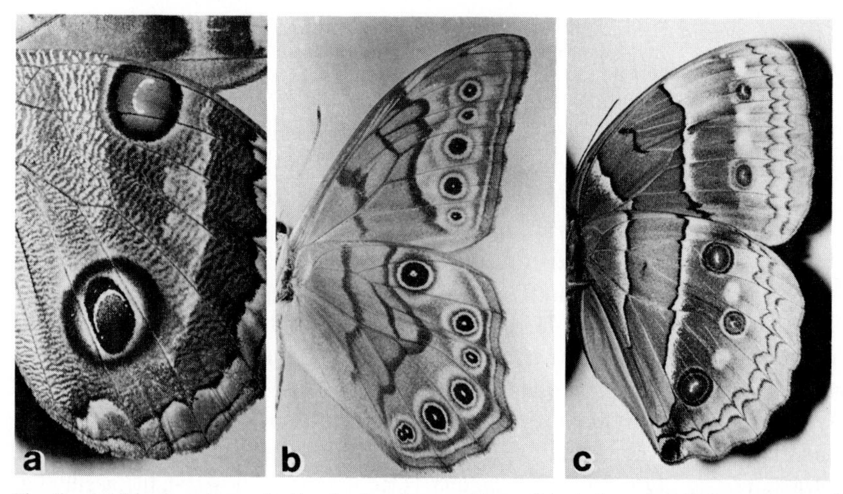

Fig. 6. (a) Ripple patterns in *Eryphanis polyxena* (Brassolidae). (b and c) Color patterns of *Enodia creola* (Satyridae) and *Stichophthalma nourmahal* (Amathusiidae), respectively, which adhere closely to the ideal of the Nymphalid Groundplan but exhibit some selectivity and a minor amount of dislocation.

background pattern upon which the elements of the Nymphalid Groundplan may develop and appear to follow the same general relations to the other elements of the wing pattern as the color fields and iridescent colors. In *Chrysiridia madagascarensis*, for instance, the broad iridescent ripple pattern is readily recognized as the inner field of the central symmetry system. Nothing is known at present about the developmental physiology of ripple patterns though it seems reasonable to assume that positive feedback mechanisms analogous to those that generate ripple patterns elsewhere in nature must be operative.

4.5 IMPLEMENTATION OF THE NYMPHALID GROUNDPLAN

4.5.1 *Selectivity*

The Nymphalid Groundplan illustrated in Fig. 5 is a hypothetical device. It is simply the maximal pattern from which the color patterns of extant species *can* be derived. Both Schwanwitsch and Süffert have emphasized that their scheme should not be interpreted as representing an ancestral or primitive pattern. Rather, the Nymphalid Groundplan is a purely formal construct that is to be used as a tool in recognizing homologies in real color patterns. No species exists whose color pattern is identical to that illustrated in the Nymphalid Groundplan. Real color patterns are "derived" from the Groundplan by selectively suppressing and/or exaggerating certain elements (e.g., see Fig. 6 and numerous illustrations in Schwanwitsch, 1924; Süffert, 1927, 1929b; and Nijhout, 1978, 1981). Thus species-specific patterns have their origin in the selective expression of only some groundplan elements or parts thereof. In addition, species may differ in the pigments they utilize in their pattern so that color patterns that are structurally very similar can have a dramatically different appearance in different species.

4.5.2 *Dislocation*

Within the framework laid out above, the most significant phenomenon involved in the generation of species-specific patterns from the Nymphalid Groundplan is *dislocation* (Schwanwitsch, 1925).The depiction of the Nymphalid Groundplan in Fig. 5 with bands of the central symmetry system running uninterruptedly from the anterior to the posterior margin of the wing and with similarly regular rows of border ocelli is an idealization. In the majority of cases such bands in fact show one or more discontinuities. These discontinuities always occur where a band crosses a wing vein, with one portion of the band shifted laterally with respect to the other. Figure 7 illustrates several instances of dislocation. Multiple dislocations are the rule and in a great many species dislocation occurs at almost every point where a

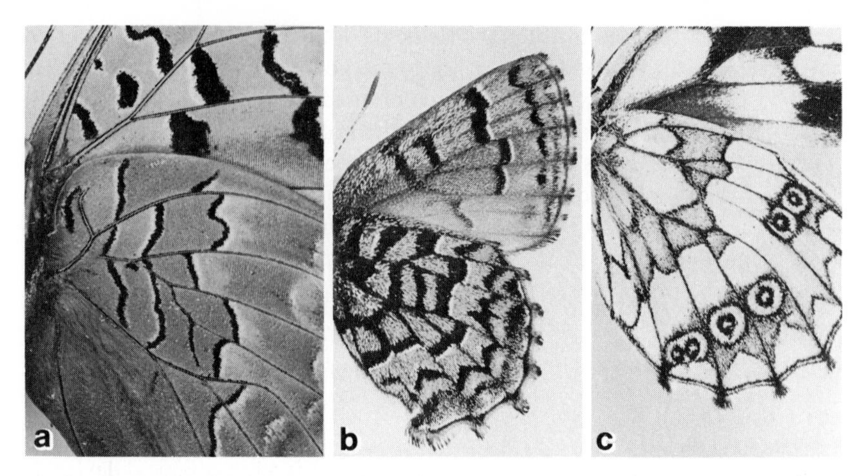

Fig. 7. Examples of dislocation in the central symmetry system. (a) *Charaxes bipunctatus* (Nymphalidae); (b) *Incisalia niphon* (Lycaenidae); (c) *Melanargia galathea* (Satyridae).

pigment band crosses a wing vein (e.g., Fig. 7a). Extreme dislocation can result in the loss of one or more segments of a band (Fig. 7c). Schwanwitsch (1924, 1925) and Süffert (1929b) have described the way in which a large number of patterns may be derived from the Nymphalid Groundplan by more or less extreme dislocation and loss of band segments.

4.6 UTILITY OF THE NYMPHALID GROUNDPLAN

Schwanwitsch (1925, 1926, 1929, 1935) is one of the few authors to have explicitly used elements of the Nymphalid Groundplan to interpret the systematics and evolution of butterflies. Unfortunately, the Nymphalid Groundplan, while known to many Lepidoptera systematists, has been used little or not at all in phylogenetic studies (the recent work of Vane-Wright, 1979, being a significant exception), and has been used only erratically in systematics. The reason that the Nymphalid Groundplan has found so little favor in systematics perhaps lies in the fact that the recognition of homologies in the color pattern is seldom straightforward; more easily quantifiable characters are preferred. The Nymphalid Groundplan is also probably too simple a model for uncritical use, and there has evidently not been a great deal of interest to adapt it to the pattern idiosyncrasies of the various taxa among the Lepidoptera.

Developmental biologists of the 1920s and 1930s have made a somewhat greater use of the Nymphalid Groundplan, but mainly as a conceptual tool to guide and focus their experimental and theoretical studies of pattern. Henke

(1928, 1933a,b, 1943, 1948) Kühn (1926), Kühn and Henke (1936) and their associates (Henke and Kruse, 1941; Kühn and von Engelhardt, 1933, 1936, 1944), and others (Catala, 1940; Wehrmaker, 1959; Schwartz, 1962) have produced a prodigious literature on the experimental manipulation of color patterns. The fact that the bulk of this work was done in a tradition that was largely descriptive and morphoplogical has resulted in its being relegated to the relative obscurity that has befallen so much of the literature on experimental developmental biology of the pre-World War II years.

It is unfortunate that a model that unifies such a truly vast diversity of patterns has received so very little attention from all but a small group of biologists.

The analysis of color pattern morphology I have presented in the preceding sections of this review reflects the insights developed by Schwanwitsch, Süffert, and others, whose approach to the problem has been largely phenomenological. Much of this view of color patterns has been rendered obsolete by new insights into the developmental physiology of color patterns that will be dealt with in the sections that follow. It will be proposed that control over pattern development occurs at the level of the wing cell (an area bounded by wing veins) rather than the wing as a whole, and that the overall pattern is built up from a number of relatively independent local patterns. This is not to say, however, that the Nymphalid Groundplan and associated phenomena have no utility for the future. Clearly, developmental phenomena may be usefully studied at many different levels of resolution and, in this case, the Nymphalid Groundplan remains an invaluable device in that it forces one to approach the comparative study of patterns in an organized fashion and, above all, it forces one to think about homologies.

5 Pattern formation

5.1 PATTERNS ARE SERIALLY HOMOLOGOUS

Perhaps the most striking feature of lepidopteran wing patterns is that in many species the color patterns of adjacent wing cells are virtually identical (Fig. 8). This is apparent even in the schematic diagrams of the Schwanwitsch/Süffert groundplan (Fig. 5) although this feature apparently did not strike either author as curious. This phenomenon is most evident in the pattern on the ventral hindwings of butterflies and the dorsal forewings of moths. These wing surfaces are generally concerned with camouflage and the color patterns they bear, for reasons not yet understood, appear to be very conservative. While we have no basis on which to critically judge which features of the color pattern are primitive and which are derived, it seems reasonable in first instance to use a democratic rule and propose that the

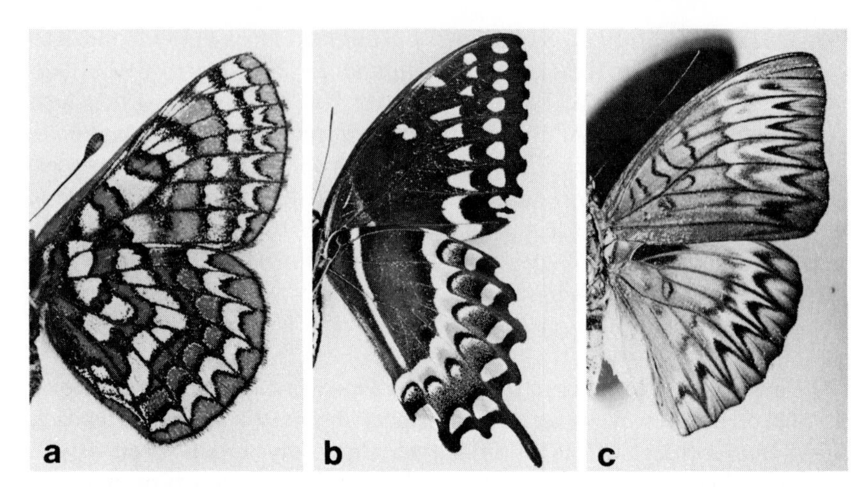

Fig. 8. Serial homology in color patterns. Each wing surface in the species illustrated bears a series of virtually identical color patterns in its wing cells. Patterns on the forewing are independent from those on the hindwing. (a) *Euphydryas editha* (Nymphalidae); (b) *Papilio palamedes* (Papilionidae); (c) *Tanaecia munda* (Nymphalidae).

most widespread features are primitive. In this case that would mean that a serial repetition of patterns, each restricted to a single wing cell, is primitive.

Because the pattern within each wing cell can be very complex, it is highly improbable that such serially repeated features arise from a single developmental mechanism that ranges over the entire surface of the wing. Instead, as will be seen in the sections that follow, there is excellent reason to believe that the color pattern in each wing cell develops independently from that in adjacent wing cells. In its most basic form, the overall color pattern is composite, made up of a serial repetition of equivalent local patterns, each restricted to a single wing cell.

5.2 ORIGIN OF THE LOCAL PATTERN

Because color patterns are simple two-dimensional distributions of chemicals, Wolpert's theory of positional information (Wolpert, 1969) has proven to be a particularly useful framework within which to develop an initial model for their origin (Nijhout, 1978). Wolpert points out that the development of spatial pattern depends critically on information that cells in a developing field receive from their cellular environment. Cells are believed to receive information about their position in a developing field. They then respond to this positional information by executing certain biochemical actions. The exact nature of a cells' response(s) depends on two things: (1) the quality of the positional signal(s), which depends on where in the field a cell is located; and (2) the prior developmental history of the cell which, through progressive compartmentalization, determination, and differentiation, restricts the types of activities that a cell is capable of performing. In terms of color patterns, the response of the scale cells consists of the synthesis of a specific pigment. As we have seen (Section 4.1) this in turn is due to the fact that a scale cell synthesizes either specific enzymes for pigment synthesis or specific transport molecules for pigments and/or their precursors. These are relatively simple molecular events whose general nature in eukaryotes is becoming increasingly well understood.

In spite of the necessary vagueness of the concept of positional information (and its generation and interpretation) some very useful ideas emerge when we examine its logical consequences. Since the position of a cell can only be determined *relative* to something else, there must exist, according to Wolpert (1969), in every developmental field one or more reference points. These reference points are called "sources" of positional information, a term that implies that they emit a signal. This signal is often assumed to be in the form of a diffusible molecule, a morphogen. If the sources of positional information are point sources, then the simplest pattern that can be generated by such a system is a ring or a circle with the source at its center. All that such a pattern requires is a *single* source that broadcasts a signal equally in all directions, and a molecular mechanism that ensures that all cells that occur at or within a critical distance from that source (and therefore experience a signal strength at or above a critical threshold) differentiate in the same way.

Circles and rings of pigment are particularly common features of the color pattern of many butterflies and some moths (e.g., Figs. 6, 10, and 11) in the form of border ocelli and eyespots derived from the discal spot. The model proposes that the development of such circular patterns depends on a specialized activity of the cell(s) at their center. This proposition can be readily tested by observing whether the development of an ocellus is affected when the cells at its center are removed or killed some time prior to, or during, its determination. In order to do such an experiment one needs to be able to identify the cells that will be at the center of the presumptive ocellus. The Buckeye butterfly, *P. coenia*, has been found to be a most useful experimental animal in this regard because it has a pigment pattern on the pupal wing cover that precisely overlies the cells that will form the center of the large eyespot on its forewing (Fig. 9a; there is excellent reason for believing that such color patterns in the pupal cuticle are organized by the same sources that later on induce development of the adult pattern; see Nijhout, 1980a). When these cells are killed by cautery a few hours after pupation the eyespot fails to develop (Fig. 9). When cautery is done 24 h after pupation a small eyespot of about half normal diameter develops. Cautery at 48 h after pupation (at 29°C) or later no longer affects eyespot development

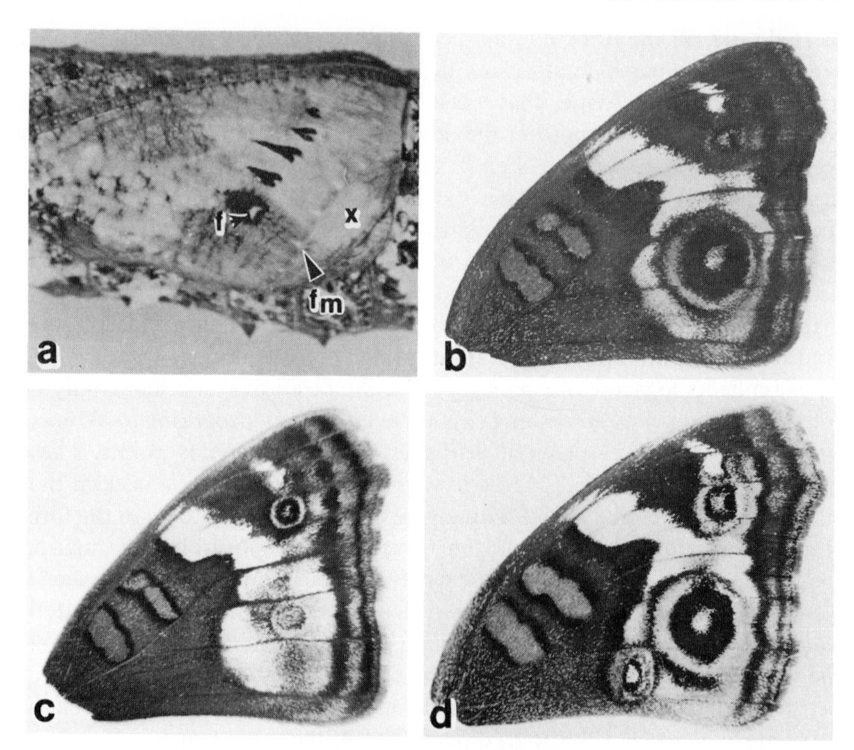

Fig. 9. Development of an ocellus in *P. coenia* (Nymphalidae). (a) Pupal forewing showing markings that correspond to foci for pattern determination. f, Focus for large eyespot; fm, possible location of a marginal focus; x, area of pupal forewing that will degenerate. (b) Normal pattern on forewing. (c) Elimination of eyespot by cautery of focus. (d) Supernumerary eyespot developed around a transplanted focus. (After Nijhout, 1980a.)

(Nijhout, 1980a). Evidently, eyespot determination occurs progressively, from the center outward. It begins about the time of pupation or shortly before and lasts until 48 h after pupation at 29°C. At 18°C eyespot determination requires about 5 days (Nijhout, 1980a; see Section 6.4.1). The small area of cells at the center of the eyespot that is responsible for its determination is called a *focus*. The exact number of cells in a focus is not known. The smallest cautery done destroyed an area of about 300 cells (Nijhout, 1980a) and was perfectly effective in abolishing development of the eyespot. Thus an effective focus consists of fewer than 300 cells. It is conceivable, of course, that a single cell serves as the focus, the source of the positional signal.

When a focus is destroyed no other cells take over its organizing function, at least not within the 2- to 5-day span required for pattern determination. That the cells of the focus are indeed specialized as pattern "organizers" can be demonstrated by transplanting them to another area of the wing where normally no eyespot would occur. Nijhout (1980b) has shown that such transplanted foci induce the formation of rings of pigment in the surrounding host cells (Fig. 9d). These pigment rings can be shown to be homologous to the outer rings of the normal eyespot because when a transplant is made close to a normally developing eyespot, the outer rings of the eyespot and the rings induced by the transplant fuse and become continuous in exactly the same way that the outer pigment rings of border ocelli from adjacent wing cells fuse. Focal transplants never induce development of very large eyespots. In fact, they induce only the development of the outer two rings and only seldom a few black scales of the inner disk. The reason for this failure to induce a fully formed eyespot probably lies in the fact that after surgery a cellular clot forms around the transplant which either prevents the establishment of the necessary cell–cell contacts between graft and host, or interferes in some other way with the transmission of the positional signal from graft to host. Hence, the surrounding host tissue probably receives the positional signal either too late or at too low an intensity to form any but the outermost elements of an eyespot before the induction-sensitive period is over.

The large eyespot of *P. coenia* is a border ocellus of the Nymphalid Groundplan and it seems reasonable to generalize the observations on this species and suggest that *all* border ocelli and similar systems of concentric rings and circles are likewise determined by some special activity of the cells at their center, the foci. Indeed, the small anterior ocellus on the forewing of *P. coenia* can be eliminated by focal cautery as can any one of the ocelli on the hindwing of *Vanessa cardui* (Nijhout, unpublished). We have already noted that in the basic groundplan of the color pattern there is one ocellus in each of the wing cells, always with its center located on the midline of the wing cell. Thus the organizing system that induces the development of a row of border ocelli consists simply of a row of foci, running anterior to posterior, with one focus per wing cell located on the wing cell midline. Species-specific differences in the distribution of border ocelli (e.g., Fig. 6a,b,c) then come about through species-specific differences in *which* foci will be active and exactly *where* along the wing cell midline they are placed. Species-specific differences in the interpretation of the positional signal(s) lead to diversity in the *fine structure* of the eyespot or ocellus.

The idea that small, locally restricted, signal sources induce pattern formation in their surroundings can be extended to the development of the central symmetry system as well. The classical work of Kühn and von Engelhardt (1933) on *E. kühniella* demonstrated that the development of the central symmetry system in this species depends on the activity of at least two "sources," one near the middle of the anterior margin of the forewing and one near the middle of the posterior margin. Furthermore, their data suggest that a third source is probably present near the middle of the wing (Fig. 20). Cautery of either of these source regions locally abolished the development of the central symmetry system. Because of the small size of the *Ephestia* wing it was impossible to pinpoint the location of the sources. There is excellent reason, however, for believing that they, like the foci of the border ocelli, are located on wing cell midlines. The reason is that the central symmetry system of many Lepidoptera is either clearly made up of a row of confluent circles or, if heavily dislocated, at least partially composed of isolated rings (Fig. 10). Such patterns are particularly widespread among the Geometridae but can also be found in selected species of almost every family of butterflies. Wherever such a makeup of the central symmetry system is evident, the component circles have their centers located on wing cell midlines. The central symmetry system thus appears to be generated from point sources just as are the border ocelli.

The central symmetry system of *Ephestia* is generated by three foci, that of *Plodia interpunctella* by two (Schwartz, 1962), while that of *Hydria undulata* (Fig. 10a) is generated by some 8–10 foci. Evidently species differ in the number of foci they employ for generating the central symmetry system, just as species differ in the number of foci that are active in their border ocelli system. Nijhout (1978) has shown that the basal symmetry system of Nymphalid Groundplan is likewise generated around discrete sources. Thus in the simplest situation the overall pattern is generated around three roughly parallel rows of foci that run from the anterior to the posterior margins of the wing (Nijhout, 1978). Each of these foci induces a system of concentric rings of pigment. Where homologous rings meet they fuse, creating an irregular

Fig. 10. Examples of circular patterns in the central symmetry system. (a) *Hydria undulata* (Geometridae); (b) *Smyrna blomfildia* (Nymphalidae); (c) *Mantaria maculata* (Satyridae).

scalloped banding pattern. This would be true, however, only if all potential foci were present, if they were arranged in reasonably straight rows (i.e., not dislocated too severely), and if each induced a similar pattern in its surroundings. There are a few cases in which these conditions are met, though such instances give comforting support for the general model outlined above. In most cases not all potential foci are expressed and those that are may be considerably dislocated (i.e., displaced laterally along the wing cell midline) so that they do not lie in an orderly row. In some cases dislocation may be severe enough to yield some confusion as to whether a focus and its surrounding pattern belongs to the border ocelli system or to the central symmetry system. Comparative studies of related species are needed in such cases and by this means the ambiguity can almost always be resolved.

This simple picture becomes complicated further by the fact that the pattern that is generated around a focus is more often than not noncircular. There exists considerable diversity in the rules or processes that govern the shape of the patterns induced by a focus. These processes will be examined in more detail below.

5.3 DIMENSIONS OF DEVELOPMENTAL FIELDS

The overall color pattern is made up of a number of fairly independent developmental fields. A developmental field corresponds to the area over which a given focus exerts its influence, that is, the area in which a focus is either solely or through interaction with another focus capable of inducing a pattern. Hence, the number of developmental fields depends on the number of "active" foci on the wing. Maximally there would be a field for every potential location of a focus; at least 3 per wing cell, for a total of between 20 and 30 per wing surface depending on the venation pattern. This maximum is, however, never observed and in most species pattern development occurs with far fewer fields. Not every portion of the wing surface need by part of a developmental field for color pattern. Many species have large patternless areas on their wing that can be considered "background" and require only a mechanism for specifying overall color. However, other apparently patternless areas, particularly those on the dorsal forewing of many butterflies, actually arise from a broadening and subsequent fusion of two or more pattern elements (Papilionidae and members of the genus *Charaxes* provide good examples; Nijhout and Wray, 1985), and thus require a set of specifications at least as complex as those that are responsible for more detailed local patterns.

In a significant sense the field corresponds to the entire area that receives the signal from a given focus, though, in practice, a developmental field can probably best be defined operationally as the area bounded by the most distal pattern elements whose development is influenced by a given focus. It is inherent in this definition that fields may overlap. Effective boundaries to a developmental field can therefore become established *passively*, by the simple fact that the signal from a given focus decays and must at some point fall below the threshold of sensitivity of the local scale-building cells. Boundaries can also be established *actively*, by the interaction of signals from two sources or where a signal is destroyed by a sink. Passive boundaries are probably impossible to detect without some knowledge about the nature of the signal and the mechanism whereby the scale cells perceive the signal. The presence of active boundaries, however, should be experimentally demonstrable. This would entail the elimination (by cautery, for instance) of one focus to observe whether the pattern determined by an adjacent focus now "expands" into the area that would have been occupied by the pattern specified by the cauterized focus. Active boundaries can also be inferred to exist wherever dislocation of the pattern is observed. Dislocations are abrupt discontinuities of the pattern along a fault line (Fig. 7) and can only be explained if at least some determinants of pattern on one side of the fault line are independent from similar determinants of pattern on the other side.

In cases of dislocation the process of pattern determination is somehow prevented from crossing an invisible boundary. This boundary usually corresponds in position to a wing vein. In a great many species almost every single wing vein corresponds to a dislocation boundary (Fig. 7). In such cases the wing cells are the units of pattern formation, and pattern determination in each wing cell is independent from that in adjacent wing cells.

Does this mean the wing veins play an active role in color pattern determination? In some cases they do; in others they do not. That they can play an active role, in principle, is demonstrated by the widespread occurrence of dependent patterns (venous stripes). The presence/absence of such dependent patterns in closely related species (e.g., *Pieris brassicoides* vs *P. rapae*, or *Acraea conradti* vs *A. caldarena*) indicates that there is significant flexibility in the use of wing veins as potential determinants of pattern. Many species (particularly in the Papilionidae) that have well-marked veins also have bands of a similarly colored pigment running along the wing cell midline (intervenous stripes, e.g., Fig. 22a). Such observations suggest that wing veins and wing cell midlines may have, at least in some cases, equivalent roles in pattern determination.

An active role for wing veins in pattern determination can also be inferred from the frequent occurrence of eyespots that are truncated or oblated at the wing veins (e.g., Fig. 11a). Such patterns are most easily explained if the wing veins act as sinks for the signal emitted from a focus. The signal cannot enter an adjacent wing cell, and the developmental field would be bounded by such sinks.

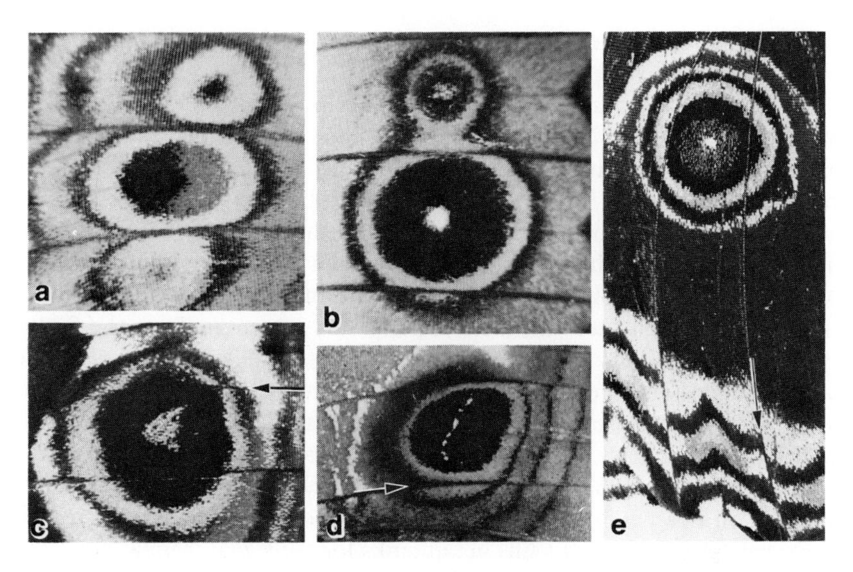

Fig. 11. Examples of different roles for wing veins in pattern determination. (a) Oblate ocelli in *Precis atlites*; ocelli remain restricted to wing cell of origin and appear to be deformed and constrained by the wing veins. (b) Ocellar bands that transgress wing vein in *Mycalesis terminus* (Satyridae). (c and d) Ocellus of *P. coenia* and *Opsiphanes quiteria* (Brassolidae), respectively, showing a change in morphology upon transgressing a wing vein (arrow) suggesting different rules of interpretation in adjacent wing cell. There is significant individual variability in this characteristic. (e) Pattern in *Morpho achilleana* in which the bands of the ocellus transgress the wing veins without alteration in shape, but the parafocal element and submarginal bands in the distal portion of the wing cell are dislocated along the veins (arrows).

There are, however, a great many cases in which the color pattern, either in part or in its entirety, develops without apparent interaction with the wing veins. Large eyespots commonly transgress wing veins (e.g., Figs. 9b and 11e), but the shape and/or coloration of the eyespot in the invaded wing cell is sometimes different from that in the wing cell of origin (the one that bears the focus). The large eyespot on the forewing of *P. coenia* (Fig. 9c) shows this phenomenon. Sibatani (1980) illustrates and discusses additional instances and suggests that when a large eyespot encompasses more than one wing cell, no matter how perfect in shape it may be, it is actually made up by integration of parts generated separately and autonomously in each of the involved wing cells. In effect, Sibatani (1980) suggests that a wing cell is a compartment, at least for purposes of color pattern formation. The focal cautery and transplant experiments of Nijhout (1980a), however, clearly demonstrate that the entire large eyespot of *P. coenia*, including the portions that lie in adjacent wing cells, is induced by a single focus. The fact that an eyespot may have somewhat different characteristics in adjoining wing cells is probably most easily explained by the fact that signal interpretation may change abruptly from wing cell to wing cell (Section 5.6).

If foci are present in two adjacent wing cells, and if they induce similar patterns, then the fact that a wing vein occurs on the boundary between such patterns could be a matter of coincidence, simply due to the fact that the wing vein is located exactly halfway between the foci. If this were the case then one would predict that when the two foci differ either in the strength or in the timing of onset of their signals, the boundary between their respective patterns would *not* coincide with a wing vein. Indeed, this is what one finds (Fig. 9b), although such instances are not very common. Thus the finding that two homologous pattern elements meet at a wing vein in itself implies nothing about the function of the vein in pattern formation. Only when homologous pattern elements are either dislocated or truncated at a wing vein can a boundary function be inferred. To ascertain whether such a boundary is a sink or a reflecting boundary would require experimental analysis. In this regard it is of interest to consider that boundaries may be temporal. That is, if wing veins act as sinks for the signal from a focus, they need not be constantly active. If they become activated rather late, after a substantial portion of the pattern has already been induced they can cause local deformations of pattern, perhaps like the one illustrated in Fig. 11d.

Thus, where wing veins act as sinks they can, in a formal sense, be regarded as signal sources that can provide positional cues within adjoining wing cells. In addition, there are cases, particularly common in *Morpho*, in which an eyespot transgresses a wing vein while certain submarginal patterns (the parafocal elements in this case) in the same wing cell show dislocation along the wing veins (Fig. 11e). Clearly, in such cases one pattern element ignores the wing veins while another is constrained by them. Whether such differences are due to a temporal pattern of activity of the wing veins (behaving as sinks only during the time that the parafocal elements are determined), or due to the fact that they are transparent to one signal but not to another, is unknown.

It appears that species differ in whether or not they use wing veins as determinants or boundaries for color pattern formation. In view of the fact that dislocation is exceedingly common it is fair to conclude that veins commonly serve as boundaries for color pattern formation but that their role in this regard would have to be established on a case by case basis.

Field boundaries *within* a wing cell, for instance between adjacent foci, are generally undetectable and may be of the passive variety mentioned above. It is probable that significant portions of the pattern within a wing cell are "controlled" by signals from two foci. For instance, the signal from the focus for the central symmetry system probably extends into the region of the border ocelli and may be partially responsible for specifying morphology in the border ocelli system (see Section 5.5.2). Furthermore, there is reason to believe that there exists a signal source at or very near the distal edge in each wing cell (see Section 6.5) which, together with the border ocellar focus, is responsible for organizing the often complex patterns near the wing margin, the parafocal elements.

5.4 DEVELOPMENTAL COMPARTMENTS

In cases in which wing veins have a boundary function, the wing cells are in effect compartments for color pattern formation, though they are not compartments in the same sense that the term is used for *Drosophila*. Developmental compartments in *Drosophila* are areas of clonal restriction (Garcia-Bellido, 1975; Garcia-Bellido *et al.*, 1973), demarcated by sharp boundaries that are respected by member cells of clones on either side. Compartment boundaries often do not follow obvious topographic features of the organ in which they occur, an observation that suggests that compartment boundaries are "transparent" to most or all of the morphogenetic signals that are involved in coordinating the development of the organ or body region in question. On the wing of *Drosophila* there are two compartment boundaries. One runs along the edge and separates dorsal and ventral compartments. The other runs in a straight line along the length of the wing and separates it into anterior and posterior compartments. The latter boundary runs just anterior to wing vein M_{1+2}. Sibatani (1980, 1983) has made an extensive study of wing homeosis in Lepidoptera and has detected what appears to be a compartment boundary between veins M_1 and M_2, a location closely similar to that of the A/P compartment boundary in *Drosophila.* Large homeotically transformed clones usually respect this boundary. In addition, Sibatani (1980, 1983) has described several instances of dorsal/ventral homeosis between two surfaces of a single wing, providing clear evidence for dorsal and ventral compartments.

Wing homeosis in Lepidoptera consists mostly of clonal transformations of small areas of the wing surface, which develop a color pattern corresponding to that of homologous areas either on the serially adjacent wing (forewing/hindwing conversion), or on the opposite wing surface (dorsal/ventral conversion). Furthermore, in species with sexually dimorphic color patterns there are many instances of gynandromorphic homeosis (Sibatani, 1980). Evidently, the homeotically transformed clones respond to local inductive signals, but they interpret these signals according to a different set of "rules" than those employed by the untransformed cells in their surroundings. The rules of interpretation that the transformed clones use are those characteristic for a different wing surface (or of a different sex).

There are no evident color pattern discontinuities at the location of the A/P compartment boundary in wing cell M_1-M_2. A number of species show pattern discontinuities at wing *veins* M_2, M_3, or Cu_1 (see Section 5.6), but there is no reason to believe that these are related to the clonal boundaries elucidated by Sibatani.

5.5 THE INTERPRETATION LANDSCAPE

5.5.1 *Noncircular patterns*

A single focus, signaling in a field of cells of homogeneous properties for signal propagation and interpretation, can only induce circular or ring-shaped patterns. When wing veins act as reflecting boundaries or as sinks for the focal signal such circles can become truncated or oblate. In order to generate patterns of any other form, for instance those illustrated in Fig. 12, the system needs more information, either in the form of additional sources or sinks or as an anisotropy in the mechanisms for signal propagation or signal interpretation.

Before considering the probable characteristics of such additional information it is useful to describe the most general properties of pattern morphology that have to be accounted for. We restrict our consideration to patterns that belong to the border ocelli system. Circular patterns are, of course, radially symmetrical around the focus. When patterns deviate from circularity they are almost always bilaterally symmetrical, with the focus on the axis of symmetry. The axis of symmetry most often corresponds exactly to the wing cell midline, though there are some cases in which it is slightly offset from, but parallel to, the wing cell midline. In a few cases there in an abrupt pattern discontinuity at the wing cell midline (Fig. 16). The majority of the patterns in the border ocelli system have only one axis of symmetry. The portion of the pattern proximal to the focus has a very different morphology from the portion distal to the focus. Thus the pattern in each wing cell is polarized along a proximo-distal axis, and bilaterally symmetrical. The field inhomogeneities that are the causes of such patterns are likely to be characterized by the same general properties. That is, there probably exists a proximo-distal gradient of some kind, as well as a gradient whose shape is symmetrical on each side of the wing cell midline. These characteristics could apply to a single gradient, or could be the combined characteristic of several. While a bilaterally symmetrical gradient seems to be essential, the polarity of the proximo-distal gradient(s) is not constrained. The majority of patterns are most readily explained by a p-d gradient with a proximal maximum (Nijhout,

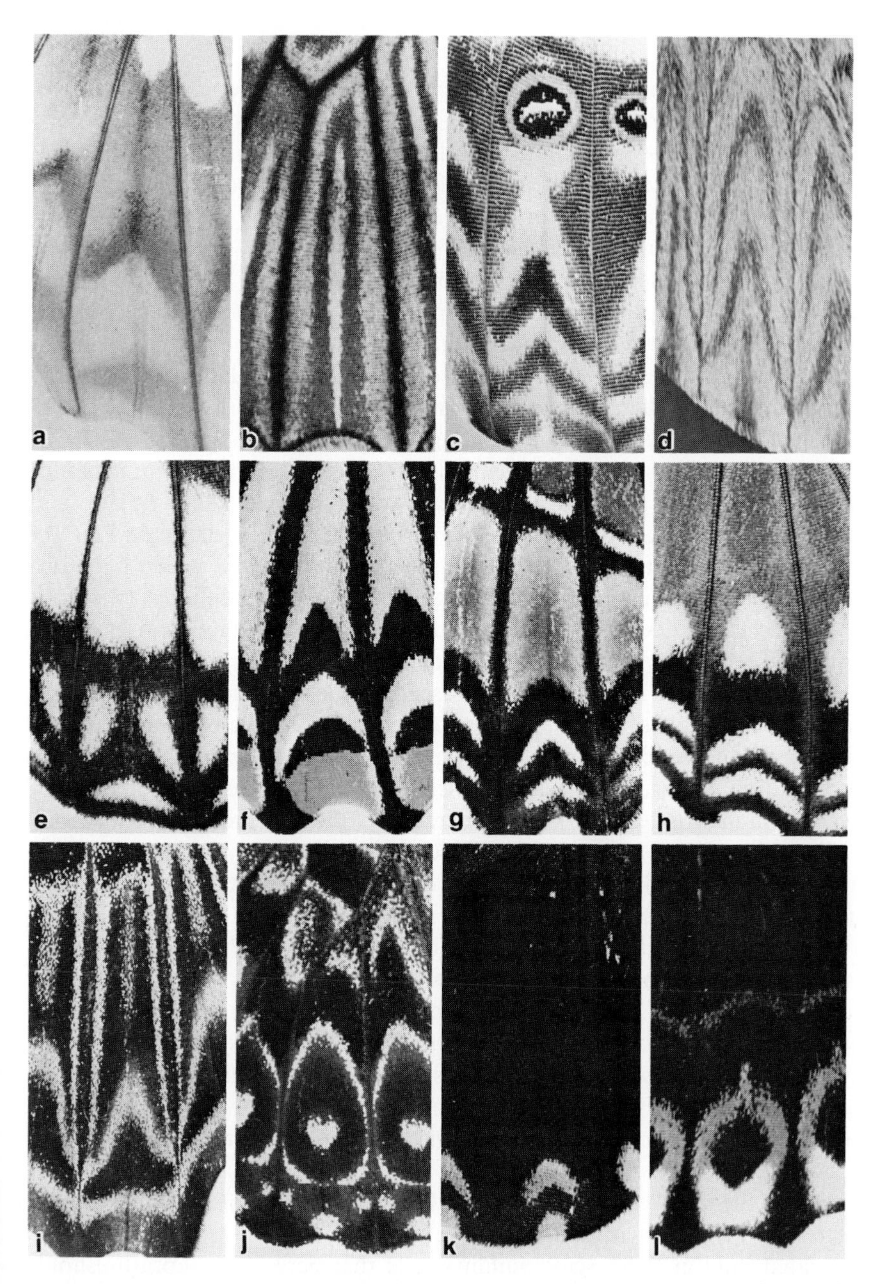

Fig. 12. Examples of wing-cell-restricted patterns. (a) *Papilio dardanus* (male); (b) *Morpho catenarius*; (c) *Eteona flava* (Satyridae); (d) *Cerura vinula* (Notodontidae); (e) *Zethera pimplea* (Satyridae); (f) *Papilio clytia*; (g) *Limenitis archippus* (Nymphalidae); (h) *Limenitis astyanax*; (i) *Parthenos sylvia* (Nymphalidae); (j) *Hamadryas amphinome*; (k) *Papilio agestor*; (l) *Symbrenthia nicea* (Nymphalidae).

1978, 1981), but there are a significant number of patterns that seem to require a distal maximum.

Nijhout (1978, 1981) has suggested that it may be possible to deduce the characteristics of the supplementary information that is required to generate noncircular patterns by a form of curve fitting, if we accept Wolpert's proposal that positional information sources (the foci) emit their signal equally in all directions and that differences in pattern are likely to be due to differences in interpretation. The supplementary information can then be described as an *interpretation landscape*, a set of local rules by which the focal signal is interpreted. In most cases such an interpretation landscape is easily visualized as a gradient whose local value (elevation) is proportional to the threshold of sensitivity of scale-building cells to the focal signal (Fig. 13).

The interpretation landscape should be viewed as a symbolism. It could represent the actual profile of a chemical gradient, but it need not. In fact, it is unlikely to be an exact description of a variable because, as drawn in Fig. 13, the interpretation landscape is referenced to a focal signal which is represented as a straight cone. If the focal signal propagates by diffusion (which it almost certainly does; Section 6.4.2) then a non-steady-state focal gradient would be exponential in character and an appropriate transform would have to be applied to the interpretation landscape to yield an accurate representation of the variable(s) it models. The point is that the slope of the interpretation landscape is *relative* to the focal gradient. Thus, by extension, if signal propagation is anisotropic within a field, this would have to be taken into account in inferring the shape and character of the interpretation landscape.

The idea of an interpretation landscape depends, of course, on the assumption that cells in different regions of a developmental field may differ in the rules by which they interpret a standard positional signal (Wolpert, 1969), and that such local rules may be nothing more than the threshold value for some response to the positional (focal) signal. It is not unreasonable to assume that such a simple situation obtains in this case, since the patterns we are attempting to account for are patterns of melanin synthesis. Melanin synthesis requires only one enzyme, so that most patterns are in reality patterns of synthesis and/or activation of a single enzyme. Such a feature could be made to respond in a stepwise manner to a graded "activator" (Lewis *et al.*, 1977).

One of the simplest models one could construct, then, for the control over patterned melanin synthesis would be that a tyrosinase is synthesized or activated only in those scale-building cells that perceive a suprathreshold focal signal, and that the tuning of this activation threshold is, in turn, a function of the strength of a second signal, described by the interpretation landscape. In most cases this is equivalent to saying that melanin synthesis occurs only where there exists a particular critical ratio between the values of

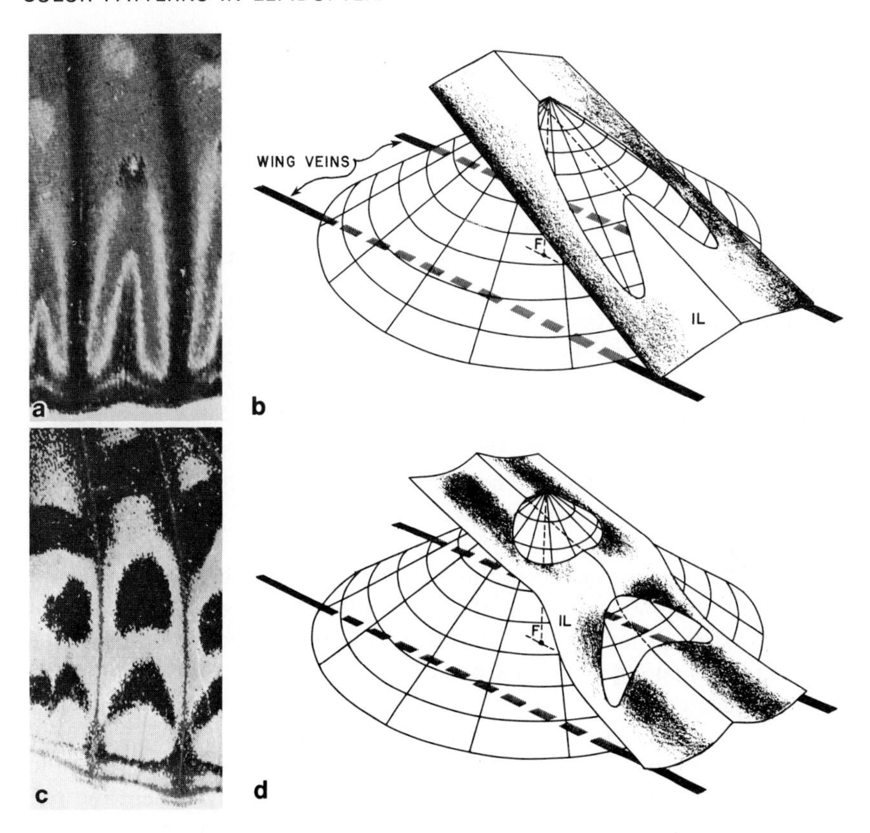

Fig. 13. Two-gradient model for the specification of patterns. Cone-shaped gradients are produced by point sources (F). The interpretation landscape (IL) is a proximo-distal gradient with a proximal high and a high along the wing cell midline. Pattern boundaries are assumed to be specified by a critical ratio (unity in this case) between the two gradients. (a and b) Pattern and model for *Dichorragia nesimachus* (Nymphalidae). (c and d) Pattern and model for *Speyeria aphrodite* (Nymphalidae). (After Nijhout, 1978, 1981.)

the focal signal and the interpretation landscape. Or, more precisely, If M represents melanin synthesis then

$$\begin{array}{llll} M, & P > T_1, & T_1 = a & \\ & P > T_2, & T_2 = bQ, & Q < c \\ & & T_2 = bc, & Q > c \end{array}$$

where P is the value of the focal signal, Q the value of the interpretation landscape, T_1 the threshold in P below which it cannot be detected by the scale-building cells, T_2 the functional threshold for P which is proportional to

Q; a, b, and c are constants. If $T_2 = f(Q)$ then the critical ratio becomes a function of Q.

5.5.2 *Origin of the interpretation landscape*

Since the focal signal is probably a diffusible substance it is likely that the interpretation landscape likewise describes the distribution of a diffusible morphogen. In fact, it is likely that the interpretation landscape is identical to the gradient that specifies the central symmetry system. As discussed above, the central symmetry system is specified by signals emanating from discrete sources and, as indicated by the high degree of dislocation that is common in the central symmetry system, there is basically one source per wing cell with the wing veins acting as active boundaries for the field of that source. Such a proximal source would be consistent with the general proximo-distal gradient slope that appears to be required to explain most patterns, while the required bilateral symmetry would be readily obtained if the wing veins served as sinks for the signal.

Experiments seem to confirm this notion. Figure 14 shows the effect on pattern formation of an ablation of the area presumed to contain the source for the proximo-distal gradient in *Speyeria aphrodite.* The most important effect is a fusion along the wing cell midline of the ocellus with the parafocal pattern element whose determination was hypothesized to depend (among others) on a focal signal (Fig. 13; and Nijhout, 1978, 1981). This pattern modification can be easily accounted for by a depression of the slope of the interpretation landscape. Evidently, at the time of the experiment the p-d gradient was already largely set up and elimination of the source interfered only with the later stages of gradient development. The pattern in adjacent wing cells is not affected apart from a substrate distortion caused by wound healing. While such an experiment lends support to the hypothesis of a

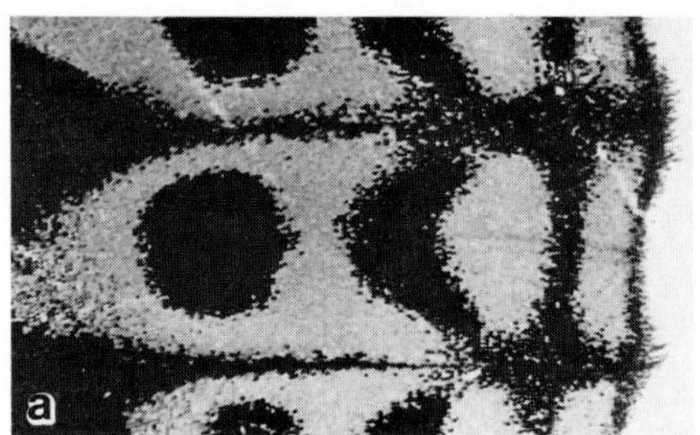

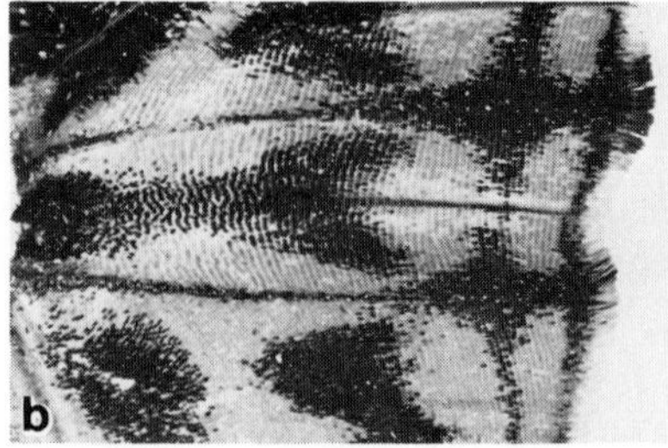

Fig. 14. Effect on color pattern of cautery at the base of a wing cell in *Speyeria aphrodite.* (a) Normal pattern; (b) pattern modification after cautery. Site of cautery is at left margin of photograph. Pattern modification consists of fusion of focal pattern and parafocal element along the wing cell midline. Patterns in adjacent wing cells are not affected, though there is some overall pattern distortion due to contraction around wound site.

proximal source for the interpretation landscape, the shape of the interpretation landscape is not fully accounted for. In particular, it is difficult to see how simple diffusion could generate the inflection in the gradient that seems to be required (Fig. 13d) for the separation of the ocellus and the parafocal element. There are basically two ways in one could generate such an inflection in a gradient. (1) A monotonic gradient of one substance can code for a gradient with an inflection in another substance if there exists a sigmoid relationship between the concentrations of the two. A process analogous to binding cooperativity or the activity of an allosteric enzyme could effect such a relationship. (2) A gradient with an inflection can also be generated by the interaction of gradients from two sources, one proximal and one distal to the focus. As will be seen below, there is good reason to believe that there exists a signaling source at the wing margin in each wing cell. Dominance of the effect of a source at the wing margin probably also accounts for those instances that appear to require a distal maximum (e.g., Fig. 12j).

5.5.3 *Global gradients*

In the majority of species the color patterns in a series of adjacent wing cells are either virtually identical to one another or can be arranged in a more or less continuous series or morphocline. *Cethosia* (Fig. 15) provides an excellent illustration of such a morphocline. Pattern formation in each wing cell evidently operates by the same "rules" because very similar patterns develop repeatedly. Yet, there is a clear and systematic alteration of the pattern as one scans a series of wing cells. The most reasonable explanation for this is that there exists a slight quantitative difference in the rules for pattern interpretation in adjacent wing cells. Whatever the variable, it evidently changes in a systematic fashion along an anterior–posterior axis. The variable appears to be smoothly graded as evidenced by the fact that in several wing cells there is a definite skew in the bilateral symmetry of the pattern with each half-pattern showing a morphology that is intermediate between that of its mirror-symmetrical other half and that of its homolog in the adjacent wing cell. It is clear then that we have a determinant of pattern that ranges over the entire wing and appears not to obey the wing vein boundaries. Yet this broadly distributed factor affects the development of patterns that are restricted to wing cells. Thus pattern determination is responsive to both global and compartment-restricted determinants. There exist species-specific (and, as Fig. 15 shows, wing-specific) differences in the manner in which this global determinant is distributed or perceived, since some species show very little pattern gradation at all, while others are known to develop gradations considerably more severe than those illustrated for *Cethosia*.

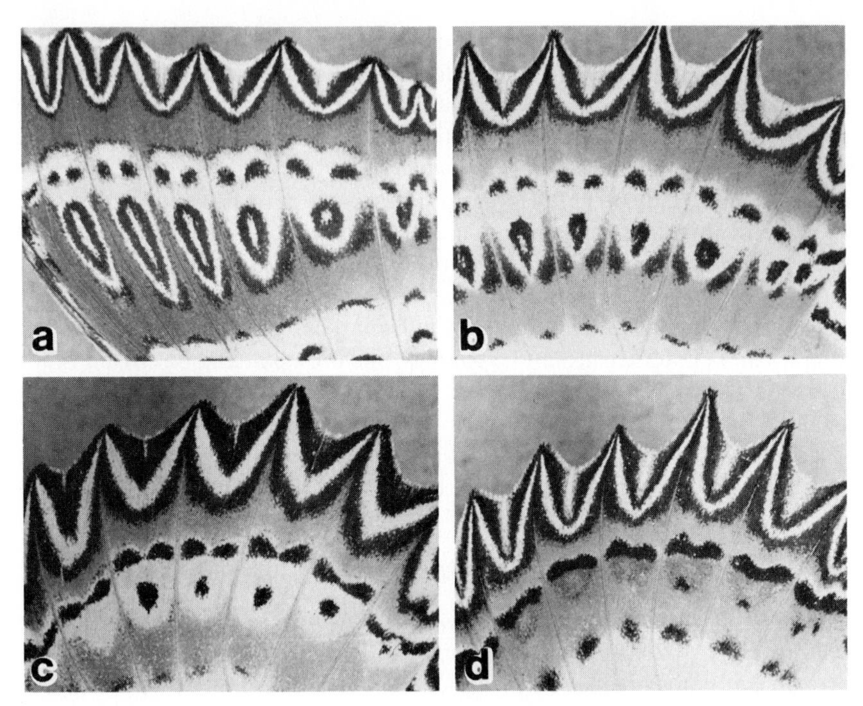

Fig. 15. Pattern series in *Cethosia*. (a and b) Forewing and hindwing, respectively, of a specimen of *C. biblis*. (c and d) Hindwings of *C. hypsea* and *C. gabina*, respectively, showing different variations of the morphological series (cf. Fig. 17).

5.6 ABRUPT DISCONTINUITIES IN SIGNAL INTERPRETATION

While pattern morphoclines that range smoothly across an entire wing are exceedingly common and give evidence of smoothly graded quantitative differences in signal interpretation in a successive series of wing cells, there are a lesser but significant and perhaps more interesting number of cases in which there are abrupt discontinuities in signal interpretation. A few such cases are illustrated in Fig. 16. *Papilio troilus* (Fig. 16a,b) has virtually identical patterns in all of its wing cells except one (the one between veins M_3 and Cu_1), which bears a dramatically different pattern, though one whose general outline corresponds reasonably well to that of the overall pattern in the other wing cells. Interestingly, one occasionally finds specimens of *P. troilus* which have a pattern in wing cell M_3-Cu_1 that is intermediate between the normal pattern for that wing cell and that characteristic for the other wing cells (Fig. 16b). It is possible to find a broad range of intermediates from nearly "normal" to nearly identical to the adjacent patterns. It seems most likely that such intermediates, as well as the normal differences in pattern between

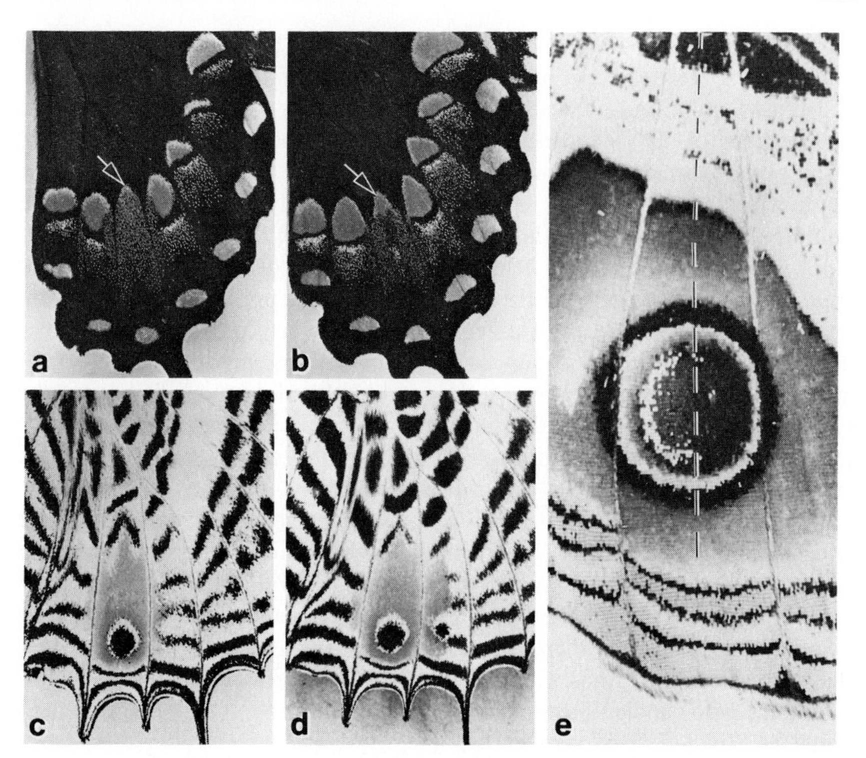

Fig. 16. Examples of abrupt changes in pattern interpretation at wing veins and on the wing cell midline. (a) Normal pattern of *Papilio troilus*; arrow points to wing cell with distinctively different pattern. (b) Naturally occurring pattern aberration in which this pattern is largely transformed to resemble that in the other wing cells. (c and d) Hindwings of *Baeotus baeotus* (Nymphalidae), showing variation in the extent of the orange field. In (c) this field ends very nearly on a wing vein; in (d) it end precisely on a wing cell midline. (e) Pattern in *Vanessa virginiensis* (Nymphalidae). Wing cell midline is indicated by dashed line; characteristics of the ocellus are different on either side of the midline.

wing cells of *P. troilus*, are attributable to differences in signal interpretation. This suggestion is strengthened by a most peculiar analogous instance in the genus *Baeotus* (Nymphalidae) shown in Fig. 16c,d.

The ventral wing surfaces of *Baeotus baeotus* have an overall pattern of black dashes on a white background. The color pattern in each wing cell of the ventral hindwing is similar except for the wing cell between veins Cu_1 and Cu_2, which bears a large orange field with a black spot at its center. Normally the orange field is confined by wing veins Cu_1 and Cu_2. In many specimens, however, the orange field "spills over" vein Cu_1 into the neighboring wing cell. Wherever this orange color occurs in the invaded wing cell the normal pattern is obliterated. The orange field never spreads across the entire invaded wing cell but seems to meet an invisible barrier at the wing cell midline (Fig. 16d). I have examined over 80 specimens of *Baeotus* in four collections and found no instance in which the orange field extended beyond the midline of the invaded wing cell. Within the orange field, pattern formation obviously follows a different set of rules, as shown by the fact that the normal black-on-white pattern stops precisely at the orange–white interface, no matter where that may lie, and by the development of a *half-black spot* where the orange field meets the midline of the invaded wing cell (Fig. 16d). The orange field therefore appears to represent a region within which signal interpretation is different from elsewhere on the wing, and the black spot that normally occurs on the wing cell midline in an orange field may well mark the position of a focus, though this cannot be known for certain without experimental intervention. Figure 16e illustrates a different example of a pattern discontinuity at a wing cell midline in *Vanessa virginiensis.*

Discontinuities of pattern series on a wing may involve more than one wing cell. In the Satyridae there are many cases in which two wing cells bear a characteristically different pattern from that of the wing cells that flank them on either side, and in *Charaxes etesippe* (and to a lesser degree in other *Charaxes* and *Polyura*) there is a boundary at wing vein M_2 that separates the wing into two regions with different pattern morphoclines [whether this boundary bears a significant relation to the compartment boundary elucidated by Sibatani (1980) is not clear].

5.7 MORPHOCLINES AND PHENOCOPIES

5.7.1 *Morphoclines*

As illustrated above, the color patterns in the wing cells of many species of Lepidoptera can be arranged in a smooth series, or morphocline. There is sufficient individual variability in most species that such a morphocline can be extended well beyond the range of patterns represented in any one specimen of a given species. Furthermore, if there are additional species in the genus it is usually possible to use their patterns to extend the endpoints of such a morphocline considerably, often with relatively few gaps. Figure 17 illustrates portions of two such morphoclines excerpted from much larger series that I have been building up over the past several years and which now contain well over 250 interconnected patterns (Nijhout, 1985). From such studies, several interesting features emerge that are of some significance in interpreting both the development and evolution of color patterns. First, it is possible to build up fairly long linear series in any one pattern element. Second, many such series can be interconnected as branching networks.

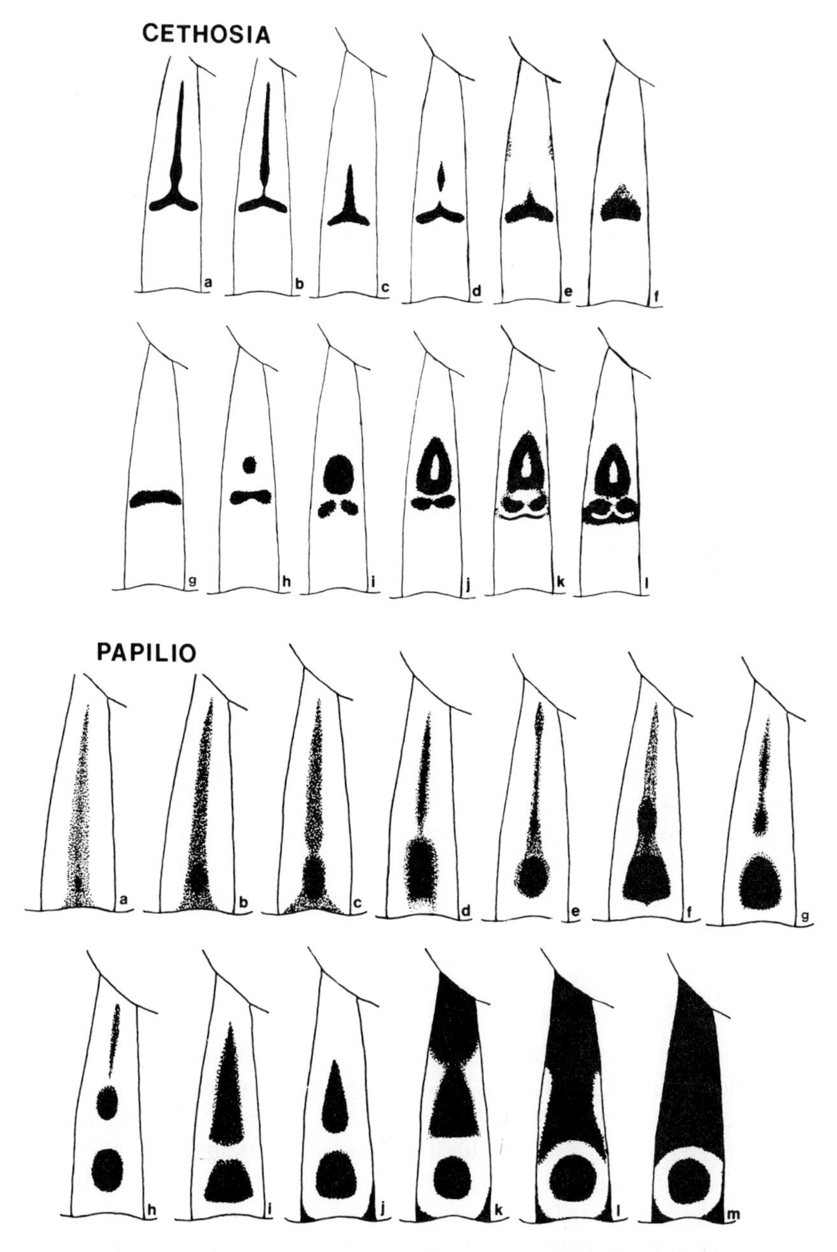

Fig. 17. Top panel, pattern morphocline for *Cethosia*. Species represented are *C. mirina* (a–g), *C. penthesilea* (h, i), *C. biblis* (j–l). Bottom panel, pattern morphocline for *Papilio memnon* (a–h) and the related *P. rumanzovia* (j–m). The forms of *P. memnon* illustrated are *retorina* (a), *butleriana* (b), *esperi* (c), *male* (d), *anceus* (e, f), *trochila* (g, i), *laomedon* (h).

Third, parallel series occur in many taxa. For instance, smooth transitions from a intervenous stripe to a round focal pattern (such as illustrated for *Papilio* in Fig. 17) are also found in *Acraea* (Acraeidae), *Idea* (Danaidae), and several genera of Nymphalidae and Satyridae. Many other pattern parallelisms occur between groups in different families. These characteristics of pattern morphoclines lead to the following hypotheses. In cases in which one finds smooth transitions from one morphology to another it is highly probable that the generating physiologies will be continuous as well. That is, adjacent patterns in a morphocline are probably generated by mechanisms that differ only quantitatively. Furthermore, any portion of a morphocline that is unbranched and unbroken probably represents a progressive change in a single variable, or in a coordinately controlled set of variables, for the simple reason that if several variables changed independently one would expect to generate either branched series, or a set of series that are noncongruent when their arrangements are based on different aspects of or portions of the pattern. Not unexpectedly, such noncongruences are rather frequent (and quite evident in Fig. 17a) and introduce a certain arbitrariness into any given arrangement of those patterns. Alternative morphoclines could therefore be used to study or emphasize different sets of variables.

Finally, while an occasional similarity of pattern in members of different taxa could be ascribed to character convergence, the finding that there are extreme *parallelisms in the morphoclines* of many taxa strongly argues for extensive similarities in the pattern-generating mechanisms of those taxa. If this is true the study of morphoclines could shed considerable light on the apparent ease with which pattern mimicry evolves among the butterflies, even though we know very little about the developmental mechanisms underlying those patterns. If primitive extremes and homoplasies can be identified, then pattern morphoclines could also provide data for phylogenetic analysis (Hennig, 1965).

5.7.2 *Phenocopies*

When organisms are subjected to sublethal trauma they may experience certain developmental abnormalities known as phenocopies. Goldschmidt (1935a) coined the term *phenocopy* as a descriptor for this phenomenon because the aberrations so produced in *Drosophila* resemble the phenotypes of several known mutants. By applying the inducing stimulus (usually some physiological stress such as heat shock, etherization, or treatment with various irritating or metabolically active chemicals) at various times it was found that specific phenocopies can be induced only during fairly brief critical periods during embryogenesis or metamorphosis (Goldschmidt, 1935a,b; Gloor, 1947; Capdevila and Garcia-Bellido, 1974; Mitchell and Lipps, 1978).

The kind of phenocopy that is induced depends only on the timing of the stress, not on the type of stimulus used. Thus *bithorax* phenocopies can be induced by ether (Capdevila and Garcia-Bellido, 1974) but also by heat shock and by treatment with phenol (Gloor, 1947). It now appears (see below) that physiological stress interferes indirectly with the expression of certain genes, and the critical period for specific phenocopy induction is probably a function of the time during which the product of the gene in question plays an essential role in the development of the wild-type phenotype.

Phenocopies induced after a given stress stimulus exhibit variable penetrance and expressivity. That is, not all individuals in a developmentally synchronous cohort develop an aberrant phenotype, and those that do, develop it to different degrees (Goldschmidt, 1935a,b; Gloor, 1947).

There exists a long tradition among lepidopterists of inducing color pattern aberrations by exposing developing pupae to abnormally high or low temperatures (see Kühn 1926, for a particularly well-documented example; and Prochnow, 1929; Goldschmidt, 1938, for reviews). Goldschmidt (1938) recognized these pattern aberrations as analogous to phenocopies in *Drosophila*. Color pattern phenocopies in Lepidoptera differ from those in *Drosophila*, of course, in that they cannot be compared with existing mutants (a case in *E. kühniella* is an exception, see below). There are, however, a large number of species of Lepidoptera, each with a characteristic color pattern, and it is not uncommon to find significant resemblances between the temperature-induced phenocopies of one species and the normal color pattern of a different species (e.g., the parafocal elements of *Precis*, Fig. 18 and *Vanessa*, Fig. 19). It has also been noted that color pattern phenocopies occasionally resemble the "normal" color pattern of certain geographic races of the species in question (Goldschmidt, 1938), suggesting that such geographic varieties may have arisen through genetic assimilation of a phenocopy (Waddington, 1953; Rachootin and Thompson, 1981). Kühn and von Engelhardt (1936) have shown that pattern phenocopies are easily induced by heat shock in *Abraxas grossulariata* (Geometridae) and that different portions of the color pattern have slightly different sensitive periods during which they are maximally altered.

Nijhout (1984) has recently shown that cold-shock-induced phenocopies in several species of Nymphalidae can be arranged in single smooth morphoclines (e.g., Fig. 19). A phenocopy morphocline for the pattern that develops around a single focus shows no noncongruences, which strongly suggests that such a series is generated by progressive alteration of a single variable. Some insight into what that variable could be is obtained in context of the model for phenocopy induction in *Drosophila* proposed by Mitchell and Lipps (1978). The Mitchell/Lipps model draws a causal relation between the *heat*

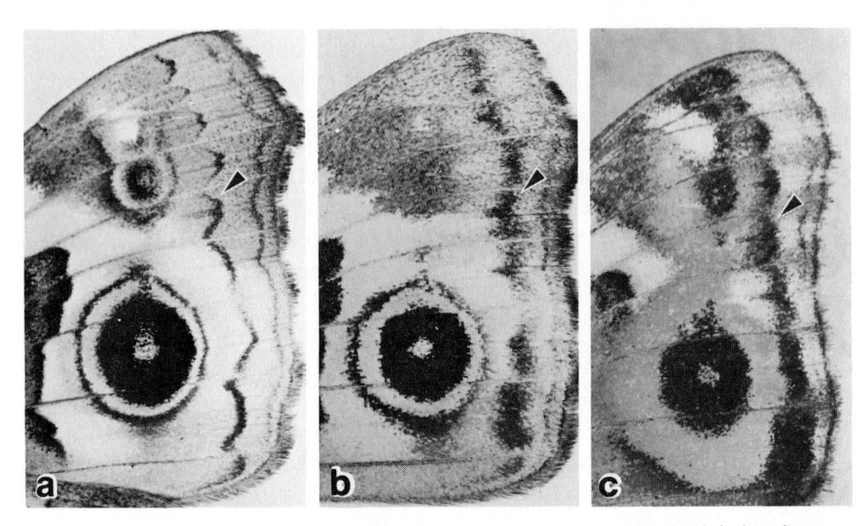

Fig. 18. (a) Normal pattern on ventral forewing of *P. coenia*; (b) cold-shock-induced pattern phenocopy in *P. coenia*. The pattern aberration consists of a loss of the small anterior ocellus and a modification of the shape of the parafocal elements. The latter are normally arcs or shallow W's and have become considerably broadened and have developed a proximally pointing peak along the wing cell midlines (arrow). (c) Normal pattern on the ventral forewing of *Precis vilida*; the morphology of the parafocal elements is almost identical to (b).

shock proteins that are synthesized by many organisms following a great diversity of environmental stresses, and the subsequent development of an abnormality that is a phenocopy. Some heat shock proteins are transcription inhibitors and cause a virtual wholesale shutdown of gene transcription. Presumably this constitutes a protection mechanism that temporarily stops development when a severe but sublethal stress is experienced. When the environmental stress is relieved, development picks up where it left off because heat shock proteins disappear and the transcription of genes that were active prior to stress resumes. Mitchell and Lipps (1978) and Chomyn *et al.* (1979) have shown that recovery is not synchronous at all genetic loci, and that some loci recover only partially and yield less product than is characteristic for an unstressed individual. Mitchell and Lipps (1978) suggest that

---

Fig. 19. Pattern phenocopies in *Vanessa*. (a) Normal pattern on the ventral hindwing of *V. cardui*. (b–f) Range of cold-shock-induced phenocopies (after Nijhout, 1984). Pattern aberration consists of the development of a proximally pointing peak on each parafocal element (arrows) and a fusion of this element with the neighboring ocellus along the wing cell midline (b, c). In more severe aberrations the fusion of parafocal element and ocellus is complete (d, e) and the ocellus vanishes (f). (g and h) Normal pattern and a cold-shock-induced phenocopy, respectively, on the ventral hindwing of *V. virginiensis* (after Nijhout, 1984). The phenocopy shows that what appears to be a submarginal band in the normal pattern (arrow in g) can be broken up,

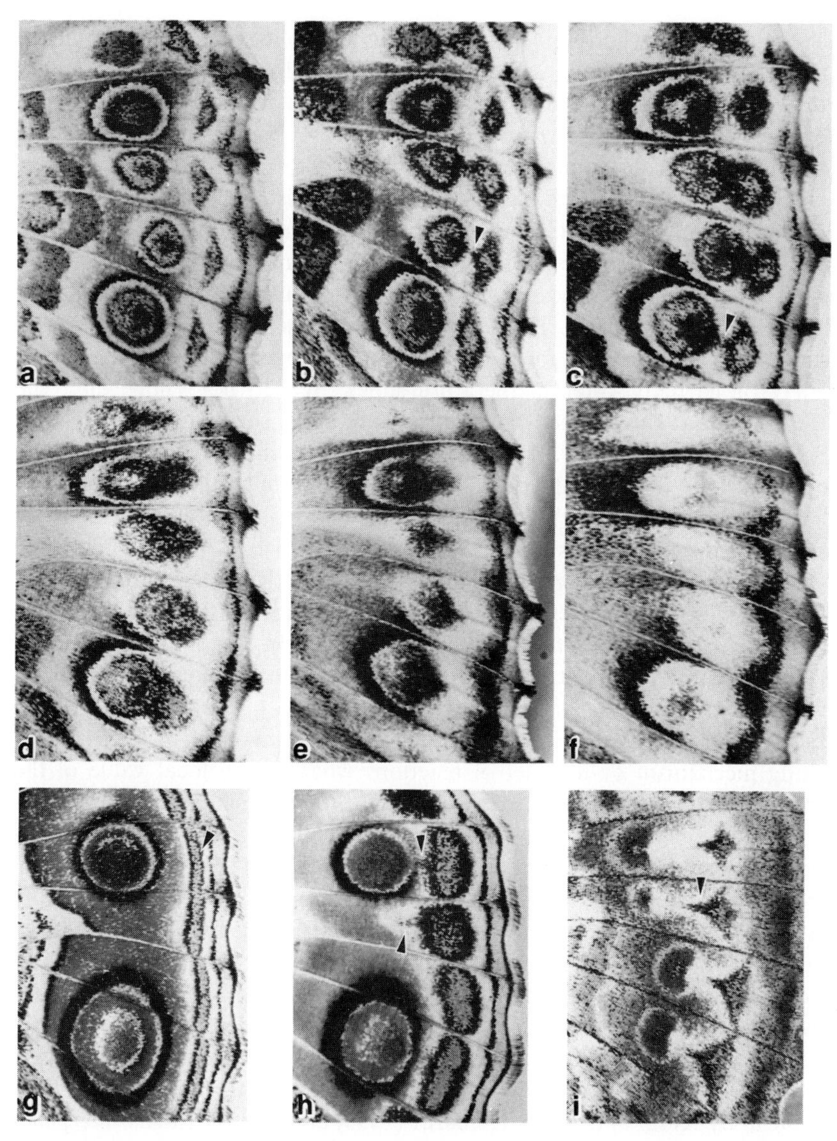

broadened, and can develop proximally pointing peaks along the wing cell midlines (arrows in h). This band is thus revealed to be a very elongate parafocal element, and the phenocopy bears considerable resemblance to the pattern in b. (i) Normal pattern of the ventral hindwing of *V. tameamea*. The parafocal elements in this species are naturally pointed toward the focus along the wing cell midline and in this feature they resemble phenocopies of the previous two species. These pattern modifications are similar in character to those found in *Precis* (Fig. 18) and also to the experimentally modified pattern of *Speyeria* (Fig. 14b), which may give a clue to the developmental parameter that is altered in the phenocopy.

phenocopies arise when a given genetic locus fails to regain normal activity in time to participate in ongoing development. While the Mitchell/Lipps model can explain the development of a phenocopy in principle, it does not, however, explain how a *specific* phenocopy can be linked to a stimulus arriving at a specific critical period.

Nijhout (1984) has shown that the frequency distribution of the various expressivities of color pattern phenocopies resembles a Poisson distribution. This finding indicates the operation of a single rare stochastic event in the origin of these phenocopies. Thus the variability in penetrance and expressivity of a phenocopy (e.g., Fig. 19) may reside in the fact that a genetic locus critical for color pattern formation failed to recover activity, either completely or partially, after temperature shock. We do not know what the function of that genetic locus is, though there are basically two possibilities. (1) We could be dealing with a gene that is involved in gradient formation. It could affect, for instance, the rate of morphogen production, the rate at which it is transmitted from cell to cell, or the rate with which it reacts with other cellular components or is degraded. Then, since there is only a finite time available for positional specification, variation in the "rate" at which this occurs would yield a series of patterns that express different temporal endpoints in one of the determinants of the pattern. In effect, the morphocline in Fig. 19 would represent a series of progressively more severe heterochronies. (2) Alternatively, it is possible that the gene in question is involved in gradient interpretation. For instance, it could be involved in a threshold-setting mechanism or in a set of reactions whereby the local value of the positional signal is translated into the synthesis of a particular pigment. If this is the case, then the morphocline in Fig. 19 would represent a continuous series of alternative ways of interpreting a given (set of) gradient(s). With the techniques presently available in this system it is impossible to critically distinguish between these two alternatives. In any case, since pattern phenocopy morphoclines appear to result from variation in a single developmental parameter they provide information about which *kinds* of patterns are developmentally adjacent.

The recognition that phenocopies form morphological series makes it possible to provide an explanation for a puzzling finding, first reported in *E. kühniella* by Kühn and von Engelhardt (1933) and later in *Plodia interpunctella* by Wehrmaker (1959) and Schwartz (1962). These investigators described a critical period in the pupal stage (between 36 and 72 h after pupation in *Ephestia*) during which cautery *anywhere* on the wing surface could dramatically alter the morphology of the color pattern that subsequently developed. A synchronous cohort of experimental animals developed a broad range of aberrations of their central symmetry system which could be arranged in a single smooth unbroken series (Fig. 20). Kühn and Henke

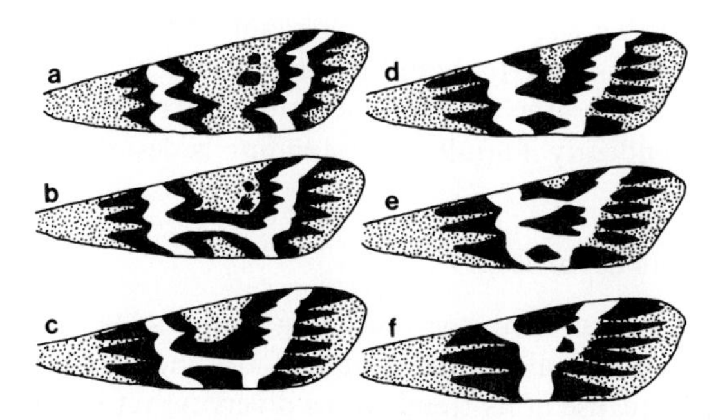

Fig. 20. (a) Normal pattern; (b)–(f) cautery-induced phenocopy series of *E. kühniella*, showing alterations of the central symmetry system. Only the central field appears to be affected and the outer extremes of the pattern remain in their normal position. (e) shows that "contraction" of the pattern is around three sources. (Redrawn from Kühn and von Engelhardt, 1933.)

(1936) have shown that pattern aberrations identical to the first half of this series can also be induced by giving pupae of *Ephestia* a brief temperature shock during the same critical period. Furthermore, portions of the morphocline (Fig. 20) also resemble the phenotype of a pattern mutant in this species (Kühn and Henke, 1936). Thus it appears that in *Ephestia* (and also in *Plodia*) simple cautery, done at the right time, can provide a sufficient stimulus for the development of a phenocopy.

6 Models and mechanisms

I do not believe that there exists another developmental system that is physically as simple and morphologically as diverse as lepidopteran color patterns. The diversity of color patterns is, initially at least, both overwhelming and confusing. Far from being an obstacle to analysis, however, this enormous diversity can be used as a tool in the analysis of the origin of color patterns because we are in effect provided with an uncommonly large number of morphologies that we can be certain are closely related to one another, and that provide an unparalleled richness of material for the development of theoretical models on the origins of spatial patterns in biology.

In this section I hope to provide a method, or at least an outline, for giving some organization to the diversity of patterns and to explore the characteristics of patterns that any model ought to be able to account for. I shall also examine, though necessarily somewhat cursorily, the potential usefulness and weaknesses of various models and mechanisms that could be evoked to explain or analyze the ontogeny of color patterns. The sections that follow deal specifically with color pattern formation in butterflies, because their patterns are the most complex. If butterfly patterns can be dealt with successfully, then we will have the terms required to account for most of the patterns of moths as well. It should be understood, though, that each taxon has pattern idiosyncrasies that do not transfer and that will have to be analyzed independently if a full understanding is desired.

6.1 COMPONENTS OF DIVERSITY

It should be evident from the preceding sections that lepidopteran color patterns present a rich and varied system of morphologies. It is a system that is complicated because of its tremendous diversity, yet it appears to be built on a fairly simple plan of design which makes this a system ideally suited for the study of comparative development and evolution. In order to lay a basis for such studies it is necessary to recognize that the *diversity of patterns* has three quite distinct components that, depending on the level at which one wishes to conduct an analysis, can each be dealt with independently from the others. (1) There is taxonomic or phylogenetic diversity in that there always exist gross and detailed differences in the patterns of any two taxa. This aspect of diversity is commonly recognized, has a strong historical component, and reflects adaptations of color pattern to different demands, internal and external. (2) Within a species, the color patterns of the four wing surfaces are always different. We know that pattern development on one wing surface is independent from that on the other surfaces (Nijhout, 1978). Whether there exist different "gene sets" for the pattern on each wing surface is unknown but not unlikely. (3) Within a wing there is a diversity of wing cell patterns. Each of these is independent from the others yet there is a strong common theme. The pattern in each wing cell gives the impression of being a permutation of a basic pattern characteristic for the wing surface. As noted above (Figs. 15 and 17), these patterns can often be arranged in a smooth morphological series, which suggests that different wing cell patterns on the same wing differ only in quantitative aspects of one or more determinants of pattern. Of course, there can, and often do, exist qualitative differences between the color patterns of adjacent wing cells, for instance in cases in which an ocellus develops in one wing cell and not in another. Such qualitative differences are in essence nothing more than the presence/absence of one of the determinants of pattern and as such they can be regarded as extremes, or endpoints, in the quantitative variation of that determinant.

Superimposed on these three components of diversity, and contributing to each, is a widespread *genetic diversity*. Many species have distinctive geographic races which, while often dramatically different in pattern morphology, are genetically compatible with each other (for instance, Clarke and Sheppard, 1971, 1972, 1975), and upon hybridization yield, as one might expect, offspring with patterns intermediate between those of the parental races. Genetic diversity also occurs within populations, as evidenced by the widespread occurrence of seasonal and sexual pattern polymorphisms.

In addition to diversity there is also a great deal of individual *variability* in the color pattern. There are individual differences in the positioning of homologous pattern elements and in their precise shapes. Within individuals there are less extensive but usually noticeable differences between left and right wings. It is often possible (at least for the willing observer) to identify individuals on the basis of details of their pattern. There are two sources of variability: (1) genetic variation, which is due to the diversity of alleles for a given gene in a population; and (2) ontogenetic variation, which refers to individual differences in morphology that arise by virtue of the noise (or freedom) that exists in developmental pathways, and of the errors that are inevitably made in the development of a multicomponent system. Furthermore, certain processes in pattern formation appear to have a stochastic component. For instance, wherever a gradient of color exists, this comes about by a gradient in the proportion of scales that are one color or another. In such a transition zone the distribution of scales of any one color appears to be random. A different situation is illustrated in Fig. 21 which shows patterns that consist simply of local variations in the proportions of scales of two contrasting colors. It seems as if different regions of a wing cell merely differ in the probability with which a particular scale color is induced. The point is that in such cases scale color determination appears to be probabilistic, not

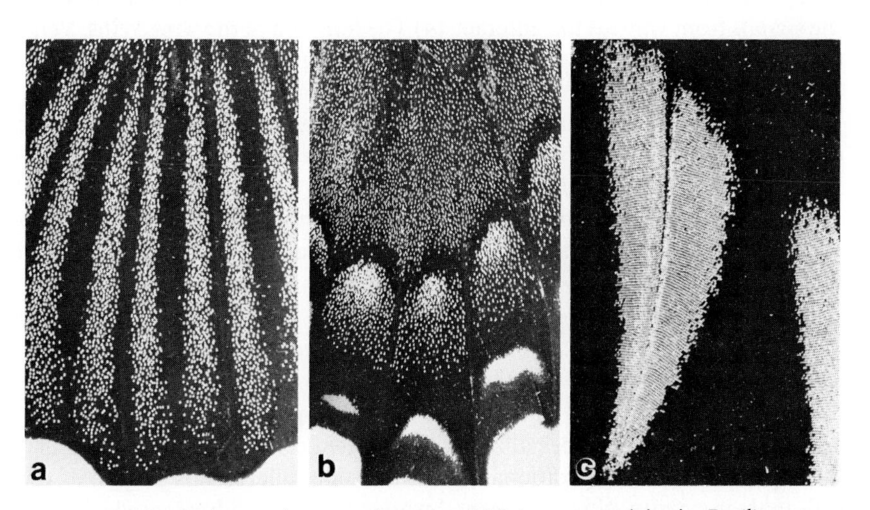

Fig. 21. Stochastic elements in pattern formation. (a) Intervenous striping in *Papilio protenor*; (b) hindwing patterns in *Papilio glaucus*, showing gradients and patterning in scale density; (c) forewing pattern in *Trogonoptera brookiana*, showing stochastic color determination in scales at what is apparently a sharp boundary.

deterministic. There is a hint of a similar stochastic component in the development of a more sharply defined patterns as well. Figure 21c, for example, illustrates a region with a sharp boundary between two colors and where an occasional scale has clearly developed an inappropriate color. Mistakes of this kind are the rule and it seems at least possible that color specification at the level of the scale-building cell is always probabilistic.

6.2 DETERMINANTS OF PATTERN

Diversity of pattern implies a corresponding diversity in the determinants of pattern. While the determinants of pattern are rather few in number, and can be readily enumerated, the preceding sections of this review illustrate that each can vary in qualitative and quantitative ways from species to species. It is this diversity in the properties and attributes of the determinants that probably accounts for most of diversity of patterns. The determinants of pattern are the following. (1) The foci on the wing cell midline. These may vary in number, in precise position (but restricted to the midline), and in shape (that is, they may be point or line sources). (2) The characteristics of the signal(s) emitted from the foci. Foci may differ in signal strength and in the timing of their activity. If the interpretation landscape indeed describes a gradient that originates from a source situated near the base and/or margin of a wing cell, as was suggested above, then it follows that foci may also be heterogeneous, differing in the kind of signal they emit. (3) The way in which the signals from various foci interact. (4) The function of the wing veins. Veins appear to commonly act as sinks for the focal signal(s), but there are many cases in which they seem to have no function at all. It is possible that wing veins may also act as signal sources. Presumably all of the above can be time dependent. (5) The way in which the signals are interpreted by the scale-building cells. This could involve a diverse array of variables from signal modification (as in reaction diffusion) to the dynamics of thresholds and (de)-repression of genes. This determinant may not always be distinguishable from 3.

6.3 THE ORIGIN OF FOCI

Nothing has been said so far about the developmental origin of the foci. The foci are themselves specialized, differentiated cells or groups of cells that are capable of emitting positional signals and that differ in this property from cells in their vicinity. No experimental work exists that addresses the question of how the characteristics and the position of a focus are specified.

The position of a focus is clearly referenced to the wing veins, in that it always occurs on the wing cell midline and this position is, with one notable exception, independent of the shape and the dimensions of the wing cell. The exception is the foci that occur in the wing cell bordered by veins Cu_2 and 2A (Figs. 22b, 6b, and 7c). In the primary venation system (Section 2.1) there is an additional vein (1A) that divides this wing cell. A double color pattern always develops, showing that the determinants of pattern (foci, etc.) are established at the time that the primary venation system exists. The two foci in wing cell Cu_2-2A are, however, not positioned precisely on the midlines of the two primary wing cells, but are shifted laterally toward the midline of the new secondary wing cell (a position formerly occupied by vein 1A). The failure to place these particular foci precisely on the midlines may be due to the fact that their determination is only weakly, or only temporarily, influenced by vein 1A, which is programmed to degenerate.

The morphocline of wing cell patterns illustrated in Fig. 17 for *Papilio* suggests that the color pattern may be affected by the simultaneous presence of point sources and of line sources along the wing cell midline. It is evident from Fig. 17 that a line source need not extend along the entire length of the wing cell and that line sources may break up (ontogenetically, at least) into several point sources. Models that attempt to mimic the origin of foci, therefore, not only must account for their position on the midline, but should also, preferably by simple quantitative changes of constants, be able to account for both the origin of line sources along part or along the entire midline, and the origin of one, two, or three point sources, spaced roughly as indicated in the patterns of Fig. 17. Any attempt to model the origin of foci

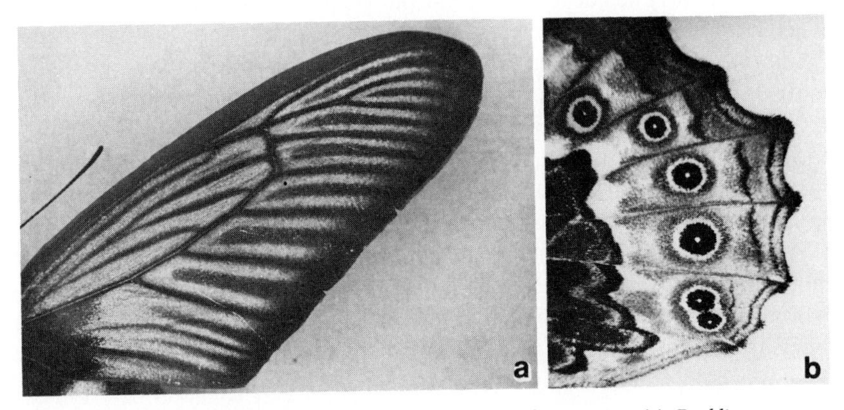

Fig. 22. Dependence of color pattern on primary venation system. (a) *Pachliopetra oreon* (Papilionidae), showing dependent patterns within the discal wing cell that reflect the former course of the base of the media vein and of vein 1A, veins that do not exist in the adult (secondary) venation pattern. (b) Hindwing of *Neope goschkewitschii* (Satyridae), showing two ocelli in one wing cell after degeneration of vein 1A. The two ocelli do not appear to be positioned exactly on the midlines of the two primary wing cells.

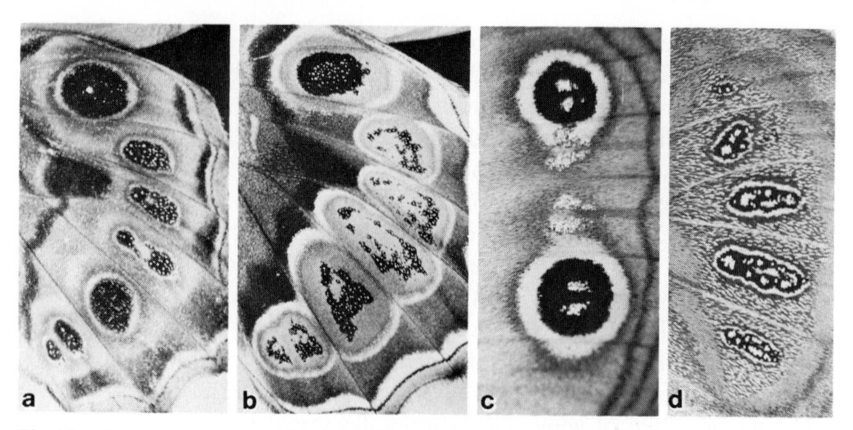

Fig. 23. Examples of unusual border ocelli with "fragmented" foci. (a) *Lethe dyrta* (Satyridae); (b) *L. europa*; (c) *Euptychia cymela* (Satyridae) with what appear to be two foci in each ocellus; (d) *E. areolatus* with much elongated ocelli that extend along the wing cell midline and suggest an elongated source or a roughly linear array of sources.

should also take note of the patterns in the genera *Euptychia* and *Lethe* (Satyridae). These genera are unique in the diversity of shapes of their border ocellar patterns (for illustrations see Fig. 23; and Seitz, 1924), and it seems that much of this is due to a corresponding diversity in the shape and arrangements of foci. In many cases there appear to be multiple foci that occur in irregular though confined clusters that may be indicative of a stochastic component in source determination.

Murray (1981) has developed a morphogen diffusion model for the origin of intervenous stripes that relies on a signal emitted from the wing veins. While the wing cell midline is a more probable source for the signal that controls intervenous stripes (Fig. 25), Murray's model may prove to be useful in investigating the origin of sources (foci) rather than of patterns, as source position is likely to be cued by the wing veins. It is of interest to note that the position of foci on the dorsal and ventral surfaces of a wing usually correspond precisely, even though subsequent development of the pattern on the two surfaces proceeds independently (Nijhout, 1978). It may be that some communication exists between the two epidermal layers during focal specification, or that foci end up in the same relative place independently because the wing cells on both surfaces are identical in size and shape and possess the same physiological determinants for foci.

6.4 CONSTRAINTS ON MODELS

Color pattern determination appears to proceed as a three-step process. First is the determination of the position and properties of the signaling sources and sinks. This is followed by an activation of these sources and the establishment of gradients in the substances or signals they produce. The third step is the "interpretation" of these gradients by the scale-building cells and their determination to form a specific pigment. There is at this time no reason to believe that these three processes interact (for instance, the signal emitted from a focus is probably not a component of the process that determines the precise location of that focus). Thus modeling and analysis at any of these three levels of pattern determination can be independent of knowledge about the other two. Performance requirements for models for source positioning were mentioned in the preceding section and will not be dealt with further. The present section deals only with constraints on models for the second step in the process, gradient formation. Throughout it will be assumed that the main element of the third step is a threshold-sensing function (Lewis *et al.*, 1977; MacWilliams and Papageorgiou, 1978).

6.4.1 *Basic requirements for models*

The simplest patterns that can be found are circles, or sets of concentric rings. It has been argued above that the majority of patterns are derivatives or permutations of such circular elements. Hence, a model for the determination of *any* pattern must also be able to account for the development of circles. Nijhout (1980a) has shown that the ontogeny of an ocellus in *P. coenia* depends on the continuous presence of the focus at its center. If the focus is removed soon after pupation no ocellus develops. If the focus is removed at progressively later times ever larger eyespots are allowed to develop (Fig. 24a–d). Thus during determination an ocellus "grows" in the sense that, if determination is allowed to proceed for a relatively brief period of time, a small eyespot is formed, and the longer the focus is active the larger the subsequent eyespot will be (see Nijhout, 1980b, for a time table of pattern determination and differentiation in *Precis*).

The temporal progress of determination of an ocellus is illustrated in Fig. 24a. This time course can be thought of as the progression of a particular threshold value of the focal signal. The data are consistent with propagation by simple diffusion and can be used to calculate an effective diffusion coefficient of about 1.5×10^{-8} cm^2/s at 29°C. Assuming that the morphogen decays while it diffuses, its real diffusion coefficient is likely to be somewhat higher. Any model for ocellus determination must mimic these dynamics. It must also be sensitive to the continuous presence of the source so that propagation of a critical value ceases very soon after elimination of the source.

An important qualitative aspect of ocellus development must be accounted for by any model, and that is that "growth" occurs by central addition, not by

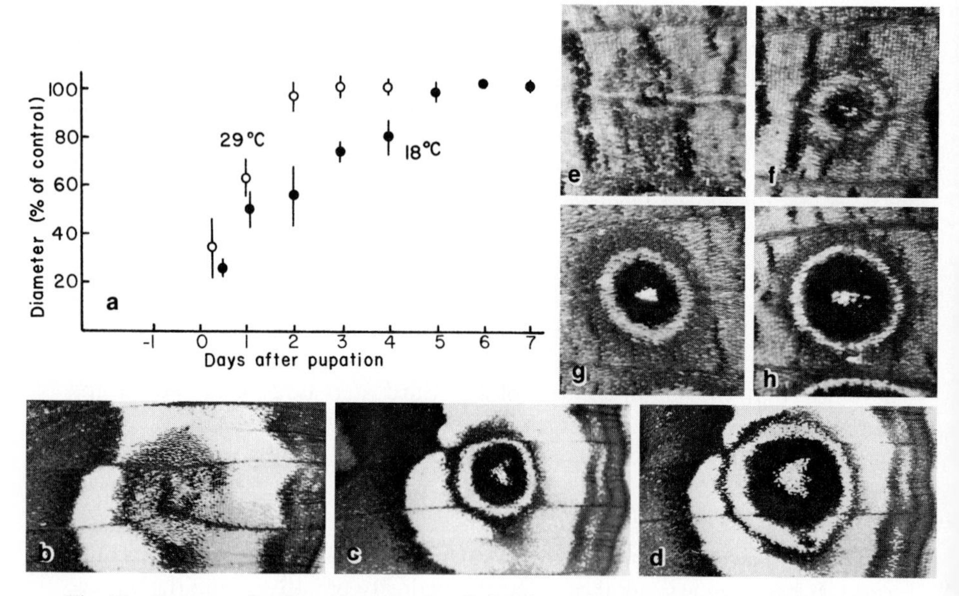

Fig. 24. Progress of pattern determination in border ocelli. (a) Graph showing progress of ocellus determination at two different temperatures in *Precis coenia.* The graph shows the diameter of the ocellus that develops after its focus is killed by cautery at various times after pupation. Cautery stops signal emission by the focus and evidently freezes the progress of determination (after Nijhout, 1980a). (b–d) Examples of ocellar morphology after cautery (in *P. coenia*) showing that the outer bands are determined before the inner portions of an ocellus. (c–h) Variation in an ocellus on the hindwing of *Cercyonis pegala* (Satyridae). This species shows considerable individual variability in the size of many of its ocelli, but in all cases the central portions of an ocellus are most variable, while the outer rings diminish only in cases where the black central disc is entirely wanting.

peripheral accretion (Fig. 24). This characteristic was noted long ago by Bateson (1894).

6.4.2 *Diffusion as a mechanism*

Simple diffusion from continuously active point sources, coupled with the assumption that a particular pigment is synthesized wherever the morphogen is above (or below) some threshold value (as, for example, in the mathematical model in Section 5.5.1), can generate circular figures whose dynamics of growth and sensitivity to persistent activity of a source satisfy the constraints mentioned above (Nijhout, unpublished results). If diffusion is the mechanism whereby a morphogen is transmitted from cell to cell then this must occur through gap junctions and not through the intercellular medium, because the latter is continuous with the hemocoel and circulating hemolymph would rapidly carry away any substance that is excreted from the cells. Using gap junctions as channels for diffusion limits the size of the signaling molecule to a maximum in the neighborhood of 1000–1400 Da (Caveney and Podgorski, 1975; Caveney, 1980; Caveney and Berdan, 1982). This would remove sizeable proteins from the candidate list of morphogens but leaves anything from small polypeptides to the majority of organics. Richard G. A. Safranyos and Stanley Caveney (University of Western Ontario) have recently measured the rate at which medium-sized organic molecules diffuse within the epidermis of *Tenebrio molitor* and calculated effective diffusion coefficients of 3.7×10^{-7} cm^2/s for 6-carboxyflourescein (MW 376) and 1.2×10^{-7} cm^2/s for lissamine rhodamine B 200 (MW 559) (personal communication). These figures are quite close to the effective D of 1.5×10^{-8} cm^2/s calculated for the putative morphogen in eyespot morphogenesis (see above) and supports the contention that diffusion through gap junctions is an adequate mechanism for signal transmission.

6.4.3 *Further constraints*

Since simple diffusion yields adequate results for the simplest morphologies, attempts to model more complex patterns might involve some forms of reaction–diffusion. However, most models that include a significant autocatalytic component for the morphogen, such as the forms proposed by Gierer and Meinhardt (Gierer and Meinhardt, 1972; Gierer, 1981; Meinhardt, 1982), will probably prove to be inadequate since they tend to make continued growth of the pattern independent of continuous activity of the initial source (models of the Gierer/Meinhardt type may, however, be quite useful for modeling earlier events, such as the distribution of foci). Other reaction–diffusion mechanisms such as the Belousov–Zhabotinsky reaction (Winfree, 1980) or the reactions involved in the formation of Liesegang patterns (which have been repeatedly contemplated as possible ways to account for periodic patterns such as those in Fig. 10a; discussions in Goldschmidt, 1938; Henke, 1948) are also inappropriate because they either are self-propagating or grow by peripheral accretion, and ocelli do not.

As argued above, to generate figures other than circles more information is needed, which will most likely be found in the form of additional sources and sinks for diffusible morphogen(s). While the possibilities for source distributions and morphologies seem quite large, there exist significant constraints: (1) Sources are restricted to the wing cell midline. (2) The maximum number of point sources appears to be three which, when present, are fairly evenly spaced. (3) Sources may also be line sources of various extents. (4) Sinks, when present, are restricted to the wing veins and wing margin. The number of permutations is finite, as are color patterns.

A wide range of choice exists in the models that one could adopt for the interaction of the morphogens produced by these sources. Here, there exist at present no a priori constraints to guide a choice, other than those posed by a reasonable biochemistry.

6.5 BASIC PATTERNS

The majority of Lepidoptera bear patterns that are permutations of the same general plan. This plan can be detected and studied at different levels of resolution. On a gross scale we have the permutations of the Nymphalid Groundplan that were worked out by Schwanwitsch (1924, 1926), Süffert (1927), Henke and Kruse (1941), and Sokolow (1936). The reason that patterns can be interpreted in such sensible ways is, as we have seen, that there are a limited number of determinants of pattern. As a rule, these determinants exert their influence at the level of the wing cell and that is, therefore, the level at which studies on the developmental physiology of pattern formation should concentrate. In this final section I want to present what I, at this time, perceive to be the basic color pattern for a "typical" wing cell of a butterfly, in the same sense as the Schwanwitsch/Süffert Groundplan (Fig. 5) represents the basic pattern for the wing as a whole. The premises on which this view is based were made explicit in the previous sections of this review.

There exist basic patterns in the distributions of sources and sinks, and basic patterns in the distribution of pigments, and the latter are related to the former in a causal manner. Figure 25 shows a scheme that relates the two and that presents a nomenclature, largely adopted from the Schwanwitsch/Süffert Groundplan. A significant addition to that classical scheme is the recognition of the parafocal pattern element as a distinct component of the pattern, at least in butterflies. Six pattern elements are recognized: (1) The intervenous stripe. (2) The venous stripe. (3) The central symmetry system, which, depending on the length of the wing cell, may be represented by two bands, or only by the distal band. (4) The border ocellus, which usually lies near the middle of the wing cell. This term is retained for historical reasons and to avoid the proliferation of unnecessary terminology, even though it is somewhat of a misnomer, as this pattern element more often than not bears little resemblance to an eyespot. (5) The parafocal element, which is morphologically the most diverse pattern element in the butterflies. (6) The submarginal bands, of which there are one or two. These are narrow bands that usually follow the outline of the wing margin faithfully. The distal submarginal band often coincides with the wing margin.

The three sources in Fig. 25 are labeled F_{css}, F_{bo}, and F_m to emphasize the fact that they are primarily responsible for controlling morphology of the

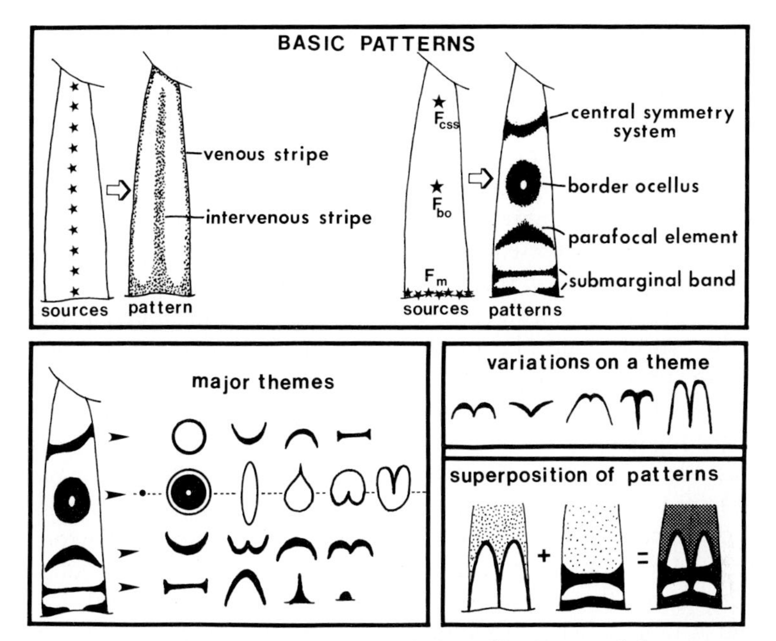

Fig. 25. A summary of basic wing cell patterns in butterflies. Top panel shows two arrangements of sources and the patterns determined by those sources. All, some, or none may be present in any one wing cell and the precise positions of sources along the wing cell midline is species specific. The bottom left panel shows the major themes of morphological diversity for each of the pattern elements. Such diversity in shapes may come about through interaction of the signals from various sources. For example, the midline peaks in the parafocal elements probably arise through interaction with a midline source or sink, while the general shape of this pattern element may be determined by the interaction of signals from various foci (see text). The way in which the shape of a pattern is influenced by adjacent sources is also seen in the morphology of the middle pattern element in Fig. 17k, l, m. The horizontal dashed line through the border ocelli themes is intended to show that this pattern element is sometimes represented only by its proximal or distal outline, just in the way that the central symmetry system may have only one or two bands within a wing cell. The parafocal elements often serve as a boundary between fields of different background coloration. On the right side are shown some variations on one of the parafocal element themes. Each of the other major themes shows a similar diversity of variations. The bottom right panel shows that some patterns come about through superposition of different elements. The marginal patterns of many Danaidae and of *Hypolimnas* appear to arise this way. Examples of pattern superposition appear to be restricted to the submarginal bands. Other pattern elements may fuse, but they never overlap.

central symmetry system, the border ocelli, and the marginal patterns, respectively. It is assumed that each focus emits a qualitatively different signal. Evidence from cautery experiments and phenocopy induction described earlier suggests, however, that patterns like the parafocal element may be affected by F_{css} and F_{bo}. Furthermore, the distal outline of the parafocal element often parallels the shape of the wing margin, so that it is possible that its morphology is also sometimes affected by F_m. The shape and location of F_m are unknown, and likely to be quite diverse. As noted above, the submarginal bands and parts of the parafocal element often parallel the outline of the wing margin perfectly, which would be most readily accounted for if a source (or sink) extended along the entire wing margin. This is the situation depicted in Fig. 25. There are other cases in which submarginal bands are absent, and more or less circular patterns develop near the wing margin, indicating the presence of a submarginal point source. Structural and pigmentary landmarks on the cuticle of many butterfly pupae coincide with F_{bo} (e.g., Nijhout, 1980a), and similar marks are frequently found on the wing cell midline at the wing edge on the pupal cuticle (e.g., Fig. 9a). These may indicate the position of F_m.

Although, in nature, the morphology of the pattern elements shown in Fig. 25 is exceedingly diverse, there are (surprisingly) relatively few *major themes*, and these are shown in the bottom left panel of Fig. 25. These basic patterns are actually very simple in form and it is fairly easy to imagine how such shapes could be controlled by the source distributions shown in the top panel of Fig. 25. Many of these patterns have "peaks" on the wing cell midline, and may come about through some sort of interaction between signals from the foci and a more extended midline source or sink, such as that normally responsible for intervenous striping. Further insights into the developmental basis for diversity in this system will likely come from the analysis, experimental and theoretical, of the origin of the patterns shown in Fig. 25. These patterns, I believe, also provide a suitable basis on which to develop a workable comparative morphology of color pattern for butterflies, and perhaps for macrolepidoptera in general.

Acknowledgments

I would like to thank Laura Williams for tireless assistance in preparing the manuscript and figures for this review and Sally Anderson for the drawings in Fig. 13. Support from the Duke University Research Council and from the National Science Foundation (Grant PCM-8214535) is gratefully acknowledged.

References

Anderson, T. F., and Richards, A. G. (1942). An electron microscope study of some structural colors in insects. *J. Appl. Phys.* **13**, 748–758.

Bateson, W. (1894). "Materials for the Study of Variation." Macmillan, New York.

Baust, J. G. (1967). Preliminary studies on the isolation of pterins from the wings of heliconiid butterflies. *Zoologica* **52**, 15–20.

Beck, S. D. (1968). "Insect Photoperiodism." Academic Press, New York.

Behrends, J. (1935). Über die Entwicklung des Lakunen-, Ader- und Tracheensystems während der Puppenruhe im Flügel der Mehlmotte *Ephestia kühniella* Zeller. *Z. Morphol. Ökol. Tiere* **30**, 573–596.

Blois, M. (1978). The melanins: Their synthesis and structure. *Photochem. Photobiol. Rev.* **3**, 115–134.

Braun, W. (1936). Über das Zellteilungsmuster im Puppenflügelepithel der Mehlmotte *Ephestia kühniella* Z. in seiner Beziehung zur Ausbildung des Zeichnungsmusters. *Roux' Arch.* **135**, 494–520.

Brown, K. S. (1981). The biology of *Heliconius* and related genera. *Annu. Rev. Entomol.* **26**, 427–456.

Capdevila, M. P., and Garcia-Bellido, A. (1974). Developmental and genetic analysis of *bithorax* phenocopies in *Drosophila. Nature (London)* **250**, 500–502.

Catala, R. (1940). Variations expérimentales de *Chrysiridia madagascarensis* Less. (Lep. Uraniidae). *Arch. Mus. Nat. Hist. Natur.* **6**, 1–262.

Caveney, S. (1980). Cell communication and pattern formation in insects. *In* "Insect Biology in the Future" (M. Locke and D. S. Smith, eds.), pp. 565–582. Academic Press, New York.

Caveney, S., and Berdan, R. (1982). Selectivity in junctional coupling between cells of insect tissues. *In* "Insect Ultrastructure" (R. C. King and H. Akai, eds.), Vol. 1, pp. 434–465. Plenum, New York.

Caveney, S., and Podgorski, C. (1975). Intercellular communication in a positional field. Ultrastructural correlates and tracer analysis of communication between insect epidermal cells. *Tissue Cell* **7**, 559–574.

Chomyn, A., Moller, G., and Mitchell, H. K. (1979). Patterns of protein synthesis following heat shock in pupae of *Drosophila melanogaster. Dev. Genet.* **1**, 77–95.

Choussy, M., and Barbier, M. (1973). Pigments biliaires des Lepidoptères: Identification de la phorcabiline 1 et de 1a sarpedobiline chez diverses espèces. *Biochem. Syst.* **1**, 199–201.

Clarke, C. A., and Sheppard, P. M. (1971). Further studies on the genetics of the mimetic butterfly *Papilio memnon* L. *Philos. Trans. R. Soc. London, Ser. B* **263**, 35–70.

Clarke, C. A., and Sheppard, P. M. (1972). The genetics of the mimetic butterfly *Papilio polytes* L. *Philos. Trans. R. Soc. London, Ser. B.* **263**, 431–458.

Clarke, C. A., and Sheppard, P. M. (1975). The genetics of the mimetic butterfly *Hypolymnas bolina* (L.). *Philos. Trans. R. Soc. London Ser. B.* **272**, 229–265.

Cockayne, E. A. (1924). The distribution of fluorescent pigments in Lepidoptera. *Trans. Entomol. Soc. London Ser. B.* 1–19.

Comstock, J. H. (1918). "The Wings of Insects." Comstock, Ithaca, New York.

Descimon, H. (1966). Variations quantitatives des ptérines de *Colias croceus* (Fourcroy) et de son mutant *helice* (Hbn.) (Lepidoptera Pieridae) et leur significance dans la biosynthèse des ptérines. *C.R. Acad. Sci. Paris* **262**, 390–393.

Descimon, H. (1975). Biology of pigmentation in Pieridae butterflies. *In* "Chemistry and Biology of Pteridines" (W. Pfleiderer, ed.), pp. 805–840. De Gruyter, Berlin.

Descimon, H. (1977). Les Ptérines des insectes. *Publ. Lab. Zool. E.N.S.* **8**, 43–130.

Enderlein, G. (1902). Eine einseitige Hemmungsbildung bei *Telea polyphemus* vom ontogenetischen Standpunkt. Ein Beitrag zur Kentniss der Entwicklung der Schmetterlinge. *Zool. Jb. Abt. Anat.* **16**, 571–614.

Ford, E. B. (1941). Studies on the chemistry of pigments in the Lepidoptera, with reference to their bearing on systematics. 1. The anthoxanthins. *Proc. R. Entomol. Soc. London Ser. A* **16**, 65–90.

Ford, E. B. (1942). Studies on the chemistry of pigments in the Lepidoptera, with reference to their bearing on systematics. 2. Red pigments in the genus *Delias* Hubner. *Proc. R. Entomol. Soc. London Ser. A* **17**, 87–92.

Ford, E. B. (1944a). Studies on the chemistry of pigments in the Lepidoptera, with references to their bearing on systematics. 3. The red pigments of the Papilionidae. *Proc. R. Entomol. Soc. London Ser. A* **19**, 92–106

Ford, E. B. (1944b). Studies on the chemistry of pigments in the Lepidoptera, with references to their bearing on systematics. 4. The classification of the Papilionidae. *Proc. R. Entomol. Soc. London Ser. A* **19**, 201–223

Ford, E. B. (1945). "Butterflies." Collins, London.

Ford, E. B. (1947). A murexide test for the recognition of pterins in intact insects. *Proc. R. Entomol. Soc. London Ser. A* **22**, 72–75.

Garcia-Bellido, A. (1975). Genetic control of wing disc development in *Drosophila. Cell Patterning. Ciba Found. Symp.* **29,** 161–178.

Garcia-Bellido, A., Ripoll, P., and Morata, G. (1973). Developmental compartmentalization of the wing disk of *Drosophila. Nature* (*London*) *New Biol.* **245**, 251–253.

Ghiradella, H. (1974). Development of ultraviolet-reflecting butterfly scales: How to make an interference filter. *J. Morphol.* **142**, 395–410.

Ghiradella, H., and Radigan, W. (1976). Development of butterfly scales: II. Struts, lattices and surface tension. *J. Morphol.* **150**, 279–298.

Ghiradella, H., Aneshansley, D., Eisner, T., Silberglied, R. E., and Hinton, H. E. (1972). Ultraviolet reflection of a male butterfly: Interference color caused by thin-layer elaboration of wing scales. *Science* **178**, 1214–1217.

Gierer, A. (1981). Generation of biological patterns and form: Some physical, mathematical, and logical aspects. *Prog. Biophys. Mol. Biol.* **37**, 1–47.

Gierer, A., and Meinhardt, H. (1972). A theory of biological pattern formation. *Kybern.* **12**, 30–39.

Gloor, H. (1947). Phänokopie-Versuche mit Äther an *Drosophila Rev. Suisse Zool.* **54**, 637–712.

Goldschmidt, R. B. (1935a). Gen und Ausseneigenschaft (Untersuchungen an *Drosophila*) I. *Z. Indukt. Abstamm. Vererb. Lehre* **69**, 38–69.

Goldschmidt, R. B. (1935b). Gen und Ausseneigenschaft (Untersuchungen an *Drosophila*) II. *Z. Indukt. Abstamm. Vererb. Lehre* **69**, 70–131.

Goldschmidt, R. B. (1938). "Physiological Genetics," McGraw-Hill, New York.

Greenstein, M. E. (1971). The ultrastructure of developing wings in the giant silkmoth, *Hyalophora cecropia.* II. Scale-forming and socket-forming cells. *J. Morphol.* **136**, 23-52.

Henke, K. (1928). Über die Variabilität des Flügelmusters bei *Larentia sordidata* F. und einigen anderen Schmetterlingen. *Z. Morphol. Ökol. Tiere* **12**, 240–282.

Henke, K. (1933a). Untersuchungen an *Philosamia cynthia* Drury zur Entwicklungsphysiologie des Zeichnungsmusters auf dem Schmetterlingsflügel. *Roux' Arch.* **128**, 15–107.

Henke, K. (1933b). Zur morphologie und Entwicklungsphysiologie der Tierzeichnung. [in 4 parts] *Naturwissenschaften* **21**, 633–640, 654–659, 665–673, 683–690.

Henke, K. (1936). Versuch einer vergleichenden Morphologie des Flugelmusters der Saturniden auf entwicklungsphysiologischer Grundlage. *Nova Acta Leopoldina* **18**, 1–37.

Henke, K. (1943). Vergleichende und experimentelle Untersuchungen an *Lymantria* zur Musterbildung auf dem Schmetterlingsflügel. *Nachr. Akad. Wiss. Gött., Match.-Physo* **Kl**, 1–48.

Henke, K. (1948). Einfache Grundvorgänge in der tierischen Entwicklung. II. Über die Enstehung von Differenzierungsmustern. [in 3 parts] *Naturwissenschaften* **35**, 176–181, 203–211, 239–246.

Henke, K., and Kruse, G. (1941). Über Feldgliederungsmuster bei Geometriden und Noctuiden und den Musterbauplan der Schmetterlinge im allgemein. *Nachr. Akad. Wiss. Gött., Math.-Phys.* **Kl**. 1–48.

Henke, K., and Pohley, H.-J. (1952). Differentielle Zellteilungen und Polyploedie bei der Schuppenbildung der Mehlmotte *Ephestia kühniella. Z. Naturforsch. Abt. B* **7**, 65–79.

Hennig, W. (1965). Phylogenetic systematics. *Annu. Rev. Entomol.* **10**, 97–116.

Inagami, K. (1954). Mechanism of the formation of red melanin in the silkworm. *Nature (London)* **174**, 1105.

Köhler, W. (1932). Die Entwicklung der Flügel bei der Mehlmotte *Ephestia kühniella* Z., mit besonderer Berücksichtigung des Zeichnungsmusters. *Z. Morphol. Ökol. Tiere* **24**, 582–681.

Köhler, W., and Feldotto, W. (1937). Morphologische und experimentelle Untersuchungen über Farbe, Form und Struktur der Schuppen von *Vanessa urticae* und ihre gegenseitigen Beziehungen. *Roux' Arch.* **136**, 313–399.

Kühn, A. (1926). Über die Änderung des Zeichnungsmusters von Schmetterlingen durch Temperaturreize und das Grundschema der Nymphalidenzeichnung. *Nachr. Ges. Wiss. Gött., Math.-Phys.* **Kl**. 120–141.

Kühn, A. (1946). Konstruktionsprincipien von Schmetterlingsschuppen nach elektronenmikroskopischen Aufnahmen. *Z. Naturforsch.* **1**, 348–351.

Kühn, A., and von Engelhardt, M. (1933). Über die Determination des Symmetriesystems auf dem Vorderflügel von *Ephestia kühniella. Roux' Arch.* **130**, 660–703.

Kühn, A., and von Engelhardt, M. (1936). Über die Determination des Flügelmusters bei *Abraxas grossulariata* L. *Ges. Wiss. Nachr. Biol.* **2**, 171–199.

Kühn, A., and von Engelhardt, M. (1944). Mutationen und Hitzemodifikationen des Zeichnungsmusters von *Ptychopoda seriata* Schrk. *Biol. Zblt.* **64**, 24–73.

Kühn, A., and Henke, K. (1936). Genetische und Entwicklungsphysiologische Untersuchungen and der Mehlmotte *Ephestia kühniella* Zeller. *Abh. Ges. Wiss. Gött., Math.-Phys.* **Kl N.F 15,** 1–272.

Kuntze, H. (1935). Die Flügelentwicklung bei Philosamia cynthia Drury, mit besonderer Berücksichtigung des Geaders der lakunen und der Tracheensysteme. *Z. Morphol. Ökol. Tiere* **30**, 544–572.

Lea, A. J. (1945). A neutral solvent for melanin. *Nature (London)* **156**, 478.

Lewis, J., Slack, J. M. W., and Wolpert, L. (1977). Thresholds in development. *J. Theor. Biol.* **65**, 579–590.

Linzen, B. (1974). The tryptophan-ommochrome pathway in insects. *Adv. Insect Physiol.* **10**, 117–246.

Lippert, W., and Gentil, K. (1959). Über lamellare Feinstrukturen bei den Schillerschuppen der Schmetterlinge vom *Urania*- und *Morpho*-typ. *Z. Morphol. Ökol. Tiere* **48**, 115–122.

Locke, M. (1967). The development of patterns in the integument of insects. *Adv. Morphogen.* **6**, 33–88.

Lounibos, L. P. (1974). The cocoon spinning behavior of saturniid silkworms. Ph.D. Thesis, Harvard University.

MacWilliams, H. K., and Papageorgiou, S. (1978). A model of gradient interpretation based on morphogen binding. *J. Theor. Biol.* **72**, 385–411.

Mason, C. (1926). Structural colors in insects. I. *J. Phys. Chem.* **30**, 383–395.

Mason, C. (1927a). Structural colors in insects. II. *J. Phys. Chem.* **31**, 320–354.

Mason, C. (1927b). Structural colors in insects. III. *J. Phys. Chem.* **31**, 1856–1872.

Meinhardt, H. (1982). "Models of Biological Pattern Formation." Academic Press, New York.

Mitchell, H. K., and Lipps, L. S. (1978). Heat shock and phenocopy induction in *Drosophila. Cell* **15**, 907–918.

Murray, J. D. (1981). On pattern formation mechanisms for lepidopteran wing patterns and mammalian coat markings. *Philos. Trans. R. Soc. London Ser. B* **295**, 473–496.

Needham, A. E. (1974). "The Significance of Zoochromes." Springer-Verlag, Berlin and New York.

Nicolaus, R. A. (1968). "Melanins." Hermann, Paris.

Nijhout, H. F. (1978). Wing pattern formation in Lepidoptera: A model. *J. Exp. Zool.* **206**, 119–136.

Nijhout, H. F. (1980a). Pattern formation on lepidopteran wings: Determination of an eyespot. *Dev. Biol.* **80**, 267–274.

Nijhout, H. F. (1980b). Ontogeny of the color pattern on the wings of *Precis coenia* (Lepidoptera: Nymphalidae). *Dev. Biol.* **80**, 275–288.

Nijhout, H. F. (1981). The color patterns of butterflies and moths. *Sci. Am.* **245**, 145–151.

Nijhout, H. F. (1984). Color pattern modification by coldshock in Lepidoptera. *J. Embryol. Exp. Morphol.* (in press).

Nijhout, H. F. (1985). In preparation.

Nijhout, H. F., and Wheeler, D. E. (1982). Juvenile hormone and the physiological basis of insect polymorphisms. *Q. Rev. Biol.* **57**, 109–133.

Nijhout, H. F., and Williams, C. M. (1974). Control of moulting and metamorphosis in the tobacco hornworm, *Manduca sexta* (L.): Growth of the last-instar larva and the decision to pupate. *J. Exp. Biol.* **61**, 481–491

Nijhout, H. F., and Wray, G. A. (1985). Homologies in the color pattern of the genus *Charaxes*. In preparation.

Overton, J. (1966). Microtubules and microfibrils in morphogenesis of the scale cells of *Ephestia kühniella*. *J. Cell Biol.* **29**, 293–305.

Paweletz, N., and Schlote, F. W. (1964). Die Entwicklung der Schmetterlingsschuppen bei *Ephestia kühniella* Zeller. *Z. Zellforsch.* **63**, 840–870.

Pohley, H. -J. (1959). Über das Wachstum der Mehlmotteflügel unter normalen und experimentellen Bedingungen. *Biol. Zbit.* **78**, 233–250.

Prochnow, O. (1929). Die Färbung der Insekten. *In* "Handbuch der Entomologie" (C. Schroeder, ed.), Vol. II, pp. 430–591 Fisher, Jena.

Prota, G., and Thomson, R. H. (1976). Melanin pigmentation in mammals. *Endeavour* **35**, 32–38.

Rachootin, S., and Thompson, K. S. (1981). Epigenetics, paleontology and evolution. *Evolution Today. Proc. Int. Congr. Syst. Evol. Biol., 2nd* pp. 181–193.

Richards, A. G. (1951). "The Integument of Arthropods." Minnesota Univ. Press, Minneapolis.

Riddiford, L. M., and Truman, J. W. (1978). Biochemistry of insect hormones and insect growth regulators. *In* "Biochemistry of Insects" (M. Rockstein, ed.), pp. 307–357. Academic Press, New York.

Riley, P. A. (1977). The mechanism of melanogenesis. *Symp. Zool. Soc. London* **39**, 77–95.

Schwanwitsch, B. N. (1924). On the groundplan of the wing pattern in nymphalids and certain other families of Rhopalocera. *Proc. Zool. Soc. London* **34**, 509–528.

Schwanwitsch, B. N. (1925). On a remarkable dislocation of the components of the wing pattern in a Satyride genus *Pierella*. *Entomologist* **58**, 226–269.

Schwanwitsch, B. N. (1926). On the modes of evolution of the wing-pattern in Nymphalids and certain other families of the Rhopalocerous Lepidoptera. *Proc. Zool. Soc. London* **33**, 493–508.

Schwanwitsch, B. N. (1929). Two schemes of the wing pattern of butterflies. *Z. Morphol. Ökol. Tiere* **14**, 36–58.

Schwanwitsch, B. N. (1935). On some general principles observed in the evolution of the wing-pattern of palearctic Satyridae. *Proc. Congr. Entomol., 6th* pp. 1–8.

Schwartz V. (1962). Neue Versuche zur Determination des zentralen Symmetriesystems bei *Plodia interpunctella*. *Biol. Zbl.* **81**, 19–44.

Seitz, A. (1924). "The Macrolepidoptera of the World. The American Rhopalocera." Kernen, Stuttgart.

Sibatani, A. (1980). Wing homeosis in Lepidoptera: A survey. *Dev. Biol.* **79**, 1–18.

Sibatani, A. (1983). A compilation of data on wing homeosis in Lepidoptera. *J. Res. Lepid.* **22**, 1–46.

Sokolow, G. N. (1936). Die Evolution der Zeichnung der Arctiidae. *Zool. Jb.* (*Anat.*) **61**, 107–238.

Stossberg, M. (1937). Über die Entwicklung der Schmetterlingsschuppen. *Biol. Zbl.* **57**, 393–402.

Stossberg, M. (1938). Die Zellvorgänge bei der Entwicklung der Flügelschuppen von *Ephestia kühniella* Z. *Z. Morphol. Ökol. Tiere* **34**, 173–206.

Süffert, F. (1924). Morphologie und Optik der Schmetterlingsschuppen. *Z. Morphol. Ökol. Tiere* **1**, 171–308.

Süffert, F. (1927). Zur vergleichende Analyse der Schmetterlingszeichnung. *Biol. Zblt.* **47**, 385–413.

Süffert, F. (1929a). Die Ausbildung des imaginalen Flügelschnittes in der Schmetterlingspuppe. *Z. Morphol. Ökol. Tiere* **14**, 338–359.

Süffert, F. (1929b). Morphologische Erscheinungsgruppen in der Flügelzeichnung der Schmetterlinge, insbesondere die Querbindenzeichnung. *Roux' Archv.* **120**, 229–383.

Süffert, F. (1937). Die Geschichte der Bildungszellen im Puppenflügelepithel bei einem Tagschmetterling. *Biol. Zblt.* **57**, 615–628.

Thomson, D. L. (1926a). The pigments of butterflies' wings. I. *Melanargia galatea. Biochem J.* **20**, 73–75.

Thomson, D. L. (1926b). The pigments of butterflies' wings. II. Occurrence of the pigment of *Melanargia galatea* in *Dactylis glomerata. Biochem. J.* **20**, 1026–1027.

Turner, J. R. G. (1971a). The genetics of some polymorphic forms of the butterflies *Heliconius melpomene* (Linnaeus) and *H. erato*. II. The hybridization of subspecies of *H. melpomene* from Surinam and Trinidad. *Zoologica* **56**, 125–157.

Turner, J. R. G. (1971b). Two thousand generations of hybridization in a *Heliconius* butterfly. *Evolution* **25**, 471–482.

Umebachi, Y., and Takahashi, H. (1956). Kynurenine in the wings of papilionid butterflies. *J. Biochem. Tokyo* **43**, 73–81.

Vane-Wright, R. I. (1979). The coloration, identification and phylogeny of *Nessaea* butterflies (Lepidoptera: Nymphalidae). *Bull. Brit. Mus.* (*Nat. Hist.*) **38**, 27–56.

Waddington, C. H. (1953). Genetic assimilation of an acquired character. *Evolution* **7**, 118–126.

Watt, W. B. (1964). Pteridine components of wing pigmentation in the butterfly *Colias eurytheme. Nature* (*London*) **201**, 1326–1327.

Watt, W. B. (1967). Pteridine biosynthesis in the butterfly *Colias eurytheme. J. Biol. Chem.* **242**, 565–572.

Watt, W. B. (1972). Xanthine dehydrogenase and pteridine metabolism in *Colias* butterflies. *J. Biol. Chem.* **247**, 1445–1451.

Watt, W. B. (1974). Adaptive significance of pigment polymorphisms in *Colias* butterflies. III. Progress in the study of the "alba" variant. *Evolution* **27**, 537–548.

Watt, W. B., and Bowden, S. R. (1966). Chemical phenotypes of pteridine colour forms in *Pieris* butterflies. *Nature* (*London*) **210**, 304–306.

Wehrmaker, A. (1959). Modificabilität und Morphogenese des Zeichnungsmusters von *Plodia interpunctella* (Lepidoptera: Pyralidae). *Zool. Jahrb. Abt. Zool. Physiol.* **68**, 425–496.

Williams, C. M. (1947). Physiology of insect diapause. II. Interaction between the pupal brain and prothoracic glands in the metamorphosis of the giant silkworm, *Platysamia cecropia. Biol. Bull.* **93**, 89–98.

Willis, J. H. (1974). Morphogenetic action of insect hormones. *Annu. Rev. Entomol.* **19**, 97–115.

Winfree, A. T. (1980). "The Geometry of Biological Time. Biomathematics," Vol. 8. Springer-Verlag, Berlin and New York.

Wolpert, L. (1969). Positional information and the spatial pattern of cellular differentiation. *J. Theor. Biol.* **25**, 1–47.

Nonspiking Interneurons and Motor Control in Insects

Melody V. S. Siegler

Department of Zoology, University of Cambridge, Cambridge, England

ADVANCES IN INSECT PHYSIOLOGY, VOL. 18

1 Introduction

The study of nonspiking interneurons in insects has arisen out of an attempt to discover how the central nervous system produces the formidable variety of motor patterns of which these animals are capable. To give a fair recounting of the many studies that have contributed to our present understanding of motor control in insects would be an immense and lengthy task, and is outside the scope of the present review. Nonetheless, some brief sketch is needed to put the study of nonspiking interneurons in perspective.

In insects, as in other animals, one of the first concerns for those interested in the neural basis of behavior was whether the patterning of motor activity arose from connections intrinsic to the central nervous system, or required the "chaining" of motor activity with sensory feedback. It is now accepted that the essential features of complex motor patterns arise centrally, whereas sensory feedback adjusts these patterns, to different degrees for different ones (Roberts and Roberts, 1983). In insects, for example, central patterns include those for flight (Wilson, 1961), walking (Pearson, 1972), and ventilation (Miller, 1965).

Once the central nature of pattern generation was recognized, the search began for the neurons involved. Some simple models proposed that direct connections among the motor neurons might be responsible. For example, mutually inhibitory interactions among antagonistic wing motor neurons could (in theory) give rise to a patterned output that approximated the flight rhythm of the locust (Wilson, 1964). When it was realized that individual motor neurons had characteristic shapes and positions within the central nervous system, it became possible to test such ideas directly with intracellular recording. Motor neurons could be treated as individuals, and the same ones studied time and time again in different individuals of a species. Recordings soon revealed that direct connections between motor neurons were the exception not the rule in insect nervous systems (e.g., Hoyle and Burrows, 1973). However, by comparing the patterns of postsynaptic potentials (PSPs) in different motor neurons during different movements, much could be inferred about the next higher level of organization, the connectivity of interneurons (Burrows and Horridge, 1974). In general, it appeared that different limb movements were achieved by activation of interneurons that selected different combinations of motor neurons. Implicit in these descriptions, however, was the idea that the premotor interneurons were spiking ones.

Our thinking about the control of motor neurons took an important step forward with the discovery of nonspiking neurons in crustacean and insect ventral ganglia (Mendelson, 1971; Pearson and Fourtner, 1975). These interneurons when depolarized with injected current could alter the rate of firing in motor neurons, and they received synaptic inputs in phase with patterned motor activity. Together, these two observations supported the idea that the nonspiking interneurons had a role in the production of motor patterns.

Since the description of nonspiking interneurons in the cockroach, studies in insects have been concerned almost solely with the nonspiking interneurons in the thoracic ganglia of the locust, and their control of leg motor neurons. The topics investigated include the synaptic transmission to motor neurons (Burrows and Siegler, 1976, 1978), the morphology of the nonspiking interneurons (Siegler and Burrows, 1979; Wilson, 1981; Wilson and Phillips, 1982), the release of transmitter by PSPs (Burrows, 1979a), the interactions between nonspiking interneurons (Burrows, 1979b), the organization of sets of motor neurons by nonspiking interneurons (Burrows, 1980), and the role of nonspiking interneurons in the control of posture (Siegler, 1981a) and in the modulation of reflexes (Siegler, 1981b).

The number of reviews that consider some aspect of nonspiking interneurons in insects approaches the number of original research papers (Pearson, 1976, 1977, 1979; Burrows, 1978, 1981; Siegler and Burrows, 1980; Wilson and Phillips; 1983; Burrows and Siegler, 1984; Siegler, 1984). By this accounting, yet another review would seem unnecessary. However, the present one is different in intent from the more general ones that have gone before. The aim here is to look in some detail at the results of the original research papers, to evaluate these results critically, and to suggest directions for further research. After a brief introduction to the organization of an insect nervous system, the physiology and then the morphology of nonspiking interneurons in insects will be considered. Finally, some comparison will be drawn with nonspiking interneurons in crustaceans.

The study of nonspiking interneurons in insects can be seen as part of a continuing investigation into the way that the interconnections among neurons in the central nervous system act to coordinate motor activity. Two attributes of nonspiking interneurons are important in placing their study in a yet wider context, however. First, nonspiking interneurons in insects are local interneurons, branching wholly within restricted anatomical regions of the nervous system. Local interneurons comprise the majority of central neurons, not only in insects and in crustaceans, but in vertebrates as well. Thus they provide us with a means of understanding the function of local interneurons and the local circuits they form, in a small yet reasonably sophisticated nervous system. Second, in nonspiking interneurons, transmitter release is initiated by graded changes in membrane potential rather than by spikes. Similarly, a wide variety of sensory receptors and central neurons in vertebrates and invertebrates operate wholly or partly by the graded spread of potentials. Yet, there are relatively few places where pre- and postsynaptic elements can readily be penetrated simultaneously, to study directly the physiology of graded synaptic transmission. Nonspiking

A

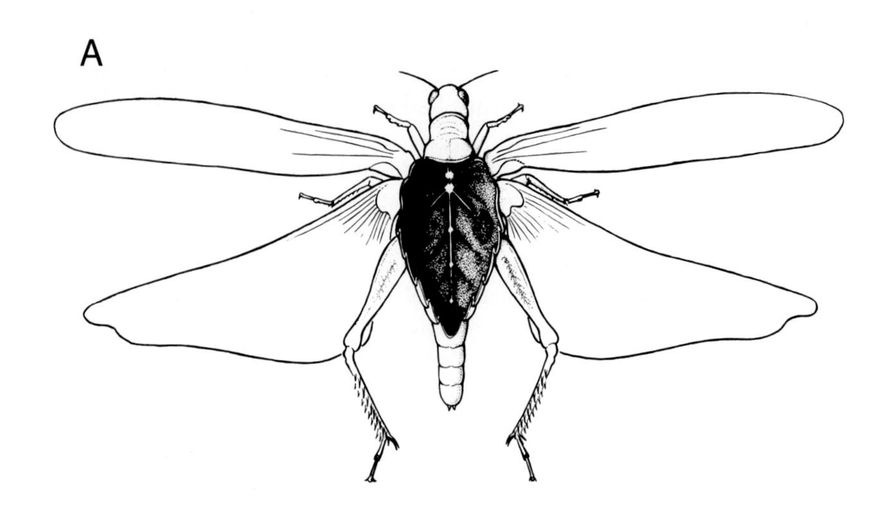

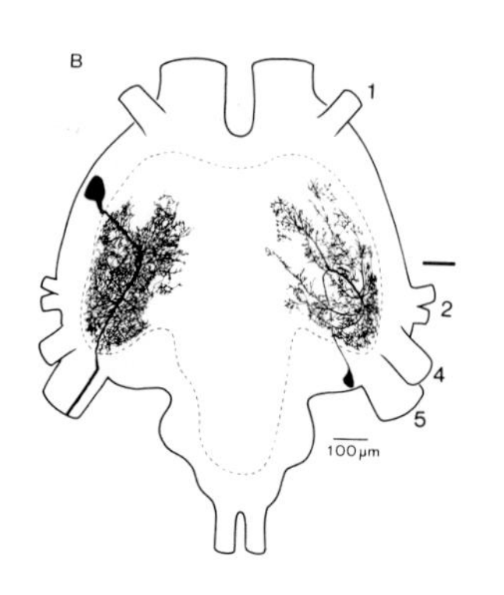

C

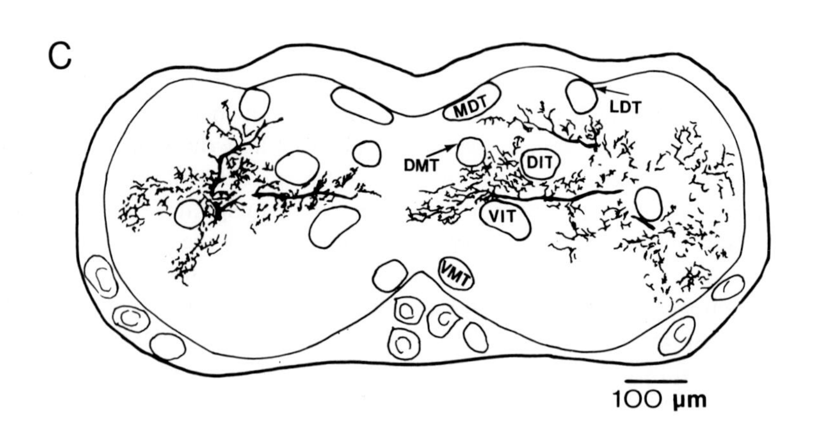

interneurons in insects afford this opportunity. In the present review, references to these latter topics are limited, inasmuch as considerable discussion of the wider significance of local and nonspiking neurons in vertebrates and invertebrates can be found elsewhere, in lengthy symposium volumes (Rakić, 1975; Schmitt and Worden, 1979; Roberts and Bush, 1981).

2 Electrophysiological properties of nonspiking interneurons

2.1 ORGANIZATION OF AN INSECT NERVOUS SYSTEM

As an introduction to the discussion of nonspiking interneurons in motor systems of insects, the features of the central nervous system of the locust, in which nonspiking interneurons have been most studied, will be summarized.

As in other insects, the central nervous system of the locust consists of a brain and a ventral nerve cord, the latter being composed of bilaterally symmetrical ganglia, joined by paired interganglionic connectives (Fig. 1A). Each of the three thoracic ganglia is about 1 mm across, and contains the cell bodies of a few thousand neurons (Fig. 1B). These include those of spiking and nonspiking local interneurons, interganglionic interneurons, motor neurons, and peripheral modulatory neurons. In addition, a small proportion of sensory neurons have centrally located cell bodies (Bräunig and Hustert, 1980), though most are in the periphery. The neuronal cell bodies are positioned around the ganglion, but are most dense on the ventral surface. They range in diameter from about 100 to 10 μm; that of a fast flexor

---

Fig. 1. (A) Dorsal view of a locust, opened to show second (mesothoracic) and third (metathoracic) thoracic ganglia, and the first three unfused abdominal ganglia. The meso- and metathoracic ganglia are associated, respectively, with the second pair of legs and the forewings, and with the third pair of legs (hind legs) and the hind wings. (B) In the metathoracic ganglion, a motor neuron and a nonspiking interneuron have been stained intracellularly with cobalt sulfide (CoS) and silver. They are drawn in plan view from cleared whole mounts of ganglia, using a camera lucida attached to a microscope. The dashed line indicates the boundary of the neuropil. On the left is a flexor tibiae motor neuron, and on the right is one of the nonspiking interneurons that when depolarized by the injection of current inhibits ipsilateral flexor tibiae motor neurons. Like all other nonspiking interneurons in thoracic ganglia, it is a local interneuron, with no branches in peripheral nerves or in central connectives. (C) Transverse section of ganglion at the level indicated in (B), showing cobalt–silver stained processes of flexor tibiae motor neuron (left) and nonspiking interneuron (right). The monopolar cell bodies of neurons occur primarily in ventral and lateral regions, at the outside of the ganglion. The section cuts across several obvious longitudinal tracts, which contain axons of intersegmental interneurons. (Abbreviations: DIT, dorsal intermediate tract; DMT, dorsal median tract; LDT, lateral dorsal tract; MDT, median dorsal tract; VIT, ventral intermediate tract; VMT, ventral median tract.) Between these longitudinal tracts and the transverse and oblique tracts are regions of neuropil where the nonspiking interneurons, motor neurons, and other inter- and sensory neurons branch profusely. Tracts, commissures, and neuropilar regions are described by Tyrer and Gregory (1982).

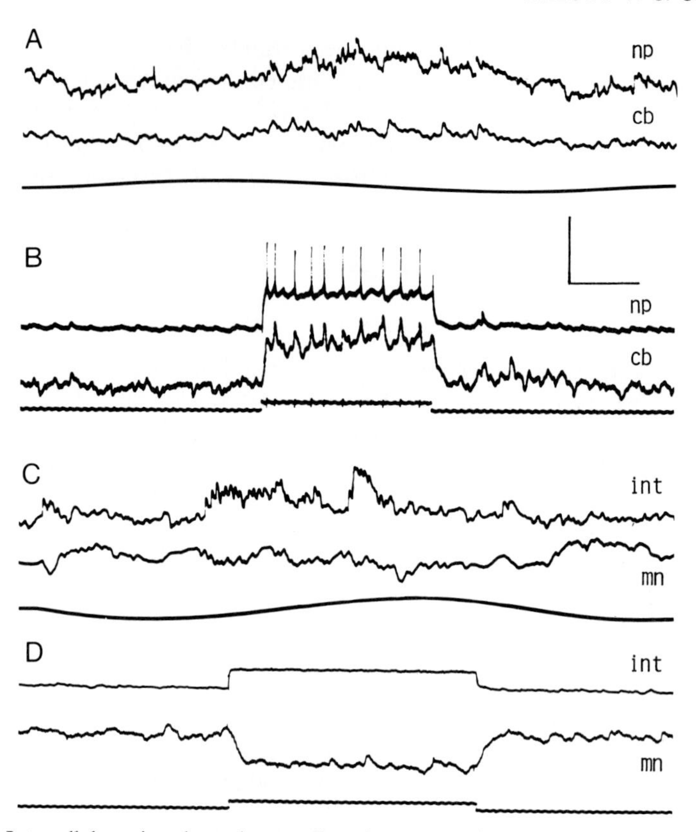

Fig. 2. Intracellular microelectrode recordings from neuropilar processes and cell bodies of neurons in locust metathoracic ganglion. (A) Synaptic events recorded simultaneously from a neuropilar process (np) and the cell body (cb) of a motor neuron of a hind leg. Third trace indicates imposed rhythmic changes in angle of femoral–tibial joint of hind leg. (B) Depolarizing current step (third trace), applied via bridge circuit to neuropilar process (np) of a hind leg motor neuron, elicits a train of spikes in the motor neuron. Spikes propagate actively in the neuropilar process, but spread passively into the cell body (cb), where they are typically 2–5 mV in amplitude. (C) Synaptic events recorded from a neuropilar process of nonspiking interneuron (int) and a hind leg motor neuron (mn). Third trace indicates imposed rhythmic changes in angle of femoral–tibial joint of hind leg. When the interneuron is depolarized with intracellularly injected current, the motor neuron is inhibited (not shown). (D) Depolarizing current step (third trace) applied via bridge circuit to a neuropilar process of a nonspiking interneuron (int) results in hyperpolarization of a postsynaptic motor neuron (mn). The motor neuron is penetrated in its cell body. The bridge circuit allows for simultaneous stimulation and recording via the intracellular microelectrode, but during stimulation the membrane potential recorded is approximate. Calibration. Vertical: for (A) and (B): np, 30 mV; cb, 6 mV. (B) current, 26 nA. (C) int, 15 mV; mn, 6 mV. (D) int, 38 mV; mn, 6 mV; current, 26 nA. Horizontal: 200 ms.

tibiae motor neuron is among the largest in the ganglion, while that of a nonspiking interneuron that affects it is among the smallest (Fig. 1B).

The morphology of these motor neurons and nonspiking interneurons can be revealed by the intracellular injection of dye (Fig. 1B,C). Motor neurons are monopolar and have a primary neurite that gives rise to a multitude of processes within the neuropil, and an axon in a peripheral nerve. Synaptic inputs and the active initiation of spikes occur in the neuropilar region, not the soma. Nonspiking interneurons also have profuse neuropilar branches, but are local interneurons, with no axons in the peripheral nerves or central connectives.

Intracellular recordings can be made simultaneously from motor neurons and interneurons, in ganglia of minimally dissected animals that have their tracheae and their peripheral and central connections with the nervous system largely intact. At the same time, the few motor neurons to each muscle of the legs or wings can be identified in extracellular myograms. It is most convenient to record from the cell bodies of the motor neurons, because of their large size and stereotyped position. However, simultaneous recording in the soma and a neuropilar process of a motor neuron shows that some events, particularly spikes, are considerably diminished at the cell body (Fig. 2A,B). For nonspiking interneurons, recordings are better made from neuropilar processes for two reasons; the cell bodies are small (10–20 μm), and it is important to be close to the neuropil region to confirm that the interneurons lack spikes. As in the motor neurons, inputs and outputs occur within neuropil regions. Simultaneous recordings can also be made routinely from a neuropilar process of a nonspiking interneuron and the cell body of a motor neuron (Fig. 2C,D). From such recordings it is clear that depolarization of the interneuron can effect changes in the membrane potential of the motor neuron without the interneuron itself spiking (Fig. 2D).

2.2 LACK OF REGENERATIVE MEMBRANE RESPONSES

The electrophysiological properties of nonspiking interneurons, which distinguish them from other types of neuron, have been described in considerable detail in the cockroach (Pearson and Fourtner, 1975) and in the locust (Burrows and Siegler, 1978).

The major property that sets these neurons apart from others within the thoracic ganglia of cockroaches and locusts is the absence of all-or-none spikes and of other graded regenerative membrane responses. The neurons do not produce spikes, either immediately upon penetration with microelectrodes, or subsequently when depolarized by the intracellular injection of depolarizing current or by summed synaptic inputs (Figs. 2D,3, and 4). Although motor neurons and spiking interneurons also do not produce spikes after microelectrode penetration unless they are damaged, they typically spike readily when depolarized by intracellularly injected current or by synaptic inputs.

Additional evidence that the interneurons do not spike comes from experiments on the locust where the neuropilar processes of individual nonspiking interneurons have been penetrated with two microelectrodes simultaneously (Fig. 3; Burrows and Siegler, 1978). This avoids the difficulty that the properties of the recording microelectrode may change during current injection via a bridge circuit. When an interneuron is depolarized by injecting a current step via one electrode, the depolarization spreads passively to the other. This depolarization has a sustained plateau, and at the offset of the current the membrane potential passively decays to its resting level. No spikes or active responses are recorded at either electrode, and no active response is seen at the onset or for the duration of the current pulse. Further evidence that this reflects the normal physiological state of the interneurons is discussed in Section 2.5.

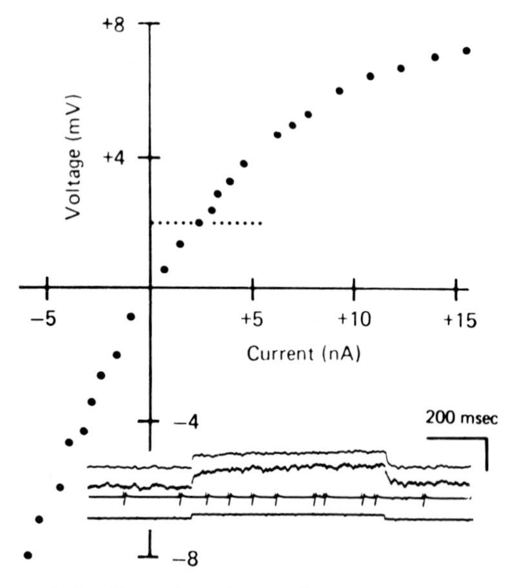

Fig. 3. Current–voltage relationship and passive membrane response of a nonspiking interneuron. In the locust metathoracic ganglion, a nonspiking interneuron was penetrated by two microelectrodes, placed some 200 μm apart in the neuropil. As shown in the inset, a sustained depolarizing current injected via a bridge circuit at one electrode (first trace) results in a sustained depolarization recorded at the other (second trace). At the same time, the frequency of spikes increases in a motor neuron, recorded extracellularly in a muscle (third trace). The current step is 2.5 nA (fourth trace). Calibration: first trace, 20 mV; second trace, 4 mV; fourth trace, 26 nA. The graph plots the change recorded at one electrode in response to current injected at the other, for a range of currents. The dotted line indicates the voltage where an effect on the motor neuron was first seen. (After Burrows and Siegler, 1978.)

2.3 MEMBRANE POTENTIAL

A second characteristic of the insect nonspiking interneurons is that their membrane potentials, recorded in quiescent animals, are significantly more depolarized than those of spiking neurons within the same ganglion. In the cockroach, Pearson and Fourtner (1975) reported that the membrane potentials of nonspiking interneurons ranged from −30 to −50 mV, whereas those of motor neurons were normally greater than −50 mV. In the locust, Burrows and Siegler (1978) found that nonspiking interneurons on the average had membrane potentials some 15 mV more positive than those of spiking neurons recorded in the same animals. The averaged value for all nonspiking neurons recorded was −48 mV (range −35 to −60 mV) and for all spiking neurons was −64 mV (range −45 to −75 mV), significantly higher. The membrane potentials must reflect the sum of synaptic inputs, which occur continually, as well as the intrinsic membrane properties of the neurons. It is not known, however, whether the two types of neurons would have significantly different resting potentials from each other in the absence of synaptic inputs, and the ionic basis of the resting potential has not been investigated. In the second-order ocellar neurons of the barnacle, which are local interneurons with graded responses, a relatively low resting potential is thought to result from a low resting conductance to K^+ (Oertel and Stuart, 1981).

2.4 SYNAPTIC EVENTS IN NONSPIKING INTERNEURONS

Reports differ as to the "characteristic" nature of the synaptic inputs to nonspiking neurons. Pearson and Fourtner (1975) reported that neuropil recordings from nonspiking interneurons in the cockroach had a characteristically high level of synaptic "noise," which was significantly different in quality from the synaptic events recorded in the neuropil from any identified motor neuron. Furthermore, in recordings from nonspiking interneurons, individual IPSPs and EPSPs usually could not be distinguished. By constrast, Burrows and Siegler (1976, 1978) found that the synaptic events recorded from the neuropilar processes of nonspiking interneurons in the locust were qualitatively no different from those recorded in the neuropilar processes of motor neurons or other spiking neurons. Although neuropilar recordings differed in the amplitude and frequency of synaptic events they revealed, this did not correlate with whether the neurons were spiking or nonspiking. Furthermore, all nonspiking interneurons showed complex fluctuations of membrane potential of at least a few millivolts in amplitude, with individual or summed excitatory PSPs (EPSPs) and inhibitory PSPs (IPSPs) occurring spontaneously, and being readily evoked in response to particular sensory inputs (see also Section 4.2). For

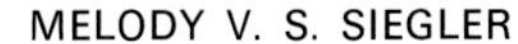

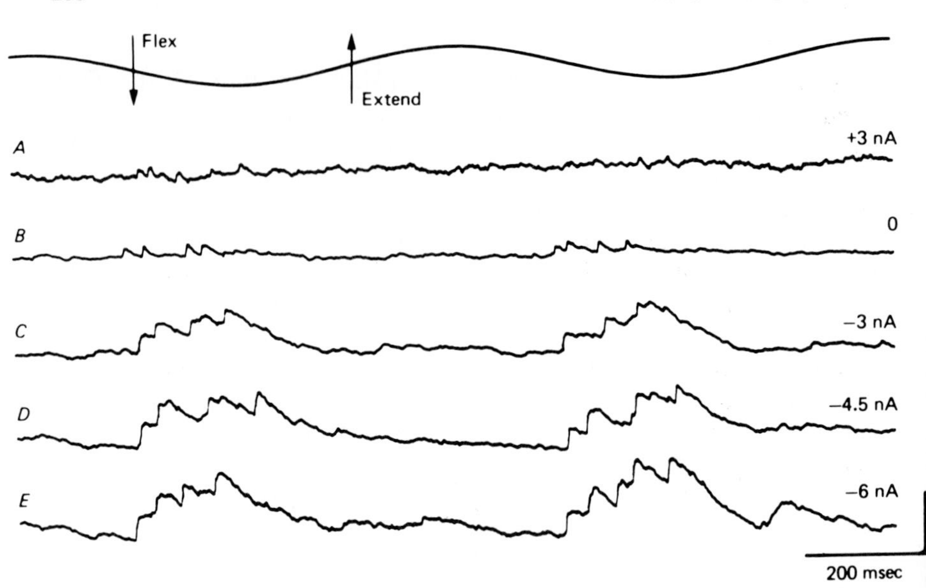

Fig. 4. Altering the membrane potential of a nonspiking interneuron fails to reveal spikes. In the locust, the tibia of a hind leg is rhythmically flexed and extended about the femur (first trace) to elicit synaptic potentials in a nonspiking interneuron in the ganglion (A–E). Current is injected through a bridge circuit to alter the membrane potential of the interneuron. The interneuron is depolarized in (A), is at its normal resting potential in (B), and is hyperpolarized in (C–E). Calibration: vertical, 16 mV and 60°. (From Burrows and Siegler, 1978.)

example, in one interneuron, depolarizing potentials a few millivolts in amplitude could be evoked by flexing the tibia about the femur (Fig. 4B). These potentials were EPSPs rather than abortive spikes or remnants of spikes, for when the interneuron was gradually hyperpolarized by injected current, the potentials increased in amplitude (Fig. 4, C–E). Conversely, they decreased in amplitude upon depolarization of the interneuron (Fig. 4A).

Nonetheless, the idea that a "noisy" membrane potential is a diagnostic feature of nonspiking interneurons has to some extent been incorporated into the literature concerning nonspiking interneurons (e.g., Pearson, 1979; Kirshfeld, 1979). Further, it has served to reinforce suspicions that nonspiking interneurons are, in fact, damaged spiking interneurons. How these two ideas are intertwined will become clear in the following section.

2.5 ARE THEY REALLY NONSPIKING INTERNEURONS?

The idea that nonspiking interneurons are damaged spiking interneurons can be traced largely to the literature investigating the giant lobula plate interneurons of flies. Each of these brain interneurons responds to particular movements in the visual field. They have been reviewed in considerable detail by Hausen (1981). Briefly, the main points of concern here are as follows. Hengenstenberg (1977) showed that some of these interneurons, thought previously to be "nonspiking," could be made to spike in response to visual stimuli if they were subjected to sustained hyperpolarizing current. The "noisy" fluctuations that otherwise occurred during visual stimuli were taken to be abortive spikes, resulting from some damage to the interneurons. Anoxia in the brain was argued to be the most likely source of damage (Hengstenberg, 1977). A further implication from this was that the "characteristic noise" reported for other nonspiking interneurons likewise resulted from damage.

However, three results of subsequent studies by Hausen (1981) on these same lobula plate interneurons argue strongly against the earlier conclusions. First, Hausen showed that the responses of the interneurons change with repeated visual stimuli. Initially, movements in the visual field elicit spikes from the interneurons. With repeated movements, the spikes gradually diminish in size, so that after some time only graded, electronically conducted potentials may be elicited. Second, spikes can be recorded from the interneurons for several hours after the tracheae of the lobula plate, and thus its oxygen supply, have been removed. Last, flies with the tracheae removed continue to perform visually guided behavior for at least a day. Thus, Hausen postulates that, normally, voltage signals may travel through the interneurons either as spikes or as graded potentials. He notes, however, that damage by microelectrodes could result in leakage currents and suppression of spikes; this might explain the earlier findings of himself and others that the lobula plate neurons were nonspiking. In short, it appears that the spike-generating mechanism of lobula plate interneurons adapts to sustained depolarization. This can occur normally, during prolonged visual stimulation, or as a result of damage upon penetration by microelectrodes.

By contrast, there are strong experimental arguments that neither microelectrode damage nor anoxia are responsible for the lack of spikes in nonspiking interneurons of the ventral ganglia of insects (Burrows and Siegler, 1978). The concern is that because these neurons are relatively small, with fine branches, they might be particularly susceptible to damage. As already noted, however, damage upon penetration of spiking neurons is indicated by a repetitive spike discharge, and this is never recorded in nonspiking interneurons; furthermore, prolonged hyperpolarizing currents do not reveal spikes in the interneurons. Perhaps the strongest argument against microelectrode damage, however, is provided by the results of paired intracellular recordings of nonspiking interneurons and postsynaptic motor neurons. A motor neuron can be penetrated first and an interneuron then penetrated, without any resulting change in the membrane potential of the postsynaptic motor neuron. If an interneuron had been damaged, and thus depolarized upon penetration, it would be expected that some change would also be seen in the membrane potential of the motor neuron. Depolarizing the interneuron with currents of small magnitude will, however, evoke the release of transmitter and alter the membrane potential of the motor neuron. Likewise, strong arguments can be made against the idea that the nonspiking interneurons lack spikes because they lack oxygen: nonspiking interneurons can be found within a few minutes of opening up an animal, when anoxia would not be expected to have developed; and in any case, the tracheal supply to the ganglion is intact throughout the experiments, and reflex and locomotory behavior continues much as normal for the duration of an experiment.

As a final argument that nonspiking interneurons are "normally" so, we can now point to the equally small, local interneurons within the same ganglion that generate "conventional" spikes (Burrows and Siegler, 1982; 1984; Siegler and Burrows, 1983, 1984). These are not "nonspiking" interneurons that have lately been converted into spiking ones through improved experimental techniques. They are sufficiently distinctive from nonspiking interneurons in their morphology and their inputs and outputs not to be confused with them, and both types of interneuron can be recorded in a given preparation.

3 Synaptic transmission by nonspiking interneurons

Pearson and Fourtner (1975) found that when nonspiking interneurons in the cockroach were depolarized by injecting current intracellularly, there was a marked change in the spike activity of motor neurons, which was recorded extracellularly. These effects were graded according to the intensity of the stimulating current, but the nature of the underlying synaptic events was not investigated. Subsequently, in the locust, Burrows and Siegler (1976, 1978) made simultaneous intracellular recordings from nonspiking interneurons and the motor neurons they affected, and thus investigated in detail the properties of synaptic transmission. There is now considerable evidence from the locust that nonspiking interneurons exert their effects by chemical transmission, and that the postsynaptic effects can be finely graded in strength and duration according to the presynaptic stimulus.

3.1 EVIDENCE FOR CHEMICAL SYNAPTIC TRANSMISSION

When a nonspiking interneuron of the locust is depolarized by injecting a current step intracellulary, the change in the membrane potential of a postsynaptic motor neuron rises slowly to reach a maximum that is sustained for the duration of the current step (Figs. 2, 3, 5, and 6). Depending upon the

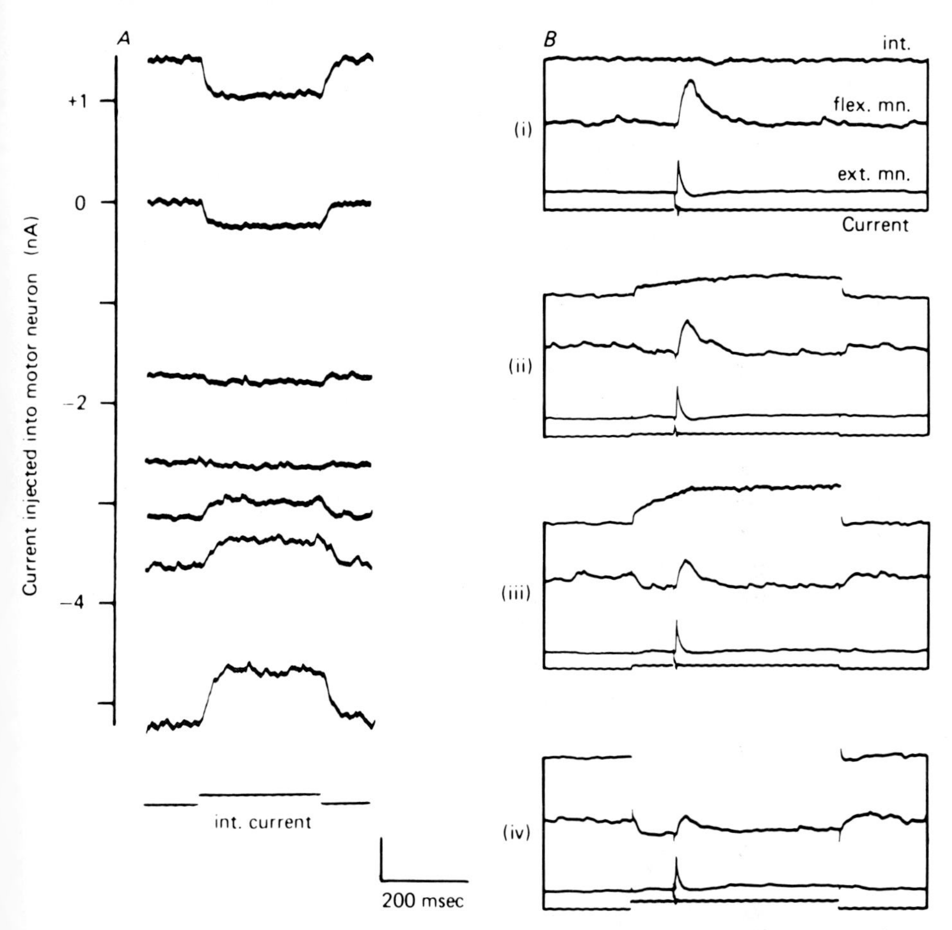

Fig. 5. Evidence for chemical synaptic transmission between nonspiking interneurons and motor neurons. Recordings from the metathoracic ganglion of locust. (A) A motor neuron is hyperpolarized when a nonspiking interneuron (not shown) is injected with depolarizing current steps of 5.6 nA, 300 ms in duration. Steady currents, of the magnitudes shown on the vertical scale, are injected into the cell body of the motor neuron. Depolarizing current increases the evoked hyperpolarizing response, while hyperpolarizing current reduces, then reverses, the response. (B) (i) A compound EPSP is evoked in a flexor tibiae motor neuron (flex. mn.) when an antidromic spike occurs in the fast extensor tibiae motor neuron (ext. mn.). (ii) When the neuropilar process of a nonspiking interneuron (int.) is injected with depolarizing current, the flexor motor neuron is hyperpolarized and the compound EPSP is decreased in amplitude. (iii, iv) Larger depolarizing currents increase the hyperpolarization of the motor neuron, and further reduce the amplitude of the EPSP. Calibration: (A) 3.6 mV; 28 nA. (B) interneuron, 33 mV; flexor motor neuron, 6 mV; extensor motor neuron, 14 mV; current, 66 nA. (From Burrows and Siegler, 1978.)

pair of pre- and postsynaptic neurons examined, the effects of an interneuron may be either depolarizing (Fig. 6) or hyperpolarizing (Fig. 7).

Evidence that the postsynaptic effects are mediated by chemical synaptic transmission has been discussed in detail (Burrows and Siegler, 1976, 1978), and can be summarized as follows. First, when the motor neuron, instead of the interneuron, is depolarized or hyperpolarized by the intracellular injection of current, there is no corresponding change in the membrane potential of the interneuron, as would be expected if there were an electrical synapse between the two. Second, when a nonspiking interneuron is depolarized the evoked postsynaptic change depends upon the postsynaptic membrane potential. As successively more hyperpolarizing current is injected directly into a postsynaptic motor neuron, an evoked hyperpolarizing potential diminishes until it reaches a reversal potential (Fig. 5A), while an evoked depolarizing potential increases in amplitude. Third, the postsynaptic potential evoked by a nonspiking interneuron "shunts" chemical PSPs of opposite polarity that occur at the same time (Fig. 5B). This also indicates that transmission involves an increased postsynaptic conductance. Last, postsynaptic potentials evoked by injecting brief pulses of depolarizing current into a nonspiking interneuron have a time course that would be expected of a chemical, but not an electrical, PSP. They last considerably longer than the stimulating pulse, and reach a peak after its offset. Results of later studies of nonspiking interneurons are consistent with these findings (Burrows, 1980; Siegler, 1981b; Wilson and Phillips, 1982), and further, show that EPSPs can effect transmitter release from nonspiking interneurons (Burrows, 1979a) and that nonspiking interneurons also affect each other by chemical synaptic transmission (Burrows, 1979b).

3.2 GRADED NATURE OF SYNAPTIC TRANSMISSION

The transmission mediated by nonspiking interneurons in the locust has been examined quantitatively by plotting the postsynaptic change in voltage, measured at the cell body of a motor neuron, against the current injected presynaptically into a neuropilar process of an interneuron (Figs. 6 and 7). The majority of interneurons examined have "resting" membrane potentials below the threshold for transmitter release, and depolarizing currents of less than 1–2 nA produce no measurable effect upon postsynaptic motor neurons. [This is true for the pair of neurons shown in Fig. 6, and one of the two pairs (▲) shown in Fig. 7.] As the current is increased, the relationship reaches a threshold region of sharply increasing sensitivity, and then becomes approximately linear. Here, transmitter release is most sensitive to the imposed current. If yet higher currents are imposed, the slope of the relationship decreases gradually, and the relationship reaches a plateau. This

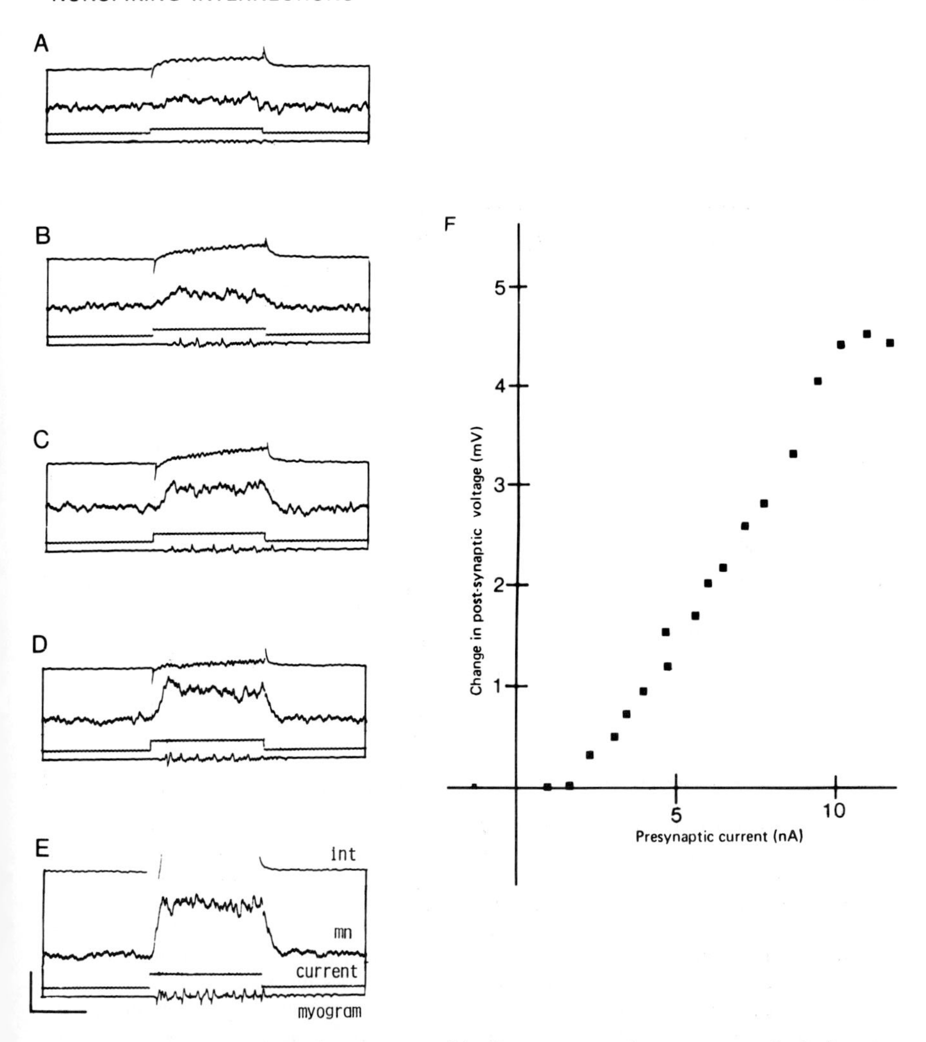

Fig. 6. Graded transmission between nonspiking interneuron and motor neuron in the locust metathoracic ganglion. (A) Depolarization of the nonspiking interneuron (first trace) with a current step (third trace) results in the depolarization of a flexor tibiae motor neuron (second trace), recorded at its cell body, and the recruitment of spikes in coxal motor neurons, recorded extracellularly (fourth trace). (B–E) Traces as in (A). Larger depolarizing currents in the interneuron evoke larger depolarizations in the flexor motor neuron. (F) The change in voltage in the postsynaptic flexor motor neuron is plotted against the current injected into the presynaptic nonspiking interneuron, for a range of currents. Calibration. Vertical: int, 38 mV; mn, 4 mV; current, 40 nA. Horizontal: 200 ms.

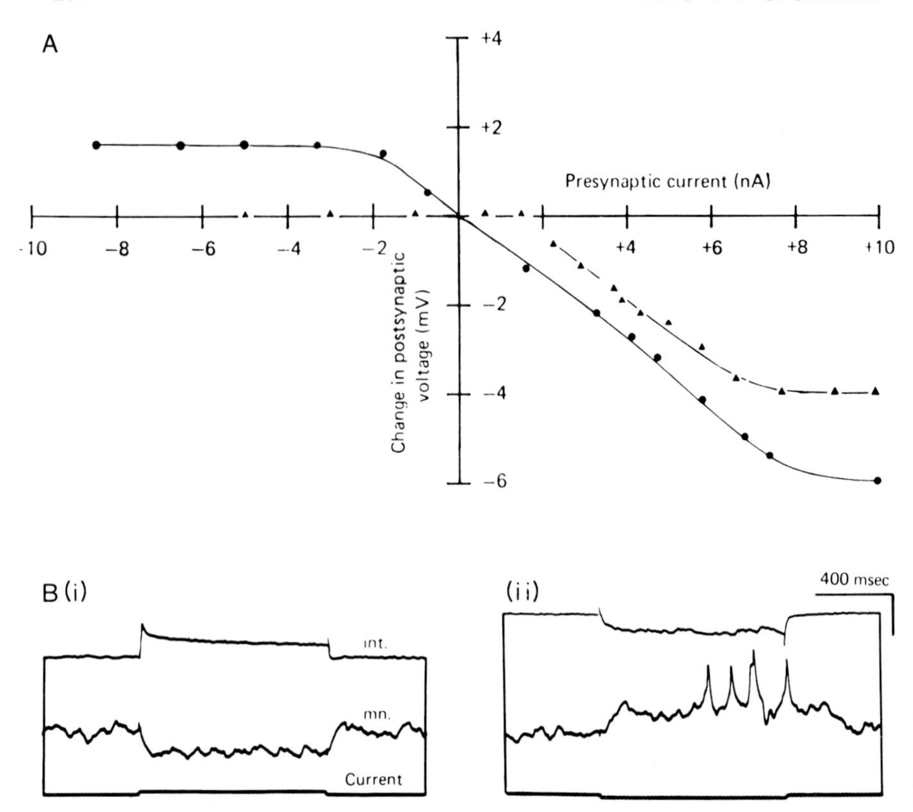

Fig. 7. Tonic release of transmitter by nonspiking interneuron. Recordings from the metathoracic ganglion of locust. (A) The relationship between current injected into a neuropilar process of the nonspiking interneuron and the change in the membrane potential of the motor neuron, recorded at the cell body are shown for two interneuron and motor neuron pairs. The interneurons are injected with current steps, about 1 s in duration, every 4 s. The change in motor neuron membrane potential is measured at its steady value, attained some 200 ms after the onset of the current step. One interneuron (●) is tonically releasing transmitter. The smallest depolarizing current hyperpolarizes the motor neuron, and hyperpolarizing current can depolarize the motor neuron by almost 2 mV. Another interneuron (▲) has a resting membrane potential more negative than its threshold for transmitter release. It must be depolarized with almost 2 nA of current before the motor neuron is hyperpolarized. (B) (i) Depolarization of a nonspiking interneuron (int) results in the hyperpolarization of a flexor tibiae motor neuron (mn). Current is injected into the interneuron via a bridge circuit, and the peak at the onset of the depolarization is an artifact. (ii) Hyperpolarization of the interneuron allows the motor neuron to depolarize and spike. Calibration: interneuron, 33 mV; motor neuron, 6 mV; current, 29 nA. (After Burrows and Siegler, 1978.)

plateau may arise partly because of rectification by the interneuron. In experiments in which two electrodes were inserted into the same interneuron (but not into a motor neuron), it was seen that at low current the current–voltage curve was linear, but currents above about 5 nA became increasingly less effective in depolarizing the neuron (Fig. 3). Presumably this rectification is due to a voltage-sensitive conductance that shunts the increasing depolarizing current. For hyperpolarizing postsynaptic effects, an additional contribution to the plateau is that the postsynaptic response approaches its equilibrium potential.

Burrows and Siegler (1978) found that in quiescent animals, some 30% of the nonspiking interneurons had their steady membrane potentials depolarized above the threshold for transmitter release. For these interneurons, the curve relating presynaptic current to postsynaptic change in voltage was of similar shape to that described for the other nonspiking interneurons, but was shifted relative to both axes (●; Fig. 7A). Even the smallest amount of depolarizing current produced a measurable postsynaptic effect, and the threshold region of the curve was revealed only when the interneuron was hyperpolarized below its steady membrane potential. Such hyperpolarizing currents had a postsynaptic effect of opposite polarity to that produced by depolarizing currents (Fig. 7B,C). The interneurons that release transmitter tonically are discussed more fully in Section 3.5.

The graded effects of the nonspiking interneurons are also shown by plotting the frequency of firing of a postsynaptic motor neuron against the current injected presynaptically. Again, for a considerable region above threshold, the relationship between presynaptic current and the postsynaptic effect is approximately linear (Pearson and Fourtner, 1975; Meyer and Walcott, 1979; Burrows, 1980; Wilson and Phillips, 1982). This is to be expected, considering the linear relationship between the frequency of spikes and the amount of current injected into motor neurons (Meyer and Walcott, 1979).

A measure of the relationship between postsynaptic voltage and presynaptic voltage, rather than current, would allow comparison with similar curves derived from other synapses. This would be technically difficult to achieve, however. It might be worthwhile to examine the effects of low presynaptic current in more detail. At currents below about 5 nA the relationship of current to voltage for the interneuron can be assumed to be linear (e.g., Fig. 3). Therefore, the curve for low currents would at least indicate the shape of the relationship between pre- and postsynaptic voltage, if not its absolute value. In addition, a more accurate plot of the postsynaptic voltages in the threshold region could be obtained by signal averaging to reduce the variation due to "background" synaptic inputs. Thus it should be possible to determine whether there is a semilogarithmic relationship between pre- and postsynaptic voltage before the plateau region, as there is at the squid giant synapse (Llinás *et al.*, 1976) and the crab nonspiking stretch receptor to motor neuron synapse (Blight and Llinás, 1980).

3.3 TIME COURSE OF POSTSYNAPTIC EFFECTS

Depolarizing the nonspiking interneurons with step current pulses of low intensity and a few hundred milliseconds in duration is a convenient way of studying the graded nature of the synaptic effects. In the course of an animal's behavior, however, the nonspiking interneurons are normally subjected to considerably more complex fluctuations in membrane potential. At one extreme, some tonically maintained stimuli may evoke steady shifts in membrane potential, whereas at the other, some phasic sensory stimuli may evoke discrete PSPs. An obvious question, then, is, how faithfully does synaptic transmission of the interneurons reflect changes in membrane potential of varying time courses?

3.3.1 *Responses to sustained changes in membrane potential*

Burrows and Siegler (1978) found that the postsynaptic effects of interneurons on some motor neurons are maintained during depolarizing current pulses 2–3 s in length. When, however, interneurons are depolarized with considerably longer pulses (<7 min) of constant current, the postsynaptic effects gradually decline. Burrows (1979b) studied the inhibitory interactions between nonspiking interneurons, and found that for one pair of interneurons the postsynaptic effect declined during pulses longer than about 200–300 ms. Nonetheless, for the longest pulses tested, 5 s, the response was still about 60% of the initial value. These reports therefore seem to suggest some variability in the time course over which nonspiking interneurons can exert their postsynaptic effects. One factor, however, which may explain this variation, is consistent in all of the published experiments in which a range of currents was tested. That is, the magnitude of the current injected presynaptically is important in determining how well the evoked postsynaptic changes are sustained. Typically, when depolarizing currents of less than 5–6 nA are injected, the postsynaptic change in voltage is undiminished throughout the current pulse (usually 400–800 ms pulses tested). At higher currents (6–12 nA), however, the postsynaptic response declines gradually during the current step (e.g., Burrows and Siegler, 1976: Fig. 3; 1978: Fig. 7; Burrows, 1979b: Fig. 1, 3, 6, 7), this effect becoming more obvious with larger currents (e.g., Burrows, 1979b: Fig. 1C–D).

Possible explanations for this decline are the following. First, transmission itself might diminish, either because less transmitter is released with time, or because the postsynaptic neuron becomes less sensitive to the transmitter released. This seems unlikely, however, since observations on tonically releasing interneurons suggest that when they are depolarized by synaptic inputs, transmission is not significantly diminished at least over the time that the recordings are made (typically tens of minutes, to as long as an hour). Second, it could result from the parallel activation of other inputs to the postsynaptic neuron, for example from sensory feedback that increases as more motor neurons are recruited by the interneuron. This possibility cannot be discounted, given the complexity of the circuitry of which the nonspiking interneurons are a part. In future experiments, peripheral feedback might be eliminated by cutting peripheral nerves, though this would make it difficult to identify the affected motor neurons. Third, the interneurons might rectify during the larger current pulses, resulting in a gradual decline in the presynaptic depolarization. This could reflect the activation of voltage-sensitive conductances. This seems the most likely explanation for the decline in postsynaptic effect, especially since higher currents are increasingly less effective in depolarizing an interneuron (e.g., Fig. 3). It would seem worthwhile to investigate this possibility, by making further recordings with two electrodes in individual nonspiking interneurons, and looking at the shape of the presynaptic voltage over time, at a range of currents. This would be technically difficult but probably the only satisfactory approach. Recordings using single electrodes presynaptically, in which current is passed through a bridge circuit, are not satisfactory for judging the presynaptic voltage, especially since acetate or citrate is typically used as the anion, to avoid the complications of injecting Cl^- ions into the presynaptic neuron. The difficulty here is that these electrodes tend to change their recording properties during the passage of larger currents. Such two electrode recordings could also be used to establish the membrane voltages at which the neurons exert their effects, an important determinant of their integrative capabilities.

3.3.2 *Responses to brief changes in membrane potential*

At the other end of the spectrum, evidence is considerable that nonspiking interneurons can mediate postsynaptic effects that faithfully reflect relatively rapid changes in membrane potential. The most convincing demonstration of this comes from Burrows (1979a), who showed that discrete EPSPs in an identified interneuron could evoke discrete IPSPs in postsynaptic motor neurons (Fig. 8A). To establish that the IPSPs were directly mediated by the EPSPs, the EPSPs were evoked repeatedly by alternately flexing and extending the tibia about the femur, and the interneuron was gradually hyperpolarized by injecting increasing amounts of current (Fig. 8B,C). Hyperpolarization increased the amplitude and the duration of the EPSPs, but reduced their peak level of depolarization (Fig. 8D,E). As a result, the

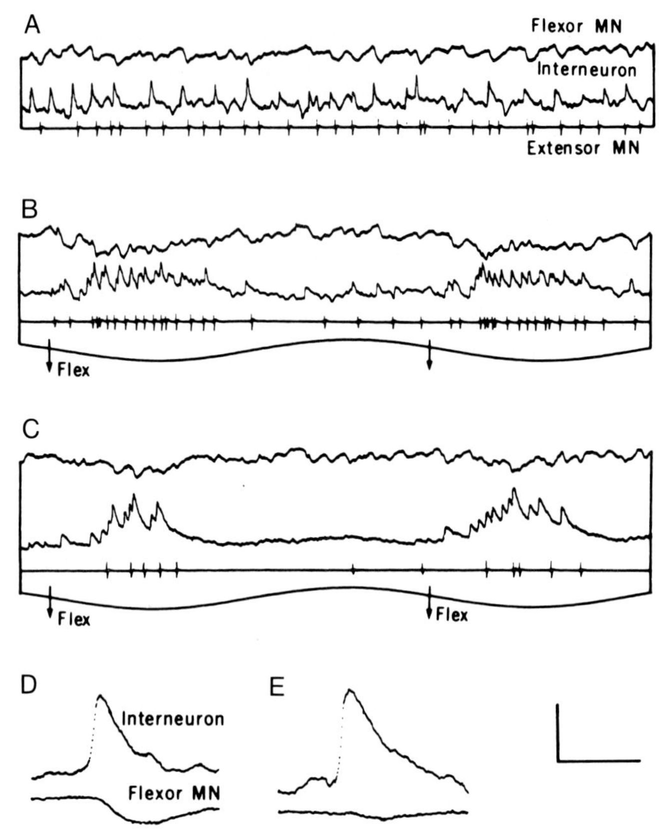

Fig. 8. Release of transmitter evoked by EPSPs in a nonspiking interneuron, recorded from the metathoracic ganglion of locust. (A) EPSPs are evoked in a nonspiking interneuron (second trace) when the tibia of the ipsilateral hind leg is fixed at 60° about the femur. Each EPSP is followed by an IPSP in a flexor tibiae motor neuron (first trace). The slow extensor tibiae motor neuron, recorded in a myogram (third trace), spikes tonically. (B) The tibia is forcibly flexed and extended about the femur sinusoidally (fourth trace; other traces as in A). Flexion evokes summed EPSPs in the interneuron, summed IPSPs in the flexor tibiae motor neuron, and an increased frequency of spikes in the extensor motor neuron. (C) The interneuron is held hyperpolarized by injecting −3.5 nA of current. Larger EPSPs occur in the interneuron with each flexion, but the hyperpolarization of the flexor tibiae motor neuron is reduced, and fewer spikes occur in the extensor tibiae motor neuron. (D) Averaged EPSPs and IPSPs, which were pretriggered from the EPSP for 32 occurrences. The EPSPs were evoked by holding the tibia flexed. The interneuron is at its normal membrane potential, and the averaged IPSP in the flexor motor neuron follows the averaged EPSP in the interneuron with a latency of 1.5 ms. (E) As for (D), except that the interneuron is held hyperpolarized by −3 nA of current. The EPSP is increased in amplitude and duration, but the IPSP is substantially reduced and occurs at a latency of about 2.5 ms. Calibration. Vertical: (A) 6 mV; (B, C) 12 mV; (D, E) 2 mV. Horizontal: (A–C) 220 ms; (D, E) 20 ms. (After Burrows, 1979a.)

corresponding IPSPs gradually diminished in size, and the latency of their onset increased from 1.5 to 2.5 ms.

When nonspiking interneurons are depolarized with brief current pulses (5–10 ms) of high intensity, the postsynaptic response also resembles a conventional, spike-mediated PSP (Burrows and Siegler, 1976, 1978). It lasts much longer than the stimulus pulse, and reaches its peak after the offset of the stimulus. The responses are graded according to the current intensity, but the currents needed to evoke transmitter release are much greater than those for longer pulses (i.e., >100 ms). The limiting factor would be the time necessary to discharge the capacitance of the presynaptic neuron enough to take the imposed voltage change above threshold.

Burrows (1979b) studied the postsynaptic response when a presynaptic nonspiking interneuron was subjected to sine wave currents of various frequencies (1–30 Hz). The response gradually diminished as the frequency increased, with frequencies higher than 25 Hz producing little postsynaptic effect. This diminution must be partly explained by the time necessary to discharge the capacitance of the presynaptic interneuron. The same would not apply when an interneuron was depolarized by chemically mediated EPSPs, and a measure based on current injection would therefore underestimate the capabilities of the interneurons to transmit patterned information. Indeed, inspection of the data of Burrows (1979a) (Fig. 8B) suggests that individual IPSPs can follow EPSPs at instantaneous frequencies as high as 100 Hz, the limiting factor being the time courses of the PSPs themselves. At higher frequencies, where the presynaptic EPSPs summate smoothly, there is a similarly sustained summed effect postsynaptically.

3.4 LATENCY AND MONOSYNAPTICITY

When neurons generate spikes, a number of physiological tests can be employed to indicate whether their effects upon other neurons are likely to be monosynaptic (Berry and Pentreath, 1976). For nonspiking interneurons, however, these tests are largely unsatisfactory (Burrows and Siegler, 1978). For example, one usual indication of a monosynaptic connection is a short and constant latency between the spikes in one neuron, and the onset of synaptic potentials in another neuron. An obvious problem with transmission from nonspiking interneurons is the absence of a presynaptic electrophysiological event that clearly indicates when the membrane potential has crossed the threshold for transmitter release. Other problems arise with tests that depend upon the differential effects of an altered ionic milieu upon monosynaptic and multisynaptic connections (briefly, for example, the greater sensitivity of multisynaptic connections to lowered Ca^{2+}). Interpretation of such tests relies on the assumption that any interposed neurons necessarily have spike-mediated synaptic transmission. But were the neurons (spiking or nonspiking) to be capable of the graded release of transmitter in the absence of spikes then the results are equivocal. Given the number of nervous systems that are now known to contain nonspiking neurons, and the increasing number of spiking neurons being shown to be capable of graded release of transmitter below the threshold for spike initiation, this poses a serious problem of interpretation, especially when synaptic delays are relatively long.

The best evidence for the monosynaptic effect of nonspiking interneurons upon motor neurons comes from experiments in which discrete EPSPs in an interneuron produce discrete IPSPs in a motor neuron (Burrows, 1979a). In this case, the shortest delay was 1.5 ms, which is comparable to that measured for other connections that are thought to be monosynaptic in insects (Burrows, 1975; Pearson *et al.*, 1976). In the majority of nonspiking interneurons investigated, however, such a measure has not been available. Instead, synaptic latency has been measured from the onset of a presynaptic current step to the onset of a response in a postsynaptic neuron (Burrows and Siegler, 1978; Burrows, 1979b, 1980; Wilson, 1981). This necessarily overestimates the synaptic latency, however, because some time is taken for the capacitance of the interneuron to be discharged. This element of the delay will decrease with stronger currents, as the onset of the voltage change becomes increasingly steep. For example, in a typical recording from an interneuron and a motor neuron, the delay decreased, from 10 ms to a minimum delay of 3 ms, as the current strength was increased (Burrows and Siegler, 1978). The shortest minimum delay measured in this way for any interneuron and motor neuron pair was 1.5 ms, though minimum delays as long as 12.5 ms could be measured. The significance of these larger delays is unknown. It is noteworthy, however, that at chemical monosynaptic connections between stomatogastric motor neurons of lobster, the latency between spikes and PSPs is 1–2 ms, whereas for the same motor neurons, graded nonspiking transmission has a latency of 50–100 ms (Graubard, 1978).

The most compelling evidence for direct connections would come from combined anatomical and physiological studies, in which nonspiking interneurons that have short latency effects upon motor neurons could also be shown at the EM level to make synapses with these motor neurons. This would involve the labeling of the putative pre- and postsynaptic elements with markers that could be distinguished from each other at the EM level. The technical difficulties are considerable, however, and it might be worthwhile in the interim to seek other ways of judging monosynapticity, for example by finding pharmacological means to elicit spikes from nonspiking interneurons, so that a more reliable measure of synaptic latency might be obtained. At present, the physiological evidence is consistent with the monosynapticity of connections, but one should be open to other possibilities that may emerge as more is discovered about the circuitry of the ganglion.

3.5 TONIC RELEASE OF TRANSMITTER

Most nonspiking interneurons are held at a membrane potential close to, but below, the threshold for transmitter release. But, as already described (Fig. 7), a proportion release transmitter tonically at their "resting" membrane potential (Burrows and Siegler, 1978; Wilson and Phillips, 1982). For example, one such interneuron had a tonic inhibitory effect upon a fast flexor tibiae motor neuron (Fig. 7), while another had a tonic excitatory effect upon the slow extensor tibiae motor neuron (Burrows and Siegler, 1978). This appeared to reflect the usual mode of functioning of the interneurons, rather than to result from damage by penetration with microelectrodes. No change occurred in the activity of postsynaptic motor neurons when the interneurons were penetrated, and the tonic effects remained unchanged for long periods (e.g., 1 h). Burrows and Siegler (1978) suggested that all nonspiking neurons are capable of such tonic release of transmitter, since all appear to be capable of exerting continuing effects when depolarized by intracellularly injected current. Which of the interneurons are "quiescent" and which are releasing significant amounts of transmitter could depend upon diverse factors such as the posture, state of arousal, and sensory inputs to the locust. Consistent with this idea, changes in the posture of a hind leg can alter the "resting" membrane potential of some nonspiking interneurons and their tonic effects upon postsynaptic motor neurons (Siegler, 1981b). The shift in membrane potential appears to result indirectly from sustained changes in the tonic sensory input provided from the joints of the leg (see also Fig. 13, and Section 4 for further discussion).

In the mesothoracic ganglion of the locust, Wilson and Phillips (1982) found two nonspiking interneurons that exert tonic effects upon motor neurons of the flexors and extensors of the tibia. They argued that these interneurons are intrinsically different (that they "characteristically release transmitter tonically") from others in the same animal that are not releasing transmitter tonically. As evidence for this, it was said that the synaptic vesicles in the tonic interneurons are "fuzzier" in appearance than those in similarly treated motor neurons and nontonic interneurons. Whether these interneurons are thought to have different membrane properties, different synaptic connections, or some other "characteristic" difference is not clear. Nonetheless, the physiological findings of Wilson and Phillips (1982) are entirely consistent with those of Burrows and Siegler (1978), and there seems no reason as yet to suppose that there are two distinct populations of nonspiking interneurons. Given that recordings are made from experiment to experiment in animals that are in approximately the same behavioral state, and in the same imposed posture, it is not surprising that particular interneurons are "typically" observed to be releasing transmitter tonically. It would seem worthwhile to investigate this problem in more detail, particularly at the EM level.

3.6 THRESHOLD FOR TRANSMITTER RELEASE AND SYNAPTIC GAIN

Two values that would be important in determining the ability of nonspiking interneurons to effect changes in the membrane potential of postsynaptic neurons are the threshold for transmitter release and the sensitivity of the postsynaptic voltage response to changes in presynaptic voltage. (This latter value is referred to here as the "synaptic gain"; see also Siegler, 1984, for further discussion.) In the absence of direct measures of presynaptic voltage, these values can only be estimated indirectly.

3.6.1 *Threshold*

Burrows and Siegler (1978) measured membrane potentials of −35 to −60 mV for nonspiking interneurons, with a mean of about −48 mV. Since the majority (70%) of the interneurons whose postsynaptic effects were studied were not tonically releasing transmitter, it seems likely that this mean value of −48 mV is somewhat higher than the threshold for release. In addition, it is assumed that the lowest membrane potentials were recorded from tonically releasing interneurons, and the highest from interneurons that were hyperpolarized beyond release threshold. Inspection of the figures from the several subsequent studies of nonspiking interneurons suggests that the largest depolarizing current needed to initiate transmitter release from an interneuron was about 3 nA. Assuming an effective input resistance of 3–5 MΩ (Burrows and Siegler, 1978) gives a maximum presynaptic depolarization of 9–15 mV. For an interneuron with the highest membrane potential −60 mV, this would put the threshold at about −51 to −45 mV. Conversely, the largest hyperpolarizing current needed to suppress transmitter release from any tonically releasing interneuron was about −3 nA. Assuming the same input resistance gives a maximum hyperpolarization of 9–15 mV, or for the interneuron with the lowest membrane potential measured, −35 mV, a threshold at about −44 to −50 mV. These different estimates place the threshold for transmitter release between −50 and −45 mV.

This is comparable to values from some other nonspiking interneurons. In ocellar interneurons in the locust, which are capable of graded, as well as spike-mediated synaptic transmission, the threshold for transmitter release is between −40 and −45 mV (Simmons, 1982); in nonspiking second-order cells of the barnacle ocellar system, the threshold is between −45 and −50 mV; and in the lobster stomatogastric ganglion, graded transmission by EX interneurons has a threshold of −47 ± 4 mV (Graubard, 1978). By contrast, in the nonimpulsive stretch receptor of the crab (Blight and Llinás, 1980) and the barnacle photoreceptor (Stuart and Oertel, 1978), the release of transmitter occurs at a membrane potential as high as −60 mV. These two sensory receptors, however, also have resting potentials that are higher (−60 mV) than those of the other nonspiking neurons.

3.6.2 *Synaptic gain*

The curves for graded transmission (postsynaptic change in voltage plotted against presynaptic current) have an approximately linear region, where the postsynaptic response shows the greatest sensitivity to a change in presynaptic current. Presumably, this is also the region of maximum synaptic gain, i.e., where changes in presynaptic voltage would be expected to produce the largest changes in postsynaptic voltage. The slope of this linear region varies considerably for different pairs of interneurons and motor neurons. Measurements of published curves yield a range of about 0.4–3.5 mV/nA for depolarizing responses, and 0.5–2.3 mV/nA for hyperpolarizing responses. The average value is about 1.25 mV/nA.

Ignoring for the present the possible sources of these differences, a very rough calculation for synaptic gain can be made, taking the average slope of 1.25 mV/nA. In doing this, at least three values must be estimated: (1) the voltage change occurring at the site of current injection; (2) the decrement from the site of current injection to the site of transmitter release; and (3) the decrement from the postsynaptic site to the recording site at the cell body of the motor neuron. Unfortunately, only indirect evidence is available concerning any of these values. First, the effective input resistance measured in a neuropilar process (presumably at a larger neuropilar process) may be on the order of 3–5 MΩ (Burrows and Siegler, 1978), so a value of 4 MΩ is assumed. This is considerably lower than that measured at the cell body of a small motor neuron (Siegler, 1982) and probably reflects the continuous bombardment of the neuropil region of the interneuron by synaptic inputs. Second, Wilson and Phillips (1982) report that output synapses occur mainly on thinner, higher order branches of nonspiking interneurons. Calculations of Rall (1981) based on anatomical data of Siegler and Burrows (1979) suggest that the decrement from a thick to a thin branch would result in a steady voltage change at a fine branch that is no larger than 70% of the initial value. Third, Watson and Burrows (1981, 1982) report that synapses on motor neurons also occur primarily on finer branches. The decrement from fine to thick processes is expected to be severe, again based on measurement by Rall (1981) on nonspiking interneurons. A decrement at the cell body to 30% of the steady-state value at synaptic sites is probably a conservative estimate.

For 1 nA injected current the estimated changes are 2.8 mV pre- and 3.3 mV postsynaptically, or a synaptic gain of 1.2. Considering the different slopes measured, the range could be from ~0.5 to ~4.2. These are probably minimum values; were the decrement either pre- or postsynaptically to be

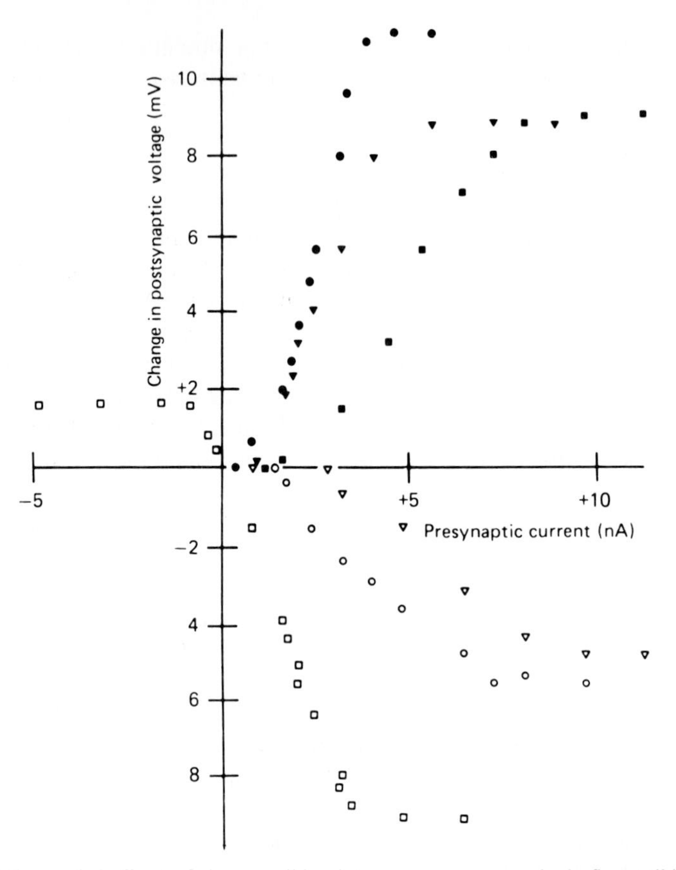

Fig. 9. The graded effects of six nonspiking interneurons upon a single flexor tibiae motor neuron in the metathoracic ganglion in one locust. Three interneurons (■, ▼, ●) depolarize the motor neuron, and three others (▽, ○, □) hyperpolarize it. One of these (□) exerts a tonic hyperpolarizing effect on the motor neuron. The postsynaptic changes in voltage are evoked by injecting the interneurons with steps of current some 500 ms in length. (From Burrows, 1980.)

any greater than estimated, then the "gain" would necessarily be higher. Nonetheless, this estimate places the value of synaptic gain within the range calculated for "conventional" synapses (0.3–3) rather than the range for the "high-gain" synapses (>50) characteristic of photoreceptors and some electroreceptors (see Shaw, 1981, for discussion). It must be stressed, however, this is only the crudest of estimates, based upon very simple assumptions, and the influence of particular neuronal geometries would be considerable on the effectiveness of graded transmission (see Graubard and Calvin, 1979, for discussion).

The variation among the different pairs of interneurons and motor neurons must reflect differences arising both pre- and postsynaptically. For example, in motor neurons with larger main processes and larger cell bodies, decrement of a given voltage from the neuropilar region to the cell body would be more severe than in the smaller motor neurons. Thus an equal synaptic input would appear to be less effective, as measured at the cell body. This cannot explain all of the variation observed, however. Burrows (1980) plotted the effect of three excitatory interneurons and three inhibitory interneurons upon a single motor neuron (Fig. 9). Measurements from the linear portion of the curves show values of 3.5, 2.2, and 1.5 mV/nA for the excitatory effects, and 2.3, 0.9, and 0.6 mV/nA for the inhibitory effects. While this could reflect actual differences in synaptic efficacy (or synaptic gain), it could also reflect differences in the site of current injection relative to the site of transmitter release, differences in the input resistance at the site of current injection, or differences in the distribution of postsynaptic sites with respect to the recording site at the cell body of the motor neuron. There is little at present to distinguish among these possibilities. An additional complication is that the apparent effectiveness of a given connection can be altered according to the synaptic inputs impinging upon the interneurons and motor neurons, for example, from sensory information that depends upon the posture and history of movement of particular joints of the hind legs (Siegler, 1981b; see also Fig. 14). It is probably not reasonable to consider the question of synaptic gain in any more detail, until we can measure the presynaptic voltages responsible for transmitter release. Even then, however, we will have the difficulty of assessing the decrement of the voltage from the site of presynaptic stimulation, to the site of postsynaptic recording.

4 Integration by nonspiking interneurons and their role in behavior

Nonspiking local interneurons are undoubtedly important elements of the local intraganglionic circuitry that controls the coordinated activity of motor neurons to limb muscles. Spiking local interneurons that may also synapse directly upon particular motor neurons have comparatively restricted motor effects, and none found so far cause the vigorous and coordinated movements that can be evoked by the nonspiking interneurons (Burrows and Siegler, 1982). Furthermore, there is little direct chemical synaptic interaction between the leg motor neurons themselves (Burrows and Horridge, 1974; Hoyle and Burrows, 1973; Wilson, 1979), and no electrical coupling has been found, except when supernumerary motor neurons are produced (Siegler, 1982).

4.1 POSTSYNAPTIC TARGETS OF NONSPIKING INTERNEURONS

The coordination of motor activity by the nonspiking interneurons is achieved in two ways, first by virtue of their particular excitatory or inhibitory postsynaptic effects upon sets of motor neurons, and second by direct inhibitory interconnections among the nonspiking interneurons themselves.

4.1.1 *Control of sets of motor neurons*

The nonspiking neurons in the metathoracic ganglion of the cockroach and locust and the mesothoracic ganglion of the locust synapse on sets of motor neurons, and control their recruitment in an orderly way (Pearson and Fourtner, 1975; Meyer and Walcott, 1979; Burrows, 1980; Wilson, 1981).

It is common in insect locomotion that excitatory motor neurons to the same muscle are recruited by size. In effect, they are activated according to the strength of the contractions they evoke: first recruited are the smaller motor neurons (those with smaller diameter cell bodies and axons), which produce weaker, slower, and more sustained contractions of the muscle. The larger motor neurons are then recruited, and these produce stronger, faster, and more phasic contractions. In the cockroach the increasing depolarization of one nonspiking interneuron recruits coxal levator motor neurons in this way (Fig. 10A,B) (Pearson and Fourtner, 1975; Meyer and Walcott, 1979). Similarly in the locust, some nonspiking interneurons recruit first the slower (smaller) motor neurons, then the faster (larger) motor neurons of a multiply innervated muscle (Fig. 10C–E) (Burrows, 1980). As a result, the strength and the speed of contraction in these muscles can be finely and precisely graded according to the depolarization of an interneuron. In the cockroach, the sequential recruitment of motor neurons by size appears to result to some extent from the properties of the motor neurons themselves, since the slower motor neurons spike more readily in response to current injected intracellularly than do the faster motor neurons (Meyer and Walcott, 1979).

This is not, however, the only way in which synergistic motor neurons are recruited by nonspiking interneurons. Burrows (1980) studied the effects of a number of different nonspiking neurons upon sets of motor neurons and found many different patterns of recruitment, even when only a slow and a fast motor neuron innervated the muscle, as for the extensor tibiae. Two interneurons excited both motor neurons, recruiting the slow one first. For one interneuron, higher currents caused the fast motor neuron to spike more rapidly than did the slow; for the other, the slow motor neuron always fired more rapidly. Still other interneurons depolarized only the slow, or only the fast, motor neuron. Further permutations in the recruitment of slow, intermediate, and fast motor neurons were found for interneurons that activated a larger pool of motor neurons, supplying more complexly innervated muscles, for example that of the flexor tibiae, which comprises at least nine motor neurons of varying efficacy.

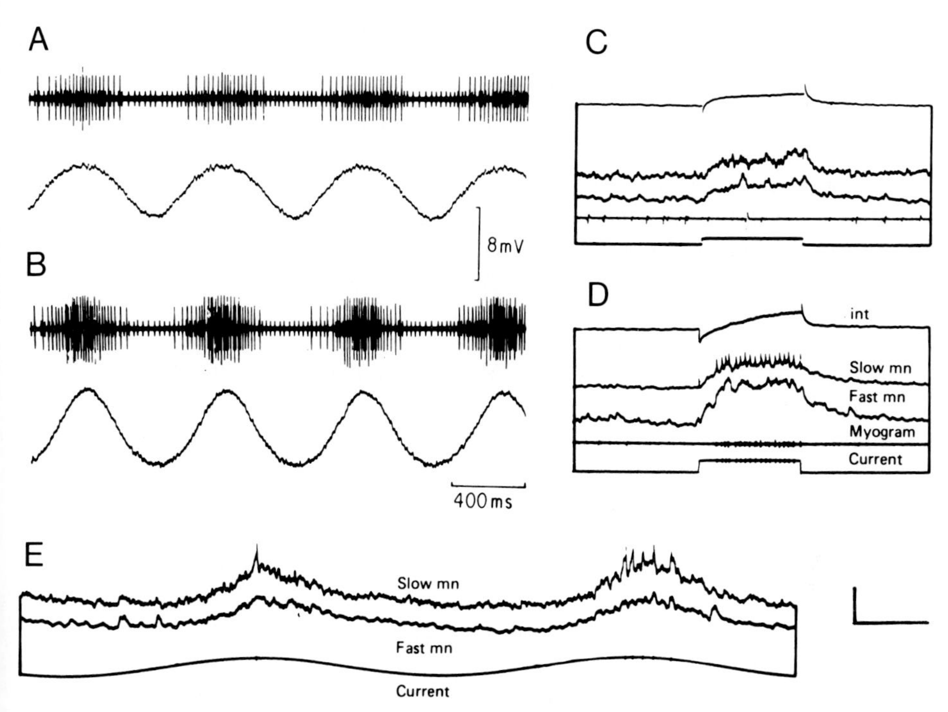

Fig. 10. Recruitment of motor neurons by size when nonspiking interneurons are depolarized by currents of increasing amplitude. (A) In cockroach metathoracic ganglion, spikes of coxal levator motor neurons 3, 5, and 6 are recorded extracellularly in nerve 6Br4 (first trace) while a nonspiking interneuron (interneuron I) (second trace) is depolarized with sinusoidally varying current via a bridge circuit. Motor neuron 3 (smallest spike) increases its frequency of firing, and motor neuron 5 (larger spike) is recruited. (B) Traces as in (A). With greater depolarization of the nonspiking interneuron, motor neuron 6 is recruited. (C) In locust metathoracic ganglion, simultaneous intracellular recordings are made from a neuropilar process of a nonspiking interneuron (first trace), a neuropilar process of a slow motor neuron (second trace), and the cell body of a fast motor neuron (third trace). Both motor neurons innervate the tergotrochanteral muscle. Depolarization of the interneuron with a step of current (fifth trace) depolarizes the motor neurons, but reduces the frequency of spikes in other coxal motor neurons recorded extracellularly in a myogram (fourth trace). (D) As in (C). A larger step of current evokes larger depolarizations, and spikes in the slow motor neuron. (E) Sinusoidally varying current injected into the interneuron (not shown) evokes continuously varying depolarization in the motor neurons and elicits spikes in the slow motor neuron. Calibration for (C and E) interneuron, 40 mV; slow mn, 7 mV; fast mn, 4 mV; current, 30 nA; time base, 400 ms. (D) slow mn, 17 mV; other traces as for (C and E). (A and B after Pearson and Fourtner, 1975; C–F after Burrows, 1980.)

In addition to ordering the recruitment of motor neurons of a particular muscle, nonspiking interneurons act in other ways to coordinate the movements of a leg. They may coordinate the excitation of motor neurons supplying different muscles that normally act together during a particular behavior (Burrows, 1980), e.g., flexor tibiae motor neurons and levator tarsi motor neurons, which would be excited together during the protraction phase of walking, or flexor and extensor tibiae motor neurons, antagonists during walking, which are excited together during the co-contraction phase of jumping or defensive kicking (Heitler and Burrows, 1977). For another, they may excite motor neurons to one muscle of a joint, while inhibiting motor neurons to another muscle that is its antagonist (Pearson and Fourtner, 1975; Burrows, 1979a, 1980; Wilson and Phillips, 1982).

The most direct evidence regarding the nature of these latter connections comes from Burrows (1980). Recordings were made intracellularly from nonspiking interneurons and two antagonistic motor neurons. One motor neuron was inhibited and the other excited by depolarization of an interneuron. These two responses had similar current thresholds, and were finely graded as the interneuron was depolarized. No discontinuities were seen in the voltage changes of the motor neurons, as might have been expected if a second interneuron were interposed and responsible for mediating the effects upon one of the motor neurons. From these results it is argued that both effects are probably direct, with one interneuron mediating opposite postsynaptic effects on the antagonistic motor neurons it supplies (Burrows, 1980).

Many motor activities require the reciprocal activation of antagonist motor neurons. Direct connections of opposite effect from a single interneuron would at first sight seem a simple way of achieving this, provided that an interneuron release two different transmitters, or that the antagonist motor neurons have receptors, or receptor–channel complexes, that respond differently to the same transmitter. The difficulty, however, is that there are a number of different nonspiking interneurons, each with different effects on the same motor neurons. Whereas one interneuron may excite a flexor and inhibit an extensor motor neuron, another may inhibit the same flexor and excite the extensor, and another may excite both (Burrows, 1979b). Thus the "simple" picture of two transmitters or two receptor types must be elaborated to account for this variety of effects. One can postulate yet more transmitters within the population of interneurons, or yet more receptor types for the motor neurons, but the hypotheses become increasingly tenuous. This problem highlights the necessity of learning something of the transmitter(s) used by the nonspiking interneurons and of obtaining anatomical evidence, at the EM level, for the connections judged to be direct from the physiological tests available.

4.1.2 *Interactions among interneurons*

The same interneurons that affect the motor neurons may also affect each other. Burrows (1979b) made recordings simultaneously from pairs of nonspiking interneurons in the metathoracic ganglion of the locust. Synaptic connections were found between the interneurons in about one-third of the paired recordings. No excitatory interactions were found; all were inhibitory, and in one direction only. The connections function in much the same way as do the inhibitory connections from nonspiking interneurons to motor neurons (Fig. 11). The effects are chemically mediated, occur with a short latency, are sustained during sustained presynaptic depolarization, and are graded according to the amount of current injected into the presynaptic neuron.

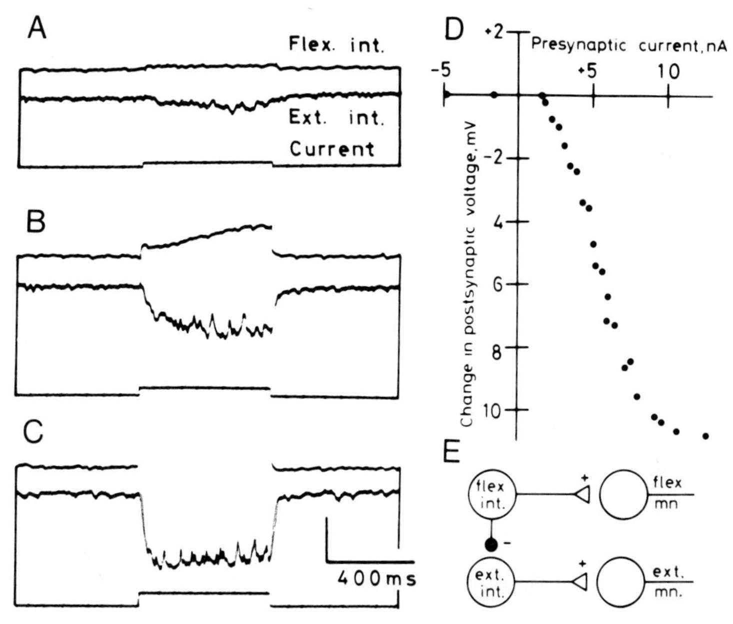

Fig. 11. Graded inhibitory interaction between two nonspiking interneurons in the locust metathoracic ganglion. One interneuron excites flexor tibiae motor neurons, and the other excites the slow extensor tibiae motor neuron. (A) Depolarization of the flexor interneuron (flex int) with a current step results in the hyperpolarization of the extensor interneuron (ext int). (B, C) As in (A). Larger depolarizing currents in the flexor interneuron evoke larger hyperpolarizations in the extensor interneuron. (D) The change in voltage in the postsynaptic extensor interneuron is plotted against the current injected into the presynaptic flexor interneuron. (E) A schematic representation of the connections between the interneurons and their motor neurons. Depolarization of the extensor interneuron had no effect on the flexor motor neurons. Calibration, (A–C) flexor interneuron, 40 mV; extensor interneuron, 8 mV; current, 30 nA. (After Burrows, 1979b.)

The inhibitory connections found within the population of nonspiking interneurons are complex, and were summarized as follows (Burrows, 1979b). Inhibitory interactions occurred (1) between interneurons that excited antagonistic motor neurons, (2) between interneurons that excited the same, or synergistic motor neurons, and (3) between interneurons that excited motor neurons to muscles moving different joints of a hind leg.

The functional significance of some of the interconnections was readily apparent. For example, inhibitory interactions between two interneurons that excited antagonistic motor neurons, such as those to flexors and extensors of the tibia, would ensure the reciprocal activation of the motor neurons during walking or certain local postural reflexes (Fig. 12A). It was

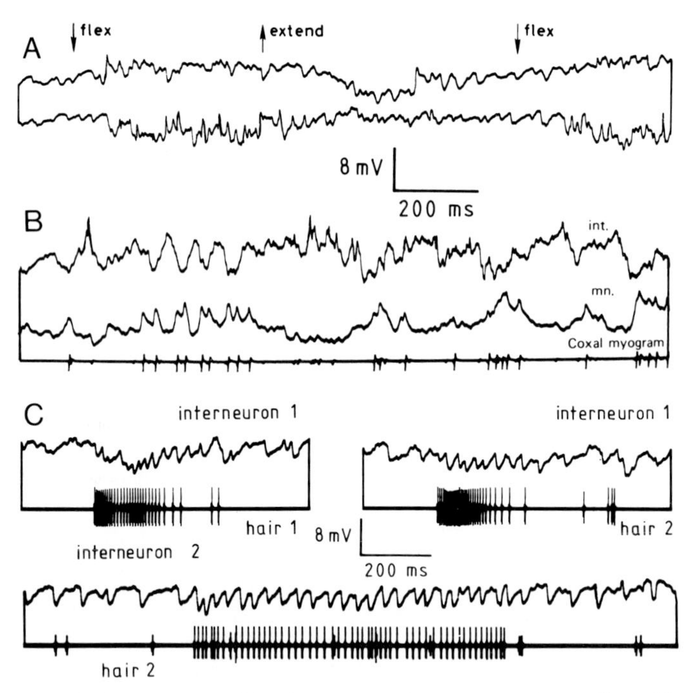

Fig. 12. Synaptic input to nonspiking interneurons in metathoracic ganglion of locust. (A) Intracellular recordings from one nonspiking interneuron (first trace) that excites flexor tibiae motor neurons and another (second trace) that excites the extensor tibiae motor neurons. When depolarized with current (not shown) the flexor interneuron inhibits the extensor interneuron. When the tibia of the ipsilateral hind leg is forcibly flexed or extended about the femur, complex changes occur in the pattern of PSPs recorded in the two interneurons. (B) Coxal motor neuron and a nonspiking interneuron that inhibits it, recorded during voluntary movements of a hind leg. (C) Touching single hairs (hair 1, 2) on the femur of a hind leg evokes trains of spikes in the primary afferent, recorded extracellularly in a peripheral nerve, and IPSPs in two nonspiking interneurons (interneurone 1,2). [Adapted from (A) Burrows, 1979b; (B) Burrows and Siegler, 1978; (C) Siegler and Burrows, 1983.]

less easy to rationalize the inhibitory interactions between other interneurons, e.g., those that excited the same motor neurons, though it was suggested that the interneurons were likely to be used in different behavioral contexts. For example, one of the interneurons might be excited during a resistance reflex while the other was inhibited.

Where the functional significance of a particular connection is not clear, this might reflect two kinds of limitations: first, incomplete knowledge of the ways in which sets of motor neurons are normally activated in the myriad of movements performed by a hind leg, and second, incomplete knowledge of the full complement of connections made by individual interneurons.

4.1.3 *Other outputs of nonspiking interneurons*

Other possible targets for nonspiking neurons have not as yet been investigated. This would seem a fruitful line of inquiry, since many nonspiking interneurons have been found without effects on leg motor neurons (Burrows and Siegler, 1978; Siegler and Burrows, 1979). Obvious possibilities would be effector neurons other than the excitatory motor neurons, such as the peripheral inhibitors (Pearson and Bergman, 1969) or the modulatory neurons of the dorsal median group (see Evans, 1980, for review); intergangliónic interneurons or local spiking interneurons (Burrows and Siegler, 1982); and the central terminals of afferent fibers, which are known from EM evidence to receive synaptic inputs (Altman *et al.*, 1980; Watson and Pflüger, 1984).

4.2 INPUTS TO NONSPIKING INTERNEURONS

Coordination of motor activity by the spiking interneurons depends not only upon the specific output connections they make with motor neurons and other nonspiking interneurons, but also upon the inputs they receive. The nonspiking interneurons in the metathoracic ganglion of the locust respond widely to sensory inputs from the ipsilateral hind leg. For example, most of the nonspiking interneurons respond to imposed changes in the angles of the joints with complex changes in membrane potential (Fig. 12A) (Burrows and Siegler, 1976, 1978; Burrows, 1979b, 1980; Siegler, 1981a,b). These responses differ from interneuron to interneuron, and also according to the speed and direction of the imposed changes in joint angle; they may be excitatory or inhibitory, and phasic or sustained. Several proprioceptors could be responsible, including chordotonal organs, multiterminal receptors, and campaniform sensilla near the joints (see Siegler, 1981b, for discussion). Another effective sensory stimulus to the nonspiking interneurons arises from the large tactile hairs arrayed over the surface of the legs (Siegler and Burrows, 1983). These hairs respond phasically to tactile stimuli, and touching hairs in different regions may excite or inhibit a particular nonspiking interneuron (Fig. 12C).

In the locust, there is no evidence that afferents synapse directly upon nonspiking interneurons, though in cockroach, stimulation of hair plate sensilla evoked EPSPs in unidentified nonspiking interneurons (Pearson *et al.*, 1976). In locust, it appears that many sensory effects are mediated indirectly, via spiking local interneurons. These spike vigorously in response to the same kinds of stimuli that affect nonspiking interneurons, though each spiking local interneuron responds in a much more restricted way. For example, the only effective stimulus for one spiking local interneuron might be an imposed flexion of the femoral–tibial joint, and for another, tactile stimuli to hairs on a small region of the hind leg (Burrows and Siegler, 1982, 1984). Some of these spiking interneurons are directly postsynaptic to afferents from hairs or campaniform sensilla on a leg, though other types of sensory receptors have not been tested (Siegler and Burrows, 1983). The pattern of spikes elicited in spiking local interneurons corresponds well to the pattern of PSPs that can be elicited from nonspiking interneurons by touching hairs individually (Siegler and Burrows, 1983) or by imposing changes in the rest angle of the femoral–tibial joint (Siegler, 1981b). This suggests that the sensory influences upon the nonspiking interneurons are mediated indirectly via spiking local interneurons. Since the nonspiking interneurons respond more widely to sensory inputs than do the spiking local interneurons, it seems probable, further, that several spiking interneurons converge on each nonspiking interneuron. Recordings can be made readily from the cell bodies of the spiking local interneurons, which are clustered in particular regions of the ganglion. Therefore, it should be feasible to record nonspiking interneurons in the neuropil, while searching among the group of cell bodies to locate putative presynaptic spiking local interneurons.

Nonspiking interneurons are also depolarized and hyperpolarized in complex ways during "voluntary" movements of the legs, which occur spontaneously or in response to touching various parts of the body such as the head or abdomen (Fig. 12B) (Pearson and Fourtner, 1975; Burrows and Siegler, 1976, 1978; Burrows, 1979b). The pathways normally involved in the recruitment of nonspiking interneurons during such voluntary movements are unknown but it is reasonably assumed that long interneurons are involved in the initiation of movements and in interganglionic coordination. Thus they are likely candidates to affect the nonspiking interneurons. At the local level, sensory feedback and inputs from other nonspiking interneurons could be important in determining the precise pattern of synaptic inputs during motor activity. Other possible inputs to the nonspiking interneurons might come from local interneurons (as yet unknown) presumed to be involved in the bilateral coordination of postural and locomotory behavior.

4.3 BEHAVIORAL ROLE OF NONSPIKING INTERNEURONS

4.3.1 *Rhythmic motor output*

Pearson and Fourtner (1975) investigated the role of nonspiking interneurons in the walking rhythm of the cockroach. This rhythm is centrally generated, with "oscillators" in individual thoracic ganglia. One nonspiking interneuron, interneuron I, strongly excited coxal levator motor neurons, and inhibited a coxal depressor motor neuron. During rhythmic leg movements interneuron I was depolarized when the levator motor neurons fired. Other interneurons, type II, were hyperpolarized during levator bursts, and when they were injected with depolarizing current they inhibited the levator. Penetration of interneuron I usually resulted in high rates of spike activity in the levator motor neurons, though in three animals, rhythmic leg movements persisted. In one of these, depolarizing pulses to the interneuron reset the rhythm of leg movements. It was argued, therefore, that interneuron I is part of the system for generating the rhythmicity, rather than being intercalated between the rhythm generator and the motor neurons.

It seems equally possible, however, that sensory feedback resulting from altered motor neuron activity contributes to the resetting. In the locust, for which the central effects of imposed leg movements have been studied in detail, sensory feedback impinges upon the nonspiking interneurons (Fig. 12A,B and Fig. 13). This, and the fact that in the cockroach depolarizing current pulses were effective only during the time that the interneuron

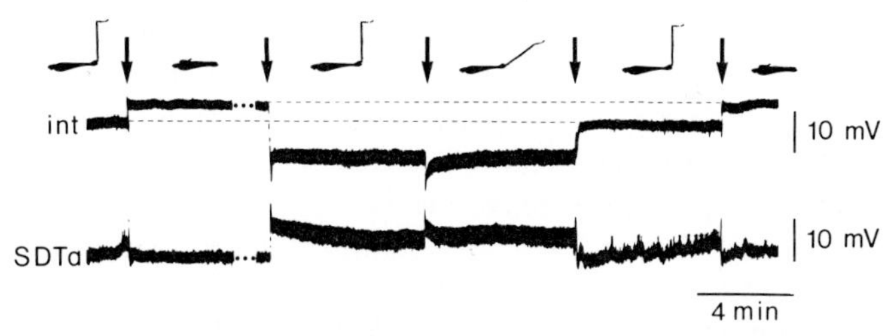

Fig. 13. Nonspiking interneuron and motor neuron in the locust metathoracic ganglion have sustained changes in membrane potential in response to imposed changes in the angle of the femoral–tibial joint of the ipsilateral hind leg. Record shows 30-min recording from a nonspiking interneuron (int) and a slow depressor tarsi motor neuron (SDTa). The recording is continuous except for short gap (dots) when the tape spool is changed. The tibia is moved every 6 min, at arrows, to the position shown in the corresponding diagrams of the hind leg (90°–0°–90°–140°–90°–0°). Leg is drawn ventral surface uppermost as positioned during the experiment. At the start of the trace, a femoral–tibial angle of 90° has been approached from 140°. The membrane potentials depend upon the angle of the femoral–tibial joint, and upon the direction from which that angle is approached. The motor neuron spikes continually at 140° and when 90° has been approached from 0°. Individual spikes cannot be distinguished at the recording speed shown. (After Siegler, 1981a.)

normally was hyperpolarized, casts some doubt upon the role of interneuron I in the central generation of the walking rhythm.

In any case, Pearson and Fourtner (1975) argue that interneuron I is not itself a pacemaker neuron, since prolonged depolarization of it did not induce rhythmic activity in a quiescent animal. Rather, the interneuron is probably part of a network of nonspiking interneurons that gives rise to the rhythmical activity.

The appealing simplicity of a rhythm-generating system with interneuron I as the key element has perhaps, and unfortunately, forestalled any further investigation of the other interneurons likely to be involved in the generation of the walking rhythm in cockroach. There are large numbers of nonspiking interneurons that control movements of the hind legs of locust, and presumably likewise a large population in cockroach. In locusts these interneurons interact with each other via inhibitory synapses, and as discussed by Burrows (1979b) these connections could give rise to rhythmical activity. It would seem that a more thorough investigation of the central basis of the generation of rhythmic leg movements and the influence of sensory feedback upon them is now warranted.

There is in addition considerable evidence that the nonspiking interneurons have other functions in the control of motor activity, namely in the maintenance of posture and in the modulation of proprioceptive reflexes.

4.3.2 *Maintenance of posture*

The capability of nonspiking neurons to sustain transmitter release suits them well for a role in the maintenance of posture. For example, some tonically releasing interneurons in the locust contribute to the sustained firing of motor neurons such as the slow extensor tibiae, and to the sustained inhibition of other phasically active motor neurons such as the fast flexor tibiae (Burrows and Siegler, 1978; Wilson and Phillips, 1982). Furthermore, nonspiking interneurons and motor neurons of the hind legs show sustained changes in membrane potential and in synaptic inputs when the angle of the femoral–tibial joint is altered, changes that are central concomitants of postural reflexes (Siegler, 1981a,b). The magnitude of these changes varies from interneuron to interneuron. Figure 13 shows one example, in which an interneuron was steadily depolarized by some 10 mV when the femoral–tibial angle of a hind leg was changed from 90 to 0°. Conversely, the interneuron was hyperpolarized by some 5 mV when the angle was changed from 90 to 140°. Upon a change to an angle of 90° the response showed hysteresis. Thus, when 90° was approached from 0°, the interneuron was some 15 mV more hyperpolarized than when 90° was approached from 140°. Motor neurons likewise showed hysteresis, and were inhibited or fired tonically when the angle of the joint was changed. These sustained changes of the membrane potential of interneurons and motor neurons and the accompanying hysteresis probably arise from the properties of sensory receptors that provide information about joint angle (Siegler, 1981b; Burrows and Field, 1982).

One consequence of an imposed change in posture was to alter the effects that nonspiking interneurons had upon motor neurons (Fig. 14). For example, for one interneuron, each of a series of depolarizing current steps

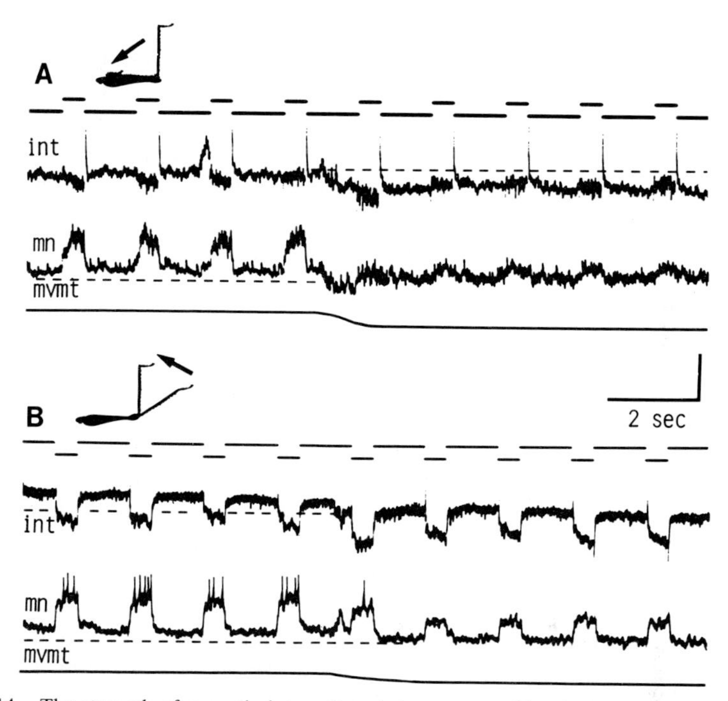

Fig. 14. The strength of synaptic interactions between nonspiking interneurons and motor neurons in the metathoracic ganglion of the locust are altered by imposed changes in the femoral–tibial angle of the ipsilateral hind leg. (A) A nonspiking interneuron (int) is repeatedly depolarized with current steps of 3 nA (first trace) to depolarize a flexor tibiae motor neuron (mn). The bridge circuit is underbalanced throughout. When the femoral–tibial angle is changed from 90 to 0° the interneuron and the motor neuron are hyperpolarized and the evoked depolarization of the motor neuron is considerably reduced. The imposed change in joint angle is indicated as a downward deflection on the fourth trace, which monitors joint movement (mvmt). The inset drawing of the hind leg shows the initial and the final position of the tibia. Hyperpolarization of the interneuron (not shown) had no effect upon the motor neuron, indicating that the membrane potential of the interneuron was below threshold for transmitter release. (B) Another nonspiking interneuron (int) is repeatedly hyperpolarized with current steps of −3 nA (first trace). A flexor tibiae motor neuron (mn) is depolarized during each current step, as it is released from tonic inhibition. Initially the femoral–tibial angle is 140°. The motor neuron spikes three to five times each time it depolarizes. When the femoral–tibial angle is changed to 90° (downward deflection on movement monitor, mvmt) the interneuron and the motor neuron are hyperpolarized and the depolarization of the motor neuron is considerably reduced. Vertical calibration: int, 8 mV, mn, in (A), 3 mV, in (B) 6 mV (After Siegler, 1981b.)

resulted in a 3-mV depolarization of a flexor motor neuron, when the angle of the femoral–tibial joint was 90° (Fig. 14A). When the angle was changed to 0°, the depolarizing response in the motor neuron was reduced to about 1 mV. For another interneuron, tonic hyperpolarizing effects were altered (Fig. 14B). When the femoral–tibial angle was 140°, each hyperpolarization of the interneuron released the motor neuron from inhibition, and allowed it to spike. When the angle was changed to 90°, the depolarization (or disinhibition) of the motor neuron was reduced, and spikes no longer occurred. These changes in postsynaptic effect probably result from complex interactions in the circuitry that governs the posture of the hind leg. Changes in the resting potential of the interneuron, for example, would only be one contributing factor, as discussed in detail elsewhere (Siegler, 1981b). Whatever the underlying mechanisms these findings emphasize that the effectiveness of central connections revealed by electrophysiological mapping may be governed by the behavioral context in which they are investigated, here for example, the posture of the animal.

4.3.3 *Modulation of reflexes*

The sustained changes evoked in nonspiking interneurons, such as those resulting from postural alterations, may also have a role in modulating reflexes of the hind legs (Siegler, 1981a,b). In one such reflex, rhythmical flexion and extension of the femoral–tibial joint of a hind leg evoke rhythmical movements at the tibial-tarsal joint. During flexion, the tarsus is levated, while during extension it is depressed (Burrows and Horridge, 1974). The effect is to keep the tarsus at a constant angle relative to the substrate. Sustained depolarization of one nonspiking interneuron enhanced the firing of the tarsal levator and the tarsal depressor motor neurons (Fig. 15A), so that during the imposed movement of the femoral–tibial joint, the movement of the tarsus was greater. Other interneurons were biased to excite one or the other of the motor neurons, so that levation or depression of the tarsus predominated (Siegler, 1981b).

Resistance reflexes at the femoral–tibial joint were also altered by the sustained depolarization of nonspiking interneurons (Fig. 15B). During a resistance reflex, motor neurons of a particular joint are activated so as to resist an imposed movement of that joint. Thus, in the example shown in Fig. 15B each extension of the tibia elicited spikes in flexor tibiae motor neurons. During each flexion of the tibia, however, the slow extensor tibiae motor neuron (recorded extracellularly) did not spike. When an interneuron was depolarized for several cycles of the imposed movement this response changed. The flexors spiked considerably less during each extension, and the slow extensor tibiae fired several times during flexion. In effect, during

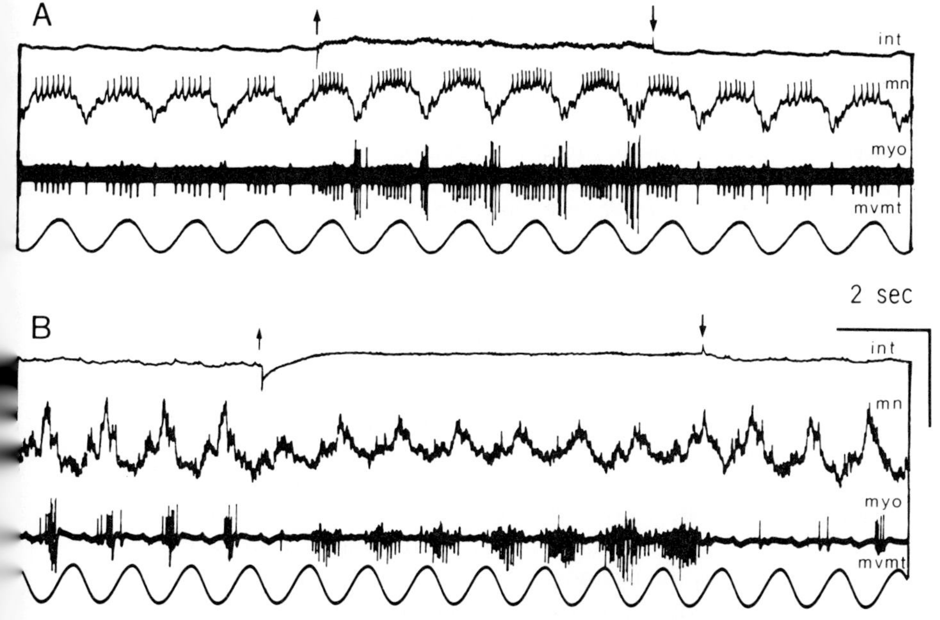

Fig. 15. Tonic depolarization of nonspiking interneurons alters compensatory and resistance reflexes of a hind leg in locust. The tibia is moved sinusoidally between angles of 30 and 60° about the femur. Flexion is downward on movement monitor (mvmt). (A) Response of a slow depressor tarsi motor neuron (mn) recorded intracellularly, and in myogram (myo). Spikes of slow depressor in myogram can be matched with those recorded intracellularly. The other, generally larger spikes in myogram are those of the slow levator tarsi, and these increase markedly in amplitude when the motor neuron fires rapidly. When the nonspiking interneuron (int) is depolarized for several seconds by 2.5 nA of current (between arrows), the frequency of spikes increases in the depressor and the levator motor neurons. (B) Depolarization of another nonspiking interneuron (int) by 1.5 nA of current (between arrows) changes synaptic input to a flexor tibiae motor neuron recorded intracellularly (mn) and alters resistance reflexes recorded in myogram (myo) from flexor and extensor tibiae motor neurons. Upward-going potentials are from flexor tibiae motor neurons; downward-going potentials are from the slow extensor tibiae motor neuron. Before depolarization, with each cycle of movement, flexor motor neuron (mn) is depolarized, and other flexor tibiae motor neurons, seen in myogram, spike in order to resist imposed extension. When the interneuron is depolarized, the maximal hyperpolarization of the intracellularly recorded flexor tibiae motor neuron is less, and other flexor motor neurons do not spike, but the slow extensor tibiae motor neuron spikes to resist imposed flexion. Vertical calibration. (A) int, 60 mV; mn, 9 mV. (B) int, 40 mV; mn, 8 mV. (After Siegler, 1981b.)

depolarization of the interneuron, the joint "attempted" to maintain a more extended angle. Other interneurons either had the reverse effect on the flexors and extensors, or intensified the spiking of both (Siegler, 1981b).

In arthropods, several types of stimuli alter reflex responsiveness. These include tactile stimuli at parts of the body remote from the moved joint, changes in angles of neighboring joints, or changes in ambient light (see

Siegler 1981b, for discussion). In principle, any source of continuous input to the motor neurons could bring this about. Nonetheless, the graded effects of nonspiking interneurons upon motor neurons and the sensitivity of these interneurons to small changes in membrane potential would seem to suit them particularly well to a role in modulating reflexes.

4.3.4 *A more general role in motor activity*

From the above discussion, it is seen that nonspiking interneurons have been implicated in the generation of rhythmic motor outputs, in the maintenance of posture, and in the modulation of postural reflexes. It seems unlikely, however, that different interneurons are specialized for these different motor activities. The same interneurons are probably involved in all of these; which facets of their activity are revealed will be limited by the ways in which the interneurons are investigated. These include the particular inputs and outputs that can be examined at any one time, and the "state" of the animal itself (e.g., the posture it assumes and its "excitability"). Furthermore, each nonspiking interneuron is only one part of the central neuronal circuitry that integrates sensory inputs and generates motor output, and the motor effects of the different interneurons are overlapping. Thus a direct and exclusive correlation between the activity of an interneuron and a particular piece of behavior seems unlikely, and is inconsistent with an integrative role.

Nonspiking interneurons were first described in the walking system of cockroaches, but subsequent work on the locust makes it clear that nonspiking interneurons have a much wider role to play in the patterning of motor activity. It is important to stress that all the evidence we have suggests that nonspiking interneurons would play a role in any behavior that involves a coordinated movement or a stereotyped posture of the legs. This, of course, includes much that an insect might do. A partial list would include walking, grooming, righting, jumping, kicking, swimming, flying, stridulation, defense, and mating. Just as the motor neurons that control muscles of the legs are involved in a myriad of behaviors, so too must be the nonspiking interneurons that control and recruit them.

5 Morphology of nonspiking interneurons

5.1 SHAPES OF INTERNEURONS

Several studies have described the morphology of nonspiking interneurons involved in the control of leg motor neurons within the thoracic ganglia of insects. Siegler and Burrows (1979) described some 16 different anatomical types of nonspiking interneuron within the metathoracic ganglion of the locust, comparing those that affected different sets of hind leg motor neurons.

Together with the anatomical descriptions of a nonspiking interneuron in the metathoracic ganglion of the cockroach (Pearson and Fourtner, 1975), and of five types of nonspiking interneuron in the mesothoracic ganglion of the locust (Wilson, 1981; Wilson and Phillips, 1982), this provides a good picture of the morphological features of the interneurons. In all of these studies, the interneurons were identified physiologically by intracellular recordings, then stained intracellularly by injecting cobalt ions via the microelectrode (Pitman *et al.*, 1972). The interneurons stained in the locust were also intensified with silver (Bacon and Altman, 1977).

All of the nonspiking interneurons are local interneurons, and all branch very extensively within one-half of a thoracic ganglion, ipsilateral to the motor neurons that they affect (Fig. 16). For the vast majority the cell body is also ipsilateral, though a few have the cell body and some neuropilar processes contralateral to their main region of branching (Fig. 17). Cell bodies are some 10–20 μm in diameter, relatively small when compared with those of the motor neurons, some 30–100 μm in diameter. From the ventrally or dorsally placed cell body, a thin process (1–3 μm diameter) travels inward to the ganglionic core, which contains neuropilar regions and tracts of neuronal processes. Within the core of the ganglion, the interneurons branch

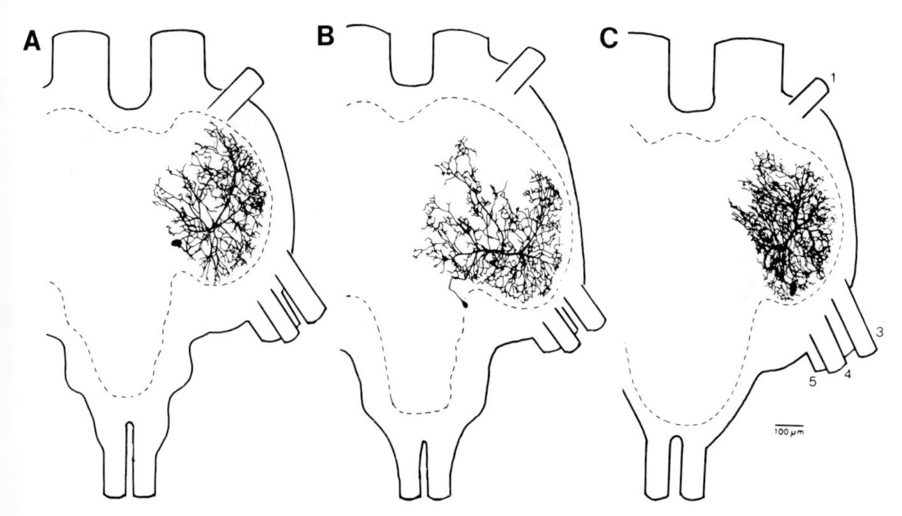

Fig. 16. Morphology of nonspiking local interneurons in the metathoracic ganglion of locust. All three when depolarized excite the slow extensor tibiae motor neuron within the ipsilateral half of the ganglion. The ganglia are viewed dorsally, and drawn in plan view from whole mounts. The dashed lines indicate the boundary of the neuropil. Interneurons in (A) and (C) have dorsal cell bodies; interneuron in (B) has a ventral cell body. Lateral nerves 1, 3, 4, and 5 are shown; others are not drawn. The interneurons were stained by the intracellular injection of cobalt ions, which were then precipitated as CoS. CoS was subsequently intensified with silver. (Adapted from Siegler and Burrows, 1979.)

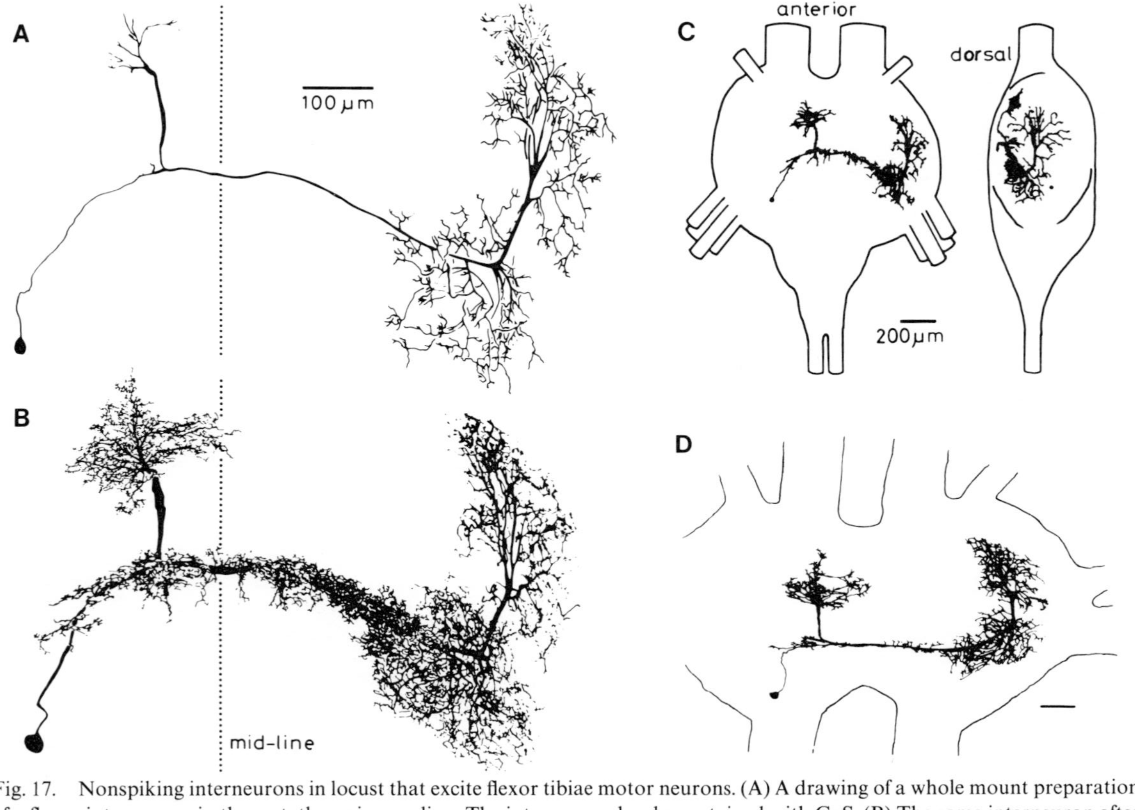

Fig. 17. Nonspiking interneurons in locust that excite flexor tibiae motor neurons. (A) A drawing of a whole mount preparation of a flexor interneuron in the metathoracic ganglion. The interneuron has been stained with CoS. (B) The same interneuron after silver intensification of the cobalt stain. The dashed line indicates the midline of the ganglion. (C) Lower magnification drawings to show the position of the interneuron in the ganglion as viewed dorsally and from the side. (D) Cobalt–silver stain of an apparently homologous nonspiking interneuron in the mesothoracic ganglion of locust. (A–C adapted from Siegler and Burrows, [illegible]

profusely. Typically, each interneuron has several main branches some 3–8 μm in diameter, and a few 100 μm in length. These give rise to a multitude of finer branches, which again branch repeatedly. The finest branches are often less than 1 μm in diameter, and only 2–15 μm in length. The main branches of a given interneuron typically differ from each other in length and diameter, and take distinctively different courses through the neuropil. Nonetheless, they generally have an equal density of fine branches, and equal frequency of secondary branching along their lengths. The density of fine branches may appear to vary from interneuron to interneuron, but this can depend upon the intensity of the staining (Siegler and Burrows, 1979). For example, in cobalt sulfide-stained interneurons with contralateral cell bodies, the process that crosses the ganglion may appear to be "axonlike," that is, lacking side branches (Fig. 17A). However, when the stain is intensified with silver, numerous fine branches become evident on this long process and elsewhere on the interneuron (Fig. 17B).

In the locust, one thoracic and three abdominal neuromeres comprise the metathoracic ganglion. Branches of the nonspiking interneurons are confined to the thoracic neuromere, and are concentrated in the region lateral to the major longitudinal tracts (Figs. 16 and 17). Typically, the branches are most profuse within the posterior two-thirds of the thoracic neuropil, and extend to the perimeter of the neuropil. Motor neurons of the hind legs have their greatest density of branches in the same region, so this is where synaptic contacts between interneurons and motor neurons are likely to occur.

This general picture of the morphology of nonspiking interneurons may be compared with that of other local interneurons within the thoracic ganglion. Two large and symmetrical groups of spiking local interneurons have been described recently in the metathoracic ganglion of the locust (Fig. 18A) (Burrows and Siegler, 1982, 1984; Siegler and Burrows, 1984). These spiking interneurons receive direct inputs from afferents, and also may have direct, but more limited motor effects than do the premotor nonspiking interneurons (Burrows and Siegler, 1982; Siegler and Burrows, 1983). The spiking local interneurons are morphologically similar to the premotor nonspiking interneurons in some respects: they branch profusely within the neuropil, and do so mainly or exclusively within one half of the ganglion; they have small cell bodies; and the range of branch lengths and diameters is comparable to that of the nonspiking local interneurons. All spiking local interneurons described so far, however, have a cell body that is contralateral to the main region of branching, and that is connected to it by a long, unbranched process which crosses the ganglion within a ventral commissure (Siegler and Burrows, 1984). Furthermore, in whole mounts of stained neurons it is obvious that the main branches are of two anatomical types. Those in the ventral region of the ganglion are thicker, and give rise to a density of very short and fine branches (Fig. 18A,i), whereas those more dorsal are thinner, varicosė, and sparsely branched (Fig. 18A,ii). The branches of two regions are connected by a single process about 100 μm in length, and of relatively large diameter (5 μm). Reconstructions of sectioned ganglia containing stained sensory afferents, stained spiking local interneurons, or stained motor neurons provide evidence that the ventral branches of the interneurons receive the sensory inputs, and the dorsal branches make the motor output connections (Siegler and Burrows, 1984). Regional anatomical and physiological specializations may also occur in the cricket prothoracic ganglion in some bilaterally branched interneurons that are involved in generating directional sensitivity to auditory inputs (Popov *et al.*, 1978; Wohlers and Huber, 1978).

In addition, a few neurons of the dorsal midline group are nonspiking local ones (Goodman *et al.*, 1980). These have small cell bodies and branches symmetrical about the midline (Fig. 18B,C). The function of these local interneurons is unknown, though in the same group other neurons with peripheral axons modulate neuromuscular transmission and muscle tension (Evans and O'Shea, 1977; Evans and Siegler, 1982).

5.2 CORRELATION OF ANATOMY WITH PHYSIOLOGY

Intracellularly stained nonspiking interneurons have been characterized physiologically mainly according to their excitatory effects upon a limited population of motor neurons (Siegler and Burrows, 1979). Clearly, this is insufficient to identify the interneurons as individuals, since several distinct anatomical types are found to excite the same motor neurons (Fig. 16). It would be convenient for further experiments, however, if each of these types could be distinguished by physiological means alone, as the different motor neurons now can be.

The nonspiking interneurons have been found to excite or inhibit different combinations of motor neurons (Burrows, 1980), and it might be possible, therefore, to characterize each one by tracing all of its outputs. It is technically feasible to record intracellularly from only a few motor neurons at a time, however, so this would be a laborious task. A more reasonable approach would be to try to distinguish interneurons with some motor effects in common according to how they respond to particular sensory stimuli. For example, imposed movements of joints will produce different patterns of synaptic activity in some interneurons (Burrows, 1980; Siegler, 1981b). A number of stimuli could be tested, and then an interneuron stained. The expectation is that gradually criteria would be developed, so that it would be possible to identify an interneuron without recourse to staining. Thus, several interneurons could be sampled in a single preparation, and particular ones

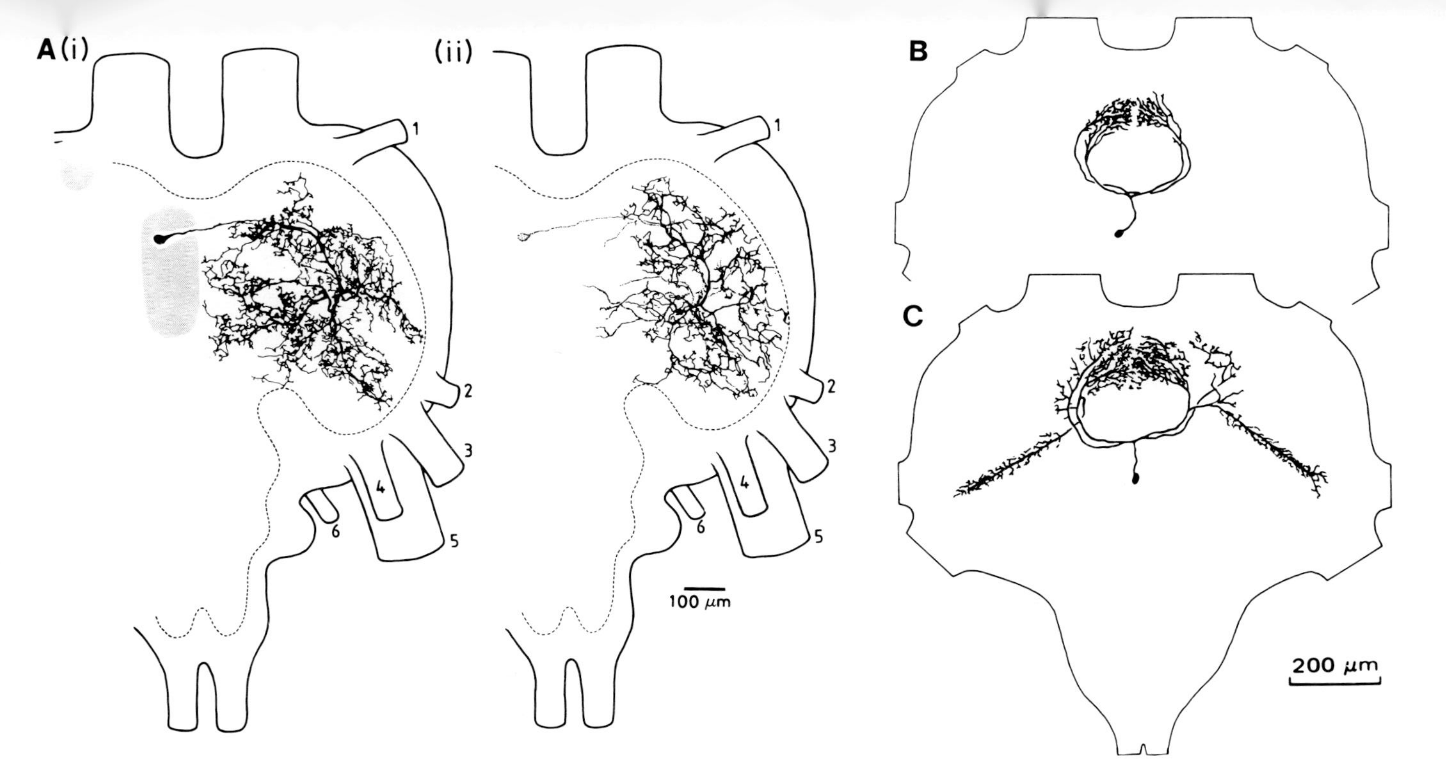

Fig. 18. Local interneurons in locust ganglia. (A) Local spiking interneuron in the metathoracic ganglion that is excited by touching hairs on the tibia of the hind leg. The interneuron is stained with CoS and silver. The dashed lines indicate the boundary of the neuropil. The ganglion is viewed dorsally. (i) Drawing of ventral region of ganglion to show the cell body, the primary neurite, and ventrally originating branches. The stippling indicates two regions in which cell bodies of spiking local interneurons have been found. (ii) Drawing of more dorsal region to show other branches of the interneuron, which arise from a single dorsal running process. The cell body and larger ventral branches are stippled. (B, C) Two nonspiking local interneurons of the dorsal unpaired median group of cell bodies, in the metathoracic ganglion. Neurons were stained by injecting Lucifer Yellow intracellularly. The function of the interneurons is unknown. (A adapted from Siegler and Burrows, 1983; B, C from Goodman *et al.*, 1980.)

selected for further physiological study. As is now possible with other identified interneurons, data from many experiments could be pooled.

5.3 NUMBERS OF NONSPIKING INTERNEURONS

One question raised in an early study was whether each anatomical type (e.g., the three different interneurons in Fig. 16) comprised a single neuron, or a few of the same shape (Siegler and Burrows, 1979). Studies of Wilson (1981) suggest that at least a few anatomical types comprise only one interneuron within each half of a thoracic ganglion (e.g., Fig. 17). All of these interneurons have contralateral cell bodies, and to confirm that the interneurons were unique to their type their primary neurites had to be traced across the ganglion for a considerable distance, in a fine tract. Unfortunately most other anatomical types of nonspiking interneuron do not appear to have shapes that are so favorable for this type of test. Nonetheless, as long as the differences in morphology reflect differences in physiological connections, it may not be too important whether there is one, or are a few, of a particular morphological type.

Another question concerns the total number of nonspiking local interneurons in a ganglion. Some 65–75% of the central neurons (i.e., those with central cell bodies) within ventral ganglia of insects and crustaceans are local, intraganglionic interneurons (Pearson, 1977; Burrows and Siegler, 1979; Reichert *et al.*, 1982). A rough idea of the absolute numbers of nonspiking neurons that we must therefore account for can be gained by considering the metathoracic ganglion of the locust. This bilaterally symmetrical ganglion contains the cell bodies of perhaps 3000 neurons (though an accurate count has yet to be made), and comprises one thoracic and three abdominal neuromeres. All nonspiking and spiking local interneurons thus far described originate within the thoracic neuromere, and branch wholly or largely within its neuropilar region (Burrows and Siegler, 1982, 1984; Siegler and Burrows, 1979, 1984). The thoracic neuromere would contain some 1800 neurons, assuming that 1200 of the estimated total in the ganglion are contributed by the three abdominal neuromeres (3 × 400 neurons, 400 being approximately the number counted for an unfused abdominal ganglion by Sbrenna, 1971). Of the 1800, about 1200, or 600 in each half, would be local interneurons, if the 65% figure holds. Interpreting our data in the most liberal way, we can account for just over one-third of these 600 pairs of local interneurons. There are at least 30 known types of nonspiking interneurons that affect mainly the motor neurons of the distal leg segments. (Sixteen types are illustrated by Siegler and Burrows, 1979, and an equal number found but not illustrated.) Physiological studies show that there are yet other nonspiking interneurons that synapse upon motor neurons of more proximal segments of a leg, or upon those of the flight muscles (Burrows and Siegler, 1978). If each of these two groups also comprises at least 30 nonspiking interneurons, then a rough guess would place 100 as a reasonable number for the nonspiking population. In addition, two groups of spiking local interneurons have been described anatomically that at most could account for 140 neurons a side, though other groups probably remain to be discovered (Siegler and Burrows, 1984). Some 360 of the 600 local interneurons within each half of the thoracic neuromere thus remain to be accounted for. These estimates force us to two conclusions about our experimental strategy. First, it is an unmanageable task to attempt to identify every local interneuron; rather, a more reasonable approach would be to find a few different types of nonspiking interneuron that we can identify from preparation to preparation, and take these as examples. Second, we need to know more about the interneurons that comprise the bulk of the central ganglionic neurons. Recent descriptions of tracts, commissures, and nerve roots within the ventral ganglia (Tyrer and Gregory, 1982) can provide us with the landmarks that we need to recognize groupings of cell bodies. Descriptions of individually stained neurons with reference to these landmarks (Siegler and Burrows, 1984) indicate that small neurons of a particular type (i.e., spiking local interneurons) have their cell bodies clustered together. Thus we may be able to generalize from individually stained neurons to those in surrounding regions with similar morphologies.

5.4 ULTRASTRUCTURE OF NONSPIKING INTERNEURONS

The ultrastructure of two identified nonspiking interneurons in the mesothoracic ganglion of the locust has been examined by Wilson and Phillips (1982). The interneurons were stained intracellularly with cobalt sulfide; then thick sections of the ganglia were intensified with silver, thin-sectioned, and examined in the EM. The primary and secondary branches of the interneurons were completely wrapped in glia, and devoid of synapses. Thus, input and output synapses were apparently confined to the smaller branches of the interneurons. The output synapses of the stained interneurons were of two types, both of which have been described previously in ultrastructural studies of cockroach thoracic ganglia (Wood *et al.*, 1977). These were bar-type presynaptic terminals with a single postsynaptic element and discrete-type dense projections, with two contiguous postsynaptic elements. Although Wilson and Phillips (1982) speculated that the two contiguous postsynaptic elements could be the processes of two different motor neurons, one of which receives inhibitory and the other excitatory inputs, there is no evidence for this idea. They also noted in passing that in these interneurons (which tonically release transmitter in a quiescent animal) the synaptic vesicles appeared to be "fuzzier" than those in similarly treated motor neurons and "nontonic" interneurons, and suggested, further, that this was related to the tonic release of transmitter.

Watson and Burrows (1982, 1983) stained identified motor neurons and interganglionic spiking interneurons by the intracellular injection of horseradish peroxidase (HRP). This is probably superior to cobalt marking, since cobalt–silver granules accumulate at synaptic densities and can obscure the synaptic structure (Phillips, 1980). By making serial EM reconstructions of selected regions of the neurons, Watson and Burrows showed that fine branches support complex synaptic arrangements. These include serial synapses, where within a few microns the stained neuron is postsynaptic to one unstained profile, and presynaptic to another; and reciprocal synapses, where input and output synapses with an unstained profile occur within a few microns of each other. Given these synaptic arrangements, it seems probable that the finer branches of the neurons can function in local circuits that would not necessarily involve more distant parts of the neuron.

An obvious next step would be to examine the nonspiking interneurons using HRP to see if they too form such serial and reciprocal synapses. Only small changes in presynaptic voltage are needed to release transmitter from these neurons, suggestive physiological evidence for such local interactions. It would also be interesting to pursue the question raised by Wilson and Phillips (1982) as to whether the synapses of different nonspiking interneurons differ from each other, and from those of motor neurons and spiking interneurons. It might be, for example, that the nonspiking interneurons that exert their physiological effects by the sustained and graded release of transmitter have synapses that are different in appearance from those of local spiking interneurons or motor neurons that exert their physiological effects by the spike-initiated release of transmitter.

6 Comparison with nonspiking interneurons of crustaceans

Nonspiking interneurons have been described in several crustacean nervous systems. They may influence the rhythmic beating of the gill bailers in lobsters and crabs (Mendelson, 1971; Simmers and Bush, 1980) and of the abdominal swimmerets in crayfish (Heitler and Pearson, 1980). In the crayfish terminal abdominal ganglion, they may also mediate directional sensitivity to water currents (Reichert *et al.*, 1983), and control the motor neurons of the uropods (Takahata *et al.*, 1981; Reichert *et al.*, 1982).

Of these nonspiking interneurons, those that affect uropod motor neurons are most comparable to the nonspiking interneurons in the thoracic ganglia of insects. They are described as anaxonic nonspiking interneurons by

Takahata *et al.* (1981), and would appear to belong to the same population of interneurons as those designated PM1-5 (unilateral nonspiking premotor interneurons) by Reichert *et al.* (1982). In common with the insect nonspiking interneurons, all have relatively small cell bodies, all branch diffusely in a fairly lateral region of the ganglion, and none have anatomically distinct regions that might be associated exclusively with inputs or outputs. They do not spike when penetrated, or in response to synaptic inputs or injected depolarizing current, or upon rebound from injected hyperpolarizing current. Furthermore, also like the insect interneurons, their membrane potentials are significantly lower than those of spiking neurons in the same ganglion. In addition, they excite or inhibit sets of motor neurons, sometimes having opposite effects upon antagonists. These effects (recorded extracellularly) can be graded according to the amount of current injected into the interneurons. The nature of synaptic transmission between the interneurons and the motor neurons is not known, however. By analogy with the locust, it might be expected that transmission would be chemical rather than electrical. Nonetheless, since electrical transmission is common in crayfish and other crustacean nervous systems (but not in those of insects), the nature of the transmission needs to be investigated directly. A final similarity between the insect and crayfish local interneurons is seen in the nature of their sensory inputs. In the crayfish, the premotor nonspiking interneurons were "not finely tuned to any (sensory) modality," and were "either excited or inhibited by most forms of sensory stimuli," whereas other, spiking local interneurons responded to distinct modalities and had small, clearly defined receptive fields (Reichert *et al.*, 1982). This parallels the different effects of sensory inputs on nonspiking and spiking local interneurons in the locust (Burrows and Siegler, 1982; see also Section 3).

The "premotor" nonspiking interneurons differ in several respects from the other type of local nonspiking interneuron described in the terminal abdominal ganglion of crayfish, the local, directionally selective (LDS) interneuron. The LDS interneurons are part of a small set of local interneurons (possibly only one pair per ganglion) in the circuitry for directional sensitivity to water currents (Reichert *et al.*, 1983). Each branches extensively within both halves of the ganglion, receiving excitatory inputs in one half from hairs on the tail fan, and making inhibitory output synapses with ascending interganglionic interneurons in the other. Their inhibitory postsynaptic effects are graded and chemically mediated, as are those of the nonspiking interneurons in insects. By contrast with these latter nonspiking interneurons, however, the LDS interneurons are not always passive in their responses to depolarizing inputs. Recordings from some interneurons show a depolarizing transient that resembles the graded and active responses recorded from other "nonspiking" or nonimpulsive crustacean neurons, such as the coxal stretch receptors of crabs and crayfish (Bush, 1981; Heitler, 1982), and the EX1 interneurons of the lobster stomatogastric ganglion (Graubard, 1978).

7 Conclusion

It is now well established that nonspiking interneurons are at the core of the premotor circuitry within the central nervous systems of insects. We know in considerable detail how their interconnections with motor neurons, and with each other, allow for the fine control of movements that underlie the complex behaviors of which insects are capable. Nonetheless, many questions remain about the form and function of the nonspiking interneurons. Some have been mentioned already; these and others are summarized here, with the intent of highlighting promising areas for future research. The research areas can be broadly divided into three categories: the synaptic physiology, the circuitry or connectivity, and the morphology of the nonspiking interneurons.

Concerning synaptic physiology, it is easy to formulate specific questions. Some may prove technically difficult to answer; many of the experiments depend upon being able to measure the presynaptic voltage, when an interneuron is depolarized or hyperpolarized by the injection of current via a microelectrode. For example, what is the gain at the synapses of nonspiking interneurons onto motor neurons or other nonspiking interneurons? Does this gain differ for different interneurons, for different regions of the same interneurons, for different postsynaptic motor neurons or interneurons, and for excitatory or inhibitory postsynaptic effects? How does the relationship between pre- and postsynaptic voltage compare with that at other synapses? Does it, for example, plot as a linear function on semilogarithmic axes, as it does at the squid giant synapse and the nonspiking crab stretch receptor? What is the threshold potential for release of transmitter? Is it the same for all of the nonspiking interneurons, or is there a difference between those interneurons that are releasing transmitter continuously, and those that are not? Are there any "historical" effects upon synaptic transmission? For example, does it show facilitation or antifacilitation when an interneuron is depolarized repeatedly? How sustained are the postsynaptic effects of an interneuron, and does this depend upon the presynaptic membrane potential? If there is a decrement with time, what is its origin? Decreased Ca^{2+} entry, decline of the presynaptic potential, depletion of transmitter, or loss of postsynaptic sensitivity are some possibilities. Finally, what transmitter(s) mediate the excitatory and inhibitory effects of nonspiking interneurons?

Much more can be learned, too, about the circuits in which nonspiking interneurons function. It is necessarily more difficult to formulate specific questions here, compared with the investigation of synaptic physiology; finding neurons that connect with the nonspiking interneurons is largely a matter of trial and error. Tracing the connectivity of neurons is, however, a process in which "positive feedback" can operate: the more that is known about the connections of a particular neuron, the easier it is to find and identify it, and the easier it is to learn still more about it. Thus the best approach will probably be to focus on a few nonspiking interneurons that can be readily identified as individuals by electrophysiological criteria, and search for the connections of these. The hope would be that they would prove to be representative. A major gap at present is our lack of knowledge of neurons presynaptic to the nonspiking interneurons, particularly those that could coordinate groups of them during complex locomotory activities. Spiking local interneurons are the most likely candidates for presynaptic neurons that we know of so far, but other types of interneurons, particularly intersegmental ones, should be investigated further.

Questions related to the morphology of nonspiking interneurons range from those of ultrastructure to those related to the intraganglionic shapes of the interneurons. Much can still be learned at an ultrastructural level, especially by examining HRP-labeled interneurons. For example, what is the distribution and relative disposition of synapses on a nonspiking interneuron? Do inputs and outputs occur primarily on different branches, as they may in spiking local interneurons, or are inputs and outputs intermingled on the same branches? Do the synaptic structures of nonspiking interneurons differ from those of spiking interneurons, sensory neurons, or motor neurons? Some specialization might be expected, associated with an interneuron's ability to sustain the graded release of transmitter. Can techniques for the differential staining of pre- and postsynaptic neurons, developed for other invertebrate preparations, be modified to examine nonspiking interneurons in insects? This could provide evidence for connections between interneurons and motor neurons, presumed from physiological evidence to interact monosynaptically, and give specific information about the way that synapses from a single presynaptic neuron are distributed on a postsynaptic target. At the level of whole neuron morphology, the geometry of a neuron will be crucial in determining how well graded potentials spread within it. Will different parts of an interneuron prove to be electrotonically remote from, or close to each other? Models based on accurate measurement of the length and caliber of the many branches of an interneuron can provide tentative answers to these questions. Ultimately, of course, the models must also take into account electrophysiological parameters which are yet unknown, such as membrane time constant and resistivity.

Nonspiking interneurons are one of two physiological types of local interneuron described so far in insect ganglia. The continued study of spiking local interneurons will undoubtedly give us additional insights into the function of the nonspiking ones. By comparing neurons of the two types, we can see what properties, anatomical and physiological, are characteristic of nonspiking local interneurons, of spiking local interneurons, and of local interneurons in general. We already know there are notable differences in the neurons of the two types, apart from the presence or absence of spikes. Anatomically, the spiking local interneurons are divided into two main regions of branching, which are probably mainly regions of input or of output. By contrast, the nonspiking interneurons cannot be readily divided into presumed input or output branches, at least based upon their appearance at the light microscope level; the prediction is that input and output synapses will prove to be intermingled on the fine branches of an interneuron, but this awaits HRP studies at the EM level. Physiologically, the spiking local interneurons are postsynaptic to some primary afferents from a leg, whereas the same afferents have no inputs to the nonspiking interneurons. However, both spiking and nonspiking local interneurons synapse upon motor neurons, though the spiking local interneurons appear to have much more restricted outputs than do the nonspiking ones.

On a more general note, the finding that there are many spiking, as well as nonspiking, local interneurons forces us to discard one rationale for the nonspiking state of the local interneurons in insect ganglia: it is no longer plausible to say that the interneurons lack spikes because of the relatively short intraganglionic distances over which intracellular potentials must spread. Spiking local interneurons have roughly the same morphological extent, so the explanation must lie elsewhere. Several suggestions have been made, including increased speed of transmission over short distances, temporal accuracy, reduced ambiguity from "noise" (Shaw, 1981), isolation of function within different parts of a neuron, sensitivity to very small signals, and the ability of single presynaptic neurons to exert finer control over postsynaptic neurons with graded rather than discrete PSPs (Pearson, 1976). Of these, the last seems the most attractive rationale, because of the behavioral requirements that postural changes and locomotion be finely graded and controlled. It is important to remember though that each motor neuron is, in fact, driven by a large number of interneurons, both nonspiking and spiking. In addition, there is an apparent paradox: discrete EPSPs in nonspiking interneurons can evoke discrete IPSPs (or, presumably, EPSPs) in motor neurons. This implies that the effects of a nonspiking interneuron can only be as finely graded as are the inputs it receives.

Neurons without spikes are by no means unique to insects. Indeed, the discovery of graded synaptic interactions in vertebrate and invertebrate visual systems long predates their discovery in insect central nervous systems (see also, Fain, 1981; Shaw, 1981; Siegler, 1984). In addition, what we know of transmission at chemical synapses indicates that whether spikes or graded potentials are the means for intracellular spread of signals, the events underlying the initiation of transmitter release are similar.

In studies of nonspiking interneurons in insects, it has been possible to span many levels: the cellular properties of individual interneurons, their synaptic transmission to other neurons, the properties of the networks in which they function, and finally, the way these networks produce and coordinate the many patterned movements that underlie the behavior of an animal. In insects our experimental enquiries can move from single neurons, to circuits, to behavior and back again, and so illuminate the many basic functions of a nervous system.

Acknowledgments

I would like to thank Drs. Malcolm Burrows, Peter Evans, Simon Maddrell, and Colin Taylor for their many useful comments on earlier versions of the manuscript, and Drs. Brian Mulloney and Kathy Radke for their hospitality during part of its writing.

References

Altman, J. S, Shaw, M. K., and Tyrer, N. M. (1980). Input synapses onto a sensory neurone revealed by cobalt-electron microscopy. *Brain Res.* **189**, 245–250.

Bacon, J. P., and Altman, J. S. (1977). A silver intensification method for cobalt filled neurons in whole mount preparations. *Brain Res.* **138**, 359–363.

Berry, M. S., and Pentreath, V. W. (1976). Criteria for distinguishing between monosynaptic and polysynaptic transmission. *Brain Res.* **105**, 1–20.

Blight, A. R., and Llinás, R. (1980). The non-impulsive stretch-receptor complex of a crab: A study of depolarization-release coupling at a tonic sensorimotor synapse. *Philos. Trans. R. Soc. London Ser. B* **290**, 219–276.

Bräunig, P., and Hustert, R. (1980). Proprioceptors with central cell bodies in insects. *Nature (London)* **283**, 768–770.

Burrows, M. (1975). Monosynaptic connexions between wing stretch receptors and flight motoneurones of the locust. *J. Exp. Biol.* **62**, 189–219.

Burrows, M. (1978). Local interneurones and integration in locust ganglia. *Verh. Dtsch. Zool. Ges.* pp. 68–79.

Burrows, M. (1979a). Synaptic potentials effect the release of transmitter from locust nonspiking interneurons. *Science* **204**, 81–83.

Burrows, M. (1979b). Graded synaptic transmission between local pre-motor interneurons of the locust. *J. Neurophysiol.* **42**, 1108–1123.

Burrows, M. (1980). The control of sets of motoneurones by local interneurones in the locust. *J. Physiol. (London)* **298**, 213–233.

Burrows, M. (1981). Local interneurones in insects. *In* "Neurones without Impulses" (A. Roberts and B. M. H. Bush, eds.) pp. 199–221. Cambridge Univ. Press, London and New York.

Burrows, M., and Field, L. H. (1982). Reflex effects of the femoral chordotonal organ upon leg motor neurones of the locust. *J. Exp. Biol.* **101**, 265–285.

Burrows, M., and Horridge, G. A. (1974). The organization of inputs to motoneurones of the locust metathoracic leg. *Philos. Trans. R. Soc. London Ser. B* **269**, 49–94.

Burrows, M., and Siegler, M. V. S. (1976). Transmission without spikes between locust interneurones and motoneurones. *Nature (London)* **262**, 222–224.

Burrows, M., and Siegler, M. V. S. (1978). Graded synaptic transmission between local interneurones and motoneurones in the metathoracic ganglion of the locust. *J. Physiol. (London)* **285**, 231–255.

Burrows, M., and Siegler, M. V. S. (1982). Spiking local interneurons mediate local reflexes. *Science* **217**, 650–652.

Burrows, M., and Siegler, M. V. S. (1983). Networks of local interneurons in an insect. *Symp. Soc. Exp. Biol.* **37**, 29–53.

Burrows, M., and Siegler, M. V. S. (1984). The morphological diversity and receptive fields of spiking local interneurons in the metathoracic ganglion of the locust. *J. Comp. Neurol.* **224**, 438–508.

Bush, B. M. H. (1981). Non-impulsive stretch receptors in crustaceans. *In* "Neurones without Impulses" (A. Roberts and B. M. H. Bush, eds.), pp. 147–176. Cambridge Univ. Press, London and New York.

Evans, P. D. (1980). Biogenic amines in the insect nervous system. *Adv. Insect Physiol.* **15**, 317–473.

Evans, P. D., and O'Shea, M. (1977). An octopaminergic neurone modulates neuromuscular transmission in the locust. *Nature* (*London*) **270**, 257–259.

Evans, P. D., and Siegler, M. V. S. (1982). Octopamine mediated relaxation of maintained and catch tension in locust skeletal muscle. *J. Physiol.* (*London*) **324**, 93–112.

Fain, G. L. (1981). Integration by spikeless neurones in the retina. *In* "Neurones without Impulses" (A. Roberts and B. M. H. Bush, eds), pp. 29–59. Cambridge Univ. Press, London and New York.

Goodman, C. S., Pearson, K. G., and Spitzer, N. C. (1980). Electrical excitability: A spectrum of properties in the progeny of a single embryonic neuroblast. *Proc. Natl. Acad. Sci. U.S.A.* **77**, 1676–1680.

Graubard, K. (1978). Synaptic transmission without action potentials: Input-output properties of a non-spiking presynaptic neuron. *J. Neurophysiol.* **41**, 1014-1025.

Graubard, K., and Calvin, W. H. (1979). Presynaptic dendrites: Implications of spikeless synaptic transmission and dendritic geometry. *In* "The Neurosciences Fourth Study Program" (F. O. Schmitt and F. G. Worden, eds.), pp. 317–331. MIT Press, Cambridge, Mass.

Hausen, K. (1981). Monocular and binocular computation of motion in the lobula plate of the fly. *Verh. Dtsch. Zool. Ges.* 49–70.

Heitler, W. J. (1982). Non-spiking stretch receptors in the crayfish swimmeret system. *J. Exp. Biol.* **96**, 355–366.

Heitler, W. J., and Burrows, M (1977). The locust jump. I. The motor programme. *J. Exp. Biol.* **66**, 203–219.

Heitler, W. J., and Pearson, K. G. (1980). Non-spiking interactions and local interneurons in the central pattern generator of the crayfish swimmeret system. *Brain Res.* **187**, 206–211.

Hengstenberg, R. (1977). Spike responses of "non-spiking" visual interneurone. *Nature* (*London*) **212**, 1242–1245.

Hoyle, G., and Burrows, M. (1973). Neural mechanisms underlying behavior in the locust *Schistocerca gregaria*. I. Physiology of identified motorneurons in the metathoracic ganglion. *J. Neurobiol.* **4**, 3–41.

Kirshfeld, K. (1979). The visual system of the fly: Physiological optics and functional anatomy as related to behavior. *In* "The Neurosciences Fourth Study Program" (F. O. Schmitt and F. G. Worden, eds.), pp. 297–310. MIT Press, Cambridge, Mass.

Llinás, R., Steinberg, I. Z., and Walton, K. (1976). Presynaptic calcium currents and their relation to synaptic transmission: Voltage clamp study in the squid giant synapse and theoretical model for the calcium gate. *Proc. Natl. Acad. Sci. U.S.A.* **73**, 2918–2922.

Mendelson, M. (1971). Oscillator neurons in crustacean ganglia. *Science* **171**, 1170–1173.

Meyer, D. J., and Walcott, B. (1979). Differences in the responsiveness of identified motoneurons in the cockroach: Role in the motor program for stepping. *Brain Res.* **178**, 600–605.

Miller, P. L. (1965). The central nervous control of respiratory movements. *In* "The Physiology of the Insect Central Nervous System" (J. E. Treherne and J. W. L. Beament, eds.), pp. 141–155. Academic Press, London and New York

Oertel, D., and Stuart, A. E. (1981). Transformation of signals by interneurones in the barnacle's visual pathway. *J. Physiol. (London)* **281**, 311, 127–146.

Pearson, K. G. (1972). Central programing and reflex control of walking in the cockroach. *J. Exp. Biol.* **56**, 173–193.

Pearson, K. G. (1976). Nerve cells without action potentials. *In* "Simpler Networks and Behavior" (J. C. Fentress, ed.), pp. 99–110. Sinauer, Sunderland, Mass.

Pearson, K. G. (1977). Interneurons in the ventral nerve cord of insects. *In* "Identified Neurons and Behavior of Arthropods" (G. Hoyle, ed.), pp. 329–337. Plenum, New York.

Pearson, K. G. (1979). Local neurons and local interactions in the nervous systems of invertebrates. *In* "The Neurosciences Fourth Study Program" (F. O. Schmitt and F. G. Worden, eds.), pp. 145–157. MIT Press, Cambridge, Mass.

Pearson, K. G., and Bergman, S. J. (1969). Common inhibitory motoneurones in insects. *J. Exp. Biol.* **50**, 445–471.

Pearson, K. G., and Fourtner, C. R. (1975). Nonspiking interneurons in walking system of the cockroach. *J. Neurophysiol.* **38**, 33–52.

Pearson, K. G., Wong, R. K. S., and Fourtner, C. R. (1976). Connexions between hair-plate afferents and motoneurones in the cockroach leg. *J. Exp. Biol.* **64**, 251–266.

Phillips, C. E. (1980). Intracellularly injected cobaltous ions accumulate at synaptic densities. *Science* **207**, 1177–1179.

Pitman, R. M., Tweedle, C. D., and Cohen, M. J. (1972). Branching of central neurons: Intracellular cobalt injections for light and electron microscopy. *Science* **176**, 412–414.

Popov, A. V., Markovitch, A. M., and Andjan, A. S. (1978). Auditory interneurons in the prothoracic ganglion of the cricket, *Gryllus bimaculatus* deGeer. I. The large segmental auditory neuron (LSAN). *J. Comp. Physiol.* **126**, 183–192.

Rakić, P. (1975). Local circuit neurons. *Neurosci. Res. Prog. Bull.* **13**, 291–446.

Rall, W. (1981). Functional aspects of neuronal geometry. *In* "Neurones without Impulses". (A. Roberts and B. M. H. Bush, eds.), pp. 223–254. Cambridge Univ. Press, London and New York.

Reichert, H., Plummer, M. R., Hagiwara, G., Roth, R. L., and Wine, J. J. (1982). Local interneurons in the terminal abdominal ganglion of the crayfish. *J. Comp. Physiol.* **149**, 145–162.

Reichert, H., Plummer, M. R., and Wine, J. J. (1983). Identified nonspiking local interneurons mediate nonrecurrent, lateral inhibition of crayfish mechanosensory interneurons. *J. Comp. Physiol.* **151**, 261–276.

Roberts, A., and Bush, B. M. H., eds. (1981). "Neurones without Impulses." Cambridge Univ. Press, London and New York.

Roberts, A., and Roberts, B. L., eds. (1983). Neural origin of rhythmic movements. *Symp. Soc. Exp. Biol.* **37**.

Sbrenna, G. (1971). Postembryonic growth of the ventral nerve cord in *Schistocerca gregaria* Forsk. (Orthoptera: Acrididae). *Boll. Zool.* **38**, 49–74.

Schmitt, F. O., and Worden, F. G., eds. (1979). "The Neurosciences Fourth Study Program." MIT Press, Cambridge, Mass.

Shaw, S. R., (1981). Anatomy and physiology of identified non-spiking cells in the photoreceptor-lamina complex of the compound eye of insects, especially Diptera. *In* "Neurones without Impulses" (A. Roberts and B. M. H. Bush, eds.), pp. 61–116. Cambridge Univ. Press, London and New York.

Siegler, M. V. S. (1981a). Posture and history of movement determine membrane potential and synaptic events in nonspiking interneurons and motor neurons of the locust. *J. Neurophysiol.* **46**, 296–309.

Siegler, M. V. S. (1981b). Postural changes alter synaptic interactions between nonspiking interneurons and motor neurons of the locust. *J. Neurophysiol.* **46**, 310–323.

Siegler, M. V. S. (1982). Electrical coupling between supernumerary motor neurons in the locust. *J. Exp. Biol.* **101**, 105–119.

Siegler, M. V. S. (1984). Local interneurones and local interactions in anthropods. *J. Exp. Biol.* **112**, 253–281.

Siegler, M. V. S., and Burrows, M. (1979). The morphology of local non-spiking interneurones in the metathoracic ganglion of the locust. *J. Comp. Neurol.* **183**, 121–148.

Siegler, M. V. S., and Burrows, M. (1980). Non-spiking interneurones and local circuits. *Trends Neurosci.* **March**, 73–77.

Siegler, M. V. S., and Burrows, M. (1983). Local spiking interneurons as primary integrators of mechanosensory information in the locust. *J. Neurophysiol.* **50**, 1281–1295.

Siegler, M. V. S., and Burrows, M. (1984). The morphology of local spiking interneurons in the metathoracic ganglion of the locust. *J. Comp. Neurol.* **224**, 463–482.

Simmers, A. J., and Bush, B. M. H. (1980). Non-spiking neurones controlling ventilation in crabs. *Brain Res.* **197**, 247–252.

Simmons, P. J. (1982). Transmission mediated with and without spikes at connections between large second-order neurones of locust ocelli. *J. Comp. Physiol.* **147**, 401–414.

Stuart, A. E., and Oertel, D. (1978). Neuronal properties underlying processing of visual information in the barnacle. *Nature* (*London*) **275**, 287–290.

Takahata, M., Nagayama, T., and Hisada, M. (1981). Physiological and morphological characterization of anaxonic non-spiking interneurons in the crayfish motor control system. *Brain Res.* **226**, 309–314.

Tyrer, N. M., and Gregory, G. E. (1982). A guide to the neuroanatomy of locust suboesophageal and thoracic ganglia. *Philos. Trans. R. Soc. London Ser. B* **297**, 91–123.

Watson, A. H. D., and Burrows, M. (1981). Input and output synapses on identified motor neurones of a locust revealed by the intracellular injection of horseradish peroxidase. *Cell Tissue Res.* **215**, 325–332.

Watson, A. H. D., and Burrows, M. (1982). The ultrastructure of identified locust motor neurones and their synaptic relationships. *J. Comp. Neurol.* **205**, 383–397.

Watson, A. H. D., and Burrows, M. (1983). The morphology, ultrastructure, and distribution of synapses on an intersegmental interneurone of the locust. *J. Comp. Neurol.* **214**, 154–169.

Watson, A. H. D., and Pflüger, H. J. (1984). The ultrastructure of prosternal sensory hair afferents within the locust central nervous system. *Neuroscience* **11**, 269–279.

Wilson, D. M. (1961). The central nervous control of flight in a locust. *J. Exp. Biol.* **38**, 471–490.

Wilson, D. M. (1964). Relative refractoriness and patterned discharge of locust flight motor neurons. *J. Exp. Biol.* **41**, 191–205.

Wilson, J. A. (1979). The structure and function of serially homologous leg motor neurons in the locust. II. Physiology. *J. Neurobiol.* **10**, 153–167.

Wilson, J. A. (1981). Unique, identifiable local nonspiking interneurons in the locust mesothoracic ganglion. *J. Neurobiol.* **12**, 353–366.

Wilson, J. A., and Phillips, C. E. (1982). Locust local nonspiking interneurons which tonically drive antagonistic motor neurons: Physiology, morphology and ultrastructure. *J. Comp. Neurol.* **204**, 21–31.

Wilson, J. A., and Phillips, C. E. (1983). Pre-motor non-spiking interneurons. *Prog. Neurobiol.* **20**, 89–107.

Wohlers, D. W., and Huber, F. (1978). Intracellular recording and staining of cricket auditory interneurons (*Gryllus campestris* L., *Gryllus bimaculatus* DeGeer). *J. Comp. Physiol.* **127**, 11–28.

Wood, M. R., Pfenninger, K. H., and Cohen, M. J. (1977). Two types of presynaptic configurations in insect central synapses: An ultrastructural analysis. *Brain Res.* **130**, 25–45.

NOTE ADDED IN PROOF: See p. 433.

Structure and Regulation of the Corpus Allatum

Stephen S. Tobe

Department of Zoology, University of Toronto, Toronto, Ontario, Canada

Barbara Stay

Department of Zoology, University of Iowa, Iowa City, Iowa

ADVANCES IN INSECT PHYSIOLOGY, VOL. 18

1 Introduction to the corpora allata

The corpora allata (CA) are endocrine glands in the posterior regions of the head, or in rare instances in the thorax, which are closely associated with the stomatogastric nervous system (Cazal, 1948). They were mistaken for part of the nervous system when first recognized in 1861 and were not shown experimentally to be endocrine organs until the 1930s (see review by Cassier, 1979). Since then much has been learned about the juvenile hormones (JH) which they secrete, the functions of the hormones, and the regulation of hormone biosynthesis. This review is concerned primarily with the regulation of the CA but it also includes a brief consideration of the embryology, innervation, and the relationships of the structure, particularly the ultrastructure, to its synthetic activity.

1.1 EMBRYOLOGY

1.1.1 *General morphology*

A critical summary of the literature on the embryology of the CA was presented by Haget (1977). The CA are of ectodermal origin, arising early in embryonic development as paired lateral groups of cells which form buds or infoldings from the intersegmental region between the maxillary and mandibular segments or thereabouts, and are evident about the time that the gnathal segments are compressed (see Haget, 1977). Some of the ectodermal buds soon develop a lumen, as in *Carausius morosus* (Haget, 1977). On the other hand, an original central vesicle resulting from an invagination may disappear in the course of development, as for example in *Oncopeltus fasciatus* (Dorn, 1975). After detaching from the epithelium, the CA migrate dorsally, sometimes in association with the tentorial arms, and become associated with the coelomic sacs of the antennae. They eventually come to rest on the stomodeum, anterior and dorsal to their site of origin. It is in this position, after dorsal closure of the embryo, that they become associated with some of the elements of the stomatogastral nervous system, namely the corpora cardiaca (CC) and the hypocerebral ganglion; this was confirmed in embryos of *Locusta migratoria* (Maltete, 1962, cited in Haget, 1977). Variation in time of detachment of the CA buds likely contributes to variation in path of dorsal migration and final destination of the glands (Haget, 1977).

1.1.2 *Evidence for synthetic activity in embryonic glands*

The possibility that embryonic CA become differentiated and synthesize JH has been proposed from histological and fine structural analyses of the CA and from titers of JH in the developing embryo.

In *Eurygaster integriceps* the CA appear to be active in the last third of embryogenesis in that they have ring-shaped nucleoli (Polivanova and Bocharova, 1974, 1982) which are found in the glands of mature females, but not in hibernating individuals (Panov and Bussurmanova, 1970). However, the accuracy of such a morphological criterion for the estimation of CA activity remains unknown (see Section 1.4).

It was recognized early in studies on titers of JH that eggs contained substances with JH bioactivity and that these were in part the product of the developing embryos. Eggs produced by allatectomized females of *Hyalophora cecropia* lacked JH in early development but showed normal amounts of hormone later, presumably the product of developed CA (Gilbert and Schneiderman, 1961). Recent chemical identification of JHs in embryos of *Manduca sexta* have demonstrated the presence of JH I, II, and the previously unknown JH 0 (see Fig. 6); JH I, the most prominent, increased in titer before the embryonic molt (day 5.5) and subsequently decreased (Bergot *et al.*, 1980, 1981b).

Changing JH titers, which suggest activity of embryonic glands, have also been observed by bioassay in other species. In *Nauphoeta cinerea*, two peaks of JH, determined by *Galleria* bioassay of hemolymph, were found between dorsal closure and hatching (Imboden *et al.*, 1978); and in *O. fasciatus*, a large increase in JH titer was shown by *Tenebrio* bioassay to occur after the CA appeared differentiated at the end of blastokinesis (Dorn, 1975). However, the extent of the titer increase is uncertain because, although Bergot *et al.* (1981b) verified the presence of JH III in *O. fasciatus* embryos, no more than 0.09 ng/g was found by physicochemical methods, whereas the bioassay suggested about 1 μg JH I equivalents/g at this stage of development. Nevertheless, Dorn presents other evidence of CA activity. The fine structure is consistent with that of active glands (Dorn, 1975) and chemical destruction or inhibition of the CA with precocene resulted in an embryonic defect (failure of dorsal closure) that was prevented in 30% of the cases by simultaneous treatment with JH (Dorn, 1982).

Ultrastructural evidence alone suggests two periods of endocrine activity by embryonic CA in *C. morosus* (Haget *et al.*, 1981). It is compelling evidence because cellular organelles were analyzed in detail at clearly staged, chronologically close intervals. The active periods coincided with dorsal closure in the prothorax and darkening of the mandibles; they were characterized in part by decreased density of mitochondrial matrix, increased extracellular space, and a sequential change in the endoplasmic reticulum (Haget *et al.*, 1981). Such morphological changes with respect to activity cycles will be discussed in greater detail later.

In the future it may be possible to assay embryonic glands by sensitive direct radiochemical assay (RCA) or radioimmunoassay (RIA), which will provide a better evaluation of the functional capacity of embryonic CA and their regulation.

The first chemical identification of JH in insect embryos was performed in *M. sexta*; in addition to JH I, two JHs were found: JH O (Bergot *et al.*, 1980) and 4-methyl JH I (Bergot *et al.*, 1981b). The titers of JH O and 4-methyl JH I I were higher that those of JH I and II; all the JHs were undetectable on day O, appeared during embryogenesis in changing titers, and, with the exception of JH I, disappeared by hatching. To demonstrate that JH biosynthesis was occurring in these embryos, these authors inhibited JH biosynthesis with fluoromevalonate (Quistad *et al.*, 1981). Not only did such treatment produce larvae with symptoms of JH deficiency but it also resulted in reduced titers of JH O, JH I, and 4-methyl JH I in the older embryos (Bergot *et al.*, 1981b). Thus it seems that not only do CA of embryos produce JH but they synthesize a form specific to embryos.

1.2 INNERVATION

1.2.1 *General morphology*

The nervous system, especially the brain, unquestionably figures largely in the regulation of the CA (see Section 4). It will clearly be necessary to identify the cells which are the direct regulators of the CA and to trace the integrative pathways through which these cells are in turn regulated. It will be some time before such quests will be accomplished unequivocally. However, simply knowing the origin, content, and termination sites of nerves which arrive at the CA is insufficient to demonstrate an interaction with the CA cells. It is possible that the CA are regulated by neurosecretions, released at a more distant site, e.g., the CC, and that the neurosecretory release sites, "synaptoids" (Scharrer, 1968), seen in CA of many insects (see Cassier, 1979), regulate functions other than those of the CA. For example, current evidence points to the release of prothoracicotropic hormone (PTTH) in neurons ramifying in the CA of larval lepidoptera (Gibbs and Riddiford, 1977; Agui *et al.*, 1980; Carrow *et al.*, 1981, 1984). Nevertheless, details of the morphology and topography of the nerve cells and neurosecretory cells (NSC) which project to the retrocerebral complex (CC and CA) are essential in designing and interpreting experiments relevant to understanding the interactions between the nervous system and the CA.

In most insects, the CA receive nerves from both the brain and the subesophageal ganglion (Cazal, 1948; Juberthie and Cassagnau, 1971). However in several primitive orders, nerves come only or at least most conspicuously from the subesophageal ganglion, e.g., in Collembola (Cassagneau and Juberthie, 1967), and Ephemeroptera (Arvy and Gabe, 1953). In *Thermobia domestica*, a fine nerve from the CC is also found (Bitsch and Lapalisse, 1984). At the other extreme are insects in which the innervation arises exclusively from the brain, e.g., in the Hemiptera, in *Leptinotarsa decemlineata* (Schooneveld, 1970) and *Hydrous piceus* (de Lerma, 1956), as well as in some Diptera (Juberthie and Cassagnau, 1971).

1.2.2 *Cells contributing nerves to the CC and CA*

The cells contributing to the nerves entering the CC and/or the CA have been established by several complementary methods of visualization including silver stains, stains for neurosecretory material, and more recently the filling of cells through cut nerves or by injection of individual cell bodies (see reviews by Raabe, 1982; Pipa, 1983). The principal nerves and their cells of origin are shown diagrammatically in Fig. 1. There are many variations on this plan; e.g., tracts of *nervi corporis cardiaci* (NCC) I and II join within the brain of *M. sexta* (Nijhout, 1975b) and *L. decemlineata* (Khan *et al.*, 1984), and the NCA I are distinct from the CC in Orthoptera (Mason, 1973) and contiguous with the CC in Dictyoptera (Fraser and Pipa, 1977). Determining with certainty the location of cells contributing to these tracts is difficult (see Pipa, 1983, for further discussion of methodological limitations), but from the several studies that have attempted to do so, it is clear that the location, number, and size vary from one group of insects to another. Table 1 shows the location of cells potentially projecting axons to the CA, as demonstrated by three different filling methods. Retrograde filling of a transected CA shows all cells projecting to that cut, and consequently does not distinguish between axons which simply traverse the CA and those which may terminate there. Also, it is not possible to distinguish between axons which terminate on the surface and those which ramify within the CA. Combined retrograde and orthograde filling and sectioning at different locations have solved some of these problems. For instance Pipa's (1978) study in *Periplaneta americana* showed that the medial and lateral cells of the protocerebrum were filled from the NCA I. However, orthograde filling of the NCC I, derived from contralateral medial cells, showed fibers traversing the CC and ramifying in the CA, whereas orthograde filling of the NCC II (derived from ipsilateral lateral cells) showed fibers only in the CC (Gundel and Penzlin, 1980). Thus, in *P. americana*, medial cells are likely those "innervating" the CA, whereas

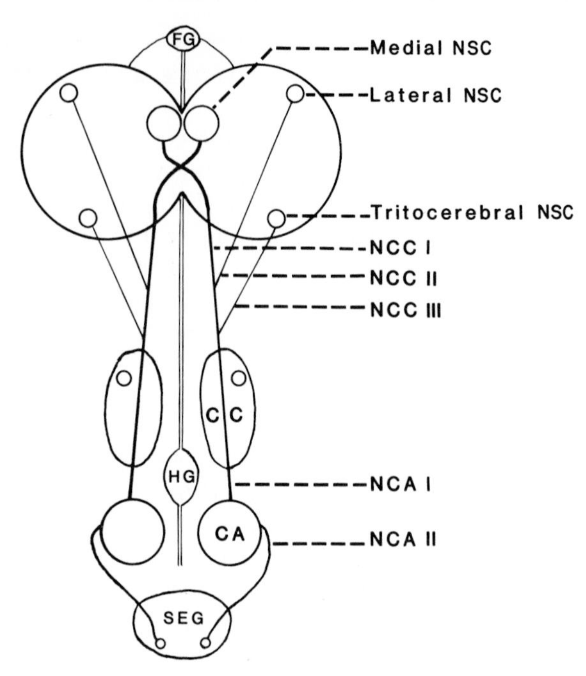

Fig. 1. A diagram of the principal features of the innervation of the corpora cardiaca (CC) and the corpora allata (CA). The three nerves which leave the brain and enter the CC are the *nervi corporis cardiaci* (NCC) I, II, and III. In the protocerebrum, medial neurosecretory cells (NSC) contribute axons to the contralateral NCC I, lateral NSC contribute to the ipsilateral NCC II, and tritocerebral NSC contribute to the ipsilateral NCC III. The nerves from the CC to the CA are the *nervi corporis allati* (NCA) I; those from the subesophageal ganglion (SEG) are the NCA II. Small nerves from the hypocerebral ganglion (HG) enter the CA; the esophageal nerve extends posteriorly from the HG; the recurrent nerve runs between HG and the frontal ganglion (FG). (Modified from Raabe, 1982.)

axons of lateral cells may terminate on the anterior surface of the CA in the extension of the CC described by Pipa and Novak (1979) as a cap of neurosecretory endings.

In *Diploptera punctata*, the location and number of cells which project to the CA and not beyond were tentatively identified by observation of the difference in perikarya visualized following section of the NCA I or postallatal nerves. This led to the conclusion that some lateral and medial cells terminate either in or on the CA (Fig. 2A) (Lococo and Tobe, 1984).

In *L. decemlineata*, only lateral cells were filled via the ipsilateral NCC II whether the filling, with horseradish peroxidase (HRP), was accomplished through the cut surface of the CA or by injecting HRP into the CA so that uncut nerves were filled more slowly (Khan *et al.*, 1984). Both medial and lateral cells were shown to project to the CA in *M. sexta* by retrograde filling

TABLE 1

Perikarya and their axons projecting to or beyond the corpora allata as demonstrated by filling techniques

Species	Perikarya		Axon	Method	Reference
	Location[a]	Number	route[b]		
Schistocerca vaga	PL (ipsi.)	30	NCC II	Co^{2+} retro. NCA I	Mason (1973)
	SG (ipsi.)	11 (2 groups)	NCA II	Co^{2+} retro. NCA II	
Periplaneta americana	PI (contra)	250	NCC I	Co^{2+} retro. NCA I	Pipa (1978)
	PL (ipsi.)	30	NCC II	Co^{2+} retro. NCA I	
	SG (ipsi)	7 (2 groups)	NCA II	Co^{2+} retro. NCA II	Pipa and Novak (1979)
	—	—	NCC I to CA	Co^{2+} (silver intensified) orthograde NCC I	Gundel and Penzlin (1980)
Diploptera punctata	PI (contra)	40	NCC I	Ni^{2+} difference between retro. NCA I and retro. post allatal nerves	Lococo and Tobe (1984)
	PL (ipsi.)	20	NCC II		
Leptinotarsa decemlineata	PL (ipsi.)	8	NCC II	HRP[c] retro. CA	Khan *et al.* (1984)
Manduca sexta	PI-Ia (contra)	4[d] large	NCC I to CA surface	HRP and Lucifer Yellow injected into perikarya in brain	Carrow (1984)
	PL-IIa (contra)	2[d] large	NCC II into CA		
	PL-IIb (ipsi.)	<5–6[d] small	NCC II into CA		
	PI-II (contra)	4–5 large 3–4 small	NCC I	Co^{2+} retro. from fused NCC I and II	Buys and Gibbs (1981)
	PL-III (contra)	2 large	NCC II		Nijhout (1975b)
	PL-I (ipsi.)	5–7 large 6 small	NCC II		

[a] PL, pars lateralis; PI, pars intercerebralis; SG, subesophageal ganglion.

[b] NCC, *nervi corporis cardiaci*, NCA, *nervi corporis allati.*

[c] HRP, Horseradish peroxidase.

[d] Determined from unstained living cells.

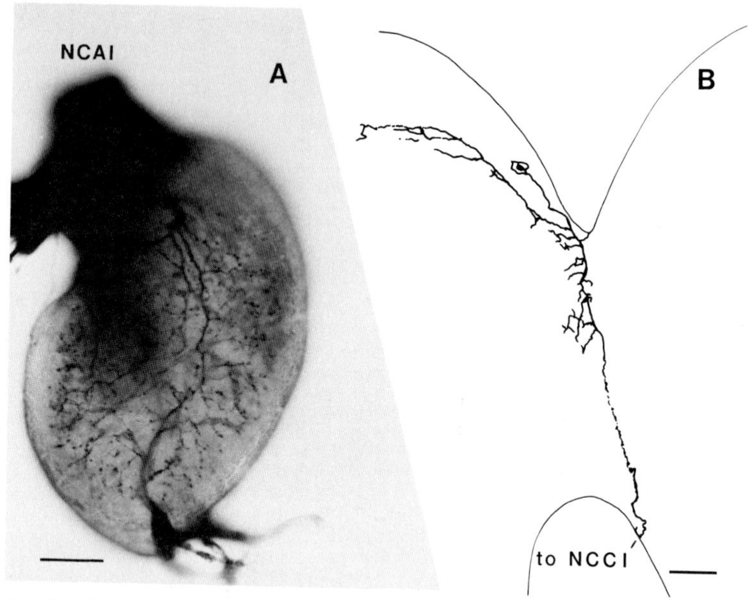

Fig. 2. (A) Nerve endings in an active CA which was filled in the orthograde direction with $NiCl_2$ and intensified with silver; photograph of whole mount. (B) Camera lucida drawing of a cell in the pars intercerebralis injected with Lucifer Yellow. Axonal collaterals branch on the ipsilateral side and the axon projects to the contralateral NCC I. From *D. punctata* females during vitellogenesis. Bars represent 0.05 mm. (Courtesy of Donald Lococo.)

with cobaltous chloride (Nijhout, 1975b; Buys and Gibbs, 1981); the injection of individual cells resolved the projections more specifically. By injecting HRP (which filled axons more readily than dendrites), or alternatively Lucifer Yellow, into brain cells that showed the opalescent blue interference colors of neurosecretory granules, Carrow *et al.* (1984) distinguished two groups of lateral cells with axons that ramified within the CA: one (group IIa) was contralateral and the other (group IIb) was ipsilateral to the CA, but both traversed the NCC II tract. In addition, contralateral medial cells (group Ia) sent axons to the surface of the CA. One of the two cells of group IIa is presumably the primary source of PTTH as identified experimentally by Agui *et al.* (1979); thus the study of Carrow *et al.* (1984) adds support to the hypothesis that in *M. sexta*, the CA are neurohemal organs for PTTH. However, direct experimental evidence for release from the CA has not been forthcoming, although PTTH bioactivity has been demonstrated by stimulating isolated CA with high K^+ (Carrow *et al.*, 1981). This study also provides evidence for potential release of other neurohormones from medial and lateral cells within and on the CA. It is tempting to speculate that one or more of these pathways represent direct delivery of neurosecretory material to the target organ—the CA. Thompson, Lococo, and Tobe (in preparation) have also demonstrated, by intracellular injection of Lucifer Yellow, contralateral medial cells that project to the CA. Zaretsky and Loher (1983) used Lucifer Yellow to fill cells in the pars intercerebralis of *Teleogryllus commodus*, which they identified as neurosecretory from their action potentials of long duration (8–50 ms). Although none of these projected to the CA, they demonstrated two morphologically and physiologically different cells with axons joining the NCC I and terminating in the anterior end of the CC. In addition, they found several cell types that have axons terminating within the brain.

1.2.3 *Arborization of axonal collaterals*[1]

Besides identifying cells that are potential regulators of the CA, the filling techniques may provide clues to the possible interactions with other pathways in the brain by defining the positions of the arborizations of the axonal collaterals or dendritic fields of these cells more precisely than is possible with other methods (Fig. 2B). For example, silver-intensified sections of brains in which cells were filled with Co^{2+} from NCC I and II (Koontz and Edwards, 1980) showed that the arborizations associated with the fibers of these nerves were not in a position to interact with ocellar fibers as had been proposed from both anatomical and experimental evidence by Brousse-Gaury (1971a,b). Also filling of single cells shows that axonal collaterals may project into both hemispheres of the brain and into proto-, deuto-, and tritocerebra (Zaretsky and Loher, 1983; Carrow *et al.*, 1984). Many more such analyses should be done, and the injection of markers into individual neurons would appear to be the method of choice, as has been demonstrated by the characterization of unique dendritic patterns for six groups of protocerebral cells in *M. sexta* (Carrow *et al.*, 1984) and for several cell types in the protocerebrum of *T. commodus* (Zaretsky and Loher, 1983).

1.2.4 *Innervation from the subesophageal ganglion*

Innervation of the CA from the subesophageal ganglion seems less important in the regulation of its activity than the brain (see Section 4). The nerve between the CA and the subesophageal ganglion, the NCA II, contains far fewer axons than the NCA I. Filling techniques have demonstrated that these axons originate in the subesophageal ganglion (see Table 1). In *P. americana*, the NCC II axons ramify in the CA and its neurohemal cap. They also extend into the CC and postallatal nerves as well as traverse to the opposite CA (Pipa and Novak, 1979; Gundel and Penzlin, 1980).

[1] Carrow *et al.* (1984) describe arborization from collaterals of axons as dendritic fields, as have other authors. Since the functional properties of these arborizations are not known, it seems of little consequence which term is used.

1.2.5 *Immunochemical staining*

Another approach to determining the functional innervation of the CA is the use of selective immunochemical staining of the CA and the central nervous system. Antisera to the brain inhibit CA activity *in vivo* in locusts (Rembold *et al.*, 1980). It may be possible to localize the cells that produce the antigens reacting with these antisera. Eckert (1977) has successfully prepared antisera to retrocerebral complex in *P. americana*, which reacted with medial NSC and the storage lobe of the CC. Using *L. migratoria*, Friedel *et al.* (1980a) prepared antisera to an isolated cystine-rich protein from brain and CC and found selective immunoreactivity in "A" medial NSC, NCC I, and the storage lobe of the CC. Although the functional significance of the immunoreactive proteins is uncertain, isolation of neurosecretory products of the insect brain will undoubtedly facilitate the localization and characterization of the cells which produce them. However, antisera to many mammalian hormones have been found to react to substances in insect neurons. We mention here only those found in the CA although, as stated previously, their presence is not necessarily related to its regulation. Immunoreactivity to some vertebrate hormones has been demonstrated in the CA of *Leucophaea maderae* including β-endorphin, luteinizing hormone-releasing factor, and substance P (Hansen *et al.*, 1982), and in *M. sexta*, enkephalin, insulin, somatostatin, substance P, vasointestinal polypeptide, and pancreatic polypeptide (El-Salhy *et al.*, 1983). This immunoreactivity to vertebrate hormones may or may not have functional significance in insects, but antisera to these hormones may be useful for mapping of NSC. Stefano and Scharrer (1981) have found high-affinity binding of an analog of an enkephalin in the brain of *L. maderae*. Since 30% more binding occurred in pregnant females than in males or in larvae of both sexes, it seems plausible that insects may utilize peptides similar to those of vertebrates in regulatory capacities.

1.3 MORPHOLOGY

1.3.1 *Location*

The general morphology of the CA and associated stomatogastric nervous system is well illustrated in the combined review and original observations of a survey of the orders of insects by Cazal (1948). Though new information will continue to be acquired and some of the observations modified (e.g., CA do exist in Collembola; Cassagneau and Juberthie, 1967), this work shows the remarkable degree of similarity in size and location of the CA among the orders of insects and yet clearly documents the variation. Cazal (1948) believed that the occurrence of paired glands situated laterodorsally to the esophagus in the posterior part of the head as is found in many orders including Orthoptera and Dictyoptera was an evolutionarily intermediate condition. A more lateroventral position of paired glands was considered to be more primitive (e.g., orders Diplura and Thysanura; in the latter the CA are embedded in the maxillae; Bitsch and Lapalisse, 1984). A more dorsal location of paired glands and fusion into a single body is the more derived condition. A single medial gland, ventral to the aorta, is characteristic of the orders Embioptera, Dermaptera, and Psocoptera and is found in some members of the Pleocoptera and Hemiptera. In higher Diptera (Brachycera and Cyclorrhapha), the CA are also fused but are dorsal to the aorta.

1.3.2 *Shape*

The characteristic shape of the CA is ovoid to round but they may be elongate as in large larvae and adults of *Libellula depressa* (Odonata) or polylobed as in *H. piceus* (Coleoptera) and *Tetegiaornia* (Homoptera, Cicadidae) (Cazal, 1948). The size of the glands is frequently about the diameter of the aorta or smaller; however, there is much variation between species and even within a species, size differs with age, sex, polymorphism, and activity cycle of the glands (see below, Section 1.3.4).

1.3.3 *Histological types (size and number of cells)*

Although only one type of glandular cell occurs in the CA, there are a variety of types of CA with respect to the number of cells per gland and the relative size of the cells. Cazal (1948) distinguished four histological types. Two of these are glands with many small cells: one, found in Paleoptera, was called "pseudolymphoid" because of the scarce cytoplasm and inactive appearance; the other, characteristic of the majority of insects, was called "the small-cell type." The latter possesses more cytoplasm and shows more "activity" than the former. Recent fine structural analysis of CA from selected Ephemeroptera at different stages of the life cycle (Kaiser, 1980) would suggest that there is not a radical difference between "active" cells of CA of Paleoptera and those of other small-celled CA. Some CA have only a few small cells which is undoubtedly correlated with the tiny size of the insect, e.g., Collembola (Cassagneau and Juberthie, 1967; Palevody and Grimal, 1976) and aphids (Elliott, 1976). Quite distinct however are glands with a few large cells, "large-cell type" (Cazal, 1948), which are characteristic of Trichoptera, Lepidoptera, Hymenoptera, and certain Diptera; cells of this type also have large, sometimes lobed nuclei. Cazal's fourth category of CA types was "epithelial type," referring to the arrangement of cells in layers in various patterns. This does not necessarily imply that the cells are arranged around a central cavity, though such occurs in *T. domestica* (Bitsch and Lapalisse, 1984) and is especially conspicuous in *C. morosus*, in which a cell product apparently forms lamellar layers within the lumen (Haget *et al.*, 1981). It could be that the complexity of the layer of cells is greater in some glands that others, since a conspicuous characteristic of most CA that have been studied ultrastructurally is the infolding of the basal lamina (see Cassier, 1979), suggesting that all CA are some form of folded epithelium. From Cazal's (1948) observations of histological material, he concluded that there was a distinct difference between an inner lumen and the spaces that occur between CA cells secondarily as in a coccid (Cazal, 1948, Fig. 95). The CA of *D. punctata* have been considered the vesicular type (Engelmann, 1970), but fine structural observation indicates that these are secondary spaces which occur after an activity cycle (Johnson *et al.*, 1985). Cassier (1965) reported that such central spaces develop when glands have reached their maximum activity.

1.3.4 *Change in gland volume with activity*

Obvious changes in volume of the CA in the course of developmental and reproductive cycles as well as differences in volume between sexes and different polymorphic forms have been recorded and frequently associated with differences in gland activity (Engelmann, 1970; Cassier, 1979). The parameters of the gland that have been estimated are overall volume, number of cells, and the relationship between these expressed as density of nuclei per volume or area of gland. Bioassays have been relied upon for estimation of activity of the glands. The use of gland size to estimate glandular activity can be evaluated now by comparing *in vitro* rates of JH biosynthesis with gland size (Table 6). One must, of course, assume that *in vitro* rates of biosynthesis are an accurate reflection of *in vivo* gland activity (see Section 3).

1.3.4.1 *Egg-laying cycles* Cazal (1948) observed frequent mitoses in CA of adults as well as larvae and hypothesized that mitotic cycles in adult CA might accompany cycles in activity of the glands. Indeed, Scharrer and von Harnack (1958) demonstrated a cyclic change not only in volume but also in nuclear number (presumably cell number) in the course of a vitellogenic cycle in *L. maderae*. However, the volume of the gland increased to a greater extent than the number of cells so that the number of nuclei per area of gland (nuclear–cytoplasmic ratio) decreased at the highest gland volume, reflecting the increase in cytoplasmic volume of the cells. The change in nuclear–cytoplasmic ratio was found to be the most conspicuous reflection of the JH-dependent egg laying cycle in this cockroach (Engelmann, 1957; Scharrer and von Harnack, 1958). Unfortunately, rates of JH biosynthesis are not available to compare with this well-documented, morphometric study. However morphometric measurements and *in vitro* synthetic activity for CA of *D.*

punctata exist. Cyclic changes in cell number occur during the vitellogenic cycle (Engelmann, 1959; Szibbo and Tobe, 1981a), but the increase in cell number does not account for the increase in synthetic capacity of the glands, i.e., activity of a pair of CA expressed as rate per cell increases, reflecting the change in rate for the entire gland pair. A comparison of morphometry (Engelmann, 1959; Szibbo and Tobe, 1981a) with rates of JH biosynthesis (Tobe and Stay, 1977; Szibbo and Tobe, 1981a) showed that the cytoplasmic -nuclear ratio began to increase before and decreased after the rates of biosynthesis. A similar discrepancy between gland volume and gland activity was demonstrated in *N. cinerea* (Lanzrein *et al.*, 1978), and between gland volume and JH titer by RIA) in *Labidura riparia* (Baehr *et al.*, 1982). Thus there are instances of reasonable, although not exact, correlation between gland activity and volume. Yet even in the same species, good correlation between volume and activity during one physiological state may not hold for another physiological state. This was clearly shown for the wasp, *Polistes gallicus*, in which the rate of JH biosynthesis reflected the volume of the CA in egg-maturing females but not in overwintering females with quiescent ovaries nor in ovariectomized females (Röseler *et al.*, 1980).

It must also be emphasized that, in some species, volume does not reflect rates of biosynthesis even for one physiological condition. For example during egg laying in *L. decemlineata*, little correlation could be found between CA volume and rates of JH biosynthesis (Khan *et al.*, 1982a). This was also true for gonadotrophic female locusts, *Schistocerca gregaria* (Tobe and Pratt, 1975b; Injeyan and Tobe, 1981) and *L. migratoria* (Ferenz and Kaufner, 1981). Nonsynchronized pulsatile biosynthesis of JH is one possible explanation for this lack of correlation (Khan *et al.*, 1982a). In all of these species a high degree of variability in rates of JH biosynthesis between individuals has been observed.

1.3.4.2 *Developmental stages* It is generally agreed that increase in volume and cell number of CA during development is associated with somatic growth rather than with increased activity of the CA. This has been confirmed by direct measurements of rates of JH biosynthesis and CA volume for selected stages in larvae of *M. sexta* (Granger *et al.*, 1979) and in more extensive measurements in the last two larval stadia of female *S. gregaria* (Injeyan and Tobe, 1981) and *D. punctata* (Szibbo *et al.*, 1982). In *D. punctata* larvae, both volume and cell number increased in the second half of the penultimate and last stadia but there were no significant changes in volume per nucleus which reflected the observed changes in rates of JH biosynthesis. In both stadia, the rate of JH biosynthesis per cell was higher early in the stadium than late in the stadium (Szibbo *et al.*, 1982).

1.3.4.3 *Polymorphism* *a. Sexual dimorphism.* Sexual dimorphism in volume of CA is common. Usually female glands are larger although this condition is sometimes reversed, most notably in the Lepidoptera (Cassier, 1979).

An example of small male CA synthesizing hormone at lower rates than the larger female CA has been demonstrated in *D. punctata* (Szibbo and Tobe, 1982). The volume and the number of nuclei of male glands are less than half those of female glands during periods of low JH biosynthesis. The male glands remain relatively constant in activity, cell number, and volume, whereas the female CA undergo large cyclic changes in both during a vitellogenic cycle. Also, denervated male glands, whether transplanted into females or left in males, maintained a low rate of biosynthesis only twice the normal rate (Szibbo and Tobe, 1982). This was a small increase relative to female CA which increase 8- to 10-fold in a vitellogenic cycle (Stay and Tobe, 1981).

Size and activity of the CA show a conspicuous sexual dimorphism in adult *H. cecropia.* The glands of the adult male, which have fivefold greater wet weight than female glands (Gilbert and Schneiderman, 1961), appear to differ from female glands not only in quantity but also quality of product. The male gland does not have the methyl transferase necessary to complete biosynthesis of JHs from the epoxy acids whereas the female gland does (Dahm *et al.*, 1981; see also Sparagana *et al.*, 1984). Although it is not certain whether female glands secrete JH acid and/or JH *in vivo*, male glands produced 100-fold greater quantity of JH acids *in vitro* than those of females (Dahm *et al.*, 1981). An *in vivo* radiochemical assay for JH accumulation using male accessory reproductive organs showed that CA of males are 10 times more active in JH acid production than are female CA (Shirk *et al.*, 1983).

b. Phases and castes. Females of the two phases of *S. gregaria* exhibit a difference in size of CA which correlates positively with activity: the solitary females have larger CA and higher rates of JH biosynthesis than gregarious females (Injeyan and Tobe, 1981). The behavioral polymorphism in egg-maturing females of the wasp, *P. gallicus*, is accompanied by correlated differences in size and rates of JH biosynthesis: dominant females have larger glands and higher rates of JH biosynthesis than do subordinate females (Röseler *et al.*, 1980). Presumptive queen larvae of honeybees have larger CA than presumptive worker larvae (Wirtz, 1973), and, although rates of JH biosynthesis are not known for these glands, the higher JH titer in the presumptive queen larvae (Lensky *et al.*, 1978) suggests that the larger glands may be more active. In the bumblebee, *Bombus terrestris*, CA from workers maintained in the presence of queens synthesized JH *in vitro* at low rates; removal of the queen resulted in an increased rate of JH biosynthesis and egg maturation by the queenless workers (Röseler and Röseler, 1978). Presumably the volume of the CA also increased, as Röseler (1977) has shown a positive correlation between CA volume and JH-dependent egg growth in queenless workers of this species. A similar phenomenon has been observed in the neotenic reproductives of the termite, *Zootermopsis angusticollis*, which upon removal of the king and queen undergo an increase in volume of the CA as well as increased rates of JH biosynthesis as measured by *in vitro* RCA (Greenberg and Tobe, 1985).

1.3.5 *Conclusion*

There are many examples of positive correlations between gland volume and activity, but from those few observations in which rates of JH biosynthesis are available for comparison with size of the glands, it can be concluded that each case must be analyzed before a positive correlation between gland size and activity can be made. Where correlation is found, it is likely that it will not be exact temporally and that the factor by which volume increases will be less than that for biosynthesis.

1.4 FINE STRUCTURE

1.4.1 *General characteristics*

Ultrastructural studies of the CA are relatively few; Cassier (1979) listed about 50, and only another dozen can now be added to that list. Yet the ultrastructural characteristics of the gland cells may provide clues to their functioning and regulation and will be summarized briefly here.

The CA are surrounded by a continuous noncellular basal lamina roughly 0.1–1 μm thick. This material occasionally projects between glandular cells into the interior of the gland, forming trabeculae which may accompany nerves and tracheae (Cassier, 1979). In *L. maderae*, collagen fibers are embedded in the matrix of the covering (Harper *et al.*, 1967); similar fibers have been seen in other cockroach species (Brousse–Gaury *et al.*, 1973) and in locusts (Odhiambo, 1966a). Thus the basal lamina maintains the integrity of the gland.

The junctional modifications of the plasma membranes abutting on the external and trabecular basal lamina are hemidesmosomes. Between cells, maculae adherens, septate desmosomes, and gap junctions occur (see Cassier, 1979, for references). The presence of gap junctions suggests that the CA cells are coupled and indeed this has been demonstrated by dye injection and intercellular recording in CA of *D. punctata* (Lococo *et al.*, 1985). However, much of the plasma membrane is apparently free of junctional modifications, and conspicuous intercellular spaces are often seen in fixed tissue. These latter may be the result of fixation procedures but occur in glands of one particular physiological state (e.g., active glands have more intercellular spaces than inactive glands; Joly *et al.*, 1968; Haget *et al.*, 1981).

The shape of the cells in many species is irregular in that coarse as well as fine microvillar-like processes project from the cell periphery. These projections interdigitate to a large extent. Sometimes the microvillar-like projections extend into conspicuous intercellular spaces (e.g., Odhiambo, 1966a,b; Papillon *et al.*, 1976; Haget *et al.*, 1981). It would seem that this architecture provides the surface area needed for exchanges with the hemolymph, as suggested by Scharrer (1964). With such irregularly shaped cells, it is not easy to establish a polarization of the cell or the distribution of organelles within the cell, though some investigators have suggested that there are local concentrations of cellular organelles. In females of *L. migratoria*, Guelin and Darjo (1974) recognized three regions of the CA cells each with a distinct distribution of organelles; Brousse-Gaury *et al.* (1973) and Scharrer (1964) also reported some localization of organelles in cockroach CA. But since cells are irregularly shaped in these species, the exact location of organelles with respect to the cell architecture, especially in cells of the gland interior, cannot be known except by reconstruction of serial sections and no such study has been made.

Of equal interest with respect to the functioning of the CA is the relative quantity of organelles in glands of different physiological states. This point has not yet been addressed with stereological procedures that would allow quantification of three-dimensional structures by extrapolation from measurements of two-dimensional electron micrographs (Weibel, 1979). However, qualitative observation of micrographs brings a consensus of opinion that mitochondria and endoplasmic reticulum, particularly smooth endoplasmic reticulum (SER), are conspicuous constituents of CA cells and that these appear in different forms as the physiological state of the gland changes. Classification of species differences according to the predominance of smooth and rough endoplasmic reticulum (RER) in the CA (Melnikova and Panov, 1975) is probably not valid because of the uncertainty in equating activity of glands.

1.4.2 *Ultrastructure and activity cycles*

1.4.2.1 *General considerations* Early ultrastructural studies of the CA, which attempted to correlate structure of the glands with changing states of activity, were hampered by lack of knowledge of the physiology of the CA and, to some extent, by the technical difficulties of preparation of this tissue for electron microscopy. The former has been ameliorated gradually; we now know the chemical identity of JH (see Section 2) and that the hormone is not stored in the gland, although glands retain hormone in proportion to the rate of hormone biosynthesis (Tobe and Pratt, 1974b; Tobe and Stay, 1977); and rates of biosynthesis and hence activity cycles can be measured directly (see Section 3).

Preservation of the tissue such that the intracellular organization is clearly visible throughout an activity cycle remains a source of difficulty in the fine structural analysis of the CA. The introduction of glutaraldehyde as a primary fixative followed by osmium tetroxide provided preservation of more cellular structure than osmium tetroxide alone. Compare for example the micrographs of cockroach glands fixed in osmium alone (Scharrer, 1964) with those fixed in glutaraldehyde followed by osmium (Brousse–Gaury *et al.*, 1973) or similarly those of locusts (Odhiambo, 1966a; Joly *et al.*, 1968, vs Papillon *et al.*, 1976; Fain–Maurel and Cassier, 1969a,b). A combination of glutaraldehyde and osmium tetroxide was used by Guelin and Darjo (1974), Melnikova and Panov (1975), and Tobe and co-workers (Tobe *et al.*, 1976; Tobe and Saleuddin, 1977). The latter group used this fixative because, as they demonstrated (Tobe *et al.*, 1976), it retained in the tissues, to some extent, radiolabeled farnesoic acid, methyl farnesoate, and JH III. Glutaraldehyde plus paraformaldehyde (Karnovsky, 1965), which penetrates tissue more rapidly than glutaraldehyde alone, has been used in modified formula by Bradley and Edwards (1979). A modification of Karnovsky's fixative containing picric acid (Ito and Karnovsky, 1968), which is reported to fix SER in steroid-secreting cells, was used by Kaiser (1980), Feyereisen *et al.*, 1981d), and Johnson *et al.* (1985) with reasonable results. The difficulty of fixing the CA may result from the lipid nature of its product and the prominence of SER, which is characteristic of most cell types that produce cholestrol or terpenoids in large amounts (Christensen, 1975). In the Leydig cells of mammalian testis, Christensen (1975) reports that SER is difficult to fix and responds best to fixation by perfusion. Since it is not possible to perfuse blood vessels in insect tissue, a fixative with rapid penetration would probably give the best results.

Cassier's (1979) review of the structure of the CA makes it clear that variation in methods of fixation is only one impediment to making generalizations regarding the relation between the structure of the gland and its cycles of activity. The major difficulty has been in establishing the criterion for activity independently of ultrastructure. Each investigator has sought clearly distinguished active and inactive states. Much experimental evidence has been brought to bear on this identification and in many cases, the bioassays used have been reasonable measures of activity, e.g., gonadotrophic cycles in adult females, or the transition from larva to pupa. Feyereisen (1985) has reviewed the limitations of bioassays, especially long-term ones, and though several ingenious test assays have been utilized, only a few short-term ones have been used in conjunction with fine structural analyses (e.g., Joly *et al.*, 1968). Since Cassier (1979) has presented a detailed review of the literature on fine structure, our comments will emphasize information which has appeared more recently and particularly that structural evidence in species for which there have been direct measurements of rates of biosynthesis by the glands or hormone titers.

An approximation of the variation in rates of JH biosynthesis among species and stages is shown by expressing rates per volume of gland tissue (Table 6). Larval glands show the lowest rates per volume and the adult females the highest. Among the females, all at a similar stage of the vitellogenic cycle, there is considerable variation between species. Undoubtedly some of this difference can be attributed to variations in methodology, but very likely the variation is real and it will be important to learn what structural and functional differences among species and stages contribute to this variation.

1.4.2.2 *Changes during developmental stages* Structure of the CA during metamorphosis has been studied in several species of the Lepidoptera, e.g., *H. cecropia* (Waku and Gilbert, 1964), *Hyphantria cunea* (Melnikova and Panov, 1975), *Diatraea grandiosella* (Yin and Chippendale, 1979b), and *M. sexta* (Sedlak *et al.*, 1983). In *M. sexta*, rates of JH biosynthesis have been determined during the last two stadia by RIA (Granger *et al.*, 1979, 1982a,b) and by RCA (Kramer and Law, 1980a). These rates reflect, though imperfectly, JH titer measurements (Fain and Riddiford, 1975; Riddiford, 1980) although precise physiochemical data remain to be published. From rate measurements, it appears that JH biosynthesis is high during the first half of the penultimate larval stadium, then declines before ecdysis to the last stadium. During the early days of the last larval stadium, JH biosynthesis is again high; it then falls in mid stadium and may rise again before the pupal molt. Rates are low in the pupa. Although Sedlak *et al.* (1983) examined CA at daily intervals during the last two larval stadia and also the first 2 days of pupal life, they were unable to discern great structural changes except between larval and pupal glands. This species deserves more detailed study of glands of known activity, not only because so much is known about its physiology but because its CA are representative of those with few large cells (Cazal, 1948).

The ultrastructure of the larval CA of the southwestern corn borer, *D. grandiosella* (Yin and Chippendale, 1979a), has also been investigated. The titers of JH, measured by bioassay (Yin and Chippendale, 1979a) and physicochemical assay (Bergot *et al.*, 1976), suggest that the activity of CA in last instar, pre-, early, and middiapausing larvae declines from high to moderate rates in the course of this sequence and reaches low rates in late diapausing and nondiapausing last instars. For each of these five stages the fine structural features were scored for their relative abundance. SER was most abundant as vesicles in prediapause (highly active glands), as stacked forms in middiapause (declining activity), and in whorled form in late diapause (inactive); SER of all forms was least abundant in nondiapausing larvae (inactive glands).

1.4.2.3 *Changes in adult male* The rates of JH biosynthesis by CA during sexual maturation in male *S. gregaria* determined by *in vitro* RCA are available (Avruch and Tobe, 1978) for comparison with fine structural observation of such glands (Odhiambo, 1966b). But as in females (see below) of *L. migratoria* (Girardie *et al.*, 1981) and *S. gregaria* (Tobe and Pratt, 1975a), the variation between animals is great, so that although rates do increase with sexual maturity, it is not possible to relate fine structural changes to rate changes with precision. Odhiambo (1966b) reported more SER and mitochondria of more complex shapes in CA of animals developing toward sexual maturity than in those just after adult emergence. Papillon *et al.* (1976) also reported the earlier appearance and greater abundance of fields of vesicles of endoplasmic reticulum in sexually mature animals (reared at 33°C) compared to animals which did not mature sexually (reared at 28°C).

1.4.2.4 *Oocyte development cycles* In *L. migratoria*, the ovarian cycles are JH dependent, and, although the successive cycles of oocyte development follow closely, each vitellogenic cycle in the basal oocytes is complete before the penultimate oocytes begin vitellogenesis (Ferenz and Kaufner, 1981). Because of the latter characteristic, it might be a good species in which to establish a correlation between ultrastructure and rates of JH biosynthesis. Many ultrastructural studies of glands of presumed different states of activity have been done (Joly *et al.*, 1968; Fain-Maurel and Cassier, 1969a,b; Guelin and Darjo, 1974). Correlation of oocyte size and/or age with CA structure has shown changes, especially in endoplasmic reticulum and mitochondria, which appear to be correlated with changing activity of the glands. According to Fain-Maurel and Cassier (1969a), the fine structural features of the glands indicate that they are most active when oocytes are relatively early in vitellogenesis. This concurs approximately with the findings of Joly *et al.* (1968), who characterized the activity of the CA by bioassay as well as by fine structural analysis. Guelin and Darjo (1974) equated active glands with the time of midcycle of oocyte development. All authors agree that there is an evolution of structural modification of the endoplasmic reticulum as the activity of the glands waxes and wanes. There is some disagreement as to whether great fields of concentric SER are indicative of glands with highest activity (Guelin and Darjo, 1974) or those with declining activity (Joly *et al.*, 1968; Fain-Maurel and Cassier, 1969a). Changes in mitochondria with the activity cycle were also noted by Joly *et al.* (1968) but more especially by Fain-Maurel and Cassier (1969b); in particular, changes in size, shape, and matrix density were observed. It is also important to point out that Guelin and Darjo (1974) found ultrastructural differences between cells within a single gland, which suggests that rates of biosynthesis may differ between cells of a gland. Fain-Maurel and Cassier (1969a) observed that as glands start a new cycle, organellar forms show active and inactive characteristics. They suggested that this was undoubtedly because the cycles follow so closely upon each other.

Theoretically these fine structural analyses can now be compared with the direct measurements of the rates of JH biosynthesis by CA of *L. migratoria* determined by RCA (Girardie *et al.*, 1981; Ferenz and Kaufner, 1981). It must be pointed out that the results obtained by these two laboratories are widely divergent, especially with respect to the rates of biosynthesis. Whereas Girardie *et al.* (1981) found mean maximal rates of 50 pmol h^{-1}, Ferenz and Kaufner (1981) reported mean rates two to three times greater. Technical problems with the assay may have led to these higher rates. In addition, Ferenz and Kaufer found that rates did not correlate with age but rather with oocyte length, whereas Girardie *et al.* found the opposite. But in both reports, there appear to be two peaks of JH biosynthesis for each gonotrophic cycle. From this finding, it is clear that the fine structural diagnosis of activity agrees only roughly with the direct measurements. The great variability between animals, which was evident in both the *in vitro* assay studies, explains to a large extent the difficulty in establishing activity cycles in this species. Such variability combined with the limited number of samples that can be used for biosynthetic rate and fine structural analysis is a great barrier to useful correlations between structure and activity in this species. Both analyses should be carried out on the same individual gland because, as Guelin and Darjo (1974) hypothesized from ultrastructural observations, there can be great variation in activity between members of a pair of glands. This was confirmed by *in vitro* measurements of JH biosynthetic rates by Ferenz and Kaufner (1981) and was originally described for *S. gregaria* (Tobe, 1977).

The alternation of JH-dependent vitellogenic cycles in the ovaries with JH-independent periods of egg guarding in *L. riparia* has been utilized by Baehr *et al.* (1982) for the fine structural analysis of active and inactive CA. Although rates of JH biosynthesis were not measured directly, the hemolymph titers of JH were determined by RIA. JH titers increased during vitellogenesis and were low during egg guarding. The CA in the active period exhibited well-developed SER, cisternae of RER, and elongate mitochondria. Apparently the SER transformed into anastomising tubules and vesicles ("structured bodies") which may be the equivalent of the lamellar bodies seen in *L. migratoria* (cf. Fain-Maurel and Cassier, 1969a). During the inactive period, these structural bodies were enclosed in autophagic vacuoles and are presumed to be eliminated during the next cycle of activity. In the inactive period, mitochondria became globular and dense, and glycogen appeared in the cytoplasm.

In adult female *D. punctata*, glands for which synthetic rates had been determined were subsequently analyzed by electron microscopy (Feyereisen *et al.*, 1981d; Johnson *et al.*, 1985). In this species, the rates of biosynthesis of individual glands in a pair are similar (Szibbo and Tobe, 1981a) and thus ultrastructural observations were made after rates were determined for a pair of glands during the first gonadotrophic cycle. Examples of glands of low activity, early in the cycle, and at near-maximal activity, in midcycle, are shown in Figs. 3 and 4, respectively. In glands approaching maximal rates of biosynthesis, the membranous structures within the cells became obscure. This was especially true of the SER, which changes from easily seen tubules (Fig. 3) to obscure tubules, anastomosing networks, and vesicles (Fig. 4). Mitochondria and RER were also somewhat obscure. However, a more outstanding characteristic of mitochondria in glands of increasing activity was a decrease in electron density of the matrix (Fig. 4), an increase in diameter, and occurrence of irregular shapes. Also in such glands the RER appeared in longer and more curved segments. Newly formed autophagic vacuoles were seen in all glands of highest activity, and these continued to be seen in glands of declining activity. Characteristic of glands with decreasing rates of biosynthesis were increased density of mitochondrial matrix, decreased width and irregular shapes of mitochondria, SER again being easily seen as tubules, larger and more conspicuous Golgi complexes, and RER remaining in curved forms. Glands maintaining low rates of biosynthesis for an extended period, such as pregnancy, were not included in this study. However, in *L. maderae*, Scharrer (1964) observed that glands of pregnant females appeared similar to those soon after ovulation; this is true also in *D. punctata* (unpublished observation).

1.4.2.5 *Ovariectomy* Before attempting to make a generalization about ultrastructure and activity of the CA, it may be instructive to consider the ultrastructure of the CA from animals experimentally treated in such a way that CA function is altered. Ovariectomy is such a treatment.

In *L. riparia*, the titer of JH, determined by RIA, is ordinarily high during vitellogenesis and low during egg guarding; ovariectomy resulted in an abnormally high titer of hemolymph JH and cessation in egg guarding. The CA of such animals did not show the normal cyclic changes in volume but rather displayed a steady increase to about twice the normal maximal volume observed during vitellogenesis. The nuclei of the gland cells were swollen, RER and structured bodies of anastomosing SER were abundant, and glycogen accumulated (Baehr *et al.*, 1982). On the basis of their appearance and the high titer of JH, these authors suggested that the CA were

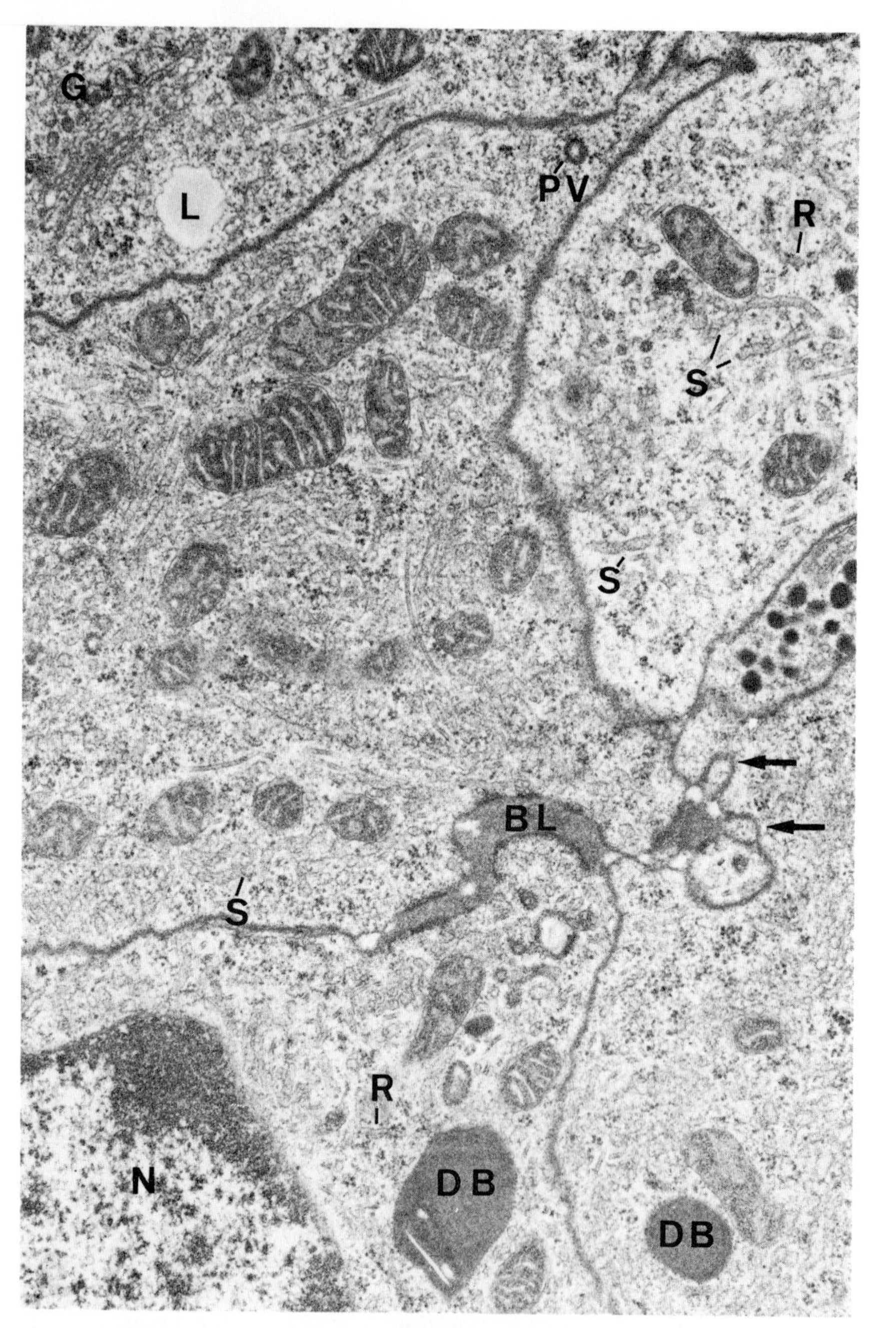

Fig. 3. Portions of CA cells from glands synthesizing JH at 9.7 pmol h^{-1} per pair from a day 1 mated female of *D. punctata* with previtellogenic basal oocytes (0.6 mm long). Note the distinct smooth (S) and short pieces of rough (R) endoplasmic reticulum. Mitochondrial matrix is similar in density to the chromatin of the nucleus (N). A neurosecretory axon terminal is present on the right, a trabecula of basal lamina (BL) appears between cells, and several microvillar-like projections can be seen (arrows). Lipid sphere, L; dense body, DB; Golgi body, G. ×25,700. (Courtesy of Genevieve Johnson.)

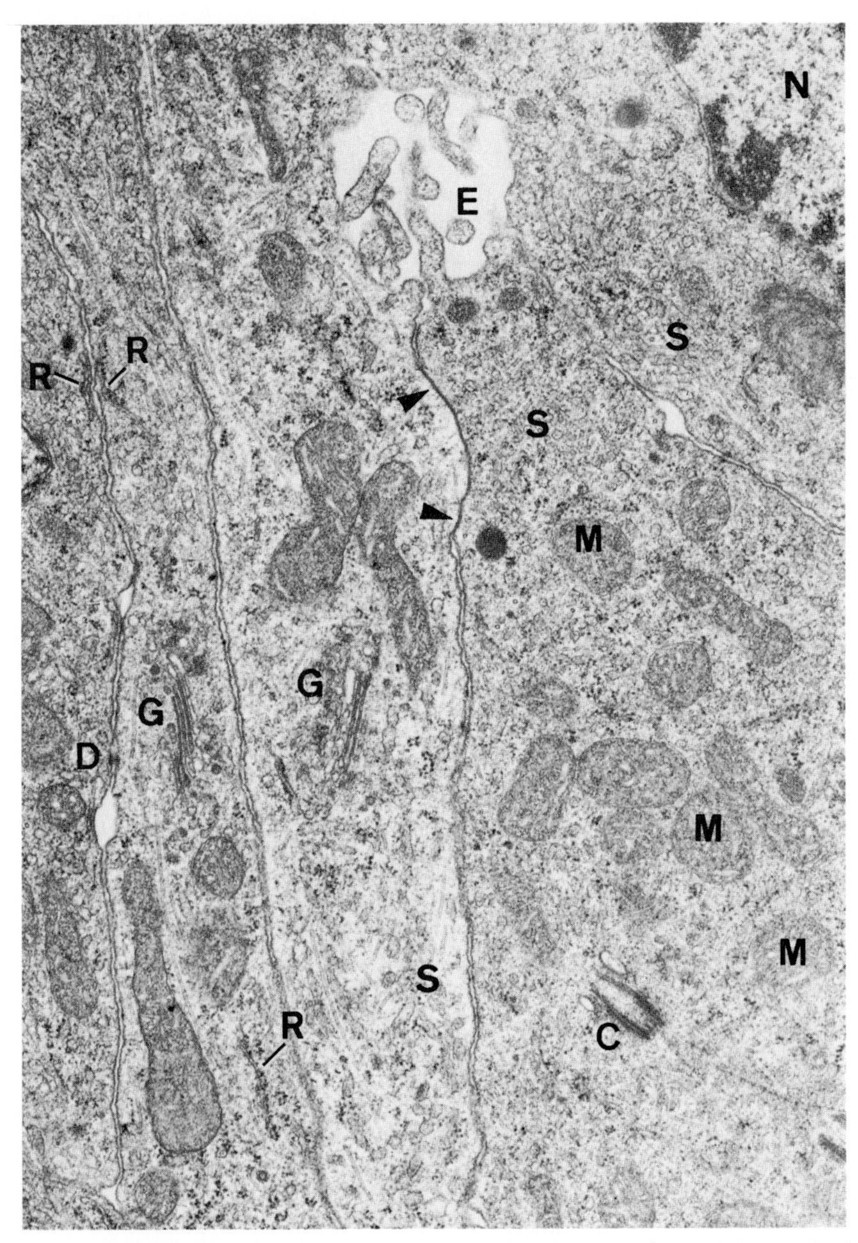

Fig. 4. Portions of CA cells from glands synthesizing JH at 140 pmol h^{-1} per pair from a mated female on day 4 with rapidly growing vitellogenic basal oocytes (1.30 mm long). Note large areas of indistinct smooth (S) and long strands of rough (R) endoplasmic reticulum. Mitochondrial matrix in cell on the right is less dense than in less active gland of Fig. 3. Microvillar-like projections of cell surface extend into a large extracellar space (E); a length of gap junction between cells is visible (arrowheads). Nucleus, N; Golgi bodies, G; centriole, C; mitochondria, M; dense body, DB. ×25,700. (Courtesy of Genevieve Johnson.)

hyperactive. However, the elevated titer could have resulted from some factor other than continued high activity by the CA, such as low JH esterase activity; alternatively, the RIA may have overestimated the titer of JH (see Section 3.2).

Following ovariectomy in *Rhodnius prolixus*, CA also appear to be hyperactive (Baehr *et al.*, 1973), although there are no assays of CA activity other than fine structure to corroborate this suggestion.

Ovariectomy in the cockroach, *L. maderae*, was also interpreted, on the basis of fine structure, to result in hyperactivity of the CA (Scharrer, 1978). In glands from such animals, there was frequently an abundance of SER, which in normal glands during a gonadotrophic cycle was not conspicuous. Neither JH titer nor rates of biosynthesis are available for ovariectomized *L. maderae*, although Engelmann (1978) has shown that JH-dependent vitellogenin biosynthesis continues in ovariectomized animals, suggesting continued activity.

The activity of the CA following ovariectomy has been demonstrated to be low in two other cockroaches, *D. punctata* and *N. cinerea* (Stay and Tobe, 1978; Lanzrein *et al.*, 1981b). The fine structure of the CA might easily be construed as indicating the contrary in that there is much distinct SER (Fig. 5). However, the mitochondrial matrix is dense and that, along with the distinctiveness of the endoplasmic reticulum, is indicative in *D. punctata* of CA about to increase in activity. This cytological evidence, together with the finding that cell numbers increase and remain elevated in ovariectomized females (Tobe *et al.*, 1984b), suggests that regulation of the synthetic activity takes place at several levels (see Section 4.6.2).

In *P. gallicus*, the synthetic rates of glands is ovariectomized animals were low, yet the volume of the CA was characteristic of that of glands with increasing activity (egg-maturing females) (Röseler *et al.*, 1980). Similarly, in *Acheta domesticus*, CA of ovariectomized females synthesized JH at lower rates than those of normal females; no structural studies accompanied these findings however (Strambi, 1981).

1.4.3 *Cellular and subcellular localization of JH biosynthesis and release*

These ultrastructural studies leave little question that SER is an important cellular compartment with respect to JH biosynthesis. Variation in abundance of SER between species may be explained by differences in their maximal rates of biosynthesis, which must be expressed as unit volume of cytoplasm. Such differences have been found for Leydig cells in vertebrates, by correlating rates of biosynthesis with quantitative morphological analysis of the surface area of SER per cell (Ewing and Zirkin, 1983). Similar studies are warranted for CA. Structural differences may also be a reflection of

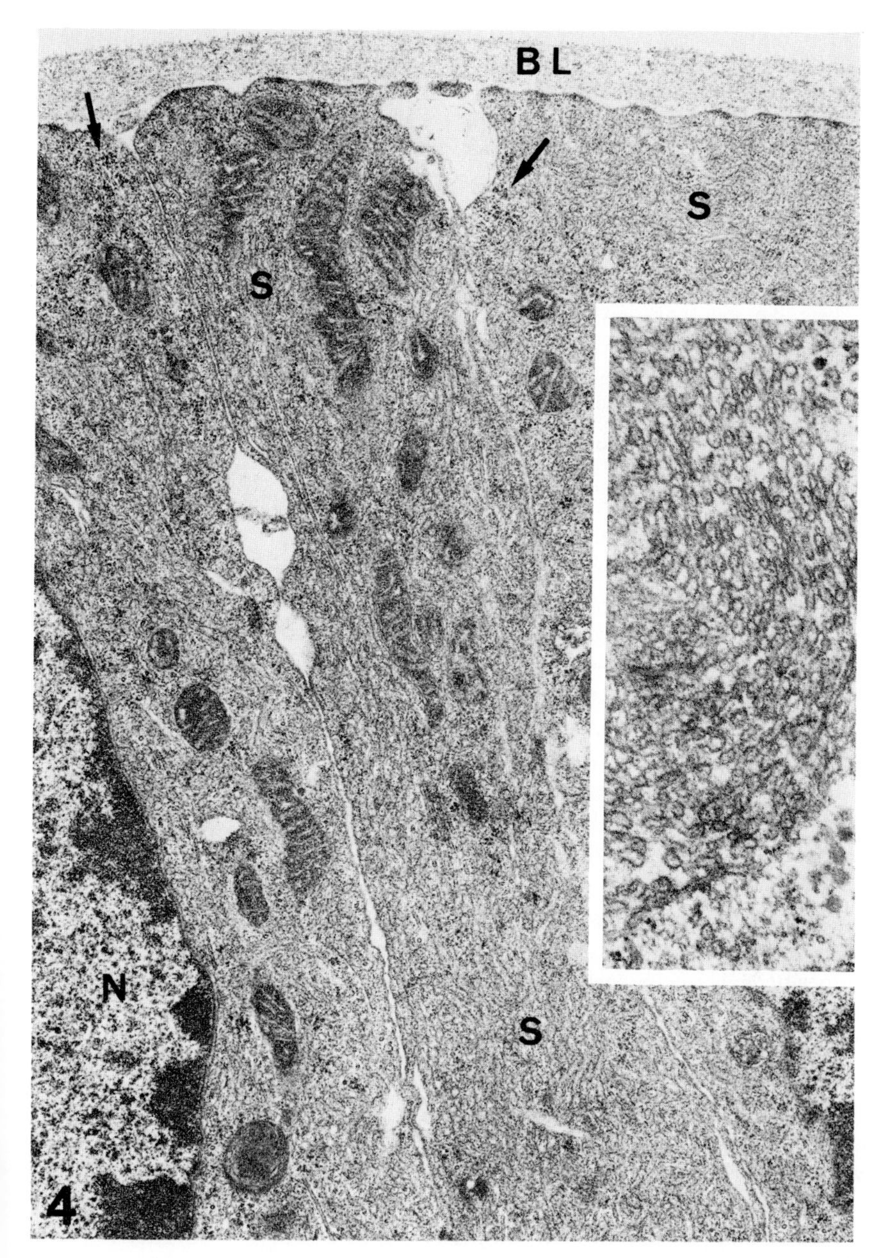

Fig. 5. Portions of CA cells from glands synthesizing JH at 11.4 pmol h^{-1} per pair from a mated 42-day-old female which had been ovariectomized as a last instar larva. Note the dense mitochrondrial matrix and the cytoplasm with abundant tubular smooth endoplasmic reticulum (S) but few ribosomes (arrows). Nucleus, N; basal lamina, BL. ×24,000. The inset shows SER in gland synthesizing JH at 8.0 pmol h^{-1} per pair, from a female ovariectomized for 102 days. ×40,000. (Courtesy of Genevieve Johnson.)

significant differences in regulation of hormone biosynthesis among species or polymorphic forms of the same species, as for example between males and females.

The structural studies also make it clear that there are no obvious storage areas for hormone other than SER. This would be expected in light of biosynthetic studies which have shown immediate release of newly synthesized hormone (see Section 2 and 3). Yet it should be mentioned that in a few species, an electron-dense exocrine product occurs. In *C. morosus*, it is released into the central cavity of the gland (Cassier, 1979; Haget *et al.*, 1981). Cassier (1979) considers this substance to be associated with RER and Golgi. Also, Scharrer (1971) observed sporadic occurrence of a dense material (cribriform inclusions) extruded from the CA cells, especially in glands presumed to be inactive.

The movement of the hormone through the cell is not revealed by the analysis of cell structure. The diffusable nature of JH and its precursors makes this difficult. In an attempt to localize the final steps of biosynthesis within the CA cells of *S. gregaria*, [^{3}H]farnesoic acid was used as a precursor of JH in a short *in vitro* incubation (Tobe *et al.*, 1976; Tobe and Saleuddin, 1977). The label was found predominantly in JH within 30 min. However, the localization of label, as demonstrated by autoradiography of thin sections, was not sufficiently concentrated over any organellar region to determine precisely the location of these last steps. Silver grains were uniformly distributed in adjacent cells (Tobe and Saleuddin, 1977) and were found in all cells of a CA (Tobe and Pratt, 1976; Tobe *et al.*, 1976). Yet it would be premature to state that all areas and cells of a gland are synthesizing at the same rate. Adjacent cells often appear quite different in electron micrographs; for example, some cells are light and some dark. Whether this is an artifact of fixation, which Christensen and Gillim (1969) believe to be the case for endocrine cells of the vertebrate testis, or whether such cells are in different activity states, as suggested for some CA (e.g., Dorn, 1973; Guelin and Darjo, 1974), we cannot distinguish at present.

Perhaps it will be possible in time to analyze the synthetic activity of individual identified cells in culture, as is now done for pituitary cells by the reverse hemolytic plaque assay (Neill and Frawley, 1983), and to determine whether there is synchronous production of JH by all cells. Asynchronous activity has been proposed for the CA cells of *Anacridium aegyptium* from fine structural analysis of a limited sample of glands during a reproductive cycle (Girardie and Granier, 1974).

The question of how the hormone leaves the CA cells also remains unresolved. Since carrier proteins for JH are known to exist in the hemolymph (see Goodman, 1983, for references), it seems likely that these may facilitate movement of hormone away from the extracellular spaces of the gland. There may be intracellular proteins which act as carriers for JH precursors. The RER in continuity with SER (Scharrer, 1964; Fain-Maurel and Cassier, 1969a) would provide for carriers within the endoplasmic reticulum. Electron-dense material can be seen within SER (Deleurance and Charpin, 1978, Fig. 2; Baehr *et al.*, 1982, Fig. 7; Bitsch and Lapalisse, 1984, Fig. 4). Segments of endoplasmic reticulum frequently appear in close proximity to the plasma membrane of CA cells (Fig. 3; also see Scharrer, 1964; Kaiser, 1980; Johnson *et al.*, 1985), suggesting that release of JH may depend upon such juxtaposition. Clearly the localization of functions within the CA cells is still in its infancy. Much more research combining cell fractionation and biochemical and fine structural analyses are required.

2 Juvenile hormone biosynthesis

2.1 THE CORPUS ALLATUM AS THE SITE OF JH BIOSYNTHESIS

2.1.1 *Historical*

The identification of JH I (C_{18}JH) [methyl (2*E*,6*E*)-(10*R*,11*S*)-10,11-epoxy-7-ethyl-3,11-dimethyl-2,6-tridecadienoate] (Fig. 6) as a JH of *Hyalophora cecropia* by Röller *et al.* (1967) and Dahm *et al.* (1968) signaled the beginning of the present era of JH physiology. This discovery was of fundamental importance since, with the chemical structure of the hormone defined, studies on the pathways of biosynthesis and degradation as well as physiological studies on the mode of action of the hormone and the regulation of physiological processes by the hormone became possible. JH II (C_{17}JH) [methyl (2*E*,6*E*)-(10*R*,11*S*)-10,11-epoxy-3,7,11-trimethyl-2,6-tridecadienoate] was subsequently identified in *H. cecropia* (Meyer *et al.*, 1968, 1970). Despite the identification of these JHs, the site of synthesis of the hormones remained to be defined conclusively. However, much indirect evidence suggested that the site of synthesis of JH was indeed the CA (see Wigglesworth, 1964, 1970; Englemann, 1970). The chemical identification of the JHs made it possible to associate precisely the synthesis of a specific hormonal product with its site of synthesis. Two nomenclatures are presently in use to identify the JHs; the original system employed carbon numbers (hence C_{16}, C_{17}, and C_{18}JH), but this system has been superceded by one in which the JHs are described numerically in order of their discovery (Dahm *et al.*, 1976) (Fig. 6) (see however, Bergot *et al.*, 1980, 1981b).

Definitive identification of the state of JH synthesis as the CA also awaited the development of *in vitro* techniques to demonstrate directly and unequivocally that these glands synthesize and release the chemically defined JH. Accordingly, it is hardly surprising that soon after the elucidation of the

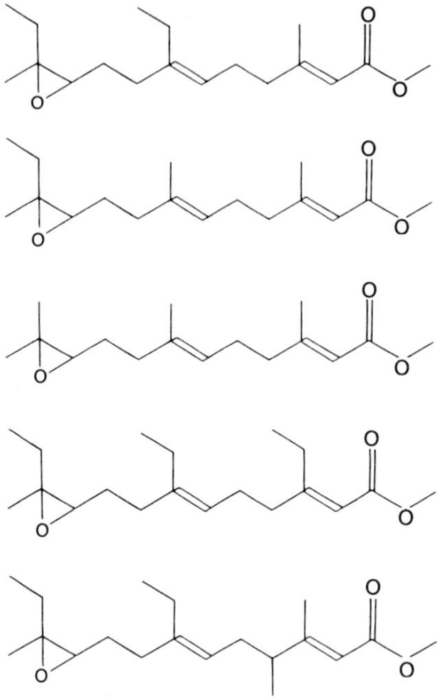

JH I (C_{18} JH)

JH II (C_{17}JH)

JH III (C_{16}JH)

JH O

4-Methyl JH I (iso-JH O)

Fig. 6. Chemical structures of identified juvenile hormones. JHs I, II, and III have been isolated not only from hemolymph and whole-body extracts but also from incubations of CA *in vitro*. JH 0 and 4-methyl JH I have been isolated from embryos of *M. sexta*.

chemical structure of JH I and II from *H. cecropia*, Röller and Dahm (1970) were able to conclude that the brain plus retrocerebral complexes of male pupae of this species produced JH I *in vitro* over a 7-day period in medium fortified with heat-treated hemolymph. It is ironic that subsequent studies have revealed that in fact the CA of adult male *H. cecropia* produce not JH but JH I acid, because they lack the *O*-methyl transferase responsible for the esterification of the JH acid; JH acid is subsequently methylated in the accessory gland (Shirk *et al.*, 1976; Peter *et al.*, 1981; Dahm *et al.*, 1981). Thus the original experiments of Röller and Dahm (1970) could not be replicated, despite the original identification of JH I from the culture medium by *Tenebrio* bioassay and gas chromatography–mass spectrometry (GC–MS). As Schooley and Baker (1984) have recently noted, the possibility of artifactual esterification of the JH I acid in the culture medium of Röller and Dahm should be considered. Transplantation of portions of the brain–retrocerebral complex confirmed that only the CA were the source of the ostensible JH (Röller and Dahm, 1970).

Although this pioneering study did not utilize radiolabeled substrates to demonstrate the incorporation into the JH molecule (and hence *de novo* biosynthesis), Röller and Dahm were able to show that neither CA extracts nor extracts of brain–retrocerebral complexes contained detectable JH I. Subsequent work from their laboratory demonstrated that the S-methyl group of [*methyl*-^{3}H]methionine was incorporated at high efficiency into JH I and II by male *H. cecropia in vivo* (Metzler *et al.*, 1971). Double-labeled JH I could also be isolated *in vivo* following injection of moths with 2-^{14}C-JH I acid and [*methyl*-^{3}H]methionine (Metzler *et al.*, 1972). Unfortunately, because of the *in vivo* nature of these experiments, it was impossible to conclude definitely that the CA were in fact the site of the biosynthetic reactions. As noted above, it is now known that the esterification of the JH acid to JH occurs not in the CA but in the male accessory gland (Peter *et al.*, 1981; Dahm *et al.*, 1981). It appears that this reaction in the accessory glands involves the intermediate formation of *S*-adenosylmethionine (SAM) (Weirich and Culver, 1979). The studies of Metzler *et al.* (1971) are particularly noteworthy because of the very high efficiencies of incorporation of the radiolabeled methyl moiety of methionine into JH (up to 0.1% of the total amount injected). This result suggested that L-methionine was not involved in many pathways of lipid biosynthesis. These studies and the fortuitous specificity of L-methionine incorporation into the JH methyl ester have been instrumental in the subsequent development of the *in vitro* radiochemical assays.

2.1.2 *In vivo JH biosynthesis*

It is appropriate at this time to consider the relative advantages and disadvantages of *in vivo* experiments for the study of JH biosynthesis. Such methodology has been utilized principally by the Röller group (see references above, as well as Rode-Gowal *et al.*, 1975). An obvious advantage of the *in vivo* approach is the relative simplicity of the methodology. In addition, there can be no doubt about the survival of experimental tissue *in vivo. In vivo* experimentation also precludes the need for the development of adequate incubation media and techniques, and obviates the question of altered biosynthetic capability of tissues maintained *in vitro*; utilization of substrates not encountered *in vivo* may also complicate interpretation of *in vitro* experimentation. Finally, the possibility of contamination of the experimental system with microorganisms and the associated nonspecific incorporation of radiolabeled substrates into apparent natural products is minimized *in vivo*. The relative advantages of the *in vivo* system can be realized *in vitro* if adequate care is taken to develop appropriate incubation media and to maintain its sterility. Experiments *in vitro* avoid several major difficulties associated with *in vivo* experimentation. (1) Generally, there are very low levels of incorporation of substrates into the desired products, principally because the biosynthetic tissue represents only a very small percentage of the total body weight. (2) It is common for injected substrates to be utilized in many metabolic pathways, both anabolic and catabolic; this results in the removal of the substrate from the precursor pool and complicates interpretation of results because of the possibility of the utilization of the modified precursors. (3) Finally, as Schooley and Baker (1984) imply, great care must be taken in the choice of radiolabeled substrates for biosynthetic studies to avoid possible elimination of the radiolabeled atoms during biosynthesis. These authors provide some pointed examples of the difficulties in interpretation when inappropriate radiolabeled substrates are employed.

2.1.3 *In vitro JH biosynthesis*

The first demonstration of *in vitro* biosynthesis of JH, as opposed to the release of previously synthesized JH, utilized the incorporation of the radiolabeled methyl moiety of methionine into the methyl ester group of JH (Judy *et al.*, 1973a), in experiments parallel to the *in vivo* approach of Metzler *et al.* (1971, 1972). These workers used long-term incubations of CA of adult female *M. sexta* in the presence of [*methyl*-^{14}C]methionine to demonstrate the biosynthesis of the previously unknown JH III [C_{16}JH; methyl (2*E*,6*E*)-(10*R*)-10,11-epoxy-3,7,11-trimethyl-2,6-dodecadienoate] as well as JH II (Judy *et al.*, 1973a). Not only did this study reveal the existence of a new juvenile hormone (which has subsequently been recognized as the principal JH of insects, with the exception of the Lepidoptera, see below), but it demonstrated the value of *in vitro* maintenance of CA in the presence of radiolabeled substrates to study JH biosynthesis. It is significant that Judy *et al.* (1973a) found only two major radiolabeled products (JH II and III) following simple extraction and chromatographic techniques; thus the *in vitro* incubation of CA resulted in surprisingly "clean" preparations of radiolabeled JHs. This realization has had several ramifications: first, it ultimately permitted the development of a simple and rapid procedure for the quantification of biosynthesized JH following short-term maintenance of CA (Feyereisen and Tobe, 1981); second, it has now proved possible to biosynthesize microgram quantities of high specific activity, enantiomerically pure JH III (using CA of *D. punctata*) requiring only minimal chromatographic purification (see Tobe and Clarke, 1985). Judy *et al.* (1973b) also used techniques similar to those employed for *M. sexta* to demonstrate the incorporation of the methyl group of radiolabeled methionine into the methyl ester of JH III by CA of *Schistocerca vaga*. However, in both cases, these workers employed long-term incubations (days and weeks) of CA of unknown activity in incubation media containing hemolymph (i.e., undefined media) and accordingly it was not possible to estimate rates of JH biosynthesis by the glands. The methodology of Judy and co-workers proved invaluable for the subsequent studies on the pathway of JH biosynthesis. The development of the RCA for JH biosynthesis by individual CA of known physiological age was occurring simultaneously and involved the use of defined incubation media and *in vitro* incubations of only short duration (Pratt and Tobe, 1974).

The *in vitro* techniques of Röller and Dahm (1970) were also adapted by Pratt and Tobe (1974) to study JH biosynthesis in *S. gregaria.* Subsequent studies by these authors demonstrated the efficacy of medium 199 as an incubation vehicle and revealed that incubation times as short as 10 min are sufficient for the recovery of detectable quantities of radiolabeled biosynthesized JH from the medium (Tobe and Pratt, 1974a; Pratt *et al.*, 1975a). These studies proved fundamental to the development of the RCA and defined the parameters for the successful deployment of this assay to a variety of insect species of different orders. The RCA was used initially to investigate the relation between JH synthesis and oocyte growth in *S. gregaria* (Tobe and Pratt, 1975a), *P. americana* (Pratt *et al.*, 1975a; Weaver *et al.*, 1975; Weaver and Pratt, 1977), and *D. punctata* (Tobe and Stay, 1977), as well as to investigate the dynamics of JH release (Tobe and Pratt, 1974b) and the relation between CA volume and JH synthesis in *S. gregaria* (Tobe and Pratt, 1975b). Numerous similar studies on other insect species have followed (see Tobe and Feyereisen, 1983, for review). The technique has widespread applicability to the study of JH biosynthesis and has revealed the basic similarity in the parameters of hormone biosynthesis in different orders and at different stages of development. The RCA has also proved invaluable in studies of pathways, precursor and substrate utilization, and enzymatic activities in JH biosynthesis, because (1) short-term incubations have allowed the isolation of biosynthetic intermediates and subsequent estimates of stoichiometry of incorporation of radiolabeled substrates—see for example, Tobe and Pratt (1974a), Baker and Schooley (1981), Baker *et al.* (1983a), Feyereisen *et al.* (1984); and (2) the use of defined incubation media precluded the possibility of incorporation of unknown precursors into the biosynthesized JH. The importance of defined media in studies on the regulation of JH biosynthesis cannot be overemphasized, because of the possibility of disruption of the biosynthetic pathway by unknown constituents of undefined media (including those containing either insect hemolymph or vertebrate sera). In addition, exogenous precursors are now well-known for their ability to alter overall rates of hormone biosynthesis (for example, farnesoic acid and its homologs: Pratt and Tobe, 1974; Tobe and

Pratt, 1976; Feyereisen *et al.*, 1981a; see also Section 2.2). In fact, simple minimal media are valuable for substrate utilization studies, e.g., those of Pratt and Tobe (1974) and Schooley and Baker (1984).

Another useful characteristic of the *in vitro* RCA, noted by Schooley and Baker (1984), is that even in the presence of relatively low concentrations of defined exogeneous precursors, the endogenous substrates are overwhelmingly diluted by the exogenous precursors, provided that there are no permeability barriers. This phenomenon can be valuable in studies of the biosynthetic pathway and points of rate limitation; e.g., utilization of possible substrates can be easily tested and rate-limiting steps assessed.

Because of the possibility of permeability barriers to certain biosynthetic precursors, cell-free homogenates of CA have been utilized. The first such study using this approach was that of Reibstein and Law (1973), demonstrating the incorporation of the methyl moiety of *methyl*-^{3}H-SAM into the ester function of JH by homogenates of CA of *M. sexta*. This study confirmed the pathway of methionine utilization in JH biosynthesis. Because the CA are probably relatively impermeable to SAM, only cell-free experiments could confirm its involvement in hormone synthesis. Although experiments utilizing cell-free homogenates of CA are tedious and difficult, because of the minute quantities of available tissue and the corresponding necessity for dissection of large numbers of animals, such experiments have provided much useful information on specific enzymes of the JH biosynthetic pathway (see for example, Baker and Schooley, 1978; Baker *et al.*, 1981; Feyereisen *et al.*, 1981a,b).

2.2 JH BIOSYNTHESIS PATHWAY

Elucidation of the pathways of JH biosynthesis has occupied the attention of a number of laboratories since the discovery of the identity of the JHs and the subsequent development of the assay systems outlined in the previous section. The JHs are unusual molecules, isolated and identified conclusively to date only in the Insecta, and it is their unusual sesquiterpenoid nature which has piqued the interest of both biochemists and chemists. In particular, investigators have been concerned with the genesis of the ethyl side chains found in JH I and II (as well in the more recently isolated JH O and 4-methyl-JH I) (see Fig. 6 for the chemical structures of the known JHs). These side chains can arise in at least two hypothesized ways: (1) the existing methyl side chains of farnesyl pyrophosphate could, through the formation of some unknown intermediate, undergo further alkylation by an appropriate methyl donor such as methionine; (2) through the substitution of a C_3 precursor for acetate, higher homologs of JH precursors could be generated. This would result in a whole series of intermediates homologous to the intermediates known to occur in the biosynthesis of farnesyl pyrophosphate.

2.2.1 *Genesis of the carbon skeleton*

At this point, it is appropriate to review briefly the biosynthetic pathway for farnesyl pyrophosphate since this compound is a known intermediate in the biosynthesis of cholesterol and other sterols. Although insects have not been demonstrated to biosynthesize cholesterol, the existence of an appreciable portion of the pathway and its subsequent divergence to give rise to the JHs has intrigued evolutionary biologists and biochemists alike.

In vertebrates, condensation of three C_2 (acetate) units give rise to a C_6 unit (3-hydroxy-3-methyl glutarate). Phosphorylation followed by decarboxylation results in a C_5 unit which, through a series of phosphorylation reactions, gives rise to a C_5 isoprenoid unit. Condensation of three C_5 isoprene units ultimately results in the formation of the C_{15} farnesyl pyrophosphate (see for example, Nes and McKean, 1977). Depending upon the mechanism for side chain generation mentioned above, farnesyl pyrophosphate could conceivably be an intermediate in the biosynthesis of JH I and II. Accordingly, some early studies did in fact investigate the incorporation of radiolabeled acetate (C_2) as well as mevalonate (C_6) into farnesol, the dephosphorylation product of farnesyl pyrophosphate, and into JH I (Metzler *et al.*, 1971). These authors were able to demonstrate incorporation of radiolabeled mevalonate into farnesol but not into JH I; however, radiolabeled acetate did incorporate into JH I. Because these studies were done *in vivo*, interpretation of results was difficult; interpretation is further compounded by the apparent elimination of radiolabeled atoms during biosynthesis (see Schooley and Baker, 1984). Thus, although these experiments proved inconclusive with respect to defining the early intermediates in the biosynthesis of JH I, Metzler *et al.* (1971) did demonstrate conclusively using microchemical degradation techniques that [*methyl*-^{3}H]methionine was incorporated only into the methyl ester function of JH I and II. This finding thus finally laid to rest the possibility of side chain alkylation for the generation of the ethyl side chains in JH I and II. JH III remained to be identified (by Judy *et al.*, 1973a), and although it could be expected that the biosynthesis of this hormone would follow the conventional farnesyl pyrophosphate pathway described in vertebrates, studies on the genesis of the carbon skeleton of the higher homologs then turned to early intermediates other than acetate.

The pioneering study of the Zoecon group on the genesis of the carbon skeleton of the JHs utilized the *in vitro* methodology of Judy *et al.* (1973a).

Schooley *et al.* (1973) were able to demonstrate conclusively the incorporation of [^{14}C]propionate (a C_3 unit) into JH II but not into JH III by CA of *M. sexta*; [^{14}*C*]acetate and [^{14}C]mevalonate on the other hand were found to incorporate into both JH II and JH III. Although these studies did not prove that the ethyl side chain of JH II arose from propionate, subsequent degradation studies were able to demonstrate that such was indeed the case (see Jennings *et al.*, 1975a; Schooley and Baker, 1984). In *H. cecropia*, the ethyl side chains of JH I have a similar origin (Peter and Dahm, 1975). The study of Schooley *et al.* (1973) was fundamental to subsequent work on the biogenesis of the carbon skeleton of the higher homologs of JH since it suggested the existence of a previously unknown pathway involving the condensation of one C_3 unit (propionate) with two C_2 units (acetate) to give rise to the higher homologs of 3-hydroxy-3-methylglutaryl-CoA (HMG-CoA) and mevalonate, i.e., 3-hydroxy-3-ethylglutaryl-CoA (HEG-CoA) (C_7) and homomevalonate (C_7). Condensation of one homoisoprenoid (C_6) unit with two isoprenoid (C_5) units via a modified isoprene pathway generated JH II, whereas condensation of two homoisoprenoid units with one isoprenoid unit generated JH I. Subsequent studies have now demonstrated the occurrence of HEG-CoA (Baker and Schooley, 1978) and homomevalonate (Lee *et al.*, 1978; Baker and Schooley, 1981) and their synthesis from propionyl-CoA (see Table 2).

2.2.2 *Enzymes of JH biosynthesis*

The enzymes involved in the biosynthesis of JH and their products are summarized in Table 2 and Fig. 7. Virtually all studies on the enzymes and their products have utilized radiolabeled substrates or cofactors. Also shown in Table 2 is the cellular location of the various biochemical reactions (where known); these are classified as either soluble or microsomal since in general there has been no further isolation of the cellular components. Clearly, these studies have utilized cell-free homogenates of CA but it should be noted that several studies, particularly those on the terminal enzymes, have used intact CA. Microsomal reactions are those which probably occur in the endoplasmic reticulum, whereas soluble reactions occur either in the cytosol or in organelles not present in the microsomal fraction. Only two of the enzymes involved in JH biosynthesis are not soluble: HMG-CoA reductase and epoxidase are both associated with the microsomal fraction of CA homogenates (see Fig. 7). Although there have been few studies on these enzymes from the CA, there is a considerable body of information on these enzymes from vertebrates (in the case of both HMG-CoA reductase and epoxidase) and in other tissues of insects (epoxidase). It appears that the CA enzymes have many characteristics in common with the enzymes from other tissues and organisms and accordingly, we will consider them briefly at this time.

2.2.2.1 *HMG-CoA reductase* This enzyme has been extensively studied in vertebrate systems, particularly in the liver; it has attracted the interest of many workers because it appears to be the rate-limiting enzyme in the biosynthesis of cholesterol and related sterols. It is susceptible to negative feedback regulation by both cholesterol and other isoprenoids (see Brown and Goldstein, 1980), and recent evidence has revealed that this feedback regulation actually functions at the level of transcription of the mRNA coding for the reductase (Luskey *et al.*, 1983). In vertebrates, the enzyme is a glycoprotein which is inserted in the membrane of the endoplasmic reticulum during its synthesis and glycosylation (see Dautry–Varsat and Lodish, 1983) and which remains there. Recently, the nucleotide sequence of HMG-CoA reductase has been described (Chin *et al.*, 1984), and it now appears feasible to begin studies on the definition of those sequences in the mRNA which mediate the negative feedback effect of cholesterol. The study of HMG-CoA reductase in insect CA remains in its infancy; however, the enzyme may be of prime importance in the regulation of JH biosynthesis. The activity of the enzyme appears to be regulated by phosphorylation/dephosphorylation reactions (Feyereisen *et al.*, 1981a; Monger and Law, 1982), as in the vertebrates (see Brown and Goldstein, 1980), and although it has not been possible to conclude that this is the only rate-limiting enzyme in JH biosynthesis (Feyereisen *et al.*, 1981a), the possibility that HMG-CoA reductase is subject to regulation by late intermediates in the biosynthetic pathway has recently been raised (Feyereisen *et al.*, 1984). In *M. sexta* CA, which synthesize both JH II and III, the enzyme presumably uses both HMG-CoA and HEG-CoA as substrates (Baker and Schooley, 1981) and in this instance, it may be more appropriate to term it HMG-/HEG-CoA reductase. In any case, the enzyme does not show great substrate specificity, and enzyme preparations from CA of both *Schistocerca nitens* and *Tenebrio molitor*, which normally only produce JH III, will efficiently convert HEG-CoA to homomevalonate, as will preparations from rat liver (Baker and Schooley, 1981).

2.2.2.2 10,11-*Epoxidase* (*monooxygenase*) This enzyme is one of the well-known series of enzymes involved in detoxification reactions called the mixed function oxidases (MFOs). In both vertebrates and insects, the epoxidase has been shown to be in the microsomal fraction of cell-free preparations and occurs specifically in the membrane of the endoplasmic reticulum (Bar-Nun *et al.*, 1980). In contrast to the HMG-CoA reductase, this enzyme lacks carbohydrate moieties (see Bar-Nun *et al.*, 1980). Epoxidase is usually linked to cytochrome P-450 and uses NADPH as cofactor (see Table 2 and Fig. 7); inhibition of the enzyme by carbon monoxide and reversal of this inhibition by white light are usually regarded as diagnostic for a P-450-linked

TABLE 2
Enzymes of JH biosynthesis

Enzyme	Cofactors	Products	Cellular location	Species	Reference
Thiolase		Acetoacetyl-CoA[a]	Soluble	*M. sexta*	Lee *et al.* (1978)
HMG-CoA synthetase		HMG-CoA (HEG-CoA)[b]	Soluble	*M. sexta*	Lee *et al.* (1978); Baker and Schooley, (1978); Schooley *et al.* (1978); Bergot *et al.* (1980)
HMG-CoA reductase	$NADPH^+$	Mevalonate (homomevalonate)	Microsomal	*M. sexta*	Lee *et al.* (1978); Kramer and Law (1980a); Baker and Schooley (1981); Monger and Law (1982)
				D. punctata	Feyereisen *et al.* (1981a)
Mevalonate kinase	ATP	Mevalonate phosphate[c] (homomevalonate phosphate)[c]	Soluble	*M. sexta*	Lee, cited in Schooley and Baker (1984)
Isomerase		Isopentenyl pyrophosphate (homoisopentenyl pyrophosphate); dimethylallyl pyrophosphate (ethylmethylallyl pyrophosphate)	Soluble	*M. sexta* *S. nitens*	Baker *et al.* (1981; and cited in Schooley and Baker, 1984)
Prenyl transferase		Farnesyl pyrophosphate	Soluble	*M. sexta*	Baker *et al.* (1981)
		Geranyl pyrophosphate		*S. nitens*	Baker *et al.* (1981)
Farnesyl pyrophosphatase		Farnesol[a]	Soluble	*M. sexta*	Reibstein *et al.* (1976); see Schooley and Baker (1984)
Farnesol dehydrogenase	NAD^+	Farnesal	Soluble	*M. sexta*	Baker *et al.* (1983*a*)
		Farnesal[a]		*D. melanogaster*	Madhavan *et al.* (1973)
		Farnesal[a]		*D. punctata*	Feyereisen *et al.* (1981*a*); Feyereisen *et al.* (1984)

Farnesal dehydrogenase	NAD^+	Farnesoic acid	Soluble	*M. sexta*	Baker *et al.* (1983a)
		Farnesoic acid[a]		*D. melanogaster*[d]	Madhavan *et al.* (1973)[d]
		Farnesoic acid[a]		*D. punctata*	Feyereisen *et al.* (1981a); Feyereisen *et al.* (1984)
O-Methyl transferase	SAM	Methyl farnesoate	Soluble	*S. gregaria*[d]	Pratt and Tobe (1974)[d]; Tobe and Pratt (1974a)
				L. migratoria	Feyereisen *et al.* (1981c)
				P. americana[d]	Pratt *et al.* (1975b)
				D. punctata	Tobe and Stay (1977)[d]; Feyereisen *et al.* (1981*a*)
		JH		*M. sexta*	Judy *et al.* (1973a); Reibstein and Law (1973); Reibstein *et al.* (1976)
10,11-Epoxidase	$NADPH^+$	JH	Microsomal	*S. gregaria*[d]	Pratt and Tobe (1974)[d]; Tobe and Pratt (1974a)
				L. migratoria	Feyereisen *et al.* (1981c)
				P. americana[d]	Pratt *et al.* (1975b)
				B. giganteus	Hammock (1975)
				D. punctata	Tobe and Stay (1977)[d]; Feyereisen *et al.* (1981b)
		Epoxyfarnesoic acid		*M. sexta*	Reibstein and Law (1973); Reibstein *et al.* (1976)

[a] Product not identified.
[b] Products in parentheses are intermediates in the biosynthesis of the higher homologs of JH III.
[c] Tentatively identified.
[d] Cellular location was not determined in these instances.

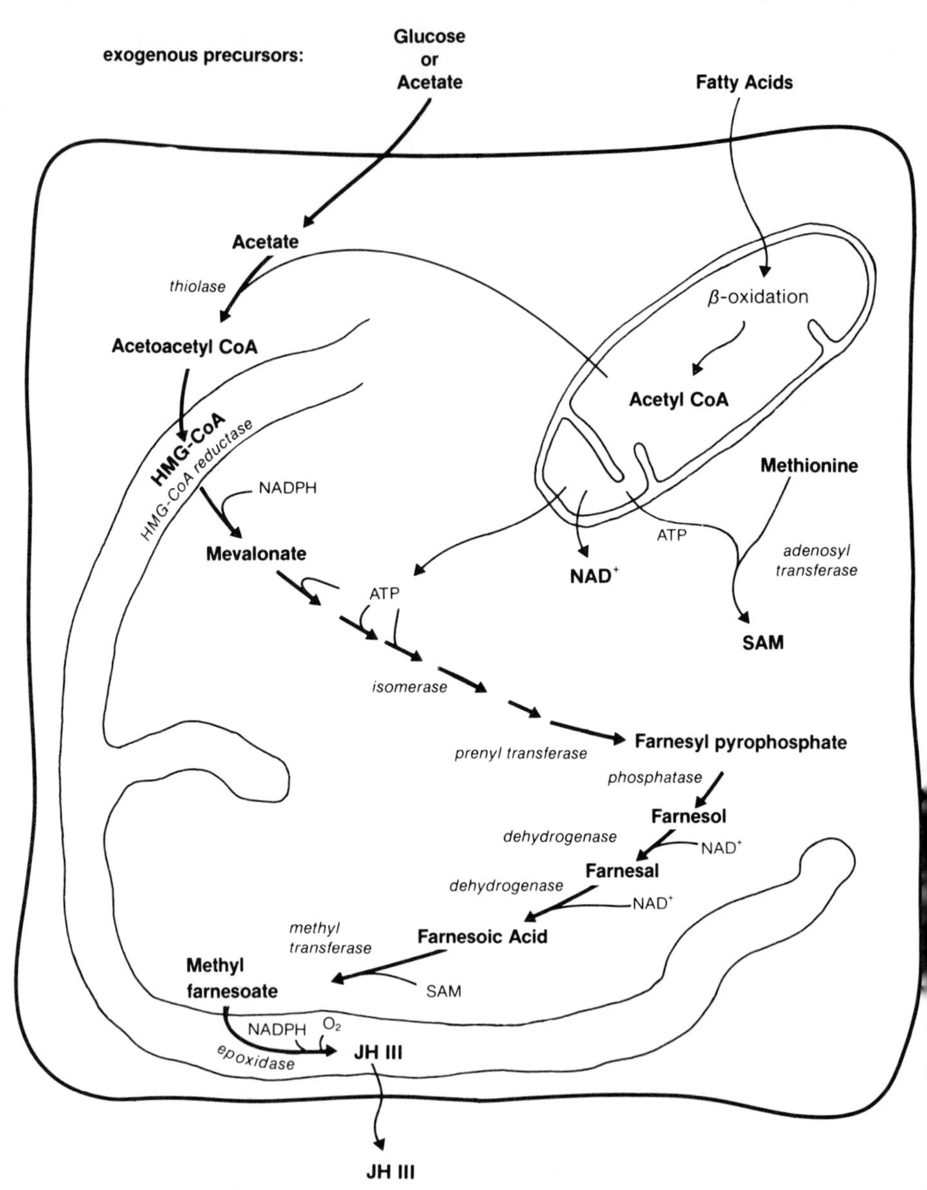

Fig. 7. Diagrammatic representation of the pathway of JH III biosynthesis within a typical CA cell. Exogenous precursors such as glucose, acetate, and fatty acids all can enter the cell and undergo subsequent conversion to acetyl CoA. Fatty acids are oxidized by β-oxidation within the mitochondria. The mitochondria are also the site of synthesis of NAD^+ and ATP. The conversion of HMG-CoA to mevalonate (by HMG-CoA reductase) and methyl farnesoate to JH III (by epoxidase) occurs within the endoplasmic reticulum, whereas all other steps appear to occur in the cytosol. Enzymes are shown in italics and known intermediates in heavy type. Cofactors are shown in light type.

epoxidase. The presence of $NADP^+$ also inhibits epoxidase activity. There have been only two conclusive studies on the epoxidase in the CA of insects (*Blaberus giganteus*: Hammock, 1975; *L. migratoria*: Feyereisen *et al.*, 1981c). In both, the authors were able to conclude that the epoxidase was a cytochrome P-450-linked enzyme, using NADPH as cofactor, and that it was found exclusively in the microsomal fraction. In fact, Feyereisen *et al.* (1981c) utilized sucrose density gradient centrifugation of CA homogenates to partially purify the enzyme, and in this case, it is quite clear that the P-450-linked epoxidase is associated with the endoplasmic reticulum. In both studies, methyl farnesoate was utilized as primary substrate and was converted with a high degree of efficiency into the corresponding epoxide (i.e., JH III). Thus, the CA of these two species contain appreciable quantities of the epoxidase. The epoxidase appears to be involved in the toxic action of the precocenes on CA (see Section 2.5.1), suggesting that the enzyme displays poor substrate specificity. In *S. gregaria* CA, it is able to epoxidize the dihomolog of farnesoic acid to JH (by way of dihomomethyl farnesoate) (Pratt and Tobe, 1974), whereas in *B. giganteus* CA, the epoxidase appears to utilize the unnatural geometric isomer (2*Z*,6*E*)-methyl farnesoate more efficiently than the natural 2*E*,6*E* isomer (Hammock and Mumby, 1978). The substrate specificity of the epoxidase in CA of *M. sexta* is unclear. It appears that the natural substrate for the enzyme in this species may be farnesoic acid and homologs, rather than methyl farnesoate (see Law, 1983), and this has led to the suggestion that the sequence of the terminal two steps in JH biosynthesis in the Lepidoptera may be reversed, i.e., farnesoic acid to epoxyfarnesoic acid to JH III, when compared to the situation in most other orders, i.e., farnesoic acid to methyl farnesoate to JH III (Fig. 7). For a more detailed discussion of the sequence of these enzymatic steps, see reviews by Law (1983), Tobe and Feyereisen (1983), and Schooley and Baker (1984).

The pathway of biosynthesis of JH III and the cellular location of the various enzymatic steps are shown in Fig. 7. Only JH III is detailed here because, to date, the higher homologs of JH have been identified only in the Lepidoptera (see Tobe and Feyereisen, 1983; Schooley and Baker, 1984; Schooley *et al.*, 1984). JH III can probably be regarded at the present as the principal JH of insects, and although identification of JH by unequivocal physicochemical techniques has been performed in only a few dozen species (see Schooley *et al.*, 1984), it is too early to conclude that all insect species, other than the Lepidoptera, use JH III as their principal hormone. The definitive identification of the JH of one of the largest orders, the Hemiptera, remains to be accomplished, despite the tentative identification of JH III (Bowers *et al.*, 1983) (see, however, Schooley *et al.*, 1984).

It is clear from Fig. 7 that the majority of the enzymatic reactions in JH biosynthesis occur in the cytosol, with only HMG-CoA reductase and 10,11-epoxidase confined to the membranes of the endoplasmic reticulum. We can only speculate on the significance of the spatial separation of the two enzymes from the cytosolic enzymes, but it is perhaps significant that the reductase may be important in rate limitation of JH biosynthesis (see Feyereisen *et al.*, 1981a), whereas the epoxidase may be involved in the exit of JH from the cell, since it is the last step in the biosynthetic pathway. Thus the availability of substrate may be regulated by way of either transport or binding proteins which facilitate the entry of HMG-CoA into the endoplasmic reticulum membrane system (e.g., Dautry-Varsat and Lodish, 1983). Similarly, the movement of methyl farnesoate into the reticular space before its conversion to JH III may facilitate the movement of the newly synthesized hormone out of the cell. It is known that newly synthesized JH moves rapidly out of the CA cells and that this movement appears to be governed by laws of physical diffusion (Tobe and Pratt, 1974b). Because diffusion rate is a function of membrane surface area (as well as factors such as concentration gradient and partition of the product between the membranes and the external milieux), proliferation of the endoplasmic reticulum as well as plasma membrane may facilitate the export of JH from the cells. CA of known high activity of *D. punctata* possess extensive endoplasmic reticulum and abundant microvilli-like projections of the plasma membrane (Johnson *et al.*, 1985; see also Section 1), whereas in inactive glands, little of the reticulum is evident. These observations are consistent with the possible role of the reticulum in export of JH from the cell. It is also notable that at least in those species which have been investigated using the RCA, JH appears to be released as soon as it is synthesized and there is no storage of hormone within the CA cells (*S. gregaria*: Tobe and Pratt, 1974b; *P. americana*: Pratt *et al.*, 1975a; *D. punctata*: Tobe and Stay, 1977). These observations are consistent with a role for the endoplasmic reticulum in the export of JH from the cell as soon as it is synthesized since the terminal step in biosynthesis is so intimately associated with these membranes.

In concluding this section, it is clear that much work remains to be done to show the significance of cellular localization of strategic enzymes in the regulation of JH biosynthesis. Biochemical studies on the isolated individual enzymes as well as their cytochemical localization will no doubt prove to be valuable in this regard.

2.3 SUBSTRATE UTILIZATION

Studies on the JH biosynthetic pathway have employed both cell-free homogenates of CA as well as intact CA isolated *in vitro* (and where noted in Table 3 *in vivo*) and from this work, we now have a body of data on the relative permeability of the CA to a range of intermediates in the pathway. Although mainly early and late intermediates have been used, such studies

TABLE 3
Incorporation of radiolabeled precursors into JH by intact CA

Precursor	JH	Species	Reference
Glucose	JH III	*Diploptera punctata*	Feyereisen *et al.* (1984)
Acetate	JH III	*Diploptera punctata*	Feyereisen *et al.* (1981a, 1984)
	JH III	*Periplaneta americana*	Dahm *et al.* (1976)
	JH II/III	*Manduca sexta*	Schooley *et al.* (1973)
	JH I/II[a]	*Hyalophora cecropia*	Metzler *et al.* (1971); Rode-Gowal *et al.* (1975)
Butyrate[a]	JH I/II	*Hyalophora cecropia*	Rode-Gowal *et al.* (1975)
Propionate	JH III	*Periplaneta americana*	Dahm *et al.* (1976)
	JH II	*Manduca sexta*	Schooley *et al.* (1973, 1976)
	JH I/II[a]	*Hyalophora cecropia*	Rode-Gowal *et al.* (1975); Peter and Dahm (1975); Peter *et al.* (1981)
	JH I/II	*Antherea pernyi*	Dahm *et al.* (1981)
	JH I/II	*Heliothis virescens*	Jennings *et al.* (1975a)
Mevalonate	JH III	*Diploptera punctata*	Feyereisen *et al.* (1981a)
	JH III	*Periplaneta americana*	Dahm *et al.* (1975)
	JH II/III	*Manduca sexta*	Schooley *et al.* (1973)
	JH I/II	*Hyalophora cecropia*	Peter *et al.* (1981)
Homomevalonate	JH II	*Manduca sexta*	Jennings *et al.* (1975b); Schooley *et al.* (1976)
Farnesoic acid	JH III	*Schistocerca gregaria*	Pratt and Tobe (1974); Tobe and Pratt (1974a, 1976)
	JH III	*Periplaneta americana*	Pratt *et al.* (1975b); Pratt and Weaver (1975)
	JH III	*Tenebrio molitor*	Weaver *et al.* (1980)
Bishomofarnesoic acid	JH I	*Schistocerca gregaria*	Pratt and Tobe (1974)
Epoxyfarnesoic acid	JH III	*Schistocerca gregaria*	Pratt and Tobe (1974)
Epoxybishomofarnesoic acid	JH I	*Schistocerca gregaria*	Pratt and Tobe (1974)
	JH I/II	*Hyalophora cecropia*	Metzler *et al.* (1972)

[a] *In vivo.*

may provide information on the relative importance of permeability of the CA as a potential factor in the regulation of JH biosynthesis. The radiolabeled precursors in Table 3 have been demonstrated to incorporate into JH, using intact CA.

It is surprising that such diverse substrates penetrate the CA and enter the substrate pools. However, several of the above-mentioned substrates (mevalonate and homomevalonate, for example) appear to undergo substantial dilution, suggesting that there are substantial permeability barriers to these substrates (see Schooley and Baker, 1984). Subsequent studies utilizing cell-free homogenates have in fact confirmed this interpretation (see e.g., Feyereisen *et al.*, 1981a). There have been numerous additional investigations on substrate utilization by intact CA but these have used unlabeled substrates in conjunction with radiolabeled methionine as the mass marker to monitor the effects of these substrates on rates of JH biosynthesis. For example, both farnesoic acid and farnesol have been found to stimulate rates of JH biosynthesis up to two- to threefold in *D. punctata* (Feyereisen *et al.*, 1981a,b). Although the CA are capable of utilizing this wide range of substrates for JH biosynthesis, it is likely that the natural endogenous substrates *in vivo* are acetate, derived from either glucose or fatty acid oxidation (to give acetyl CoA), in those species producing JH III, or propionate, derived from odd-chain fatty acids and/or amino acid oxidation (to give propionyl-CoA), in those species producing the higher homologs. The likelihood of the sesquiterpenoid acids serving as natural substrates *in vivo* is not great since these compounds could be expected to undergo rapid catabolism, probably by hydration of the double bonds and cleavage of the ester function. The high degree of efficiency with which the CA of *D. punctata* are able to utilize radiolabeled acetate in the absence of glucose for the generation of the carbon skeleton of JH III indicates that at least under these very special conditions, the CA of this species can use only acetate for the genesis of the skeleton (see Feyereisen *et al.*, 1984).

In vitro studies on the utilization of exogenous substrates by CA for the biosynthesis of JH have proved valuable in the investigation of the rate-limiting steps in hormone biosynthesis. For example, the early investigation of JH biosynthesis by CA of *S. gregaria* (Tobe and Pratt, 1974a) demonstrated that the terminal steps in JH biosynthesis (conversion of farnesoic acid to methyl farnesoate and thence to JH III; see Fig. 7) were not rate limiting in hormone biosynthesis because up to a 10-fold stimulation of biosynthesis was observed in the presence of farnesoic acid relative to spontaneous unstimulated values. The addition of farnesoic acid to incubation media apparently resulted in the bypass of rate-limiting steps. Subsequent investigations have revealed that the CA of other species are able to utilize exogenous farnesoic acid, with a concomitant stimulation in rates of JH biosynthesis (see above and Table 3). These studies allow us to conclude that the final two steps in JH biosynthesis are not rate limiting in any of the species studied and demonstrate the value of experiments with exogenous substrates. It is not necessary to use radiolabeled precursors provided that there is some means to monitor rates of JH biosynthesis. Most experiments have utilized the incorporation of radiolabeled methionine into the methyl ester function of JH as a mass marker to monitor overall rates of JH biosynthesis. In this way, it has been possible to assess the effects of exogenous precursors such as mevalonate, farnesol, and farnesoic acid on overall rates of JH biosynthesis (e.g., Feyereisen *et al.*, 1981a). A stimulation in the rate of JH biosynthesis in the presence of a given precursor would suggest that all enzymatic steps following the formation of the product in question are not rate limiting. Of course, it must be shown that the precursor does not encounter permeability barriers with respect to entering the endogenous precursor pool. This can only be demonstrated by examining the degree of dilution of the endogenous pool by the precursor. Finally, that the exogenous precursor is in fact a natural intermediate in the biosynthetic pathway should be confirmed, either by isolation of the endogenous compound from CA (as in the case of farnesoic acid from CA of *L. migratoria*: Hamnett *et al.*, 1981) or by demonstration of incorporation of the radiolabeled exogenous precursor into JH (Table 3).

Experiments involving the use of exogenous precursors have also contributed appreciably to our understanding of substrate specificity of the various enzymatic steps in biosynthesis (e.g., Pratt *et al.*, 1978, 1981). This approach allowed Baker and Schooley (1981) to examine the conversion of substrates by the HMG-CoA reductase in cell-free homogenates of CA of *M. sexta*, *S. nitens*, and *T. molitor* and to conclude that in fact the same enzyme is responsible for the conversion of HEG-CoA to homomevalonate as well as HMG-CoA to mevalonate. Similarly, Pratt *et al.* (1981) assessed the conversion of a wide range of substrate analogs by the *O*-methyl transferase of *L. migratoria* CA. From studies such as these, it appears that many of the enzymes of the JH biosynthetic pathway display relatively loose substrate specificities, although in a number of instances, preference for the correct geometric isomers and enantiomers does exist (see Peter *et al.*, 1979; Baker and Schooley, 1981; Pratt *et al.*, 1981).

2.4 RATE LIMITATION IN JH BIOSYNTHESIS

Exogenous substrates can clearly be utilized by CA or cell-free homogenates of CA for the biosynthesis of JH. Systematic studies using known or presumed intermediates in JH biosynthesis as the exogenous substrates have thus been important to our understanding of rate limitation in hormone biosynthesis. Although no single enzymatic step has been definitely established as the single point of rate limitation in the biosynthetic pathway, there is now evidence to suggest that more than one step may be involved in overall rate limitation, and this will be reviewed briefly below. The elucidation of the points of rate limitation in the biosynthetic pathway is of obvious physiological importance, since it is at such points that humoral and/or nervous signals regulating CA activity must ultimately act. Intracellular messengers such as cyclic nucleotides and ions may serve as transducers of the humoral and nervous signals impinging on the CA cells, and accordingly, it will be such intracellular messages which affect directly the strategic enzymatic steps regulating JH biosynthesis. Research in this area of CA physiology should prove most fruitful in the coming years.

The stimulation of JH biosynthesis in the presence of exogenous precursors such as farnesol (Feyereisen *et al.*, 1981a, 1984; Baker *et al.*, 1983a) and farnesoic acid (Pratt and Tobe, 1974; Tobe and Pratt, 1974a, 1976, and references in Table 3) suggests strongly that rate limitation occurs prior to the entry of these compounds into the biosynthetic pathway. In addition, the apparent suppression of synthesis of such precursors in the presence of micromolar concentrations of the exogenous precursors suggests a complex feedback loop operating on rate-limiting enzymes prior to this point in the pathway. Such suppression was inferred from a molar incorporation ratio approximating 1, with respect to incorporation of [^{3}H]farnesoic acid and [^{14}C]methionine into JH III (see Tobe and Pratt, 1974a: *S. gregaria*; Pratt and Bowers, 1977: *P. americana*; Weaver *et al.*, 1980: *T. molitor*). Stimulation of JH biosynthesis has also been observed in the presence of mevalonate and homomevalonate (Dahm *et al.*, 1976; Schooley *et al.*, 1976; Feyereisen *et al.*, 1981a), although the degree of stimulation does not appear to be as appreciable as that observed in the presence of the later precursors. Nonetheless, these data do suggest that a major point of rate limitation occurs before or at the formation of mevalonate.

Sterol biosynthesis in the vertebrate hepatic system appears to be regulated primarily through the activity of HMG-CoA reductase (see Rodwell *et al.*, 1976) although this regulation may be complex, involving a variety of sterols (see Brown and Goldstein, 1980; Kandutsch *et al.*, 1978). Enzyme activity appears to be subject to multivalent feedback regulation (Brown and Goldstein, 1980). A similar mechanism has been suggested to regulate in part JH biosynthesis in the CA of *D. punctata* (Feyereisen *et al.*, 1981a). There is also recent evidence that nonsterol isoprenoid intermediates (such as farnesol and farnesoic acid) can influence the activity of the earlier rate-limiting steps, as determined from carbon flow using [^{14}C]acetate (Feyereisen *et al.*, 1984), although the precise site of action remains to be defined. In vertebrates, HMG-CoA reductase is interconvertible between a phosphorylated, inactive form and a unphosphorylated, active form (Keith *et al.*, 1979), and there is evidence that a similar mechanism may be operative in insect CA (Feyereisen *et al.*, 1981a; Monger and Law, 1982) to control the catalytic activity of this enzyme. Although the data are limited it appears that rate limitation in JH biosynthesis occurs early in the pathway (prior to mevalonate) and that it cannot be strictly associated with a single enzymatic step. Feyereisen *et al.* (1981a) have suggested that CA activity and hence overall JH biosynthesis are therefore regulated by the interaction of at least three different mechanisms: (1) control of CA cell number; (2) coordinated control of several successive enzymes; and (3) control of carbon flow rate by multivalent feedback. More recent studies on regulation of CA cell number (Szibbo and Tobe, 1981a,b, 1983; Szibbo *et al.*, 1982; Tobe *et al.*, 1984b) and of carbon flow (Feyereisen *et al.*, 1984) support this interpretation.

2.5 INHIBITION OF JH BIOSYNTHESIS

Inhibitors of JH biosynthesis have been of major interest to insect endocrinologists/biochemists during the past decade because of their obvious potential importance as insect control agents. Studies on the biosynthetic pathway, on the individual enzymatic steps, and on the substrate specificities of these enzymes have all provided a sound framework on which to base subsequent work on specific inhibitors of JH biosynthesis. There have been several recent reviews on specific inhibitors of JH biosynthesis (see Bowers, 1983), and no attempt will be made here to review the topic in detail. Rather, we wish to mention only two specific groups of inhibitors of JH biosynthesis because of their established mechanisms of action on the CA and the insights which they provide on the physiology of the CA.

2.5.1 *Precocenes (ageratochromenes)*

These compounds were originally isolated from the house plant *Aegeratum houstonianum* by Alertsen (1955) and Kasturi and Manithomas (1967) and were subsequently demonstrated to cause precocious metamorphosis in *O. fasciatus* (Bowers *et al.*, 1976). For this reason, Bowers renamed these compounds *precocenes*, and they have now been shown to cause similar disruptive metamorphic effects, particularly in selected members of the Hemiptera and Orthoptera. Pratt and colleagues (Pratt and Bowers, 1977; Pratt *et al.*, 1980) have provided the bulk of the data on the mode of action of the precocenes. The species of choice for such studies has been *L. migratoria* because of its apparent sensitivity to precocene treatment (despite the use of very high doses for *in vivo* experimentation; see Pener *et al.*, 1978). Using CA maintained *in vitro*, Pratt *et al.* (1980) demonstrated that radiolabeled precocenes are rapidly converted to the 3,4-dihydrodiols; the macromolecules of the CA also become extensively labeled. These findings led Pratt and colleagues to suggest that the precocenes serve as substrate for the epoxidase within the CA. The formation of highly reactive precocene epoxides results in extensive alkylation of macromolecules, followed by cell death and ultimately the destruction of much of the CA tissue (e.g., Feyereisen *et al.*, 1981d). Thus, the precocenes appear to be "suicide substrates" for the epoxidase within the CA—the formation of the product ultimately results in the destruction of the enzyme catalyzing its formation (as well as other macromolecules).

Since the precocenes appear to exert their effects directly on the CA, by disrupting JH biosynthesis (Pratt and Bowers, 1977; Pratt *et al.*, 1980), they should be regarded as *proallatocidins* (Pratt *et al.*, 1980; Feyereisen *et al.*, 1981d) and not as antiallatotropins (Bowers and Martinez-Parado, 1977) since the latter term implies a compound which interferes with the action of allatotropins, i.e., stimulators of CA activity (see Section 4). It is unfortunate that the term *antiallatotropin* has now entered the literature; such usage is inappropriate and should be discontinued.

As noted above (Section 2.2.2.2), the epoxidase may show loose substrate specificity and it is clearly for this reason that precocenes can be converted to the reactive epoxide intermediates. The ultimate value of the precocenes as insect control agents remains speculative, but these precocene studies do demonstrate the possibility for the rational development of suicide substrates and other novel approaches to the inhibition of JH biosynthesis. Such studies also reveal conclusively the value of basic biochemical knowledge in the development of rational insect control methods.

2.5.2 *Compactin (mevastin)*

Compactin is a fungal metabolite (Endo, 1981), known for its ability to competitively inhibit HMG CoA reductase in vertebrate systems. Compactin and its homolog mevinolin have been used extensively in humans as hypocholesterolemic agents, although the precise effect of these compounds in reducing cholesterol levels remains unclear. Compactin is an effective inhibitor of JH biosynthesis in intact CA of *M. sexta* (Monger *et al.*, 1982) and *P. americana* (Edwards and Price, 1983), with an ID_{50} of 1-10 nmol. Recent studies with CA of *D. punctata* also have demonstrated that compactin inhibits JH biosynthesis (Tobe and Clarke, unpublished) although the ID_{50} appears to be somewhat higher. However, incubation parameters appear to be quite important in assessing the effects of compactin. For example, the time course of inhibition of hormone biosynthesis probably varies between species; in *D punctata*, a 2-h preincubation is necessary to obtain high degrees of inhibition and this may result from low permeability of compactin and the size of endogenous pools of both mevalonate and later biosynthetic intermediates. In this latter respect, compactin and related HMG-CoA reductase inhibitors may prove useful in studies on rate limitation and carbon flow in JH biosynthesis. However, compactin appears to be relatively ineffective *in vivo* following either injection or topical application for "chemical allatectomy" (Monger *et al.*, 1982) and repeated injections were necessary to obtain any anti-JH effect in *M. sexta.* Thus the potential usefulness of compactin and related compounds as *in vivo* inhibitors of JH biosynthesis remains to be established. Nonetheless, it should prove to be useful biochemical probe.

3 Assays for juvenile hormone biosynthesis

3.1 RADIOCHEMICAL ASSAY: BACKGROUND, THEORY, AND PRACTICAL DEPLOYMENT

The pioneering work from Röller's laboratory provided much of the impetus for the development of the RCA (Section 2). In particular, the observation that radiolabeled methionine served as the principal methyl donor for the methyl ester function of JH *in vivo* proved to be of fundamental importance (Metzler *et al.*, 1971), and subsequent studies confirmed that *in vitro*, the *S*-methyl function of [*methyl*-^{14}C]methionine was incorporated without appreciable dilution into the methyl ester moiety of JH (Judy *et al.*, 1973a, 1975; Pratt and Tobe, 1974; Tobe and Pratt, 1974a).

3.1.1 *Molar incorporation ratio*

Two different approaches have been employed to establish this. One, used by Judy and colleagues, involved the analysis by mass spectrometry of JH biosynthesized *in vitro* in the presence of [*methyl*-^{14}C]methionine at high isotopic abundance; in this way, the percentage contribution of the radiolabeled methionine to the methyl ester function in the isolated JH could be estimated directly. In incubations of CA from *M. sexta*, 90–100% of the methyl ester was found to derive from the radiolabeled methionine in the medium (Judy *et al.*, 1973a), whereas in incubations of *T. molitor* CA, about 70% of the methyl ester was derived from the radiolabel (Judy *et al.*, 1975). The low incorporation percentage for *T. molitor* glands is probably a reflection of the long duration of the incubations (1–2 weeks) in suboptimal medium (see below). Alternatively, because of the very low productivity of the CA, endogenous methionine, present within the CA at the time of extirpation, may have served as methyl donor for an appreciable portion of the biosynthesized JH.

The second approach utilized the observation that CA of *S. gregaria* can efficiently incorporate exogenous farnesoic acid into JH III, even under circumstances in which the CA are virtually inactive (Pratt and Tobe, 1974). Using CA from female *S. gregaria* showing only very low rates of spontaneous JH biosynthesis (see Tobe and Pratt, 1975a) and [^{3}H]farnesoic acid as the mass marker to monitor overall rates on JH biosynthesis, the relative incorporation of [methyl-^{14}C]methionine was determined by double-label liquid scintillation spectrometry. The ratio of incorporation approached 1:1 within 2 h of incubation, demonstrating that the CA use methionine almost exclusively as the universal methyl donor, despite the presence of other potential carbon donors in the incubation medium (Tobe and Pratt, 1974a). In these experiments, CA of low spontaneous activity were employed to minimize endogenous production of farnesoic acid. It is noteworthy that both farnesol and farnesoic acid can suppress incorporation of [^{14}C]acetate into JH III by the CA of *D. punctata* (Feyereisen *et al.*, 1981a, 1984), and if a similar situation pertains to *S. gregaria*, it may not prove necessary to adhere strictly to the requirement for CA of low activity in order to establish stoichiometry.

The results of Tobe and Pratt (1974a) thus established that radiolabeled methionine could be used to monitor JH biosynthesis in short-term incubations, provided that appropriate allowance was made for the time required for equilibration of the isotope within the CA. Subsequent studies have demonstrated that an equilibration time of 30–60 min is sufficient in most species (see below). Table 4 is a compilation of those studies which have demonstrated JH biosynthesis by CA *in vitro* in different species.

The short equilibration times noted above imply that the endogenous methionine pool within the CA is small and is rapidly diluted by exogenous radiolabeled methionine in the incubation medium. This interpretation is supported by the observation that radiolabeled JH III can be detected in the medium within 10 min of the addition of radiolabeled methionine (Tobe and Pratt, 1974a: *S. gregaria*) and a similar rapid release of radiolabeled JH III also occurs in *D. punctata* (Tobe and Stay, 1977, and unpublished). Estimates

TABLE 4
Identity and quantity of JHs produced by CA *in vitro*

Species	JH produced	Stage	Range (pmol h^{-1} CA pair^{-1})	Reference
Lepidoptera				
Hyalophora cecropia	JH I	L, P	N.D.[a]	Dahm *et al.* (1976)
	JH II	L	N.D.	Dahm *et al.* (1976)
	JH I	P	N.D.	Röller and Dahm (1970)
	JH I acid	A♂	N.D.	Peter *et al.* (1981)
	JH II acid	A♂	N.D.	Peter *et al.* (1981)
Manduca sexta	JH I	L	N.D.	Dahm *et al.* (1976)
	JH II	L	N.D.	Dahm *et al.* (1976)
	JH III	L	1.04–2.6	Granger *et al.* (1982a)
	JH I/II/III	L	0–2.1	Kramer and Law (1980a)
	JH I/II/III	L	<0.1–1.3	Sparagana *et al.* (1984)
	JH III	L	0.06–0.4	Granger *et al.* (1979)
	JH II/III	A♀	0.1–6	Kramer and Law (1980a)
	JH II	A♀	N.D.	Judy *et al.* (1973a)
	JH III	A♀	N.D.	Judy *et al.* (1973a)
Heliothis virescens	JH I	A♀	N.D.	Jennings *et al.* (1975a)
	JH II	A♀	N.D.	Jennings *et al.* (1975a)
Galleria mellonella	JH II	L	N.D.	Dahm *et al.* (1976)

TABLE 4 (*continued*)

Species	JH produced	Stage	Range (pmol h^{-1} CA pair^{-1})	Reference
Dictyoptera				
Periplaneta americana	JH III	A♂	N.D.	Muller *et al.* (1974)
		A♂	N.D.	Dahm *et al.* (1976)
		A♀	4–25	Pratt *et al.* (1975a)
		A♀	0.2–2.6	Weaver and Pratt (1977)
Periplaneta fuliginosa	JH III	A	N.D.	Dahm *et al.* (1976)
Blaberus discoidalis	JH III	A♀♂	N.D.	Dahm *et al.* (1976)
Diploptera punctata	JH III	L	<0.1–2.8	Szibbo *et al.* (1982)
		A♀	5–120	Tobe and Stay (1977) Tobe (1980)
		A♂	2–7	Tobe *et al.* (1979)
Nauphoeta cinerea	JH III	A♀	0.1–3	Lanzrein *et al.* (1978)
Leucophaea maderae	JH III	A♀	N.D.	Koeppe *et al.* (1980)
Isoptera				
Macrotermes subhyalinus	JH III	A	N.D.	Lanzrein *et al.* (1977)
Zootermopsis angusticolla	JH III	A	0.1–0.6	Greenberg and Tobe (1985)
Orthoptera				
Schistocerca gregaria	JH III	L	0.5–4.5	Injeyan and Tobe (1981)
		A♀	N.D.	Pratt and Tobe (1974)
		A♀	<0.1–25	Tobe and Pratt (1975a)
		A♂	<0.1–12	Avruch and Tobe (1978)
Schistocerca vaga	JH III	A♀	N.D.	Judy *et al.* (1973b)
Locusta migratoria	JH III	L	0.2–0.5	Couillaud *et al.* (1984)
		A♀	N.D.	Hamnett and Pratt (1978)
		A♀	1–50	Girardie *et al.* (1981)
			4–35	Couillaud *et al.* (1984)
Gastrimargus africanus	JH III	A♀	N.D.	Judy *et al.* (1973b)
Melanoplus sanguinipes	JH III	A♀	1–20	McCaffery and McCaffery (1983)
Coleoptera				
Leptinotarsa decemlineata	JH III	A♀	N.D.	Dahm *et al.* (1976)
		A♀	<0.1–3	Kramer (1978a)
		A♀	<0.1–6	Khan *et al.* (1982a)
		A♂	2–4	Khan *et al.* (1982a)
Tenebrio molitor	JH III	A♀	N.D.	Judy *et al.* (1975)
			0.1–4	Weaver *et al.* (1980)
Hymenoptera				
Bombus terrestris	JH III	A♀	2.6–17.4	Röseler and Röseler (1978)
Polistes gallicus	JH III	A♀	0.1–3.5	Röseler *et al.* (1980)

[a] N.D., Not determined.

of the endogenous methionine pool size in CA of *D. punctata*, using preincubation of glands in medium containing only [*methyl*-^{14}C]methionine, i.e., no unlabeled methionine, followed by a chase in methionine-free medium, revealed that the endogenous pool size is probably no greater that 15 pmol per pair (Feyereisen *et al.*, 1984). In methionine-free medium, JH biosynthesis ceased within 1 h, indicating that the intraglandular pool of methionine was exhausted within this time. A similar pool size can be implied from the quantity of JH biosynthesized by *S. gregaria* CA in methionine-free medium in the presence of farnesoic acid (see Tobe and Pratt, 1974a).

3.1.2 *Methionine concentration for JH biosynthesis*

The RCA has routinely employed L-methionine concentrations of 100–300 μM. This range of values was derived from the original study of Tobe and Pratt (1974a) which demonstrated that under conditions of farnesoic acid-stimulated JH biosynthesis, maximal rates of biosynthesis were observed over this entire range. Thus, L-methionine concentration appears to be relatively uncritical with respect to JH biosynthesis, and subsequent studies have shown a similar situation in *T. molitor* (Weaver *et al.*, 1980), *D. punctata* (Feyereisen *et al.*, 1981b), and *L. decemlineata* (Khan *et al.*, 1982a). However, medium 199 contains 100 μM L-methionine and in those studies employing this medium, it was clearly impossible to ascertain the effect of L-methionine concentrations of less than 100 μM. Recent studies utilizing L-methionine-free medium 199 supplemented with L-[*methyl*-^{3}H]methionine have revealed that maximal rates of biosynthesis by CA of *D. punctata* can be realized at much lower concentrations of L-methionine (Tobe and Clarke, 1985), with values of 30–40 μM being sufficient. These experiments demonstrated that the overall rate of JH biosynthesis did not appreciably change the methionine requirement—both high and low activity glands displayed similar dose responses for L-methionine. Typical dose–response curves are shown in Fig. 8. Much reduced L-methionine requirements for optimal JH biosynthesis have also been recently described for the termite, *Zootermopsis angusticolla* (Greenberg and Tobe, 1985).

The necessity for only a low concentration of labeled L-methionine to monitor JH biosynthesis is noteworthy for several reasons. In conjunction with the use of methionine-free medium supplemented with [*methyl*-^{3}H]methionine, such a system permits the production of radio-labeled JH III of high isotopic enrichment. This can improve the sensitivity of the assay appreciably (by one to three orders of magnitude, depending upon the specific radioactivity of the radiolabeled methionine). Second, the cost of the assays can be dramatically reduced as a result of both the lower cost of the tritiated isotope and the reduced requirements for methionine. Finally, it is

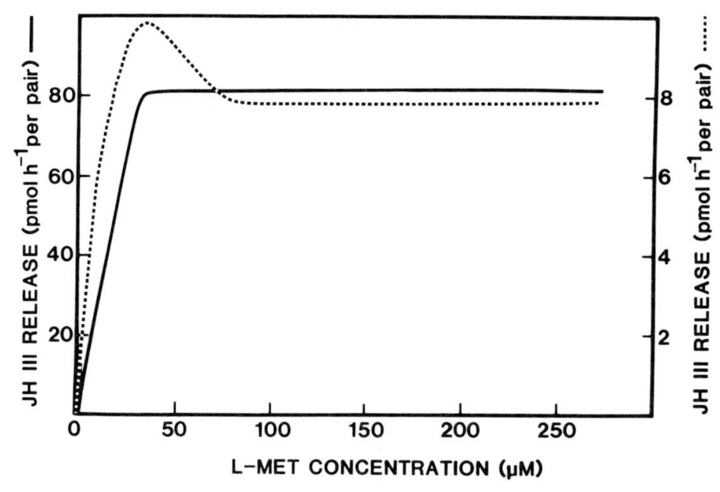

Fig. 8. Dose–response relationship for JH release *in vitro* as a function of L-methionine concentration in the incubation medium for CA of day 5 (solid line) and day 11 (dotted line) of female *D. punctata*. (Data from Tobe and Clarke, 1985.)

possible to biosynthesize JH of high specific activity, using high specific activity [*methyl* -^{3}H]methionine for studies on both receptors and binding proteins. This biosynthesized JH is of the natural 10*R* enantiomer exclusively (Tobe, Baker, and Schooley, unpublished). Despite the very high specific radioactivity of the methionine, it is possible to realize final methionine concentrations which support maximal rates of JH biosynthesis, provided that the volume of the incubation medium is somewhat reduced.

3.1.3 *Sensitivity of the radiochemical assay*

The final specific radioactivity of JH biosynthesized in the presence of radiolabeled methionine should approximate the specific radioactivity of the precursor radiolabeled methionine. Assuming that the endogenous methionine pool in the CA is very small and that this pool is rapidly diluted by the exogenous radiolabeled precursor, the final specific radioactivity of the JH can safely be assumed to be identical to that of the starting compound. The sensitivity of the RCA is directly related to the specific radioactivity of the precursor methionine. Because its final specific radioactivity is a function not only of the initial specific radioactivity of the exogenous methionine but also of the methionine concentration in the medium, to achieve maximum sensitivity in the assay it is clearly necessary to utilize medium which does not contain any exogeneous methionine. Under these circumstances, the sensitivity of the RCA, using [*methyl*-^{14}C]methionine of specific radioactivity of 60 mCi/mmol (2GBq/mmol), is about 0.1–0.2 pmol. This value assumes that the minimum significant level of radioactivity that can be detected is twice that of background. [*Methyl*-^{3}H]methionine is available at much higher specific radioactivity, and accordingly its use can dramatically improve the sensitivity of the RCA. Use of [^{3}H]methionine of the highest specific radioactivity currently available, 80 Ci/mmol (3.1 TBq/mmol), improves the sensitivity of the assay by greater than three orders of magnitude to about 0.1 fmol.

3.1.4 *Parameters for deployment of the radiochemical assay*

In order to use the RCA for the monitoring of JH biosynthesis by CA *in vitro*, several parameters must be adequately defined before the rates of JH biosynthesis can be assumed to be accurate. For each species to be assayed, all the following criteria should be met.

3.1.4.1 *Identification of biosynthesized JH in vitro* *a. Isomers*. Because CA in general produce only small quantities of JH *in vitro* (and presumably *in vivo*), generally in the range of 0.1–100 pmol per pair CA per hour, it is essential to demonstrate that the biosynthesized JH in the medium is of genuine biosynthetic origin *in vitro* rather than a contaminant of either chemical or biological origin within the CA or in the incubation medium. The identification procedures employed by Zoecon have established unequivocally that the JHs detected in *in vitro* incubations of CA are of biological origin. Essentially, the Zoecon group has shown that the biosynthesized JHs are specific geometric and optical isomers rather than a mixture. The 2*E*,6*E* geometric isomer has been established to be the natural isomer of the known JHs (Judy *et al.*, 1973a,b, 1975; Jennings *et al.*, 1975a,b), and the natural enantiomer has similarly been established to be 10*R* in the case of JH III (only one asymmetric carbon) or 10*R*,11*S* in the case of the higher homologs (two asymmetric carbons). Elaborate precautions are taken (Schooley, 1977; Schooley and Baker, 1984) and hence the chirality of the products from *in vitro* incubations of CA establishes their biosynthetic origin. The identification of the natural enantiomers of JH in incubations of CA *in vitro* has been performed in only a limited number of species, including *M. sexta*, *S. vaga*, *Heliothis virescens*, *T. molitor*, and *L. migratoria* (Judy *et al.*, 1973a,b, 1975; Jennings *et al.*, 1975a; Hamnett *et al.*, 1981). However, it does not seem unreasonable to assume that the natural JHs, in those species known to produce any of the presently identified JHs, are the 10*R* or 10*R*,11*S* enantiomers.

Despite the relative dearth of information on the chirality of JHs biosynthesized by other species, the chemical identity of biosynthesized JHs has been established using a combination of high-performance liquid chromatography (HPLC) and mass spectrometry (MS), sometimes in conjunction with chemical derivatization, for a much wider range of species. The description of the chemical procedures necessary for such identification is beyond the scope of the article, and the reader is referred to the reviews by Schooley (1977) and Schooley and Baker (1984) and the detailed procedure of Bergot *et al.* (1981a).

b. Detection limit. The detection limit of the methodology used should also be considered for identification. From Table 4, it is apparent that with the exception of the Lepidoptera, JH III is the only JH detected in either hemolymph or whole bodies. However, the limit of detection for JHs other than JH III by conventional techniques is a function not only of the concentration of the other JHs but also of JH III: in samples in which the JH III concentration is high, the possibility of detecting small quantities of other JHs is compromised because of limitations in dynamic range and signal-to-noise ratio. Hamnett and Pratt (1978) incubated CA of *P. americana* in medium containing high-specific-radioactivity methionine, and used subsequent analysis by gas–liquid chromatography (GLC) with a radioactivity monitor to establish that JH III is produced by these CA *in vitro* at a level of at least 10^5 times greater than other JHs. These authors were able to conclude that JH III is the exclusive product of the CA *in vitro.* A different approach has recently been utilized to demonstrate that JH III is the exclusive JH in the hemolymph of *D. punctata* (Tobe *et al.*, 1985). Adult females of this species have extremely high titers of JH III in the hemolymph (see Section 5) (up to 6 μM). By using an appropriate internal standard (see Bergot *et al.*, 1981a) and separation of the JH III from the putative JH I and II by HPLC, followed by analysis by derivatization, HPLC, and MS, it was possible to demonstrate that the higher JH homologs were present at levels at least 5×10^4 times lower than that of JH III (Tobe *et al.*, 1985). Accordingly, it is safe to assume that at least in these two species, JH III is the exclusive product of the CA, *in vitro* and *in vivo.*

c. Verification of in vivo and in vitro products of CA. Although the identification of the products of CA incubation *in vitro* has provided no surprises with respect to the known JHs, it can be argued that CA incubated *in vitro* might biosynthesize JHs different from those biosynthesized *in vivo.* This may occur as a result of early substrate availability, causing a modification in the biosynthesized product. In fact, experiments utilizing early precursors of JH such as propionate and mevalonate have revealed that the ratio of JH I: JH II acid produced *in vitro* by CA of *H. cecropia* can be modified, depending upon the relative concentration of these precursors (Peter *et al.*, 1981) (see Schooley and Baker, 1984; and Section 2.2). For this reason, it is also essential that the identity of JHs *in vivo*, isolated from the hemolymph or whole bodies, be established. To date, such identification has been performed only in a very few species (see Table 5). It is clear from Table 5 that the exclusive JH in species other than those of the Lepidoptera is JH

TABLE 5
JH identity and titer, determined by physicochemical methods

Species	JH	Stage[a]	Titer[b]	Reference
Lepidoptera				
Hyalophora cecropia	JH 0	E	0.02–0.13 ng/g	Bergot *et al.* (1981b)
		A	24 ng/g	Schooley *et al.* (1984)
	JH I	E	0.02–0.50 ng/g	Bergot *et al.* (1981b)
		L	0.15 ng/g	Dahm *et al.* (1981)
	JH II	E	0.02–0.05 ng/g	Bergot *et al.* (1981b)
		L	0.53 n*M*	Dahm *et al.* (1976)
		L	0.06 n*M*	Dahm *et al.* (1981)
Manduca sexta	JH 0	E	1.6–2.7 ng/g	Bergot *et al.* (1981b)
	JH I	E	0.05–1.47 ng/g	Bergot *et al.* (1981b)
		L	0.07–1.7 n*M*	Bergot *et al.* (1981a)
			<0.01–2.1 n*M*	Peter *et al.* (1976)
		A	0.02–0.34 n*M*	Peter *et al.* (1976)
	JH II	E	N.D.–0.013 ng/g	Bergot *et al.* (1981b)
		L	0.04–3.9 n*M*	Bergot *et al.* (1981a)
			<0.01–3.9 n*M*	Peter *et al.* (1976)
		A	0.14–0.59 n*M*	Peter *et al.* (1976)
	JH III	L	N.D.–0.15 n*M*	Bergot *et al.* (1981a)
			<0.01–0.45 n*M*	Peter *et al.* (1976)
		A	<0.08–0.41 n*M*	Peter *et al.* (1976)
	4-Methyl JH I	E	1.20–1.95 ng/g	Bergot *et al.* (1981b)
Diatraea grandiosella	JH I	L	N.D.–9.5 n*M*	Bergot *et al.* (1976)
		A	0.3 ng/g	Bergot *et al.* (1976)
	JH II	L	N.D.–14 n*M*	Bergot *et al.* (1976)
		A	0.9 ng/g	Bergot *et al.* (1976)
	JH III	L	1.9–5.3 n*M*	Bergot *et al.* (1976)
		A	0.9 ng/g	Bergot *et al.* (1976)
Attacus atlas	JH II	A	4–17 ng/g	Schooley *et al.* (1984)
Trichoplusia ni	JH II	L	—	Cited in Schooley *et al.* (1984)
Dictyoptera				
Diploptera punctata	JH III	L	132–210 n*M*	Tobe *et al.* (1984a, 1985)
		A	190–6000 n*M*	Tobe *et al.* (1984a, 1985)
Blatta orientalis	JH III	A	3.1 ng/g	Trautmann *et al.* (1976)
Leucophaea maderae	JH III	A	3.5 ng/g	Trautmann *et al.* (1976)
Nauphoeta cinerea	JH III	A	6.1 ng/g	Trautmann *et al.* (1976)
Orthoptera				
Schistocerca gregaria	JH III	A	0.5 ng/g	Trautmann *et al.* (1976)
Locusta migratoria	JH III	L	1.5–3.4 n*M*	Bergot *et al.* (1981c)
		A	N.D.–90 ng/g	Rembold (1981)
		A	150 n*M*	Bergot *et al.* (1981c)

TABLE 5 (*continued*)

Species	JH	Stage[a]	Titer[b]	Reference
Teleogryllus commodus	JH III	E	0.05 ng/g	Loher *et al.* (1983)
	JH III	L	N.D.–9 n*M*	Loher *et al.* (1983)
	JH III	A	11 n*M*–241 n*M*	Loher *et al.* (1983)
Teleogryllus oceanicus	JH III	A	26 n*M*	Loher *et al.* (1983)
Taeniopoda eques	JH III	A	327–387 n*M*	Loher *et al.* (1983)
Coleoptera				
Leptinotarsa decemlineata	JH III	L	N.D.–9 n*M*	de Kort *et al.* (1981, 1982)
		P	N.D.–0.5 ng/g	de Kort *et al.* (1981, 1982)
		A	4.5–330 n*M*	de Kort *et al.* (1981, 1982)
		A	11.1 ng/g	Trautmann *et al.* (1976)
Tenebrio molitor	JH III	A	N.D.–7.4 ng/g	Trautmann *et al.* (1976)
Melolontha melolontha	JH III	A	4.1 ng/g	Trautmann *et al.* (1976)
Diabrotica undecimpunctata	JH III	A	—	Schooley *et al.* (1984)
Hymenoptera				
Apis mellifera	JH III	A	2.8 ng/g	Trautmann *et al.* (1976)
		L	N.D.–50 ng/g	Rembold and Hagenguth (1981)
		P	N.D.–50 ng/g	Rembold and Hagenguth (1981)
		A	1–10 ng/g	Rembold and Hagenguth (1981)
Diptera				
Drosophila hydei	JH III	L	0.3–0.5 ng/g	Klages *et al.* (1981)
		P	0.08 ng/g	Klages *et al.* (1981)
		A	2.3 ng/g	Klages *et al.* (1981)
Aedes aegypti	JH III	L	0.3 ng/g	Baker *et al.* (1983b)
		A	1.7–3.3 ng/g	Baker *et al.* (1983b)
Thysanura				
Thermobia domestica	JH III	L	0.3 ng/g	Baker *et al.*, cited in Schooley *et al.* (1984)
		A	0.4–1.4 ng/g	Baker *et al.*, cited in Schooley *et al.* (1984)
Hemiptera				
Megoura viciae	JH III	A	0.15 ng/g	Hardie *et al.*, cited in Schooley *et al.* (1984)

[a] Stages: E, embryos; L, larvae; P, pupae; A, adults.
[b] N.D., Not detected.

III. Although JH III may prove to be the predominant JH in many species, we must bear in mind that the JHs in fewer than 20 species have been identified by unequivocal physicochemical methodology and in view of the million-plus species of insects, it is dangerous to extrapolate. In addition, the unequivocal identification of the JH of the Hemiptera, one of the larger orders, remains to be determined. Nonetheless, in those species in which the JHs biosynthesized *in vitro* and the JHs present *in vivo* have been identified, there is no discrepancy and we can probably assume that the JHs biosynthesized *in vitro* do in fact reflect the JHs found *in vivo*.

3.1.4.2 *De novo biosynthesis vs storage of JH* It is essential to demonstrate that JH released into the incubation medium has been biosynthesized *de novo* rather than stored and released. This does not pose a problem when using the RCA because the incorporation of radiolabeled methionine is monitored and accordingly, all radiolabeled JH must have been biosynthesized *de novo* during the incubation. JH does not appear to be stored by the CA of several species (*S. gregaria*: Tobe and Pratt, 1974b; *P. americana*: Pratt *et al.*, 1975a; *D. punctata*: Tobe and Stay, 1977; Feyereisen *et al.*, 1981b) and is released as soon as it is synthesized. However, this phenomenon must be confirmed for each species, particularly if RIA is used to monitor JH biosynthesis since RIA measures total quantity of JH in the incubation medium, irrespective of radiolabel.

The possibility of storage of JH within the CA is important not only in the context of confirmation of the CA as the site of JH biosynthesis but also in terms of the role of the CA and JH in metamorphosis and reproduction and in understanding how the CA are regulated, i.e., whether regulation of release is an important component of CA regulation. JH may be retained in CA by certain fixation regimes (see Section 1.4.3), but ordinarily it is not retained as a visible product in fixed tissue at the level of either light or electron microscopy. Bioassay techniques have been employed to determine the JH content of CA of several species, including *H. cecropia* (Röller and Dahm, 1970) and *L. migratoria* (Johnson and Hill, 1973). Only in the latter case was JH detected in the CA (of adult males), but these authors concluded that there was no appreciable storage of the hormone. Using the RCA, it has been demonstrated that newly biosynthesized JH is rapidly released from the CA, irrespective of the overall rates of JH biosynthesis (*S. gregaria*: Tobe and Pratt, 1974b; *P. americana*: Pratt *et al.*, 1975a; *D. punctata*: Tobe and Stay, 1977). This observation arose from the observed precise correlation between the quantity of JH retained in the CA and that released into the medium, and permitted the conclusion that at least in these species, there is no appreciable storage of JH in the CA. The study of Tobe and Pratt (1974b) also provided some insights into the dynamics of JH release. These authors concluded that the rate of release of JH was first order with respect to the intraglandular content of JH and that the CA release almost three times their steady-state intraglandular JH content each hour (this value is the "release coefficient" described by Tobe and Pratt, 1974b). The available evidence all suggests that the CA do not store appreciable quantities of JH, and accordingly, the vast majority of JH isolated from *in vitro* incubations of CA must represent newly biosynthesized hormone, rather than stored product. We still know very little about the mechanism of JH release from the CA, but as noted in Section 2.2.2, the final stage in release of newly synthesized JH may be intimately associated with the epoxidation of methyl farnesoate in the endoplasmic reticulum to yield JH III. The presence of JH within the endoplasmic reticulum may facilitate its exit from the CA cells.

3.1.5 *Incubation media*

The suitability of the incubation medium and its ability to support rates of JH biosynthesis that are an accurate reflection of those *in vivo* are prime considerations for all *in vitro* experimentation. To this end, it is necessary to establish a number of parameters including optimal pH, salt concentrations, and methionine concentrations, among others. Different workers have employed different media, including Grace's, Landereau's, TC 199, and MEM (Minimum Essential Medium), but with few exceptions, there has been little rationale for the selection of a given medium. However, the study of Weaver *et al.* (1980) did establish conclusively that for CA of *T. molitor*, TC 199, MEM, and fortified Hank's saline provided maximal rates of biosynthesis, whereas the media of Grace, Schneider, or Lender–Duveau–Hagege were consistently inferior. It is unfortunate that no systematic survey of incubation media has been conducted for other species. However, the observations from farnesoic acid stimulation may be instructive in this regard; we have noted previously that the addition of farnesoic acid to the incubation medium consistently results in the stimulation in rates of JH biosynthesis by CA *in vitro*. Depending upon the species and stage, the degree of stimulation (relative to spontaneous rates of biosynthesis) is between 1.2 and 10. In *S. gregaria* and *D. punctata* adult females, the lowest degree of stimulation is in the 1.2–1.5 range, this being observed in CA synthesizing JH spontaneously at near-maximal rates (Tobe and Pratt, 1974a, 1976; Feyereisen *et al.*, 1981b). Thus in these CA, JH biosynthesis is only slightly stimulated by farnesoic acid. Farnesoic acid is believed to stimulate JH biosynthesis because it enters the biosynthetic pathway after the rate-limiting steps; accordingly, the rates of biosynthesis observed in the presence of farnesoic acid probably represent the maximal rates of which the CA are capable. Thus, if spontaneous rates approach those observed in the presence of farnesoic acid (which is clearly the case in the above two examples), it is likely that the CA are functioning optimally *in vitro*. In both of the above examples, medium 199 was employed. It would be unwise to extrapolate from these data that medium 199 is suitable for all species; nonetheless, it has been employed successfully for the CA of *S. gregaria* (Tobe and Pratt, 1975a, 1976), *P. americana* (Pratt *et al.*, 1975a), *D. punctata* (Tobe and Stay, 1977), *T. molitor* (Weaver *et al.*, 1980), *M. sexta* (Kramer and Law, 1980a), *L. decemlineata* (Kramer, 1978), *L. migratoria* (Hamnett and Pratt, 1978; Girardie *et al.*, 1981), *Melanoplus sanguinipes* (McCaffery and McCaffery, 1983), *B. terrestris* (Röseler and Röseler, 1978), *N. cinerea* (Lanzrein *et al.*, 1978), and *Z. angusticolla* (Greenberg and Tobe, 1985).

Most workers have relied, for CA maintenance, upon media used in the culture of other insect cells. These media were soundly based on the pH, salt, and organic acid concentrations of the hemolymph of the species in question (e.g., Grace's) and it is likely that such media would prove suitable for the maintenance of CA of that species (and related species presumably). However, the use of media specifically formulated for Lepidoptera for maintenance of CA from the Coleoptera is unwise. Grace's medium was employed for the maintenance of CA from both *S. vaga* and *T. molitor* (Judy *et al.*, 1973b, 1975), and we now know, at least in the case of *T. molitor*, that the rates of JH biosynthesis in Grace's medium were significantly less than those observed in media such as 199 and MEM, using the RCA (Weaver *et al.*, 1980).

Defined media, rather than media supplemented with either vertebrate or insect serum, should be chosen for maintenance of CA, if accurate rates of JH biosynthesis are to be realized, because serum may contain factors capable of modulating JH biosynthesis. pH and osmolarity optima should be determined and will no doubt closely reflect hemolymph values. pH has been shown to influence rates of JH biosynthesis in several species although a certain degree of latitude is apparent (Tobe and Pratt, 1974a; Weaver *et al.*, 1980). Optimal methionine concentrations must also be established (see above), although JH biosynthesis appears to be maximal over a wide range of concentrations (Tobe and Pratt, 1974a; Weaver *et al.*, 1980; Feyereisen *et al.*, 1981b; Khan *et al.*, 1982a; Tobe and Clarke, 1985) (see Fig. 8). Oxygenation of the incubation medium may also be important, although no enhancement of JH biosynthesis was observed in an enriched O_2 environment; however, biosynthesis was curtailed in an N_2 environment (Tobe and Pratt, 1974a).

3.1.6 *Time course of JH synthesis and release*

It must be demonstrated that JH biosynthesis and release occur at a constant rate over the defined incubation interval. Deviations from linearity preclude the use of the RCA to measure rates of JH biosynthesis. Therefore the time required for dilution of the endogenous methionine pool by the radiolabeled substrate must be determined (typically 30–60 min; Tobe and Pratt, 1974a; Pratt *et al.*, 1975a; Tobe and Stay, 1977), as well as the interval over which JH biosynthesis is linear. In most cases, an interval of 3 h suffices for a standard incubation period. JH biosynthesis was linear over a 5-h period for most of those species listed in Table 4, in experiments which utilized the RCA for measurement of rates of JH biosynthesis/release. However, there is an appreciable degree of variability in these measurements, with some CA showing increased rates over time and others, decreased rates. At present, the reasons for this variability are unknown. The CA of certain species show linear rates of biosynthesis for periods much greater than 5 h (e.g., *P. americana*: Pratt *et al.*, 1976; *D. punctata*: Stay and Tobe, 1977), but in all cases, rates of biosynthesis eventually declined. This decline in biosynthesis by CA maintained *in vitro* for long intervals was originally reported by Judy *et al.*, (1973a) for *M. sexta* CA, and was accurately documented by Pratt *et al.* (1976) for *P. americana* and Stay and Tobe (1977) for *D. punctata.* The decline in rates of JH biosynthesis over time can be attributed to a number of factors: (1) the absence of signals *in vitro* which normally modulate hormone biosynthesis *in vivo*; (2) depletion of substrates or cofactors necessary for biosynthesis; this depletion could occur either in the medium or within the CA; (3) cell death and an absence of proliferation of CA cells; (4) contamination of incubation medium with microorganisms, as a result of either dissection or maintenance techniques; (5) destruction of cellular organelles involved in biosynthesis; (6) turnover of enzymes of the biosynthetic pathway with reduced synthesis of these and other proteins; and (7) deterioration of CA cells due to accumulation of toxic compounds and/or lack of O_2. At present, we do not know which of these factors are most important but it should be noted that farnesoic acid stimulation of CA of *P. americana* still occurred after 44–56 h *in vitro*, indicating that at least the enzymes catalyzing the final two steps in JH biosynthesis and the associated cofactors (NADPH) must be present and functional (Pratt *et al.*, 1976). Nonetheless, several of the above factors could be attributable to suboptimal incubation media for long-term maintenance of CA.

Two final points should be made with regard to the linearity of JH biosynthesis *in vitro*. First, sufficient JH must be produced in the incubation time employed to ensure that it can be accurately quantified. This interval will be dictated by activity of the CA. The use of high-specific-radioactivity methionine and methionine-free incubation medium (see Section 3.1.2) should ensure sufficient sensitivity so that prolonged incubations should not prove necessary, even for quantification of CA of low activity. Extended *in vitro* incubations should be avoided to ensure that linearity is maintained over the entire period. Second, CA of both high and low activities must show linearity over the entire incubation period; at least in the case of *P. americana* and *S. gregaria*, low-activity CA appear to show a longer period of linearity than high-activity glands (Pratt *et al.*, 1975a; Avruch and Tobe, 1978). Accordingly, in defining the time course of JH biosynthesis for any given species, CA producing JH at both high and low rates should be assayed.

3.1.7.1 *Do rates of JH biosynthesis in vitro reflect those in vivo?* Until it is possible to measure JH biosynthesis *in vivo*, this question will not be resolved. However, there are several reasons which support the view that *in vitro* rates are an accurate reflection of *in vivo* rates. (1) There is an excellent correlation between rates of JH biosynthesis *in vitro* and JH titer in hemolymph and whole bodies (Tobe *et al.*, 1984a, 1985; Fig. 10). (2) Physiological manifestations of changing JH titer, such as oocyte growth, correlate well with changes in JH titer and biosynthesis (Tobe *et al.*, 1984a, 1985) (see Fig. 9). (3) Biosynthesis of JH occurs at linear rates for at least several hours following removal from the donor insects. In addition, using the RCA, radiolabeled JH can be detected in the incubation medium within 30 min or less of addition of radiolabel (Tobe and Pratt, 1974a). This apparent continuity in the performance of the CA following removal from the donors suggests that their activity simply reflects that *in vivo*. Rapid initial changes in biosynthesis do not occur, as might be expected from a system in which the modulators of endocrine activity are relatively slow acting (see Tobe and Pratt, 1976). There is no evidence that CA activity can be modified rapidly, and even operations such as denervation of the glands, which are known to cause an increase in the rates of JH biosynthesis in *D. punctata* (Stay and Tobe, 1977), appear to require at least 7 h for the manifestation of the effects (Tobe *et al.*, 1981). (4) There are dramatic changes in rates of JH biosynthesis as a function of age, particularly in females maturing oocytes (see Table 4 for references). These changes may occur over a 24-h period or less (e.g., see Fig. 9). Thus CA incubated under identical conditions show age-dependent JH biosynthesis. (5) The responses of CA to denervation *in vivo* are independent upon the species, and rates of JH biosynthesis determined by RCA reflect data obtained by transplantation bioassay. Thus, denervation of CA in virgin *D. punctata* results in a stimulation of JH biosynthesis (Stay and Tobe, 1977; Tobe *et al.*, 1981) and a dramatic growth in oocytes (Stay and Tobe, 1977). On the other hand, denervation of CA in females of *S. gregaria* and *L. migratoria* results in a dramatic decline in JH biosynthesis (Tobe *et al.*, 1977; Couillaud *et al.*, 1984). Transplantation bioassay of CA of *S. gregaria* suggests that transplanted CA lose activity (Pener, 1965, 1967). Accordingly, the CA of different species show different responses to denervation. Data from the RCA are thus consistent with the observed biological and transplantation effects.

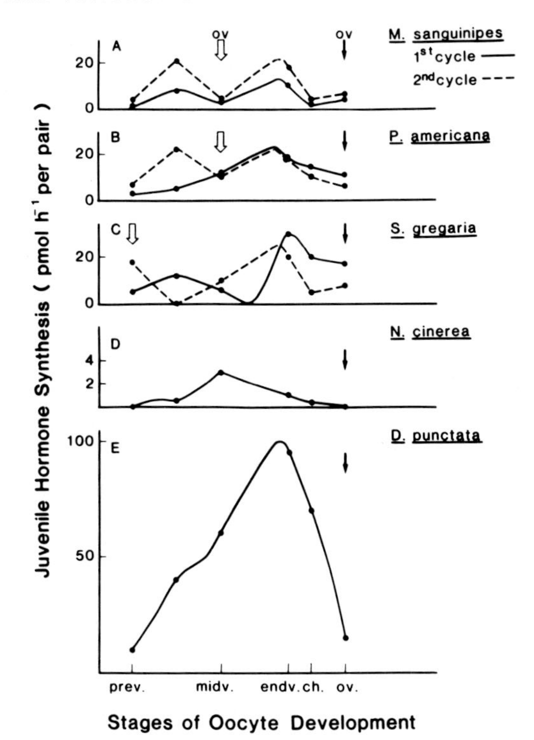

Fig. 9. Rates of JH biosynthesis during the first (solid line) and second (dashed line) vitellogenic cycles in different species of insects, showing differences in maximal rates and in rates at similar stages in the vitellogenic cycles. Previtellogenesis (prev.), midvitellogenesis (midv.), end of vitellogenesis (endv.), chorionation (ch.), and oviposition at end of first and second cycles (ov., solid arrow). The open arrow indicates the point of ovulation of first cycle of oocytes with respect to the second cycle of vitellogenesis. Sources: (A) McCaffery and McCaffery (1983); (B) Weaver *et al.* (1977); (C) Tobe and Pratt (1975a); (D) Lanzrein *et al.* (1981b); (E) Tobe (1980); Feyereisen *et al.* (1981b). Note difference in scale in (D) and (E). (D) and (E) show only first cycle since cycles are widely separated by pregnancy.

3.1.7.2 *Theoretical vs measured rates of JH biosynthesis* The view has been advanced that *in vitro* incubations may not permit the full expression of CA activity, i.e., that rates of JH biosynthesis *in vitro* are less than those *in vivo* possibly because of suboptimal incubation conditions (see Kramer, 1978; De Kort *et al.*, 1981). A comparison of rates of JH biosynthesis *in vitro* by CA of *L. decemlineata* with the theoretical rates, calculated from measured hemolymph JH titer, revealed that the theoretical rates were considerably higher than the observed rates. These calculations assumed steady-state conditions, with rates of biosynthesis equal to rates of degradation of hormone, and a half-life of JH of about 25 min, derived from studies using exogenous radiolabeled JH I (Kramer *et al.*, 1977) (note that JH III is the naturally occurring JH of *L. decemlineata*; Table 5). Similar calculations, using half-life data from Rotin and Tobe (1983), have also been performed for *D. punctata* and revealed that theoretical rates of JH biosynthesis are considerably higher than those observed *in vitro* (see Section 5) (Tobe *et al.*, 1984a, 1985, and unpublished). Although the difference between actual and theoretical rates could be attributable to suboptimal incubation conditions, it is more likely that the theoretical rates are overestimates, in part because the calculations have assumed a steady-state condition (which is unlikely to prevail in view of the rapid changes in measured JH titer) and in part because the actual half-life of endogenous JH may be much greater than previously estimated from breakdown of exogenous hormone (Tobe *et al.*, 1984a, 1985). These factors will be explored in greater detail in Section 5, but the observation that rates of biosynthesis by CA of *D. punctata*, even in the presence of farnesoic acid (Feyereisen *et al.*, 1981b), do not approach the theoretical values for spontaneous JH biosynthesis supports our suggestion that the theoretical values are overestimates.

3.1.8 *Stoichiometry of incorporation*

In order to determine precisely the quantity of JH biosynthesized in a given time interval (and hence determine the rate of biosynthesis), it is necessary to demonstrate that an exact 1:1 molar incorporation ratio exists between the incorporation of the S-methyl function of radiolabeled methionine and the carbon skeleton of the JH molecule, that is, for each mole of radiolabeled methionine incorporated into the methyl ester function of JH, one mole of farnesoic acid (the presumed substrate for this reaction) should be utilized. Demonstration of this 1:1 molar incorporation ratio must be regarded as a cornerstone of the RCA and deviations from this stoichiometry preclude its use. A 1:1 molar incorporation ratio implies that the specific radioactivity of the biosynthesized JH is identical to that of the methionine in the incubation medium, and it is this latter value which is used in the calculation of rates of JH biosynthesis.

Several procedures have been utilized to establish stoichiometry of incorporation. Judy *et al.* (1973a) employed mass spectral analysis of JH biosynthesized by CA of *M. sexta* in medium containing [*methyl*-^{14}C]methionine at high isotopic abundance to demonstrate that 90–100% of the methyl ester function derived from the radiolabeled precursor; thus the molar incorporation rate was between 90 and 100%. Pratt and Tobe (1974) and Tobe and Pratt (1974a) used [^{3}H]farnesoic acid and [*methyl*-^{14}C]methionine in the incubation medium to demonstrate a 1:1 incorporation into JH III by the CA of *S. gregaria*. A similar technique has been employed to demonstrate stoichiometry with CA of *P. americana* (Pratt and Bowers, 1977) and *T.*

molitor (Weaver *et al.*, 1980). This procedure is less technically demanding than that employed by Judy *et al.* (1973a) but, unfortunately, [^{3}H]farnesoic acid and its homologs are not commercially available. For these reasons, Feyereisen *et al.* (1981b) used a different procedure to demonstrate a 1:1 molar incorporation ratio. JH III, biosynthesized by CA of *D. punctata* in the presence of [*methyl*-^{14}C]methionine, was analyzed using HPLC and the fractions containing the hormone were assayed for radioactivity. By using different quantities of biosynthesized product, these authors were able to generate a plot of radioactivity in JH III as a function of quantity of JH III biosynthesized (i.e., specific radioactivity). Comparison of the slope of the regression line relating these parameters (i.e., specific radioactivity) with that from synthetic radiolabeled JH III of known specific activity indicated that the radiolabeled methionine was the exclusive methyl donor in the biosynthesized JH III. This latter method is usable by any laboratories equipped with HPLC and liquid scintillation capabilities and hence may be of more widespread interest. Molar incorporation ratios have been demonstrated for only a limited number of species: *M. sexta* (Judy *et al.*, 1973a), *S. gregaria* (Tobe and Pratt, 1974a), *T. molitor* (Judy *et al.*, 1975; Weaver *et al.*, 1980), *P. americana* (Pratt and Bowers, 1977), and *D. punctata* (Feyereisen *et al.*, 1981b), and these measurements should be extended to all species in which the RCA is routinely employed.

3.1.9 *Selection of biological material*

To define the preceding parameters for the RCA, it is advantageous to use CA exhibiting near-maximal rates of hormone biosynthesis, particularly because the larger quantities of JH biosynthesized by these glands facilitate its accurate quantification (Table 6). Although Table 4 lists the species in which the RCA has been employed for the measurement of rates of JH biosynthesis, in only some cases has a time course of JH biosynthesis with respect to a physiological event been defined. The most commonly used parameter has been oocyte growth and maturation, particularly vitegellogenesis and subsequent chorion formation; in general, rates of JH biosynthesis are higher during vitellogenesis and lower during previtellogenesis and chorion formation (see Fig. 9). Unfortunately, this is only a generalization and cannot be applied universally without further verification. In *S. gregaria* and *L. migratoria*, plots of JH biosynthesis as a function of basal oocyte length (covering the period of previtellogenesis, vitellogenesis, and chorion formation) have revealed little, if any, correlation (see Fig. 9; and Tobe and Pratt, 1975a: *S. gregaria*; Girardie *et al.*, 1981, and Couillaud *et al.*, 1984: *L. migratoria*). On the other hand, a much more precise relation between basal oocyte length and JH biosynthesis appears to exist in the cockroaches and

TABLE 6
Quantity of JH produced per volume of CA tissue

Species	Stage[a] and sex	Volume of one CA (nl)	JH biosynthesis (pmol h^{-1} nl^{-1})	References
Manduca sexta	L(IV–1)		0.4	Granger *et al.* (1979)
Diploptera punctata	L(III–1)	1.5[b]	0.5	Szibbo *et al.* (1982)
	A♀	9[b]	9.0	Szibbo and Tobe (1981a)
	A♂	1[b]	2.5	Tobe *et al.* (1979); Szibbo and Tobe (1982)
Nauphoeta cinerea	A♀	9	0.2	Lanzrein *et al.* (1978)
Schistocerca gregaria	A♀	30	0.5	Injeyan and Tobe (1981)
Locusta migratoria	A♀	45	1.1	Ferenz and Kaufner (1981); Girardie *et al.* (1981)
Leptinotarsa decemlineata	A♀	0.5	6.6	Khan *et al.* (1982a)
Bombus terrestris	A queenless worker	2[b]	2.5	Röseler (1977); Röseler and Röseler (1978)

[a] L, Larvae (stadium, day); A, adults (♀♀ at maximum rates of vitellogenic cycles).
[b] Data from fixed CA, the volume may be half the fresh volume (see Szibbo and Tobe, 1981a); other volumes known or assumed to be of fresh CA.

this relationship is particularly striking in *D. punctata* (Tobe, 1980; Feyereisen *et al.*, 1981b; Rankin and Stay, 1984). In this latter instance, oocyte length is an accurate indicator of JH biosynthesis and hence can be of enormous predictive value in selecting animals exhibiting known rates of JH biosynthesis. Implicit in this statement is the suggestion that there is a precise relationship between oocyte length and JH titer and this does in fact appear to be the case (see Fig. 10 and Section 5). Similarly, there is a precise dose–response relationship between oocyte growth and exogenous JH or JH analog dose in allatectomized females of *D. punctata* (Tobe and Stay, 1979), *N. cinerea* (Lanzrein, 1979), and *L. maderae* (Engelmann, 1979). However, these species appear to be more the exception than the rule and it is likely that other species will show much less precise relationships between oocyte growth and hormone biosynthesis; this may be related to the particular mode of reproduction of the species in question (see Tobe, 1980). Nonetheless, in surveying JH biosynthesis in a previously uninvestigated species, the greatest likelihood for high rates of biosynthesis would be realized in females with oocytes in midvitellogenesis. Larvae would not appear to provide CA showing high rates of biosynthesis since in those few species studied, JH biosynthesis is considerably less than the maximal rates observed in adult females (*D. punctata*: Szibbo *et al.*, 1982; *S. gregaria*: Injeyan and Tobe, 1981; *M. sexta*: Kramer and Law, 1980a).

3.1.10 *Contamination of CA*

The CA of species from several different orders are intimately associated with the corpora cardiaca (CC) (see Section 1.1); included in this group are the CA from some of the more intensely studied species such as *D. punctata, P. americana, N. cinerea, M. sexta, L. decemlineata*, and *T. molitor*. Because of this association, it is difficult, if not impossible, to obtain CA which are not contaminated to some degree with CC tissue. This contamination is a potential source of difficulty, not only because of the possibility that the CC may contain modulators of CA activity but also because it precludes the identification of the CA as the sole site of JH biosynthesis. These problems were addressed by the study of Stay and Tobe (1977); these authors were able to demonstrate that CA of *D. punctata* that had as much CC removed as possible showed rates of JH biosynthesis similar to those observed by CA plus CC. In addition, the time course of biosynthesis over a 96-h period was similar for both types of CA. These observations suggest that at least in *D. punctata*, contamination of the CA with CC tissue does not affect rates of JH biosynthesis. Similar studies should be extended to other species.

Contamination of CA with fat body may also pose difficulties and every attempt should be made to ensure that dissected CA are free of any fat body tissue. The fat body is the probable source of the JH-specific esterase (see Section 5.2) and contamination of CA with this tissue may result in artificially low rates of JH biosynthesis, because of the cleavage of the radiolabeled methyl ester function of newly synthesized JH by the esterase. In this regard, it is noteworthy that the CA of *M. sexta* do show appreciable JH esterase activity, in both homogenates and intact glands (Reibstein *et al.*, 1976; Granger *et al.*, 1979; Kramer and Kalish, 1984; Sparagana *et al.*, 1984), although it is uncertain how much of this activity can be attributed to contamination of CA with fat body. The presence of such esterase activity complicates interpretation of data obtained using the RCA, particularly at times when the esterase activity is high, and for this reason, Kramer and Kalish (1984) used a modified assay system in which the incorporation of radiolabeled propionate, rather than methionine, into both JH and JH acid (the product of the esterase) was monitored. Such an assay system should be of value in those species which biosynthesize the higher homologs of JH but clearly is of no use in those species which produce only JH III. The presence of esterase activity in incubations of CA producing JH III would dictate the use of some other readily available radiolabeled precursor, but at present we know little about the utilization of such precursors for JH biosynthesis (see Section 2). Although the incorporation of a radiolabeled substrate such as acetate may prove satisfactory, it will be necessary to demonstrate the stoichiometry of acetate incorporation. Acetate may also be incorporated into a wide range of products, and extensive purification of biosynthesized JH may prove necessary. Such extensive cleanup procedures may militate against the use of such precursors for the routine assay of JH biosynthesis.

Great care must also be exercised when removing extraneous tissue and we have repeatedly observed that violation of the integrity of the CA, including the basement lamina, invariably results in reduced rates of hormone biosynthesis (unpublished). Any damage to the CA must be avoided and damaged CA should not be employed for the RCA. The effects of physical damage on the CA are unknown but the basement membrane may serve as a permeability barrier, regulating the interior composition of the CA. Damaged CA may also show reduced cell coupling (Lococo *et al.*, 1985), resulting in the disruption in integration of the glands.

3.2 RADIOIMMUNOASSAYS: BACKGROUND, THEORY, AND DEPLOYMENT

The enormous growth in the field of vertebrate endocrinology during the past 25 years must be attributable in part to the advent of the sensitive and reproducible methods to quantify hormones in biological tissue. Probably the most widely used and adapted method is radioimmunoassay (RIA) and since its original development by Berson and Yallow (1958) for quantification of insulin, RIAs for most known vertebrate hormones have appeared. Insect endocrinology has not fared quite so well and although a RIA for ecdysteroid was developed over a decade ago (Borst and O'Connor, 1972), only in the past 5 years has this technology been employed by a large number of laboratories for quantification of ecdysteroids in biological samples. The history of RIAs for JHs has been even less auspicious and only a limited number of laboratories have ventured into this potentially difficult but valuable area of insect endocrinology. RIAs in general are known for their sensitivity (limits of detection in the pico- to nanogram range), reproducibility, and simplicity and hence could be of enormous value in the quantification of JH *in vivo* as well as *in vitro*. They are also characterized by straightforward protocols and a minimum of sample preparation. Unfortunately, these generalizations are not necessarily applicable to RIAs for JH, in part because of the unusual nature of the JH molecule and in part because of the existence of JH homologs and their respective degradation products. Because we are concerned here with the CA and the assay of its secretory product, we shall consider briefly the parameters for successful deployment of the RIAs and some of the problems encountered in its use. For more complete considerations of the RIA, see Tobe and Feyereisen (1983) and Granger and Goodman (1983).

Many of the parameters for use of the RIA are identical to those outlined for the RCA, and strict adherence to the definition of these parameters is equally important to the RIA and the RCA. Thus, unequivocal identification of biosynthesized JHs, optimization of incubation media, and demonstration of a constant rate of JH biosynthesis or release during the incubation period must all be rigidly established. There are several additional parameters which must be defined for the RIA only and which are related to the generation and characterization, including specificity, of the antibody. Characterization of the antibody is a fundamental requirement for the successful utilization of any RIA.

3.2.1 *Establishing new biosynthesis*

One major advantage of the RIA over the RCA in terms of the parameters for deployment is that there is no necessity to establish stoichiometry in biosynthesis of JH because the RIA does not use a radiolabeled precursor to monitor synthesis. However, it is necessary to establish the quantity of JH within the CA at the beginning and end of the incubation, in order to determine the quantity of hormone synthesized during incubation. Any JH within the CA at the time of dissection and subsequently released *in vitro* will be detected by RIA; accordingly, there is no apparent delay in the release of JH and hence no assurance that new JH is biosynthesized. This contrasts with the RCA in which a lag time in release of newly synthesized hormone is apparent because of the time required for equilibration of the radiolabeled methionine with endogenous methionine within the CA (see Section 3.1).

3.2.2 *Generation of antibodies*

The JH molecules are poor immunogens and there has been considerable difficulty in raising antibodies to these molecules at high titer and specificity. For this reason, virtually all studies have utilized conjugation of the JHs with immunizing proteins, particularly human serum albumen (HSA) (Lauer *et al.*, 1974; Baehr *et al.*, 1976; Strambi *et al.*, 1981). Granger and Goodman (1983) note that the use of other immunizing proteins, for example bovine thyroglobulin, may improve the specificity of the generated antibodies but such studies remain to be reported. These authors also point out that changes in immunization schedules and quantities of injected antigen can profoundly influence the generation and specificity of antibodies, and again, this area demands further research. Generation of antibodies is influenced not only by the conjugated immunizing protein and the immunization protocol but also by the molecular location of conjugation and by the conjugating agent. The initial studies of Lauer *et al.* (1974) employed conjugation to HSA through the C-1 position of JH, using the *N*-hydroxysuccinimide ester, and a similar approach was pursued by Baehr *et al.* (1976). Although the antibody raised by Lauer *et al.* (1974) showed very low binding affinity for JH III (33 μg/ml assay for 50% displacement), in the hands of Baehr *et al.* (1976), the affinity of an independently generated antibody for JH I was improved considerably (1.5 ng/ml for 50% displacement). Sensitivity of this RIA was also improved dramatically by conjugation of the JH acids to histamine through the C-1, permitting the iodination of the immunogen. Granger *et al.* (1979, 1982a), using antisera from the same source, have reported affinities for JH I and III of 1.2 and 1.0 ng/ml, respectively, for 50% displacement. Strambi *et al.* (1981) utilized a different approach to the coupling of JH to the immunizing protein, conjugating JH I diol to HSA by succinylation through the C-10 position. This yielded antibody with a higher affinity (0.6 ng/ml for 50% displacement of the JH I immunogen, ^{125}I-Suc < JH-Gly-Tyr) and at very high titer. The approach of Baehr and colleagues has permitted the generation of separate antibodies to JH I, II, and III, although each of the antibodies show considerable cross-reactivity to other JHs (> 1%) (see below); this limits their usefulness. On the other hand, the main advantages of the approach of Strambi *et al.* (1981) are (1) increased immunoreactivity of the diol; (2) increased aqueous solubility of the diol; and (3) equal immunoreactivity of the antibody to all JH homologs. This latter point is moot unless the individual JHs are separated before assay. There is no advantage associated with the measurement of the sum of the JHs.

The physical characteristics of the JH molecules make them difficult to handle—they have low aqueous solubility, particularly the higher homologs (see Giese *et al.*, 1977), a phenomenon which can result in problems because the RIA is clearly aqueous based. However, the advantages associated with the RIA of Strambi *et al.* (1981) minimize, although by no means overcome, these problems.

3.2.3 *Characterization of antibody*

3.2.3.1 *Cross-reactivity* Characterization of antibodies used in RIA with respect to cross-reactivity can be a tedious procedure; cross-reactivity refers to the specificity of the antibody and the ability of substances other than the immunogen to bind with the antibody. These substances include metabolites and precursors of JH as well as totally unrelated compounds. Antiserum with high cross-reactivity is unable to discriminate between the three-dimensional structure of the hapten and other compounds and is of limited usefulness in the RIA. The case of JH can be particularly difficult because of the existence of up to five different hormones and their associated metabolites and precursors, at least in the Lepidoptera. The situation is less complex in other orders because of the presence of only JH III. Nonetheless, the metabolites and precursors of JH III, as well as non-JH molecules, must also be examined for cross-reactivity in members of these orders. There has been some attempt to characterize the antibodies presently in use for the JH RIA and the most notable of these are the studies of Baehr *et al.* (1976) and Granger *et al.* (1979). Conjugation of the JHs to the immunizing protein through the C-1 position generated antisera specific to each JH, theoretically able to discriminate between the original immunogen and the other JH homologs. However, these studies demonstrated that there was appreciable cross-reactivity; for example, for a JH I antiserum, cross-reactivities of JH II and JH III were 13 and 2%, respectively (Granger *et al.*, 1979). Typically, antiserum showing a cross-reactivity of less than 1% is regarded as specific. It is questionable if the above JH I antiserum would be of use in RIA without prior separation of the JH homologs. However, Granger *et al.* (1982b) have employed this antiserum to quantify JH I production by CA of larval *M. sexta in vitro*, without prior separation of the homologs.

Although the antisera utilized by the Baehr group and the Granger group clearly showed some degree of specificity, that employed by the Strambi group displayed very little specificity for the JH homologs (i.e., it was unable to discriminate between the JH homologs), no doubt because of the C-10 linkage used to conjugate the hormone to the immunizing protein. However, this was intentional and permitted these workers to utilize a single antiserum to quantify the diols of the JH homologs (Strambi *et al.*, 1981). Clearly, in order to quantify the individual homologs, derivatization and separation of the homologs prior to assay are necessary. Although it is unclear if the Baehr antibodies are able to discriminate between the enantiomers of the JH homologs (see below), this is irrelevant in the Strambi assay because the chiral center no longer exists following derivatization to the diols. Granger and Goodman (1983) suggest that the inability of the Strambi antiserum to distinguish between the JH homologs may prove useful for the assay of total JH in a sample. However, the biological meaning of such measurements is moot, principally because the JH homologs are well known to display widely different biological activities (see Staal, 1975; Sehnal, 1976).

The foregoing discussion indicates that although there have been some attempts to characterize the antibodies currently in use for the RIAs for JH, these studies have concentrated on cross-reactivity of JH homologs. The question of cross-reactivity of JH precursors and metabolites as well as non-JH molecules remains to be adequately explored. Granger *et al.* (1982a) have examined the cross-reactivity of JH III acid, diol, and methyl farnesoate with JH III antiserum and observed cross-reactivities of 77, 12, and 8%, respectively. The high value for JH acid is to be expected, because of the conjugation through the C-1 position. Methyl farnesoate appears to be contained only with the CA cells and is not released into the medium (Tobe and Pratt, 1974a; Pratt *et al.*, 1975a,b; Tobe and Stay, 1977; Weaver *et al.*, 1980), at least for those species examined, and accordingly, the relatively high degree of cross-reactivity of the JH III antibody with this compound may be relatively unimportant. However, both JH III acid and JH III diol are metabolites of JH III, JH III acid being the product of the JH esterase and JH III diol the product of the epoxide hydratase (Slade and Zibitt, 1972), and these compounds can be expected to be present in either hemolymph or tissues. Because of the high degree of cross-reactivity of these compounds with the JH III antiserum, it is clearly impossible to utilize the RIA on biological samples without prior separation of the JH homologs from the metabolites. The question of cross-reactivity of antisera with non-JH molecules remains largely unexplored, although methylepoxy stearate has been reported to have virtually no cross-reactivity with JH III antiserum (Granger *et al.*, 1982a)—the rationale for the choice of this compound is moot, since it is probably not an appreciable component of insect tissue and its three-dimensional structural relation to JH was not ascertained. Until such time as the cross-reactivity of a range of naturally occurring non-JH molecules with the JH antisera has been determined, it would be unwise to employ the RIA on biological tissue without extensive prior purification of the JHs. It seems likely that much of the JH-immunoreactive material reported by Baehr and colleagues to occur in species in which only JH III has been unequivocally demonstrated by physicochemical methods can be attributed to non-JH compounds (see Baehr *et al.*, 1979, 1981; Deleurance *et al.*, 1980; Papillon *et al.*, 1980). It is unfortunate that appropriate "biological blanks," i.e., samples from which the endogenous hormones were separated, were not included in these studies. For a consideration of the possible reasons for the discrepancy between RIA data and physicochemical data, see Tobe and Feyereisen (1983). Granger and Goodman (1983) have noted that RIA values are often 10- to 100-fold greater than LC-MS data but offer no explanation for these discrepancies other than possible differences between populations and rearing regimes. It is unlikely that such enormous differences can be attributed simply to variations in these parameters.

The chiral centers in the JH molecules may also pose problems with respect to specificity. As noted in Section 2.1, the C-10 position in JH III and the C-10 and C-11 positions in the higher JH homologs are chiral centers, with the natural enantiomers being 10*R* for JH III and 10*R*,11*S* for the higher homologs (Faulkner and Peterson, 1971; Nakanishi *et al.*, 1971; Judy *et al.*, 1973a; see also Hamnett *et al.*, 1981). It is unknown at present if antisera raised against racemic mixtures of JH contain subpopulations of antibodies specific for each diastereomer but if such is the case, it is essential to utilize only the appropriate natural enantiomer as immunogen in the generation of the antisera. In addition, the construction of standard displacement curves must also be performed with only the natural enantiomer. This problem awaits investigation but it should be noted that measurement of both diastereomers in the preparation of displacement curves, as probably occurs with the antisera employed by both the Baehr and Granger groups, may result in overestimates of the JH titers.

The RIA of Strambi *et al.* (1981) does not discriminate between diastereomers because linkage to the immunizing protein occurs through the C-10 position, resulting in the destruction of the chiral center. This may be regarded as advantageous since the JH diols employed in the generation of antisera and in the displacement curves can be derived from racemic mixtures of the JHs. However, conversion of JHs to diols on a microscale may be whimsical (Schooley, 1977), and it is essential to include some type of internal standard in each sample to monitor conversion and recovery.

3.2.4 *Internal standards and recovery of JH*

Measurements of conversion and/or recovery in the RIA of Strambi *et al.* (1981), using internal standards, have yet to be published but it should be noted that the use of a "correction factor," derived from occasional parallel standard incubations, may be unreliable (Strambi *et al.*, 1981). The use of internal standards for the other RIAs for JH should also be encouraged so that recoveries of JH from extracts of hemolymph or whole bodies can be accurately monitored. Unless recoveries are accurately monitored in the above situations, the total quantity of JHs in a given sample cannot be assumed to be equal to the sum of the individual JH homologs. This relation will also be influenced by any sources of interference for the individual RIAs; the most accurate method for the determination of interference would require the separation of the JHs from the biological sample followed by assay of both the JH and the non-JH fractions for each individual sample. Recoveries must also be determined using internal standards. Such measurements would also be appropriate for the assay of JH in CA incubations *in vitro*. Biological blanks utilizing only incubation media without CA (Granger and Goodman, 1983) are not sufficient; rather, assay of the JH and non-JH fractions of media following incubation with CA should be performed, in conjunction with determination of recovery using internal standards.

3.2.5 *Conclusions*

A final point regarding the RIA concerns the identity for the immunoprecipitable products. In no case have the immunoprecipitable products of biological extracts or incubations been identified unequivocally (using physiochemical methods) as JH or JH diols (depending upon the particular RIA). Such information may also provide insights into the identity of interfering substances.

It is clear from the foregoing discussion that much work on the generation and characterization of antisera remains to be done before all of the data generated from RIA can be accepted unequivocally. Great care must be taken in the preparation of samples and in most cases, separation of the JHs from the extracts or media, using chromatography (preferably HPLC), is the method of choice. The recovery of the JHs should be monitored routinely using internal standardization, and to establish the validity of the RIA, a direct comparison of RIA values with those obtained from physiocochemical methods should also be performed for each system studied. Although the difficulties of the JH RIA preclude its routine use in many laboratories, it is nonetheless a most valuable technique and should prove particularly useful in monitoring JH titers in hemolymph and whole-body extracts and in measuring release of JH homologs in incubations of lepidopteran CA *in vitro*.

4 Regulation of the corpora allata

4.1 THE NERVOUS SYSTEM AS A REGULATORY CENTER OF CA ACTIVITY

As should be clear from the consideration of the variation in morphology of CA among insects species and of their polymorphic forms within a species, the genetics and the developmental history of the glands set limits to the quality and quantity of their product. But within these constraints there is obviously change in the synthetic activity of the glands which includes changes in cell number and in cell configuration as well as in the control of the synthetic pathway. The nervous system provides a mechanism for integrating regulatory signals effecting these changes from both the internal and external environments. In the following section, the evidence for regulation of the CA by way of the nervous system is presented. This includes evidence that the nervous system is the source of both stimulation and inhibition of the CA and that this regulation reaches the glands by way of the hemolymph and axon tracts. The elements of the nervous system which are responsible for this regulation undoubtedly involve NSC and possibly conventional neurons. If the NSC function only by direct delivery of their products to the glands, it is difficult to distinguish their effect from that of excitatory or inhibitory neural terminals of the conventional sort by the simple experimental procedure of interrupting axonal pathways to the CA, because this would prevent the transmission of nerve impulses required for the release of either neural or neurosecretory messages. In the discussion that follows we will consider experimental evidence for stimulatory neurosecretions acting by way of the hemolymph under the heading of allatotropins (Scharrer, 1958) and then cases in which intact nerves are required for increased rates of biosynthesis. A similar distinction will be made for the downward modulation of CA activity (see below), though it is not known at this time whether the regulation transmitted via the fiber tracts involves factors which are distinct from those traveling via the hemolymph.

4.2 ALLATOTROPINS

4.2.1 *Techniques for demonstrating allatotropins*

Most evidence for the existence of allatotropins comes from studies of CA in adult insects, particularly with respect to their regulation of the female reproductive cycles and periods of diapause. Correlations of changes in the staining patterns of NSC with the gonotrophic cycles was one of the first indications that neurosecretory factors from the brain might be involved with regulation of the CA (see Girardie, 1983, for references). Experiments utilizing the extirpation of NSC from the brain of the blow fly (Thomsen, 1948) began the work demonstrating the importance of neurosecretion in regulating reproduction, and such procedures, with more sophisticated tools and assay methods, continue to be used in the search for brain factors which regulate the CA.

Unequivocal demonstration of the existence of allatotropins is only now becoming possible as substances can be isolated from extracts of neural tissue by HPLC (Girardie *et al.*, 1983; Hayes and Keeley, 1984) and their stimulatory effect on rates of JH biosynthesis assayed *in vitro*. However, most of the evidence for allatotropins comes from experiments in which CA and brains were manipulated in the whole animal and bioassays were used to estimate the resulting activity of the CA. Interpretation of the results of such experiments depends upon the sensitivity of the bioassay and cannot exclude the possibility of indirect effects on the CA. Yet such experiments form a basis for choosing brains from the stages in which allatotropins might be expected to be found and also for choosing the CA which would be most suitably used to assay the putative allatotropins.

4.2.2 *Allatotropin in immature stages*

Galleria mellonella. Last instar larvae of this species undergo an extra larval molt following implantation of several brains from donors of similar age early in the stadium (Granger and Sehnal, 1974; Sehnal and Granger, 1975). Presumably the JH titer is elevated by increased activity of the CA. This assumption is supported by finding that an inhibitor of JH biosynthesis, fluoromevalonate (Quistad *et al.*, 1981), suppressed the action of an implanted stimulatory brain (Cymborowski and Bogus, 1983). The pars intercerebralis (PI) is required for this stimulatory action: implanted PI-cauterized brains are ineffective (Granger and Sehnal, 1974). The M-2 cells of the PI have been suggested as the source of allatotropin based on the allatotropic effect of brains in which only the M-2 cells appeared to retain neurosecretory activity (Granger and Borg, 1976; Borg and Granger, 1977). Temporal changes in the production of allatotropin are suggested by the observation that brains early in the stadium were more effective in stimulating extra molts than were brains later in the stadium (Granger and Sehnal, 1974; Granger *et al.*, 1981), but this was not confirmed by Pipa (1977) and Bogus and Cymborowski (1984) found nearly constant activity during the whole stadium except for higher activity between 0–4 h. However, chilling the brain donors enhanced the production of extra larval molts (Cymborowski and Bogus, 1976) if chilling occurred early but not late in the stadium (Bogus and Cymborowski, 1984).

Manduca sexta. Last instar larvae of *M. sexta* have also been analyzed for allatotropin. A regimen of 3 days of starvation followed by feeding early in last stadium results in an extra larval molt, presumably as a result of stimulation of CA by an allatotropic factor (Bhaskaran and Jones, 1980). JH biosynthesis by CA, as determined by RCA, was not above normal in the second day of starvation but was much higher on the third day and on the first day of feeding (Bhaskaran, 1981). JH biosynthesis remained low, however, in starved animals in which the median NSC (MNSC) had been cauterized (Bhaskaran, 1981). The bioassayed titer of JH was also high during starvation and declined gradually in the first days of refeeding (Cymborowski *et al.*, 1982). By cautery of different groups of NSC, both medial and lateral, the source of allatotropin was localized to a group of medial cells designated II by Nijhout (1975b) and Buys and Gibbs (1981). This group corresponds to group I of Carrow *et al.* (1984). Animals deprived of MNSC could respond to starvation and feeding following implantation of brains but not CC or CA (Bhaskaran and Jones, 1980). This suggests that the allatotropin is either not stored in these organs or not released from them following denervation. On the basis of experiments involving implantation of brains from animals of various stages into test hosts, the allatotropin appeared to be present in newly emerged penultimate instars, but not in normal last instars; however, in starved and subsequently fed last instars, it was present on day 1 of starvation and days 1 and 3 of subsequent feeding (Bhaskaran and Jones, 1980). The persistence of the allatotropin in the brain for this considerable period suggests that CA are dependent upon it for continued activity and this notion is corroborated by the observation that active CA do not remain so when implanted into MNSC-cauterized, starved–fed hosts (Bhaskaran, 1981). The CA do not, however, require intact innervation to respond to the allatotropin (Bhaskaran and Jones, 1980). The allatotropin is released and presumably produced in these experimental animals in response to starvation. A brain can be conditioned to produce allatotropin by implantation into a suitable host and hence the brain cells must respond to factors in the hemolymph (Bhaskaran, 1981). Because the titer of trehalose is low during starvation (Jones *et al.*, 1980), these authors suggested that the allatotropin center is activated in response to low trehalose (Bhaskaran, 1981). In normal development, a threshold body size appears to be the signal which presumably inactivates the CA (Nijhout, 1975a). It seems clear that this is overridden by the starvation induced allatotropin.

An *in vitro* assay for JH biosynthesis was used to test for allatotropins in extracts of brains of larval *M. sexta* (Granger *et al.*, 1981, 1982b, 1984). By incubating two left CA with brain extract and two right CA without, a three-fold activation (i.e., ratio of JH produced by stimulated gland to JH produced by control gland) was observed within 2 h (Granger *et al.*, 1984). JH I and III in the medium were assayed by RIA which detects both the hormones and the corresponding acids; both JH I and III were found but only the JH III production was stimulated by the brain extract (Granger *et al.*, 1981). Extract of abdominal ganglia elicited 25% of the response obtained with brain extract (Granger *et al.*, 1984). Preliminary characterization of the JH III-allatotropic factor suggests that it is a peptide (Granger *et al.*, 1984). These studies are an important step in the demonstration of allatotropins. Further studies should demonstrate that activation can occur in glands monitored before and after the addition of extracts and that the degree of activation of the CA can account for titer changes *in vivo*. In addition it will be important to know whether JH III can be identified in the hemolymph of *M. sexta* at the stages under study because the relative titer of the JHs change with development (Schooley *et al.*, 1984).

Rhodnius prolixus. An allatotropic effect of the brain was demonstrated in penultimate larval instars of *R. prolixus* (Baehr, 1976). The allatotropin was bioassayed using the degree of larvalization of the next instars after the following experimental protocol: animals were given a blood meal, and the PI was then cauterized; an injection of ecdysone was also provided to compensate for failure of PTTH release after cautery. This complicated assay system showed that the PI is required for 20 h for "activity" of the CA to be sufficient for a larval molt; thereafter, the PI can be cauterized and a larval molt still ensues. The allatotropic function of the PI probably acts via a humoral route because the CA may be denervated just before the blood meal and a larval molt still occurs. The loss of allatotropic function caused by cautery of MNSC cannot be restored by implantation of brain tissue. This may be because the implanted brain functions for only a short time, whereas the brain is required for 20 h (as opposed to 1 h in the adult; Baehr, 1973). In this blood-feeding species, the brain must remain connected to the ventral nerve cord in order for feeding to activate the brain (Wigglesworth, 1934). Based on correlations in changes of NSC with physiological events in adult females, Baehr (1973) proposed that "A" cells in the PI produce allatotropin; unfortunately the size of the brain of *R. prolixus* precludes selective cautery to test this hypothesis.

There is also some evidence for an allatotropic effect of the brain in larval stages of *L. migratoria*. Selective cautery of "C" cells in the PI, which resulted in a few cases of premature metamorphosis in *L. m. cinerascens*, suggested to Girardie (1967) that the PI contains a stimulatory factor and that it acts humorally, since denervation of the CA did not alter metamorphosis. However, Couillaud *et al.* (1984) have shown that intact nerves to CA are required for their activation in larvae of *L. migratoria* (see Section 4.3.1) and following CA denervation, although metamorphosis was delayed, it was normal.

4.2.3 *Evidence for allatotropins in adults*

Locusts. Studies of adults of several species of locusts have provided much evidence for the existence of brain allatotropins. In *L. migratoria*, loss of reproductive functions followed cautery of the PI in newly ecdysed adult females and males (Girardie, 1966), and JH biosynthesis was severely reduced following cautery of the brain tract of the NCC I (Girardie *et al.*, 1981). Similarly, following cautery of the PI in young adults, McCaffery (1976), Lazarovici and Pener (1978), and Goltzene and Porte (1978) showed loss of CA function as determined by the absence of vitellogenesis. Pratt and Pener (1983) confirmed by RCA that rates of JH biosynthesis do not increase after PI cautery of newly ecdysed adults. Poras *et al.* (1983) suggested that the CA require stimulation from the brain by way of NCC II before they can be activated and also that the CA require humoral stimulation from the PI for continued activity during vitellogenesis. Thus in *L. migratoria* the PI is implicated as a source of allatotropin.

In *S. gregaria*, cautery of selected areas of the brain also prevented the CA from attaining mature size and appearance, with consequent failure of oocyte maturation (Strong, 1965b). Because the cautery of lateral NSC produced this effect, it appears that the lateral cells rather than medial ones are responsible for production of allatotropin. This was further supported by the observation that unilateral cautery affected the ipsilateral CA and that following transection of the NCC II ($n = 2$), the CA did not "mature" (Strong, 1965b). This is compatible with evidence from cobalt fills of nerves, which indicate that the CA are innervated by lateral cells (Mason, 1973). But the possibility remains that there are axons from the PI to the CA of such diameter that are not easily filled with cobalt. If such axons from the PI regulate the CA, then the effect of lateral cell cautery could result from interactions between the PI and lateral cells. Thus it is premature to conclude that there is a major difference in control centers between *Schistocerca* and *Locusta.*

The presence and source of allatotropins in the brain have also been demonstrated, utilizing the electrical stimulation of local brain areas, followed by an assay of CA activity. The precision of this technique is limited by the size and identity of the area stimulated. In *L. migratoria*, stimulation of the PI promoted oocyte growth and presumably CA activity (Moulins *et al.*, 1974). In *A. aegyptium*, the stimulatory effect was observed following denervation of the CA and thus a humoral route for the allatotropin is implicated in this species (Moulins *et al.*, 1974). In an attempt to show that such stimulation is indeed affecting JH biosynthesis in *L. migratoria*, Tobe *et al.* (1982) measured rates of JH biosynthesis *in vitro* over an interval of 6 days and found an increase in rates 2–3 days after stimulation. This suggests either a long-term stimulus or a delayed response. Stimulation of the PI following cautery of the MNSC tracts resulted in low rates of synthesis in both stimulated and control animals. Thus in this species, intact nerves from the PI to the CA are required for stimulation of the CA. Stimulation of lateral cells also showed a slight allatotropic effect after 1–2 days; this effect was enhanced somewhat after cautery of the MNSC. Thus, from these experiments the role of medial and lateral cells in the production of allatotropins is not clear, but it is important to recognize that there is a potential interaction between these groups of cells because of the ramifications of their axonal collaterals (see Section 1.2.3).

The isolation of the brain allatotropin in locusts has also been attempted. Although there is strong evidence for the existence of allatotropin in these insects, the variability in activity of the CA (see Section 1.4.2.4) is a barrier to the development of a reliable assay for it. A preliminary report by Ferenz and Diehl (1983) suggests that the CA may be stimulated *in vitro* by extracts of CC and subesophageal ganglia. Gadot and Applebaum (1984) make a similar claim for the effect of extracts of brain tissue on *in vitro* activity of CA. It will be particularly important to establish the proper conditions for the RCA (see Section 3) in the quest for identification of allatotropins in this species.

An allatotropic function of the brain has been surmised from experiments utilizing antibodies to brain tissue in *L. migratoria* (Ulrich *et al.*, 1984). Injection of these nonspecific antibodies blocked vitellogenesis, and these authors concluded that the antibodies were eliminating allatotropin needed to stimulate CA activity (Rembold *et al.*, 1980; Ulrich *et al.*, 1984). If it can be demonstrated that the antibodies are reacting specifically with hemolymph-transmitted allatotropin and such material can be identified immunohistochemically in medial and or lateral cells in the brain, it might then be possible to demonstrate that the function of the nerve fibers to the CA, in those cases where innervation is required (see below), is indeed delivery of neurohormone to the target organ.

Leptinotarsa decemlineata. Normally long-day conditions elicit elevated titers of JH and oviposition, whereas short-day conditions result in lowered titers and diapause (de Wilde and de Boer, 1969). The original evidence for a humoral stimulatory factor in this species comes from the demonstration that denervated CA can be activated by changing the animals from a short-day to long-day photoperiod (de Wilde and de Boer, 1969). Khan *et al.* (1983) have shown by RCA that denervated CA under short-day conditions do not show the high rates of JH biosynthesis characteristic of glands in long-day animals. Yet under conditions of starvation, denervated CA, whether in animals kept in short- or long-day conditions, show elevated JH biosynthesis. This may indicate that an allatotropic factor is released during starvation or that an allatostatin ceases to be produced. Further evidence is required to demonstrate an allatotropin in *L. decemlineata.*

In *R. prolixus*, cautery of the PI after a blood meal prevented vitellogenesis but application of a JH analog restored it (Baehr, 1973). Baehr (1973) thus postulated that allatotropins are produced in this region of the brain. Furthermore, by cauterizing the PI at various times after the blood meal, he demonstrated that the PI is necessary for only 1 h after the blood meal, suggesting that the allatotropin acts as a trigger. Evidence for the possible triggering effect of the putative allatotropin can also be found in experiments using *Drosophila melanogaster* (Postlethwaite and Shirk, 1981). In this species, for oocyte development to occur, the head is needed for only 10 min after eclosion, whereas the thorax is required for 16 h. Both of these parts of the body can be removed before these critical periods and oocyte development restored by application of a JH analog (Handler and Postlethwaite, 1978). Because removal of CA by surgery (Bouletreau-Merle, 1974) also resulted in failure of oocyte development, it appears that the brain is triggering the CA (Postlethwaite and Shirk, 1981). Similar experiments in *Aedes aegypti* indicated that the brain is needed for between 1 and 24 h after adult eclosion for the occurrence of CA-dependent ovarian growth (Hagedorn *et al.*, 1977).

In *L. riparia*, an allatotropic factor has also been postulated. Parsectomy by cautery resulted in altered volume and fine structure of the CA characteristic of an inactive state (Baehr *et al.*, 1982).

In *N. cinerea*, decapitation experiments suggested that the brain is required for only a short period after adult emergence for functional capacity of the CA during the reproductive cycle (Barth and Sroka, 1975). Similarly in *D. punctata*, there is some suggestion of brain-mediated stimulation of the CA. Using female CA implanted into males, rates of JH biosynthesis by female CA can be stimulated. Such increase can be reduced by cautery of the PI and partially restored by implantation of female brains into the operated males (Tobe *et al.*, 1981). This could indicate that the implanted brain released an allatotropin; however, in females, the increase in JH biosynthesis associated with the gonadotrophic cycle can occur in the absence of the brain (head) (Rankin and Stay, 1983). Thus the increase in CA activity is not dependent upon the brain in females and perhaps there are nerve cord or ovarian allatotropic factors responsible for the stimulation. These have yet to be isolated.

4.3 STIMULATION OF CA BY WAY OF INTACT NERVES

In the foregoing discussion of allatotropins we have indicated the instances in which brain factors function by way of the hemolymph. If innervation from the brain to the CA must be maintained in order for the CA to respond to brain signals, then either direct delivery of neurosecretion or conventional neural signaling from the brain to the CA is required. These alternatives have not as yet been distinguished. However in some instances, it is known that the CA must retain innervation to become and remain "active." Some of these have already been mentioned, e.g., *L. migratoria.* The source of the stimulation may be indicated by identification of the axonal pathway required for activation. Loss of CA activity following severance of NCA I suggests brain–CC influence, whereas such a loss following NCA II severance suggests subesophageal ganglion influence. Although Engelmann (1957) implicated the NCA II as the pathway for stimulation of the CA in *L. maderae*, this has not been confirmed in other studies. Severence of the NCA II did not alter the function of the CA in *P. americana* (Pipa, 1982) nor did it prevent activation of the CA in *D. punctata* (Stay and Tobe, 1977). Also, signals reaching the CA by way of the NCA II are not responsible for the basal level of JH biosynthesis by CA in *L. migratoria*, as there was no difference in activity of CA following transection of NCA I and transection of both NCA I and II (Couillaud *et al.*, 1984). Thus the regulatory route to the CA seems to be principally by way of NCA I and axons of the NCC I, II, and III from the brain in this species, in *S. gregaria*, and in *L. decemlineata* (see Raabe, 1982, for references).

4.3.1 *Immature stages*

Using allatectomy, CA implantation, and JH titer measurements, Yin and Chippendale (1979a) have shown in *Diatraea grandiosella* that active CA maintain diapause and that CA must have intact nerves to the brain (NCA I) to maintain their activity. Since four to eight brains from prediapausing larvae could not induce diapause in nondiapausing larvae, they were not able to demonstrate that brains have an allatotropic effect; however, they did observe neurosecretory endings in the CA (Yin and Chippendale, 1979b) and it is possible that an allatotropin must be delivered to the CA cells in order to be effective.

The necessity of intact nerves to the CA for maintenance of CA activity has also been demonstrated in the larval earwig *Anisolabis maritima*; severance of the NCA resulted in precocious adults (Ozeki, 1962, cited in Bhaskaran, 1981). A brain or CC factor acting through intact nerves to the CA was also shown in early third and fourth instars of *Pyrrhocoris apterus* by experiments involving implantation of brains in complex with CC and CA into test larval hosts and evaluation of juvenilization at the subsequent molt; CA implanted alone had no such effect (Hodkova, 1979b).

It has been clearly demonstrated in young penultimate instars of *L. migratoria* that the NCA I must be intact for the normal increase in JH biosynthesis in this stadium (Couillaud *et al.*, 1984). Measurements by RCA showed that NCA I severance resulted in low rates of JH biosynthesis compared to controls (about fourfold difference); the subsequent molt was normal but the duration of the stadium was extended. From these studies, no evidence is provided for the source of the stimulatory factor.

4.3.2 *Adult stages*

We have referred already to the instances in adults in which intact nerves to the CA are required for the stimulatory regulation in *L. migratoria* (Poras *et al.*, 1983) and *S. gregaria* (Strong, 1965b; Tobe *et al.*, 1977). Additional evidence for the importance of intact innervation for CA activity was provided by Strong's (1965a) observation that implanted CA, whether immature or mature, did not resume activity following implantation into allatectomized females. This author used oocyte growth and CA volume as criteria for CA activity. Cautery of NCC I tracts from the PI of *L. migratoria* showed that these axons must remain functional for the normal increase in rates of JH biosynthesis by CA (Girardie *et al.*, 1981; Tobe *et al.*, 1982).

In adult female *P. apterus*, the volume of the CA remained small after NCA I severance although oocyte development was not significantly different from controls (Hodkova, 1977). She suggested, however, that the CA were not as active as normal and that a different assay would be required to show this. Couillaud *et al.* (1984) have shown in *L. migratoria* that the development of oocytes is scarcely affected by denervation of the CA, yet the rates of JH biosynthesis are at least 20-fold lower than the maximal rates during the normal gonadotrophic cycle. Also in *S. gregaria*, direct measurements of rates of JH biosynthesis confirmed that CA require innervation for continued activity in this locust (Tobe *et al.*, 1977).

4.4 ALLATOHIBINS/ALLATOSTATINS

The downward modulation of biosynthetic activity of the CA may be regulated by factors which travel in the hemolymph and possibly also by way of intact nerves to the CA. These have been called allatohibins (Williams, 1976) if they act to reduce rates of biosynthesis and allatostatins (Tobe, 1980; Friedel *et al.*, 1980b) if they preserve a status quo, i.e., the unstimulated rates of biosynthesis. The cessation of release of allatotropins and the coincidental release of allatohibins/allatostatins would ensure the decline in JH biosynthesis, which is particularly important for metamorphosis to pupal and adult forms and for maintaining periods of vitellogenic inactivity. Definitive demonstration of such substances would require the same approach as that for demonstrating allatotropins (see Section 4.2.1), but as yet no attempt has been made to demonstrate the action of inhibitory substances *in vitro.*

4.4.1 *Allatohibins in immature stages*

An allatohibin was postulated to be part of the regulatory system for the CA in last instar larvae of *M. sexta* (Williams, 1976). Bhaskaran *et al.* (1980b) have shown that by day 1 after ecdysis, the CA of last instar larvae can no longer respond to the allatotropin of the starved–fed assay larva (see Section 4.2.2); presumably the CA have become irreversibly unresponsive to an allatotropin in the hemolymph. By exposing active CA to an interim host for a day, they found that an allatohibin was present in the hemolymph of last instar larvae only on day 0–1 (not after day 2). Cautery of large areas of the brain suggested that the source of the allatohibin is the MNSC region of the brain (group II and possibly some of group I cells of Nijhout, 1975b). Thus medial cells of the brain may be a source of both allatotropin and allatohibin.

In *L. migratoria* larvae of the penultimate stadium, cauterization of the center of the PI ("A and B" cells) inhibited metamorphosis (lengthened the stadium) in only a very few animals; addition of PI to such animals shortened the stadium, suggesting to Girardie (1967) that the PI restrained the CA. Yet destruction of lateral cells also seemed to remove a CA restraint and this could be reinstated by implantation of either medial or lateral neurosecretory regions of the brain (Girardie, 1974). Thus, it is not clear if either medial or lateral NSC produce allatohibins in these larvae.

In last instar females of *D. punctata*, JH biosynthesis is low in the last quarter of the stadium (Szibbo and Tobe, 1983). This appears to be due, at least in part, to humoral inhibition because larval glands removed from neural inhibition by denervation continue to show low rates of activity. In addition, CA transplanted from gonadotrophic females show reduced activity in such hosts (Szibbo and Tobe, 1983). Alternatively, it could be suggested that the allatotropin was no longer present in the larvae. However, there is no strong evidence for the occurrence of allatotropin in either larval or adult females of *D. punctata*. In fact, there is no loss of CA activity in gonadotrophic females following cautery of medial or lateral neurosecretory regions of the brain (Ruegg *et al.*, 1983) or decapitation shortly after adult emergence (Rankin and Stay, 1983).

4.4.2 *Allatohibins/allatostatins in adult stages*

Experiments with pregnant females of *D. punctata* suggest that the brain may be a source of allatostatin because removal of the brain by decapitation resulted in more activity by "loose" CA implanted into the abdomen than by "loose" CA in starved controls or headless animals into which protocerebra from pregnant females were implanted (Rankin and Stay, 1985a). Protocerebra, optic lobes, or deutocerebra from vitellogenic females did not restrain the CA (Rankin and Stay, 1985a).

Females of *L. decemlineata* kept in long-day conditions were used as donors of active CA which were implanted into animals kept under short-day conditions; JH biosynthesis by these CA was shown to be reduced, indicating that they were inhibited (Khan *et al.*, 1983). Thus a humoral factor is implicated in this downward regulation of the CA, but the source and the time of release of the allatostatin remain unknown. It has, however, been shown that CA from long-day animals are never as fully inhibited as are transplanted glands from short-day animals, suggesting that prediapause conditions provide a stepwise inhibition of the CA (Khan *et al.*, 1983).

In *P. apterus*, Hodkova (1979a) demonstrated an inhibitory factor operating in adult females under short-day conditions (in which reproduction ceases presumbly as a result of lack of JH). However, demonstration of humoral inhibition was complicated by the fact that the CA were also inhibited by way of intact nerves (see Section 4.5). Nevertheless, inhibition from the brain was reduced (oocyte production resumed) by removal of the PI and enhanced (oocyte production depressed) by the introduction of an additional isolated brain and pair of CA. One interpretation of these results is that in short-day conditions, the PI produces a factor which inhibits the CA by way of the hemolymph.

In *L. migratoria* Girardie (1966, 1967) presented evidence that the centrally located A and B cells in the PI inhibit the CA; but the observations of MacCaffery and Highnam (1975a) on the quantity of neurosecretory material in A cells following extirpation and implantation of CA suggest that A cells are stimulatory. It is not possible at present to distinguish between these.

4.5 INHIBITION REQUIRING NERVES

Whereas there are few demonstrations of humorally transmitted inhibition of the CA, there is much evidence that the CA are restrained by way of intact nerves from the brain. The pioneering observations of Wigglesworth (1948) and Scharrer (1952) revealed increased bioactivity of the CA after severance of the nerves between the CA and the brain. This is the only technique which has been used to demonstrate a neural route for inhibition. It remains to be demonstrated whether postsynaptic changes occur in the CA in response to stimulation of nerve cells in the brain and what type of nerve cells might be eliciting such a response, if it exists.

4.5.1 *Immature stages*

Analysis of the downward regulation of CA in last instar larvae of *M. sexta* suggests an initial decrease in activity exerted by way of the hemolymph followed by inhibition by way of neural tracts (Bhaskaran *et al.*, 1980b). Using an assay presumably sensitive to low levels of JH (namely allatectomized penultimate instar larvae), they demonstrated a declining activity of CA from day 1 to 3 of the last stadium. The final inhibition was shown to require nerves since denervation of the CA before day 3 resulted in continued CA activity as measured by the penultimate larval assay (Bhaskaran *et al.*, 1980b). Cautery of the MNSC did not destroy this neural inhibition, and it appears that the neural inhibition was reversible only if it had not been preceded by initial (day 1) humoral inhibition (Bhaskaran *et al.*, 1980b; Bhaskaran, 1981).

In early last larval instars of *G. mellonella*, the CA appear to be inhibited by way of their nervous connection. The inhibition can be partially removed (7% of experimental animals) by severance of the nerves to the CA and to a greater extent (20%) by implantation of stimulatory larval brains. The assay for activity in these experiments was induction of extra larval molts (Sehnal and Granger, 1975; also Sehnal and Granger unpublished, in Granger *et al.*, 1981).

CA of penultimate instar larvae of *R. prolixus* may be neurally inhibited to a small extent because destruction of the PI just after a blood meal results in more juvenilization at the next molt than does allatectomy. Baehr (1976) interprets these results to mean that CA may be neurally inhibited in the penultimate instar larvae.

In larvae of cockroaches the CA are unquestionably restrained by way of nervous connections. Early studies on *L. maderae* showed that denervation of the CA resulted in apparent increased CA activity, as indicated by changes in volume and the induction of supernumerary larval molts (Scharrer, 1946, 1952, Lüscher and Engelmann, 1960). In last instar female larvae of *D. punctata*, denervation of CA resulted in a supernumerary molt in 95% of the animals but only 18% in sham-operated controls (Szibbo and Tobe, 1983). This alteration in metamorphosis resulted from much higher rates of JH biosynthesis by the denervated CA than by sham-operated ones as demonstrated by RCA (Szibbo and Tobe, 1983). The route of this inhibition appears to be by way of the NCA I (Lüscher and Engelmann, 1960) and more specifically by way of the NCC I. In *P. americana* larvae, neural inhibition of the CA probably occurs to a lesser extent than in *L. maderae* and *D. punctata*. Although denervation of the CA resulted in supernumerary molts in 100% of the animals, 60–85% of sham-operated controls responded similarly (Pipa, 1980). Thus injury also appears to release CA from inhibition.

4.5.2 *Adults*

A brain inhibitory center which affects the CA by way of a neural route has been demonstrated in *R. prolixus*: the mean length of vitellogenic oocytes was greater in fed females with denervated CA than in control females (Baehr, 1973). Electrocautery of the PI, after the critical period for allatotropin release, gave similar results as denervation of the CA, and Baehr (1973) does not exclude the possibility that the PI may be the source of the inhibition.

In *L. migratoria* Kazalinsk strain, the inhibition of the CA induced by long-day conditions may also require intact nerves (Darjo, 1976). A similar condition occurs in the Savio strain which is photoperiod sensitive and diapauses under long-day conditions (Poras *et al.*, 1983). Implanted CA can induce oocyte development in diapausing animals and differential severance of nerves to the CA indicates that the NCC II are responsible for the inhibition, because severance of this nerve after the CA are mature appears to result in an increase in the titer of JH as measured by RIA and oocyte growth (Poras *et al.*, 1983). Thus in this instance, lateral NSC are implicated in the restraint of the CA.

Short days result in adult diapause and low rates of JH biosynthesis in *L. decemlineata* (Khan *et al.*, 1983). Glands from short-day females implanted into similar hosts show much higher rates of biosynthesis for several days, indicating that neural inhibition was reduced by denervation; a humoral inhibition ensues under short-day conditions (Khan *et al.*, 1983). Under long-day conditions, the normally high rates of biosynthesis can be reduced by starvation; denervated (implanted) glands escape this inhibition, thus indicating that the restraint of CA as a result of starvation is imposed through intact nerves (Khan *et al.*, 1983).

Neural inhibition of the CA has also been proposed in several other species of adult insects in which starvation prevents the normal increase in activity of the CA. Johansson (1958) induced oocyte production and presumably increased CA activity in starved *O. fasciatus* by severance of nerves to the CA. In *P. americana*, Pipa (1982) showed by oocyte growth bioassay for CA activity that severance of NCA I removed inhibition whereas severance of NCA II did not. In this species, Weaver and Pratt (1981) had shown previously by RCA of CA activity that starvation lowered rates of JH biosynthesis as well as oocyte growth.

Suppression of CA activity and oocyte development is maintained by neural inhibition in cockroaches when enbryos are carried in the brood sac. Denervation of CA at the level of NCC I appeared to release this inhibition of CA activity in *L. maderae* (Scharrer, 1952; Engelmann and Lüscher, 1957). Denervated CA also appeared to resume activity during pregnancy in *D. punctata* (Engelmann, 1959; Roth and Stay, 1961). RCA of rates of JH biosynthesis after denervation of the CA during pregnancy showed a twofold increase in rates compared to sham-operated controls (Rankin and Stay, 1985a). This neural inhibition of the CA decreases in the course of pregnancy and appears to come from a different source than that of the inhibition transmitted by way of the hemolymph (Rankin and Stay, 1985b).

The process of mating enhances reproductive capacity in many insects (see Engelmann, 1970, for references). Since denervation of the CA mimics mating in several species of insects, it would appear that in these, the CA are restrained by way of nerves. In the harlequin bug, *Dindymus versicolor*, Friedel (1974) showed that the oocyte development normally associated with mated females occurred in virgins following section of the allatal nerve or extirpation of an area around the MNSC; extirpation of MNSC had no effect. Thus, in *D. versicolor*, the source of this inhibition appears to be outside the MNSC. In young virgin *D. punctata*, the rates of JH biosynthesis remain low unless the females are mated or the CA are denervated (Stay and Tobe, 1977). Bilateral cautery of the MNSC was also followed by an increase in biosynthesis of both CA (Ruegg *et al.*, 1983). Unilateral cautery of the area adjacent to the medial cells or cautery of the lateral cells on one side of the brain resulted in activation of the ipsilateral CA (Ruegg *et al.*, 1983). Since unilateral severance of the NCC I or NCC I and II also activated the ipsilateral gland (Tobe *et al.*, 1981), it appears that inhibition of the CA involves an area lateral to the MNSC which very likely interacts with the contralateral medial region of the protocerebrum.

The foregoing descriptions indicate unquestionably that the nervous system, particularly the protocerebrum, is involved in the modulation of activity of the CA. The route of action may be directly by way of nerves to the CA or by way of the hemolymph. The results available from *in situ* experimentation and bioassays of CA activity and the more reliable *in vitro* RCAs of gland activity provide the background information necessary for rigorous demonstrations of the existence of allatotropins and allatohibin/allatostatins. Once this is accomplished, it should be possible to locate the sources and to elucidate the modes of action of these substances.

4.6 RESPONSES OF THE CA TO REGULATORY SIGNALS

4.6.1 *Asymmetry between CA*

Measured rates of JH biosynthesis have shown that in normal gonadotrophic activity cycles, the members of a pair may be well matched in rates as in *P. americana* (Weaver, 1979) and *D. punctata* (Szibbo and Tobe, 1981a) or show considerable difference in rates as in *S. gregaria* (Tobe, 1977) and *L. migratoria* (Ferenz and Kaufner, 1981). Weaver (1979) attributed this difference to differences in the pathway for stimulation of the CA: locusts require intact nerves for increased CA activity (Strong, 1965a; Tobe *et al.*, 1977, 1982; Poras *et al.*, 1983) whereas cockroach CA can be stimulated humorally (Scharrer, 1952; Engelmann, 1959; Stay and Tobe, 1977). In locusts, the neural stimulation of the CA must be specific and different for each CA. It should be pointed out that, in cockroaches, a neural inhibitory signal must be removed before the glands can be stimulated humorally; thus the normal neural regulation must be symmetrical.

Asymmetrical activation of the CA has been achieved experimentally in species in which CA are inhibited by way of intact nerves. Scharrer's (1952) supposition that a single CA was activated by denervation in *L. maderae* was confirmed by RCA in virgin female *D. punctata* (Tobe *et al.*, 1981; Ruegg *et al.*, 1983) and by bioassay in starved *P. americana* (Pipa, 1983). In these instances, the disinhibited gland underwent a normal cycle of biosynthesis that was regulated by way of the hemolymph and the other member of the pair remained uninfluenced by the factors which regulated the cycle in the denervated gland. This leads to the conclusion that the neural inhibition takes precedence over all of the "humoral" signals and that in the normal animal either both CA are released from inhibition symmetrically, or, if there is slight asymmetry in the disinhibition, it is overcome perhaps by signals from the ovary. If there is humoral ovarian stimulation of the CA in locusts, it does not take precedence over the neural stimulation.

4.6.2 *Steps in the regulation of the CA*

In a normal gonadotrophic cycle of CA activity, it can be hypothesized from the changes in morphology of the glands (see Section 1.4.2.4) that the regulation of synthetic activity takes place at several levels: proliferation of cells, surface membrane and organellar proliferation, and proliferation of enzymes for JH biosynthesis, followed by regression of all of these parameters. It is clear from the discussion of demonstrated regulatory factors that in any one species, there are multiple regulators for the gonadotrophic cycle (see Tobe and Pratt, 1976). It is possible that each factor regulates one of these parameters. Experimental uncoupling of the processes involved in the activity cycle of the CA during the gonadotrophic cycle in *D. punctata* suggests that the regulation does indeed occur at different levels. Normally there is an increase in number of cells in the CA as rates of synthesis increase. This proliferation of cells was recorded in the absence of increase in JH biosynthesis in females that were mated (CA disinhibited) but ovariectomized (Tobe *et al.*, 1984b). The cell number remained at that of maximally active glands yet in such animals, JH biosynthesis remained low (Stay and Tobe, 1978; Stay *et al.*, 1983). In ovariectomized females with CA disinhibited by severance of the NCA I (Stay and Tobe, 1977) or electrocautery of the brain (Ruegg *et al.*, 1983) rather than mating, JH biosynthesis also remains low. Although the cell organelles were not quantitated in ovariectomized mated females, the SER appears to be enriched (Fig. 4). Subsequent reduction in cell number was effected by the injection of 20-hydroxyecdysone (Tobe *et al.*, 1984b). This suggests that mitosis and organellar proliferation may be stimulated by "neural" disinhibition and that proliferation or activation of enzymes involved in JH biosynthesis requires stimulation emanating from the developing ovary, as does the decrease in cell number. In these ovariectomized females, JH biosynthesis by the CA can be somewhat stimulated *in vitro* by the addition of farnesoic acid (Tobe, unpublished), which indicates that at least the terminal enzymes in JH biosynthesis are present. At other normal stages of inactivity such as last instar larvae, these enzymes appear to be lacking, as the gland cannot be stimulated by addition of farnesoic acid (Tobe, unpublished).

An additional demonstration that proliferation of cells and rates of JH biosynthesis are regulated independently can be found in experiments utilizing unilaterally allatectomized females of *D. punctata.* Following this operation, the cell numbers undergo normal cyclic changes but the rates of JH biosynthesis per cell are doubled throughout the cycle (Szibbo and Tobe, 1981b).

4.7 FEEDBACK LOOPS REGULATING THE CA

There is clearly much to be learned about how the nervous, including neurosecretory, system regulates the CA. How this system is in turn regulated is even more of a challenge, not only with respect to the demonstration of pathways of action for the internal and external environmental signals, but also how they are integrated in order to modulate appropriately CA activity. The possibility that factors in the internal milieu act directly on the CA also exists and this too needs to be investigated further.

There are a few studies which have explored the CA-modulating factors in the internal milieu by direct measurement of CA activity and it is principally these which are outlined below.

4.7.1 *Negative feedback from JH*

The regulation of hormonal titer by negative feedback is a common phenomenon (Goldsworthy *et al.*, 1981). Thus, it is not surprising that JH biosynthesis by the CA is decreased at times of elevated titer. This phenomenon was predicted by the experiments of Slama *et al.* (1974) in which application of a JH analog resulted in morphological characteristics of the CA that indicated glands of low activity in *P. apterus.* Since these experiments, others have, in part, verified this conclusion by direct radiochemical measurements of CA activity after implantation of supernumerary CA or treatment of animals with exogenous JH or JH analogs. For a review of the biological effects of JH and JH analogs see Slama *et al.* (1974) and Sehnal (1976, 1981).

In *D. punctata*, following implantation of a supernumerary pair of CA into a gonadotrophic female, rates of JH synthesis by the implanted pair and host pair (whether denervated or innervated) were lower than was normal for a single pair of glands in such a female; however, the combined rates were slightly higher than normal (Tobe and Stay, 1980; Stay and Tobe, 1981). Although the JH titer was not measured directly in these experiments, the growth of oocytes served as an indirect measure of JH titer and it was found that with supernumerary CA, the oocytes grew only slightly faster than normal. Thus the rates of biosynthesis by the CA appear to be roughly adjusted to maintain near "normal" hormone titer. It should be noted that this was not obvious when rates of biosynthesis were measured at one time point during the gonadotrophic cycle (Stay and Tobe, 1978). In *L. decemlineata*, a single supernumerary pair of CA sufficed to reduce rates of biosynthesis of the host glands (Schooneveld *et al.*, 1979). However, Khan *et al.* (1982b) found that two to three supernumerary pairs were required to reduce sharply the rate of JH biosynthesis by the host glands.

Treatment of animals with exogenous JH or a JH analog also sharply reduced the rates of JH biosynthesis by CA. This was shown in gonadotrophic female *D. punctata* following application of a JH analog (ZR512, 25 or 100 μg/animal) (Tobe and Stay, 1979) and in gonadotrophic female *L. decemlineata* following five daily topical applications of 30 μg JH I per animal (Schooneveld *et al.*, 1979) or injection of 50 μg JH III per animal (Khan *et al.*, 1982b). In the last case, JH synthesis by CA, whether innervated or denervated, was suppressed by the added hormone. Buschor *et al.* (1984) probably also showed negative feedback by injected JH III in early pregnant females of *N. cinerea* from which the egg cases were removed because although oocytes grew in response to this treatment, rates of JH biosynthesis remained low. Thus growing oocytes were unable to stimulate CA activity (see Section 4.8).

4.7.2 *Positive feedback of JH*

A positive feedback of JH on the CA of saturniid pupae was suggested originally by Siew and Gilbert (1971). These authors observed an increased nuclear RNA synthesis in the CA in response to JH treatment although the relation between RNA synthesis and JH synthesis is unknown. In *D. punctata*, a low dose (2.5 μg/animal) of the JH analog ZR 512 produced a marked brief stimulation in JH synthesis by CA of gonadotrophic females early in oocyte development (Tobe and Stay, 1979). A similar stimulation in rates of JH biosynthesis was observed in pregnant females of *N. cinerea* from which egg cases were removed; low doses of JH III (2 or 10 μg/animal) stimulated not only oocyte growth but also JH synthesis (Buschor *et al.*, 1984).

Evidence for positive feedback may also be found in the experiments utilizing the procedure of unilateral allatectomy which results in compensation by the remaining gland, such that its rate of biosynthesis is increased above normal. In *D. punctata*, this resulted in a doubling of activity (Stay and Tobe, 1978) that was accompanied in part by an increase in number of cells (Szibbo and Tobe, 1981b). Unilateral allatectomy in *L. decemlineata* also resulted, as shown by RCA, in a doubling in rates of biosynthesis by the remaining gland and, as shown by bioassay, an apparently normal JH hemolymph titer (Schooneveld *et al.*, 1979). This compensation mechanism indicates that titer of JH probably plays an important role in regulating the synthetic activity of the CA.

4.7.3 *JH feedback pathway*

The pathway for the JH feedback appears not to be directly on the CA because CA incubated *in vitro* with JH are not inhibited (Pratt and Finney, 1977; Kramer and Staal, 1981; Khan *et al.*, 1982b). On other hand, there is no evidence that this feedback operates through the brain even though JH does appear to influence NSC of the brain. Thomsen and Lea (1969) demonstrated that allatectomy shortly after adult emergence prevented the normal increase in an index of activity (increase in nuclear volume) in the MNSC of meat-fed female *Calliphora erythrocephala* whereas implantation of CA restored this activity; this suggests that JH acts on the MNSC. McCaffery and Highnam (1975a,b) found that either implantation of supernumerary CA or JH injections stimulated activity (increase in nuclear volume and release of stainable material) in A cells of the MNSC in brains of adult female *L. migratoria*. These studies were done *in vivo* and consequently, the effect of the JH could be indirect. In addition, it remains to be rigorously demonstrated that the brain cells in question produce an allatotropin. Nevertheless, demonstrations such as these provide strong impetus for further investigation of the CA–neurosecretory cell axis. In such studies, it will be important to define the hemolymph titers of hormones that elicit the negative and positive feedback responses in the intact animal. Utilization of the JH RIAs would be most useful in this endeavor.

4.8 OVARIAN INFLUENCE

Ovarian development in many adult insects is dependent to a large extent on JH, which plays a dual role of stimulating the fat body to synthesize vitellogenin and of stimulating ovarian competence for vitellogenesis (Engelmann, 1979; Davey, 1982).

In other insects, the ovary may be less dependent on JH. For example in mosquitos, JH is required only for growth of oocytes to the stage of previtellogenesis, i.e., competent to undergo vitellogenesis (Gwadz and Spielman, 1973; Hagedorn *et al*., 1977; Rossingnol *et al*., 1981).

CA activity during vitellogenic cycles has been measured directly by RCA in several species. In Section 1.4.2.4, we discussed those for which fine structural studies of the CA were available. Those that correlate reasonably well with oocyte development are shown in Fig. 9,A–E. Rates of biosynthesis are plotted for approximately similar stages in the vitellogenic cycle of all species (see legend for sources of data). In *S. gregaria*, *N. cinerea*, and *D. punctata*, a vitellogenic cycle occurs only in the basal follicles of the ovary and in these species, one major peak of JH biosynthesis occurs per cycle: at about midcycle in *N. cinerea* (Fig. 9D) and toward the end of the cycle in *S. gregaria* (Fig. 9C) and *D. punctata* (Fig. 9E). However in *S. gregaria*, there is a small peak before the major peak of the first cycle. In *P. americana* and *M. sanguinipes*, while the basal follicles are still vitellogenic, the penultimate ones also enter vitellogenesis. In *P. americana*, the first cycle has only one major peak of JH biosynthesis but in subsequent cycles, there is a decline in JH biosynthesis at midvitellogenesis (Fig. 9B). In *M. sanguinipes*, such a decline occurs in the first as well as the subsequent cycle (Fig. 9A). Note that the decline in the second cycle in both of these species occurs at ovulation of the first mature oocytes.

The close correlation of stage of oocyte development with rates of JH biosynthesis shown in Fig. 9 suggests that the ovary may have a changing influence on CA activity, especially in view of the fact that no cycle of JH biosynthesis at midvitellogenesis (Fig. 9B). In *M. sanguinipes*, such a decline 1978; Stay *et al*., 1983) and *N. cinerea* (Lanzrein *et al*., 1981a). The changing influence of the ovary on the CA has been analyzed in *D. punctata*. Ovaries in various stages of development were grown for short periods (24 and 48 h) in previously ovariectomized females; RCA of the hosts' CA showed that the size of the ovary was positively correlated with the JH biosynthesis elicited in the test period. This relationship was observed until shortly before chorionation, at which stage this stimulation of JH biosynthesis ceased (Rankin and Stay, 1984). The ovary not only ceases to be stimulatory, but becomes inhibitory as the basal oocytes grow. This was first demonstrated in experiments involving the implantation of female CA into males of *D. punctata* (Stay *et al.*, 1980). In males, the presence of an ovary is not required for elevated JH synthesis by the implanted female CA, but in its presence, JH biosynthesis declines at the end of vitellogenesis. That a similar phenomenon occurs in females is more difficult to demonstrate. Lanzrein *et al.* (1981a) showed that there is a slower decline in JH synthesis following removal of the ovaries at the stage of near-maximal growth. To demonstrate that ovaries actively inhibit the CA as opposed to ceasing to stimulate them in *D. punctata*, Rankin and Stay (1985b) implanted presumably inhibitory ovaries along with stimulatory ones into ovariectomized host females and observed rates of JH biosynthesis intermediate between those found when each ovary was implanted alone.

The ovary of *A. aegypti* appears to inhibit JH biosynthesis following its JH-dependent growth to the previtellogenic stage (Rossignol *et al.*, 1981). Pupal ovaries transplanted into adults failed to grow in the hosts' previtellogenic ovaries were present but grew in ovariectomized hosts; however, the implanted pupal ovary did grow in the presence of previtellogenic ovaries following treatment of the host with JH analog (Rossignol *et al.*, 1981). Thus the previtellogenic ovary appears to lower JH titer, presumably by inhibiting CA activity.

How the ovary exerts it dual influence, both stimulatory and inhibitory, over the CA is unclear. In *D. punctata*, it is known that only the vitellogenic follicles (i.e., oocytes and surrounding follicle cells) of the ovary are required to elicit the normal cycle of JH biosynthesis (Rankin and Stay, 1984), that only one-quarter of the normal compliment of ovarioles per animal suffices, and that the ovary need not be innervated (Stay *et al.*, 1983). Thus, either the oocytes or the follicle cells appear to produce a regulatory substance(s). It has been suggested that the inhibitory function of the ovary is effected by ecdysteroids (Stay *et al.*, 1980), because mature ovaries contain ecdysteroids (Hoffmann *et al.*, 1980) and may release ecdysteroids during *in vitro* incubation (e.g., Hanaoka and Hagedorn, 1980; Koeppe, 1981). In *A. aegypti* the ovary releases ecdysteroids at the start of vitellogenesis (Hagedorn *et al.*, 1975). Also ecdysteroids, administered in sequential doses, are known to inhibit the CA *in situ* in some species (Stay *et al.*, 1980; Friedel *et al.*, 1980b; Lanzrein *et al.*, 1981b) though not in others (Khan *et al.*, 1982b) and to reduce CA cell numbers after they have increased in the absence of the ovary (Tobe *et al.*, 1984b; also see Section 1). In addition, titers of ecdysteroids are reduced following ovariectomy in *A. domesticus* (Renucci and Strambi, 1981), *R. prolixus* (Ruegg *et al.*, 1981), and *L. riparia* (Baehr *et al.*, 1982) but not in *D. punctata* (Stay *et al.*, 1984). Accordingly, it has not been demonstrated unequivocally that the ovarian ecdysteroids regulate the CA.

We do not know whether the putative ovarian regulatory factor acts directly or indirectly on the CA. In *D. punctata*, exposure of CA to ecdysteroids *in vitro* did not alter the rates of JH biosynthesis (Friedel *et al.*, 1980b), suggesting that the influence was indirect. Action through the brain, however, was excluded by decapitation experiments in the same species: a normal cycle of JH biosynthesis occurred in the absence of the head (Rankin and Stay, 1983).

4.9 ECDYSTEROIDS

As mentioned above, treatment with ecdysteroid has been shown to result in decreased rates of JH biosynthesis *in vivo* in adult insects. If ecdysteroids are potential regulators of JH biosynthesis in adults, it seems even more probable that they play this role in larval stages, in which interactions of JH and ecdysteroids appear to be important in the process of metamorphosis (Gilbert and King, 1973; Riddiford, 1980; Laufer and Borst, 1983).

In last instar larvae of female *D. punctata*, rates of JH biosynthesis decline (Szibbo *et al.*, 1982) as titers of ecdysteroids reach a peak (Stay and Tobe, unpublished), suggesting that there is a relationship between elevated titers of ecdysteroids and the decline in JH synthesis.

As the measurements of JH and ecdysteroid titers become more accurate and detailed and available for more species, it may be possible to choose the system best suited for definitive demonstration of the role of ecdysteroids in the regulation of the CA.

4.10 OTHER INFLUENCES ON CA REGULATION

Clearly there must be numerous factors in the internal environment in addition to those mentioned above that influence the regulation of the CA, either directly or by way of the nervous system. Feeding, for example, may be an integral part of the regulatory system, and starvation may modify this in larval development, e.g., in *M. sexta* (Bhaskaran, 1981; Cymborowski *et al.*, 1982) and *L. decemlineata* (Khan *et al.*, 1983), and in gonadotrophic females, e.g, in *P. americana* (Weaver and Pratt, 1981), *O. fasciatus* (Johansson, 1958), and *S. gregaria* (Tobe and Chapman, 1979). Starvation could affect the nervous system either directly or by changing the composition of the hemolymph. The chilling that elicits release of a presumptive allatotropin from the brain of *G. mellonella* larvae appears to act through the ventral nerve cord (Bogús and Cymborowski, 1981). Severance of the ventral nerve cord resulted in a significant loss of the ability of chilled larvae to undergo JH-induced supernumerary molts (Bogús and Cymborowski, 1981).

The specific JH titers and their precise timing which provide for normal development require that CA be regulated in response to the progress of development. Just what information about development is important in this regard and how it is relayed to the CA are largely unknown. Hodkova (1979b) suggests that the brain counts instars in *P. apterus*. However, an absolute or relative body size may also be assessed, as in *M. sexta* (Nijhout and Williams, 1974) and *O. fasciatus* (Blakley and Goodner, 1978; Nijhout, 1979). In the latter case, stretch receptors are implicated as the sensory receptors (Nijhout, 1981). Numerous external conditions have also been shown to affect the rates of JH biosynthesis. Of course these must eventually be manifested as internal signals.

Social interactions between insects are among the external influences affecting CA activity. These include mating behavior and interactions between castes in social insects, as well as the less specific interactions such as the degree of crowding of individuals. Environmental conditions such as temperature and day length are, as has been mentioned previously, important in the regulation of the CA in some species.

The reader is directed to other reviews, for example, those of Engelmann (1970), Doane (1973), and Nijhout and Wheeler (1981), for references on these subjects. They are important for a complete understanding of the regulation of the CA in that the pathways for the relevant incoming information need eventually to be described. In some cases, the sensory receptors have been implicated and it will be important to identify them unequivocally so that the complete regulatory pathway can be followed. This may not be an unrealistic goal for the near future. The regions of the brain which exert influence over the neurosecretory cells that presumably influence the CA have been pointed out in Sections 1.2 and 4.1–4.5. At the present time their identification rests on experiments involving destruction of relatively large areas of the brain. But the technology is now at hand to selectively destroy identified neurons (Miller and Selverston, 1979). Utilization of such techniques as well as selective stimulation of such neurons will undoubtedly be an essential part of understanding the regulation of the CA, at least that part of it which is directed through the nervous system.

5 Regulation of juvenile hormone titer

Although a review of the mode of action of JH is beyond the scope of this article, there is ample evidence to indicate that JH is a vital component in the regulation of both metamorphosis and reproduction in virtually all insect species, with the apparent exception of selected Lepidoptera and Diptera. It seems likely that changes in the titer of the hormone regulate specific processes such as synthesis of specific cuticular mRNAs and proteins in metamorphosing Lepidoptera (Riddiford, 1980; Riddiford and Hori, 1981), the synthesis and uptake of vitellogenin in female Dictyoptera and Orthoptera (Engelmann, 1979, 1981, 1984a,b; Chen *et al.*, 1976, 1979; Dhadialla and Wyatt, 1981, 1983), and the synthesis and utilization of oothecins by the left colleterial gland in female Dictyoptera (Weaver, 1981b). Although the absolute dependence of these processes on JH appears clear, how changes in JH titer control these processes remains unknown, in part because changes in JH titer have not been accurately documented by unequivocal measurements and in part because absolute dose dependence of these processes on JH *in vivo* remains to be determined. Nonetheless, changes in JH titer appear to be fundamental to the regulation of the processes which it controls. In addition, changes in JH titer may regulate JH biosynthesis by the CA, either through modulation of the allatotropin/allatostatin system (Section 4) or by action on the target organs which themselves feedback on the CA (e.g., ovaries; Section 4.8). We will consider briefly here how titer may be regulated *in vivo* and how *in vitro* measurements of rates of JH biosynthesis may provide useful information on the regulation of JH titer.

It is obvious that JH titer is ultimately regulated by the interaction between biosynthesis and release of the hormone from the CA on the one hand, and its degradation and clearance from the hemolymph by tissue uptake and excretion on the other hand. JH titers will be in steady state only when the rate of biosynthesis and release of JH (V_s) equals the rate of degradation and clearance of hormone from the hemolymph (V_d). There are several factors involved in degradation, including esterase activity in the hemolymph and tissues as well as clearance of JH from the hemolymph by uptake and/or sequestration of JH in the tissues and excretion of the hormone, presumably by the Malpighian tubules (Erley *et al.*, 1975). It is unlikely that steady-state conditions often occur at stages in which changes in JH titer are regulating specific physiological processes, and accordingly, it may be incorrect to assume steady-state conditions in general. Nevertheless, at any given time, the titer of JH must be regulated by an interaction of biosynthetic and catabolic processes. There has been some debate concerning the relative contribution of these two processes to the determination of JH titer, some authors suggesting that in larval Lepidoptera, catabolism of JH by esterases is the major mechanism controlling JH titer (Akamatsu *et al.*, 1975; Sparks and Hammock, 1980; Wing *et al.*, 1981; Sparks *et al.*, 1983; Sparks and Rose, 1983), whereas other authors have suggested that in adult insects, changes in JH biosynthesis are a major factor in the regulation of JH titer (Stay and Tobe, 1978; Tobe and Stay, 1980; de Kort *et al.*, 1978; Kramer, 1978; Kramer and de Kort, 1976; de Kort and Granger, 1981). We shall briefly examine the evidence supporting each of these views.

5.1 IMPORTANCE OF JH BIOSYNTHESIS

Ample evidence has accumulated to indicate that JH biosynthesis is a major regulator of JH titer. However, much of this evidence has been indirect, as summarized by Tobe (1980). The most compelling indirect evidence concerns the large and predictable changes which occur in JH biosynthesis as a function of both age and basal oocyte length (Tobe, 1980; Feyereisen *et al.*, 1981b). Patterns of CA activity appear to be specific to each species studied, but in general, rates of JH biosynthesis are high during vitellogenesis and low during previtellogenesis and chorion formation; some examples of the changes in JH biosynthesis which occur during the gonotrophic cycle are shown in Fig. 9. It is only in *D. punctata* and *N. cinerea* that oocyte length or ovarian dry weight can be precisely associated with known rates of JH biosynthesis, providing a useful parameter for predicting JH biosynthesis (Tobe, 1980; Feyereisen *et al.*, 1981b; Lanzrein *et al.*, 1978; Rankin and Stay, 1984). Nonetheless, changes of at least 5- to 10-fold have been observed in most species during the gonotrophic cycle (Fig. 9), and in view of the dose dependence of oocyte growth on JH (e.g., Tobe and Stay, 1979; Lanzrein, 1979), it is likely that these changes in JH biosynthesis are reflected in changes in JH titer. Other manipulations which appear to alter JH titer by way of changes in JH biosynthesis, such as nerve transection, addition or removal of CA, exogenous JH treatment, or treatment with known inhibitors of JH biosynthesis, all have a profound effect upon oocyte growth, and again assuming that oocyte growth is directly related to JH titer (see Fig. 10), we must conclude that these directly measured changes in JH biosynthesis and oocyte growth are an accurate reflection of JH titer (see Tobe, 1980). This conclusion is based largely on work with adult female *D. punctata* and should be extended to other species.

The question of the relation between JH biosynthesis and JH titer is of course best answered by direct measurement of these two parameters. Unfortunately, such determinations have been performed in very few species, the principal examples being *L. decemlineata* (de Kort *et al.*, 1981, 1982), *L. migratoria* (Rembold, 1981), and *D. punctata* (Tobe *et al.*, 1984a, 1985). In only the latter example has sufficient data been accumulated to permit their analysis by linear regression but in this case, there is an excellent correlation ($r = 0.97$, $n = 14$; y intercept = 0.2 μM) (Tobe *et al.*, 1984a, 1985). Thus, at least in *D. punctata*, changes in JH biosynthesis result in precise and predictable changes in JH titer. Two points are of particular interest here. (1) The y intercept for the plot of JH biosynthesis and release vs JH titer is

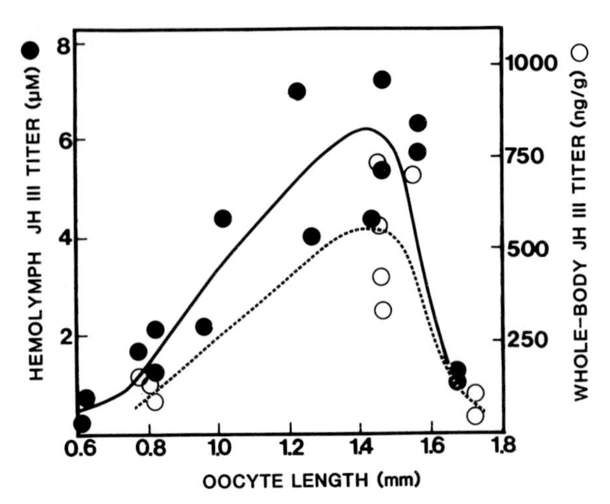

Fig. 10. JH III titer, as determined by physicochemical methods, as a function of basal oocyte length in *D. punctata*. JH III titer in both hemolymph (●) and whole-body extracts (○) are shown. Oocytes enter vitellogenesis at 0.8 mm and complete chorion formation at 1.6–1.7 mm. (Data from Tobe *et al.*, 1984a.)

0.2 μM—in other words, when JH biosynthesis and release is 0, the JH titer is 0.2 μM. This suggests that the tissues of this species are quite insensitive to JH and that a JH titer of 0.2 μM may essentially represent the JH threshold. Alternatively, if JH is in fact a "metabolic" hormone, regulating general metabolic events in insects, as has been suggested by numerous authors (see Steele, 1976), this titer of JH may represent the basal level necessary for the maintenance of "metabolic" effects but not specific reproductive effects. In either case, a JH titer of 0.2 μM has been observed in newly emerged females, before appreciable oocyte growth, in pregnant females, in late last instar females, and in adult males (Tobe *et al.*, 1984a). In the last three instances, allatectomy does not affect the ability of the animals to complete gestation (Stay, unpublished), to undergo metamorphosis (Szibbo and Tobe, 1983), or to fertilize females, respectively (Tobe *et al.*, 1979), supporting the notion that this JH titer is below the threshold. (2) The changes in JH biosynthesis and release which occur during the gonotrophic cycle (5- to 20-fold) (see Fig. 9) are similar in magnitude to the changes in hemolymph JH titer. The changes in hemolymph JH titer are also reflected in changes in whole-body JH titer (see Fig. 10; Tobe *et al.*, 1984a), again supporting the validity of this physicochemical methodology for the determination of JH titer (Bergot *et al.*, 1981a).

Although the relation between JH titer and biosynthesis and release is apparent in *D. punctata*, additional information is required to establish the validity of this relationship in other species. Hence, it is not possible to comment on the universality of this relationship at present. It would be particularly useful to explore the relation between JH titer and biosynthesis in the larval Lepidoptera since there is considerable evidence to indicate that JH esterases play a role in the regulation of JH titer in these animals. For the present, both biosynthesis and degradation of JH must be regarded as the major regulators of JH titer in the Lepidoptera.

5.2 ROLE OF JH ESTERASE

5.2.1 *Classification of esterase*

Esterases (E.C. 3.1.1) capable of hydrolyzing the methyl ester of JH have been reported in the hemolymph of many species (see Hammock and Quistad, 1981). Carboxylic ester hydrolysis appears to be the primary pathway of JH degradation, giving rise to the product JH acid. An alternate pathway involves the hydrolysis of the 10,11-epoxide by an epoxide hydratase (equivalent to hydrase and hydrolase; E.C. 3.3.2.3) (Slade and Zibitt, 1972) to produce JH diol. JH acid and JH diol may serve as substrates for the hydratase and esterase, respectively, resulting in the production of the final degradation product JH diol acid. JH acid, diol, and diol acid all appear to have low biological activity relative to JH (Hammock and Quistad, 1981), although recent reports have suggested that JH acid may exert biological effects in the Lepidoptera (Bhaskaran *et al.*, 1980a; Sparagana *et al.*, 1984). Epoxide hydratase activity is confined largely to the tissues, and accordingly, the primary pathway of inactivation of JH in the hemolymph is by ester hydrolysis (see Wing *et al.*, 1981).

5.2.1.1 *Esterase inhibitors* JH esterases have been classified largely on the basis of substrate specificity and effects of inhibitors. They appear to be poorly inhibited by such classical esterase inhibitors as diisopropyl phosphorofluoridate (DFP) and eserine and their resistance to DFP inhibition has been used extensively to distinguish a priori between specific and nonspecific esterases (Sanburg *et al.*, 1975a,b). Nonspecific esterases have been proposed to be capable of hydrolyzing JH (when unbound to proteins) (Sanburg *et al.*, 1975a). However, little difference in JH esterase activity has been observed in DFP-treated and untreated hemolymph of larval *M. sexta* (Vince and Gilbert, 1977; Beckage and Riddiford, 1982; Sparks *et al.*, 1983). Sparks *et al.* (1983) have also determined that in fact the "general," nonspecific esterases in the hemolymph of *M. sexta* did not appreciably hydrolyze JH, relative to the JH esterase. It thus appears that the nonspecific esterases make little contribution to the hydrolysis of JH in the hemolymph in larval *M. sexta*.

JH esterases are inhibited by many organophosphates, particularly certain phosphoramidothiolates (Pratt, 1975; Hammock *et al.*, 1977; Sparks and Hammock, 1980; Sparks *et al.*, 1983), but are poorly inhibited by carbamates, *p*-hydroxymercuribenzoic acid (PHMB), and phenylmethyl sulfonylfluoride (PMSF) (Sparks and Hammock, 1980; McCaleb *et al.*, 1980; Roe *et al.*, 1982; Sparks *et al.*, 1983), all well-known esterase inhibitors. However, the JH esterase of Dipteran species displays a different sensitivity to these inhibitors (see Sparks and Hammock, 1980) and these authors have suggested that the JH esterase may be of less importance in the regulation of JH titer in this order. McCaleb *et al.* (1980) classified the JH esterase of *G. mellonella* as an acetylesterase (E.C. 3.1.1.6) but this seems inappropriate on the basis of its substrate specificity. For this reason, Sparks, Hammock, and co-workers proposed that the JH esterase of *Trichoplusia ni* and *M. sexta* be classified as a carboxylic ester hydrolase (E.C. 3.1.1.1) (Sparks and Hammock, 1980; Sparks *et al.*, 1983).

5.2.1.2 *Substrate specificity* The JH esterase appears to be highly selective for the JHs and substrate studies using competitive inhibitors of the esterase in the presence of JH I or III have indicated that the JHs are the preferred substrate for the esterase (Weirich and Wren, 1973, 1976; Hammock *et al.*, 1977; Sparks and Hammock, 1980; Sparks *et al.*, 1983). However, little is known of the enantioselectivity of the enzyme, although recent reports have demonstrated that in *L. migratoria*, the 10*R* enantiomer is preferentially hydrolyzed (Peter *et al.*, 1983), whereas in *L. decemlineata* and *M. sexta*, the esterase displays no enantioselectivity (de Kort *et al.*, 1983; Peter *et al.*, 1981). Clearly, more research in this area is required. The studies of enantioselectivity of the JH esterase utilized derivatization and analysis of the diastereomer products following incubation of the esterase with racemic mixtures of radiolabeled JH; such studies should now be complemented by measurement of hydrolytic rates using the natural JH enantiomers as substrates in these and other species.

5.2.2 *Role in regulation of JH titers*

The role of JH esterase in the regulation of JH titers may differ between species and perhaps between stages. However, there is evidence to support the view that at least in larval Lepidoptera, the JH esterase is an important component in the regulation of titer. (1) Known JH esterase inhibitors (phosphoramidothiolates) can disrupt metamorphosis. Treatment of larval *T. ni* with EPPAT (*O*-ethyl-*S*-phenyl phosphoramidothiolate) resulted in delayed pupation and larval/pupal intermediates (Sparks and Hammock, 1980). These effects are symptomatic of JH or JH analog treatment (Sehnal, 1976) and Sparks and Hammock (1980) accordingly suggested that JH titers was elevated in EPPAT-treated insects. Unfortunately, this is the only report of metamorphic disruption by a JH esterase inhibitor. Nonetheless, this important paper provides compelling indirect evidence for the role of JH esterase in the regulation of JH titer in larval Lepidoptera. (2) Induction of JH esterase activity by treatment of insects with JH or JH analogs has been reported for numerous species; this induction has been demonstrated to occur in a dose-dependent fashion (Whitmore *et al.*, 1972, 1974; Kramer, 1978; Reddy *et al.*, 1979; Sparks and Hammock, 1979a,b; Wing *et al.*, 1981; Hammock *et al.*, 1981; Rotin *et al.*, 1982; Rotin and Tobe, 1983). Thus, the hormone appears to be regulating its own breakdown. The validity of such an experimental paradigm remains to be demonstrated but such a mechanism may function to reduce JH titer following peaks in hormone synthesis and titer. Steady-state JH titers would have no effect on JH esterase activity. It is noteworthy that JH treatment does not regulate the absolute appearance of the JH esterase but rather modulates the level of esterase activity, and this action is consistent with the role of the esterase in reducing high JH titers. Similarly, in last instar *T. ni*, there are two peaks of JH esterase activity, the first of which appears to be JH independent, regulated by a factor from the head, whereas the second peak is JH inducible and unaffected by the presence of the head (Sparks and Hammock, 1979b; Jones *et al.*, 1981; Hammock *et al.*, 1981). This differential regulation of the JH esterase also supports the notion that induction of esterase activity may be important in the regulation of titer. The inducible and noninducible forms of the esterase appear to be identical (Sparks and Hammock, 1979b). (3) JH esterase activity undergoes large changes during metamorphosis and reproduction in many species (e.g., Sanburg *et al.*, 1975a; Kramer and de Kort, 1976; Sparks *et al.*, 1979, 1983; Rotin *et al.*, 1982). The magnitude of these changes ranges from about twofold in adult female *D. punctata* (Rotin *et al.*, 1982), to 100-fold or more in adult female *L. decemlineata* (Kramer and de Kort, 1976) and last instar *M. sexta* (Sanburg *et al.*, 1975a) and *T. ni* (Sparks *et al.*, 1979). Unfortunately, there are little data correlating JH esterase activity with JH titer (determined by physicochemical methods), but clearly if these changes in JH esterase activity can be correlated with changes in JH titer, a role for the JH esterase in the regulation of JH titer is implied. In adult female *D. punctata*, the only species for which data on JH biosynthesis, esterase, and titer are available (Tobe *et al.*, 1984a, 1985), maximal levels of JH esterase activity occur at times when the JH titer is low or declining. Although there is an apparent correlation in this species, it should be noted that the JH esterase activity changes by a factor of less than two during these times. We do not know if a change of such magnitude is sufficient to effect the decline in JH titer and additional studies on the kinetics of the enzyme will be necessary, utilizing the natural enantiomer. Nonetheless, the apparent change in the level of JH esterase activity in many species suggests a role for the esterase in the regulation of JH titer. (4) The JH esterase shows high specificity for JH (Weirich and Wren, 1973; Kramer and de Kort, 1976; Hammock *et al.*, 1977; Sparks and Hammock, 1980; Slama and Jarolim, 1981; Coudron *et al.*, 1981; Sparks and Rose, 1983). This suggests that JH is in fact the natural substrate of the esterase and on the basis of this assumption, Sparks and Rose (1983) have hypothesized that the JH esterase is important in the regulation of JH titer, at least in last instar *T. ni*.

From the preceding discussion, it appears that the JH esterase may be important in the regulation of JH titer in some species, particularly the larval Lepidoptera, but may be less important in other orders. In general, the catabolic activity of the hemolymph, in terms of JH esterase activity, is far greater than the *in vitro* rate of JH biosynthesis, by a factor of several thousand (see *D. punctata*: Tobe *et al.*, 1984a, 1985; *L. decemlineata*: Kramer and de Kort, 1976; Kramer, 1978a,b; de Kort *et al.*, 1981, 1982; Khan *et al.*, 1982a), and at least in these species, this argues against a role for the esterase in JH titer regulation. However, these observed levels of JH esterase activity may be artificially high because (1) these studies have utilized racemic mixtures of JH rather than the natural enantiomer; the esterase may show enantiomeric specificity; and (2) preparations of crude diluted hemolymph have been used to estimate JH esterase activity. In view of the existence of hemolymph proteins with high affinity for JH (see below) and the apparent enhancement of JH esterase activity in diluted hemolymph (see de Kort *et al.*, 1983), crude hemolymph preparations may represent an unsatisfactory system for the investigation of JH esterase activity. Isolation of the JH esterase would clearly be of value in future studies on the kinetics of the esterase (Rudnicka and Hammock, 1981; Yuhas *et al.*, 1983; Rudnicka and Kochman, 1984). Accordingly, a reinvestigation of JH esterase activity, using isolated purified esterase and enantiomerically pure authentic JH, should be useful in clarifying the role of the JH esterase in the regulation of JH titer.

5.3 ROLE OF JH BINDING AND SEQUESTRATION

Binding of JH to proteins, particularly lipoproteins and JH-specific carrier proteins, may play an important role in the regulation of JH titer. While it is clear that JH can associate both with high capacity, low affinity, low specificity lipoproteins (e.g., Whitmore and Gilbert, 1972; Bassi *et al.*, 1977) and with low capacity, high affinity, high specificity proteins (e.g., Kramer *et al.*, 1974, 1976; Kramer and Childs, 1977; Peterson *et al.*, 1977) in the hemolymph of some species, it is uncertain if all species possess JH-specific binding proteins and the role of such proteins remains speculative. JH-binding proteins were originally suggested to afford protection to JH from hydrolysis by the nonspecific hemolymph general esterases but not the JH-specific esterase in larval *M. sexta* (Kramer *et al.*, 1974; Sanburg *et al.*, 1975a,b), although the importance of protection from the nonspecific esterase has recently been challenged (Sparks *et al.*, 1979, 1983; Wing *et al.*, 1981). Because the JH-specific binding proteins do not appear to be universally distributed in insects, their role in the regulation of JH titer may be less important than originally believed. Rather, these proteins may ensure the rapid and uniform distribution of JH within the insect, perhaps as a result of reduced partitioning of JH into lipophilic compartments and decreased sequestration by the tissues (Hammock *et al.*, 1975; Nowock *et al.*, 1976; Kramer and Law, 1980b; Wing *et al.*, 1981). Both Gilbert *et al.* (1978) and Wing *et al.* (1981) have noted that the concentration of the JH-specific binding protein varies with the hemolymph protein concentration and this observation led Wing *et al.* (1981) to suggest that these relatively small changes in titer of the binding protein (about two- to threefold relative to JH esterase) militate against a major role for the binding protein in the regulation of JH titer in certain larval Lepidoptera. However, the study of Hartmann (1978) suggests a direct role for JH-binding protein in the regulation of JH titer in female *Gomphocerus rufus*. Injections of antibody to the JH-binding protein into females mimicked the effect of allatectomy in terms of suppression of oocyte development. Unfortunately, it is uncertain that this author was in fact dealing with the authentic binding protein because only JH I was employed in the binding studies. In addition, the effect of injection of heterogeneous antiserum remains unclear.

Recently, the existence of an enantioselective high affinity JH III binding protein in the hemolymph of *L. migratoria* and *L. decemlineata* has been proposed (Peter *et al.*, 1979; de Kort *et al.*, 1983), extending the observations on the binding proteins of *M. sexta* (Goodman *et al.*, 1978; Schooley *et al.*, 1978). The unnatural 10*S* enantiomer is degraded more rapidly than the 10R form in both species, leading these authors to conclude that the 10*S* enantiomer is not afforded protection from the nonenantioselective JH esterase (Peter *et al.*, 1979; de Kort *et al.*, 1983). It should also be noted that in *L. decemlineata*, *L. maderae*, and *L. migratoria*, the JH-specific binding protein has much higher affinity for JH III than JH I (de Kort *et al.*, 1983; Kovalik and Koeppe, 1983; Peter *et al.*, 1979), whereas in *M. sexta*, the protein has a higher affinity for JH I and II than JH III (Peterson *et al.*, 1982). It therefore appears that the JH-binding proteins show higher affinity for the natural JH(s) of the species in question. These studies emphasize the importance of using only the appropriate species-specific JH in investigations of JH-specific binding proteins and it will no doubt be necessary to reinvestigate the occurrence of JH-specific binding proteins in other species, using the natural JH enantiomers. On the basis of the occurrence of the enantioselective JH III-binding protein in *L. decemlineata*, de Kort *et al.* (1983) speculated that this protein was important in the protection of JH III from the hemolymph JH esterase. This implies a role for this protein in the regulation of JH titer in this species, but until the kinetics of JH, JH-binding protein, and JH–esterase associations are investigated, the role of the binding protein in the regulation of JH titer remains uncertain. One such study has recently been completed, using semipurified JH III-binding protein and esterase from the hemolymph of vitellogenic females of *L. maderae* (Engelmann, 1984a,b; Gunawan and Engelmann, 1984). The authors found that two hemolymph proteins were able to bind JH at high affinity ($K_d = 10^{-8}\ M$) and, surprisingly, that both of these proteins were in fact inhibitors of the JH esterase. These proteins had more than an order of magnitude greater affinity for JH III than the esterase. These results suggest that the so-called protection of JH from the esterase by binding protein may occur by inhibition of the esterase. It is uncertain if the JH esterase is able to hydrolyze JH bound to the binding proteins and how degradation of this complex may be effected. However, these data also suggest that the binding protein–JH interactions may be able to influence JH titer.

Sequestration of JH into lipid compartments within the tissues may also afford a degree of protection for the hormone from the esterase and may provide a depot from which JH can reflux. The occurrence of such a process could clearly influence JH titer and binding proteins may facilitate this uptake and reflux. At present, virtually nothing is known of the existence of such sequestered pools but it should be noted that reflux of JH has been reported in the fat body of *M. sexta* maintained *in vitro* (Hammock *et al.*, 1975; Nowock *et al.*, 1976). These authors assumed that much of the JH uptake could be attributed to specific binding although it is equally likely that a sequestration of JH into protected pools was also occurring (see below).

5.4 HALF-LIFE OF JH

We have noted previously that JH titer is regulated by an interaction between hormone synthesis and degradation. For steady-state conditions, the rate of synthesis is equal to the rate of degradation. Although degradation includes several different parameters including JH esterase activity, removal of JH from the hemolymph by tissue uptake, and sequestration into protected lipophilic pools as well as excretion and binding of JH, the entire degradation component can be conveniently described by the half-life of the hormone in the hemolymph. Accordingly, direct measurement of the half-life of JH should provide an estimate of the overall rates of either biosynthesis or degradation under steady-state conditions. Such measurement has been used to estimate the theoretical rates of JH biosynthesis necessary to maintain a given JH titer (e.g., Fain and Riddiford, 1975; Kramer, 1978; see also de Kort *et al.*, 1981); invariably, observed rates of JH biosynthesis determined using the RCA *in vitro* were considerably less than the theoretical rates, thus raising the possibility that the RCA consistently underestimated rates of JH biosynthesis (see Section 3.1).

JH half-life determinations have been mainly based on the rate of disappearance of exogenous radiolabeled JH from the hemolymph (e.g., Kramer et al., 1977; Kramer and Childs, 1977), although some early studies utilized the decline in material detectable by JH bioassay at selected times following allatectomy or neck ligation to estimate half-life (Johnson and Hill, 1973; Fain and Riddiford, 1975). JH half-life has been estimated to be 1.5 h in adult *L. migratoria* (Johnson and Hill, 1973) and fourth instar *M. sexta* (Fain and Riddiford, 1975), < 4 h in larval *G. mellonella* (Reddy and Krishnakumaran, 1972), 1 h in larval *Paramyelosis transitella* (Kramer and Childs, 1977), and 1.6 h in adult *D. punctata* (Rotin and Tobe, 1983). A shorter half-life (25 min) has been reported in *L. decemlineata* (Kramer *et al.*, 1977) and fifth instar *M. sexta* (Nijhout, 1975c). However, recent measurement of half-life in adult *D. punctata* using physiocochemical determination of endogenous JH titer following allatectomy has demonstrated that the rate of disappearance of JH is not linear but rather declines after about 4 h (Tobe *et al.*, 1984a, 1985). This unexpected finding indicates that the decay in JH titer following allatectomy does not obey first-order kinetics and although the initial estimate of half-life using only 1- and 2-h titer values is about 1.2 h, in good agreement with values obtained from exogenous JH decay in this and other species, half-life values obtained in this way seriously overestimate the rate of disappearance of hormone. Although this phenomenon has been described only in *D. punctata* to date, it is clear that it will be necessary to extend these observations to other species. Such overestimates in V_d clearly would result in overestimates in V_s, for steady-state conditions and, accordingly, theoretical rates of JH biosynthesis derived in this way may be inaccurate. An example of these calculations may prove instructive, using the *D. punctata* system.

Using the equation $k = 0.693/t_{0.5}$ derived from $x = x_0 e^{-kt}$ for a first-order reaction (where $x =$ titer of hormone at time t, and $x_0 =$ titer of hormone at $t = 0$), the rate constant k was calculated to be $0.58\ h^{-1}$ for the initial half-life value of 1.2 h cited above. This value was then used to calculate the rate of JH biosynthesis necessary at this stage to maintain a steady-state quantity of JH per animal, using the equation $V_s = kx$ [where $V_s =$ rate of JH biosynthesis (ng/h), $k =$ rate constant (h^{-1}), and x is the presumed state-state titer of hormone, and assuming that $V_s = V_d$, the rate of JH degradation]. Using the observed JH titer value for this age (Tobe *et al.*, 1984a, 1985), the theoretical rates of JH biosynthesis were calculated to be 1.5–2 times higher than those observed *in vitro* using the RCA. This overestimate stems largely from the high calculated value for the rate constant and in view of the demonstrated non-first-order decay in JH titer, the use of this value is questionable. Accordingly, half-life values for other species should be determined using measurement of endogenous JH titers after allatectomy. The determination of accurate half-life values is essential to our understanding of the factors which regulate JH titer in insects.

The reasons for the observed slower decay in JH titer deserve consideration since such processes clearly impinge upon the regulation of JH titer. On the basis of differences between hemolymph JH titer and whole-body JH titer and non-first-order decay in JH titer, Tobe *et al.* (1984a, 1985) proposed that in *D. punctata*, there was a pool of JH outside of the hemolymph in equilibrium with the hemolymph JH pool. Measurement of growth of basal oocytes after allatectomy support this interpretation because the stimulatory effect of JH on oocyte growth persisted long after allatectomy (Barth and Sroka, 1975; Tobe *et al.*, 1985). This sequestered pool of JH is thought to occur in the tissues of *D. punctata*, but attempts to identify the sites of sequestration, by direct measurement of JH titer in the fat body, ovary, and accessory glands, proved only partially successful (Tobe *et al.*, 1985). Although substantial quantities of JH were found in these tissues, the quantities were insufficient to account for the difference between hemolymph and whole-body JH titers. It therefore appears likely that all the sites of sequestration have yet to be identified. For a complete discussion of other possible explanations for this discrepancy, see Tobe *et al.* (1984a, 1985). Although the existence of the extrahemolymph JH pool appears likely, the function of such a pool remains uncertain. Both uptake and reflux of JH from the fat body of *M. sexta* have been reported (Nowock *et al.*, 1976); significantly, most of the product released from the fat body was JH and not metabolites, indicating that the JH had been protected from catabolic enzymes known to occur in the fat body (e.g., Wing *et al.*, 1981). Thus, the compartmentalized JH may represent a protected pool, in equilibrium with JH in the hemolymph, which functions to maintain the hemolymph JH titer. The primary function of this pool may therefore be homeostatic, perhaps preventing rapid changes in titer. It will be important to determine if such pools occur in other species and other stages since this may provide insights into the function of extrahemolymph JH with respect to the maintenance of JH titer. At present, we may only speculate that this pool functions to maintain JH titer under adverse environmental conditions which are known to suppress JH biosynthesis by the CA, for example starvation, so that normal reproductive or metamorphic processes can continue.

6 Prospects

There is now sufficient information of the *in situ* regulation of the CA in a variety of species to embark on the search for and identification of allatotropins and allatostatins/allatohibins. Isolation and identification of these substances are of fundamental importance in advancing our understanding of CA regulation. The advent of *in vitro* techniques for the monitoring of CA activity (JH biosynthesis) now permits the direct assessment of putative CA regulators; however, preliminary attempts at the demonstration and isolation of such factors have met with only limited success and it will be important to elucidate the reasons for the apparent lack of responsiveness of CA to these regulators *in vitro*. It will also be essential to heed the many precautions in the deployment of the *in vitro* assays. Our abilities to demonstrate and identify putative regulators will no doubt be limited by our experimental ingenuity and it is likely that factors such as the critical periods of activity and sensitivity of CA to regulators will further complicate our experimental protocols.

The unequivocal identification of known JHs in additional species and the elucidation of the structures of yet undescribed JHs will also prove to be important tasks in the coming years. This will probably be of particular significance in the orders Hemiptera and Diptera, whose members include many serious agricultural and medical pests, and it will be necessary to ascertain if the few reports on the occurrence of low levels of JH III in selected species can in fact be extended to other species of these orders. In conjunction with the identification of the JHs, unequivocal information on the titers of the JHs, both in normal animals and in animals subjected to experimental manipulation, will prove essential to our understanding of the regulation of the CA and to elucidating the mode of action of JH.

In tandem with the demonstration and identification of CA regulators, investigations on the control of JH biosynthesis at the cellular and molecular levels should also proceed. Elucidation of the points of rate limitation in the biosynthetic pathway is an obvious first step, but it will probably not prove possible to establish unequivocally the role of specific enzymes in the regulation of hormone biosynthesis until such enzymes have been isolated to purity and their kinetics studied *in vitro*. Such isolation will of course facilitate the cloning of the genes for these enzymes—only then will it be possible to establish the precise roles of CA regulatory factors on the modulation of JH biosynthesis. However, studies on likely cellular intermediates such as the cyclic nucleotides will provide useful insights into the molecular mechanisms regulating JH biosynthesis. At the cellular level, investigations on cell proliferation and destruction within the CA will also prove essential, particularly because JH biosynthesis appears to be regulated at a number of different levels. Similarly, studies on the electrical coupling of CA cells and its modulation by neurotransmitters and by direct nervous input should provide much exciting information.

Understanding of the regulation of the CA will not be complete until the integration of regulatory pathways, at the level of both the CA and the central nervous system, has been defined. This should prove to be a stimulating challenge for the future.

References

Agui, N., Granger, N. A., Gilbert, L. I., and Bollenbacher, W. E. (1979). Cellular localization of the insect prothoracicotropic hormone: *In vitro* assay of a single neurosecretory cell. *Proc. Natl. Acad. Sci. U.S.A.* **76**, 5694–5698.

Agui, N., Bollenbacher, W. E., Granger, N. A., and Gilbert, L. I. (1980). Corpus allatum is release site for insect prothoracicotropic hormone. *Nature (London)* **285**, 669–670.

Akamatsu, Y., Dunn, P. E., Kezdy, F. J., Kramer, K. J., Law, J. H., Reibstein, D., and Sanburg, L. L. (1975). Biochemical aspects of juvenile hormone action in insects. *In* "Control Mechanisms in Development" (R. Meints and E. Davis, eds.), pp. 123–149. Plenum, New York.

Alertsen, A. R. (1955). Ageratochromene, a heterocyclic compound from the essential oils of *Ageratum* species. *Acta Chem. Scand.* **9**, 1725–1726.

Arvy, L., and Gabe, M. (1953). Données histophysiologiques sur la neurosecretion chez les Paleoptères (Ephemeroptères et Odonates). *Z. Zellforsch.* **38**, 591–610.

Avruch, L. I., and Tobe, S. S. (1978). Juvenile hormone biosynthesis by the corpora allata of the male desert locust, *Schistocerca gregaria*, during sexual maturation. *Can. J. Zool.* **56**, 2097–2102.

Baehr, J. C. (1973). Contrôle neuroendocrine du fonctionnement du corpus allatum chez *Rhodnius prolixus*. *J. Insect Physiol.* **19**, 1041–1055.

Baehr, J.-C. (1976). Etude du contrôle neuro-endocrine du fonctionnement du corpus allatum chez les larves du quatrième stade de *Rhodnius prolixus*. *J. Insect Physiol.* **22**, 73–82.

Baehr, J.-C., Cassier, P., and Fain-Maurel, M.-A. (1973). Contribution expérimentale et infrastructurale a l'étude de la dynamique du corpus allatum de *Rhodnius prolixus* Stål. Influence de la nutrition, de l'activité ovarienne, de la pars intercerebralis et de ses connections. *Arch. Zool. Exp. Gen.* **114**, 611–626.

Baehr, J.-C., Pradelles, P., Lebreux, C., Cassier, P., and Dray, F. (1976). A simple and sensitive radioimmunoassay of insect juvenile hormone using an iodinated tracer. *FEBS Lett.* **69**, 123–128.

Baehr, J.-C., Pradelles, P., and Dray, F. (1979). A radioimmunological assay for naturally occurring insect juvenile hormones using iodinated tracers: Its use in the analysis of biological samples. *Ann. Biol. Anim. Biochem. Biophys.* **19**, 1827–1836.

Baehr, J.-C., Caruelle, J. P., Porcheron, P., and Cassier, P. (1981). Quantification of juvenile hormones using a radioimmunological assay in the analysis of biological samples. *In* "Juvenile Hormone Biochemistry: Action, Agonism and Antagonism" (G. E. Pratt and G. T. Brooks, eds.), pp. 47–57. Elsevier, Amsterdam.

Baehr, J.-C., Cassier, P., Caussanel, C., and Porcheron, P. (1982). Activity of corpora allata, endocrine balance and reproduction in female *Labidura riparia* (Dermaptera). *Cell Tissue Res.* **225**, 267–282.

Baker, F. C., and Schooley, D. A. (1978). Juvenile hormone biosynthesis: Identification of 3-hydroxy-3-ethylglutarate and 3-hydroxy-3-methylglutarate in cell-free extracts from *Manduca sexta* incubated with propionyl- and acetyl-CoA. *J. Chem. Soc. Chem. Commun.* **7**, 292–293.

Baker, F. C., and Schooley, D. A. (1981). Biosynthesis of 3-hydroxy-3-methylglutaryl-CoA, 3-hydroxy-3-ethylglutaryl-CoA, mevalonate and homomevalonate by insect corpus allatum and mammalian hepatic tissues. *Biochim. Biophys. Acta* **664**, 356–372.

Baker, F. C., Lee, E., Bergot, B. J., and Schooley, D. A. (1981). Isomerization of isopentenyl pyrophosphate and homoisopentenyl pyrophosphate by *Manduca sexta* corpora cardiaca-corpora allata homogenates. *In* "Juvenile Hormone Biochemistry: Action, Agonism and Antagonism" (G. E. Pratt and G. T. Brooks, eds.), pp. 67–80. Elsevier, Amsterdam.

Baker, F. C., Mauchamp, B., Tsai, L. W., and Schooley, D. A. (1983a). Farnesol and farnesal dehydrogenase(s) in corpora allata of the tobacco hornworm moth, *Manduca sexta*. *J. Lipid Res.* **24**, 1586–1594.

Baker, F. C., Hagedorn, H. H., Schooley, D. A., and Wheelock, G. (1983b). Mosquito juvenile hormone identification and bioassay activity. *J. Insect Physiol.* **29,** 465–470.

Bar-Nun, S., Kreibich, G., Adesnik, M., Alterman, L., Negishi, M., and Sabatini, D. D. (1980). Synthesis and insertion of cytochrome P-450 into endoplasmic reticulum membranes. *Proc. Natl. Acad. Sci. U.S.A.* **77**, 965–969.

Barth, R. H., and Sroka, P. (1975). Initiation and regulation of oocyte growth by the brain and corpora allata of the cockroach, *Nauphoeta cinerea*. *J. Insect Physiol.* **21**, 321–330.

Bassi, S. D., Goodman, W., Altenhofen, C., and Gilbert, L. I. (1977). The binding of exogenous juvenile hormone by the haemolymph of *Oncopeltus fasciatus*. *Insect Biochem.* **7**, 309–312.

Beckage, N. E., and Riddiford, L. M. (1982). Effects of parasitism by *Apanteles congregatus* on the endocrine physiology of the tobacco hornworm *Manduca sexta*. *Gen. Comp. Endocrinol.* **47**, 308–322.

Bergot, B. J., Schooley, D. A., Chippendale, G. M., and Yin, C.-M. (1976). Juvenile hormone titer determinations in the southwestern corn borer, *Diatraea grandiosella*, by electron capture-gas chromatography. *Life Sci.* **18**, 811–820.

Bergot, B. J., Jamieson, G. C., Ratcliff, M. A., and Schooley, D. A. (1980). JH zero: New naturally occurring insect juvenile hormone from developing embryos of the tobacco hornworm. *Science* **210**, 336–338.

Bergot, B. J., Ratcliff, M., and Schooley, D. A. (1981a). Method for quantitative determination of the four known juvenile hormones in insect tissue using gas chromatography-mass spectroscopy. *J. Chromatogr.* **204**, 231–244.

Bergot, B. J., Baker, F. C., Cerf, D. C., Jamieson, G., and Schooley, D. A. (1981b). Qualitative and quantitative aspects of juvenile hormone titers in developing embryos of several insect species: Discovery of a new JH-like substance extracted from eggs of *Manduca sexta*. *In* "Juvenile Hormone Biochemistry" (G. E. Pratt and G. T. Brooks, eds.), pp. 33–45. Elsevier, Amsterdam.

Bergot, B. J., Schooley, D. A., and de Kort, C. A. D. (1981c). Identification of JH III as the principal juvenile hormone in *Locusta migratoria*. *Experientia* **37**, 909–910.

Berson, S. A., and Yallow, R. S. (1958). Isotopic tracers in the study of diabetes. *Adv. Biol. Med. Phys.* **6**, 349–420.

Bhaskaran, G. (1981). Regulation of corpus allatum activity in last instar *Manduca sexta* larvae. *In* "Current Topics and Insect Endocrinology and Nutrition" (G. Bhaskaran, S. Friedman, and J. G. Rodriguez, eds.), pp. 53–81. Plenum, New York.

Bhaskaran, G., and Jones, G. (1980). Neuroendocrine regulation of corpus allatum activity in *Manduca sexta*: The endocrine basis for starvation-induced supernumerary larval moult. *J. Insect Physiol.* **26**, 431–440.

Bhaskaran, G., DeLeon, G., Looman, B., Shirk, P. D., and Röller, H. (1980a). Activity of juvenile hormone acid in brainless, allatectomized diapausing *Cecropia* pupae. *Gen. Comp. Endocrinol.* **42**, 129–133.

Bhaskaran, G., Jones, G., and Jones, D. (1980b). Neuroendocrine regulation of corpus allatum activity in *Manduca sexta*: Sequential neurohormonal and nervous inhibition in the last-instar larva. *Proc. Natl. Acad. Sci. U.S.A.* **77**, 4407–4411.

Bitsch, J., and Lapalisse, J. (1984). Modifications volumetriques et ultrastructurales des corpora allata du Lepisma *Thermobia domestica* (Pacard) (Thysanura: Lepismatidae) au cours des cycles biologique. *Int. J. Insect Morphol. Embryol* **13**, 37–49.

Blakley, N., and Goodner, S. R. (1978). Size-dependent timing of metamorphosis in milk weed bugs (*Oncopeltus*) and its life history implications. *Biol. Bull. Woods Hole* **155**, 499–518.

Bogus, M. I., and Cymborowski, B. (1981). Chilled *Gallaria mellonella* larvae: Mechanism of supernumerary moulting. *Physiol. Entomol.* **6**, 343–348.

Bogus, M. I., and Cymborowski, B. (1984). Induction of supernumerary moults in *Galleria mellonella*: Evidence for an allatotropic function of the brain. *J. Insect Physiol.* **30**, 557–561.

Borg, T. K., and Granger, N. A. (1977). Ultrastructure of the protocerebral neurosecretory cells of larval *Galleria mellonella, in situ* and after culture of the brain *in vitro. Tissue Cell* **9**, 645–652.

Borst, D. W., and O'Connor, J. D. (1972). Arthropod molting hormone: Radioimmune assay. *Science* **178**, 418–419.

Bouletreau-Merle, J. (1974). Stimulation de l'ovogenèse par la copulation chez les femelles de *Drosophila melanogaster* privées de leur complexe endocrines retrocerebral. *J. Insect Physiol.* **20**, 2035–2041.

Bowers, W. S. (1983). The precocenes. *In* "Endocrinology of Insects" (R. G. H. Downer and H. Laufer, eds.), pp. 517–523. Liss, New York.

Bowers, W. S., and Martinez-Pardo, R. (1977). Antiallatotropins: Inhibition of corpus allatum development. *Science* **197**, 1369–1371.

Bowers, W. S., Ohta, T., Cleere, J. S., and Marsella, P. A. (1976). Discovery of insect anti-juvenile hormones in plants. *Science* **193**, 542–547.

Bowers, W. S., Marsella, P. A., and Evans, P. H. (1983). Identification of an Hemipteran juvenile hormone: *In vitro* biosynthesis of JH III by *Dysdercus fasciatus. J. Exp. Zool.* **228**, 555–559.

Bradley, J. T., and Edwards, J. S. (1979). Ultrastructure of the corpus cardiacum and corpus allatum of the house cricket *Acheta domesticus. Cell Tissue Res.* **198**, 201–208.

Brouse-Gaury, P. (1971a). Description d'arcs reflexes neuro-endocriniens partant des ocelles chez quelques orthoptères. *Bull. Biol. Fr. Belg.* **105**, 83–93.

Brousse-Gaury, P. (1971b). Influence de stimuli externes sur le comportment neuroendocrinien de Blattes. I. Les organes sensoriels cephaliques point de depart de reflexes neuro-endocriniens. *Ann. Sci. Nat. Zool.* **13**, 181–332.

Brousse-Gaury, P., Cassier, P., and Fain-Maurel, M. A. (1973). Contribution expérimentale et infrastructurale a l'étude de la dynamique des corpora allata chez *Blabera fusca.* Influence du groupement visuel, des afferences ocellaires et antennaires. *Bull. Biol. Fr. Belg.* **107**, 143–169.

Brown, M. S., and Goldstein, J. L. (1980). Multivalent feedback regulation of HMG CoA reductase, a control mechanism coordinating isoprenoid synthesis and cell growth. *J. Lipid Res.* **21**, 505–517.

Buschor, J., Beyeler, P., and Lanzrein, B. (1984). Factors responsible for the initiation of a second oocyte maturation cycle in the ovoviviparous cockroach *Nauphoeta cinerea. J. Insect Physiol.* **30**, 241–249.

Buys, C. M., and Gibbs, D. (1981). The anatomy of neurons projecting to the corpus cardiacum from the larval brain of the tobacco hornworm, *Manduca sexta* (L.). *Cell Tissue Res.* **215**, 505–513.

Carrow, G. M., Calabrese, R. L., and Williams, C. M. (1981). Spontaneous and evoked release of prothoracicotropin from multiple neurohemal organs of the tobacco hornworm. *Proc. Natl. Acad. Sci. U.S.A.* **78**, 5866–5670.

Carrow, G. M., Calabrese, R. L., and Williams, C. M. (1984). Architecture and physiology of insect cerebral neurosecretory cells. *J. Neurosci.* **4**, 1034–1044.

Cassagnau, P., and Juberthie, C. (1967). Structure nerveuse, neurosecretion et organes endocrines chez les Collemboles. II. Le complex cerebral des Entomobryomorphes. *Gen. Comp. Endocrinol.* **8**, 489–502.

Cassier, P. (1965). Contribution a l'étude du comportement phototropique du criquet migrateur (*Locusta migratoria migratoroides* R. and F.). *Ann. Sci. Nat. Zool. Biol. Anim.* [12] **7**, 213–412.

Cassier, P. (1979). The corpora allata of insects. *Int. Rev. Cytol.* **57**, 1–73.

Cazal, P. (1948). Les glandes endocrines retro-cerebrales des insectes. *Bull. Biol. Belg. Suppl.* **32**, 1–227.

Chen, T. T., Couble, P., De Lucca, F. L., and Wyatt, G. R. (1976). Juvenile hormone control of vitellogenin synthesis. *In* "The Juvenile Hormones" (L. I. Gilbert, ed.), pp. 505–529. Plenum, New York.

Chen, T. T., Couble, P., Abu-Hakima, R., and Wyatt, G. R. (1979). Juvenile hormone-controlled vitellogenin synthesis in *Locusta migratoria* fat body. Hormonal induction *in vivo. Dev. Biol.* **69**, 59–72.

Chin, D. J., Gil, G., Russell, D. W., Liscum, L., Luskey, K. L., Basu, S. K., Okayama, H., Berg, P., Goldstein, J. L., and Brown, M. S. (1984). Nucleotide sequence of 3-hydroxy-3-methyl-glutaryl coenzyme A reductase, a glycoprotein of endoplasmic reticulum. *Nature* (*London*) **308**, 613–617.

Christensen, A. K. (1975). Leydig cells. *Handb. Physiol. Sect.* 7 **5**, 57–94.

Christensen, A. K., and Gillim, S. W. (1969). The correlation of fine structure and function in steroid-secreting cells, with emphasis of those of the gonads. *In* "The Gonads" (K. W. McKerns, ed.), pp. 415–488. Appleton, New York.

Coudron, T. A., Dunn, P. E., Seballos, H. L., Wharen, R. E., Sanburg, L. L., and Law, J. H. (1981). Preparation of homogeneous juvenile hormone specific esterase from the haemolymph of the tobacco hornworm, *Manduca sexta. Insect Biochem.* **11**, 453–461.

Couillaud, F., Girardie, J., Tobe, S. S., and Girardie, A. (1984). Activity of disconnected corpora allata in *Locusta migratoria*: Juvenile hormone biosynthesis *in vitro* and physiological effects *in vivo. J. Insect Physiol.* **30**, 551–556.

Cymborowski, B., and Bogus, M. I. (1976). Juvenilizing effect of cooling on *Galleria mellonella. J. Insect Physiol.* **22**, 669–672.

Cymborowski, B., and Bogus, M. I. (1983). The allatotropic activity of the brain of *Galleria mellonella* larvae. *Int. Symp. Invertebrate Reprod., 3rd, Tübingen* Abstract, pp. 571.

Cymborowski, B., Bogus, M., Beckage, N. E., Williams, C. M., and Riddiford, L. M. (1982). Juvenile hormone titres and metabolism during starvation-induced supernumerary larval moulting of the tobacco hornworm, *Manduca sexta. J. Insect Physiol.* **28**, 129–135.

Dahm, K. H., Röller, H., and Trost, B. M. (1968).The juvenile hormone. IV. Stereochemistry of juvenile hormone and biological activity of some of its isomers and related compounds. *Life Sci.* **7**, 129–137.

Dahm, K. H., Bhaskaran, G., Peter, M. G., Shirk, P. D., Seshan, K. R., and Röller, H. (1976). On the identity of the juvenile hormone in insects. *In* "The Juvenile Hormones" (L. I. Gilbert, ed.), pp. 19–47. Plenum, New York.

Dahm, K. H., Bhaskaran, G., Peter, M. G., Shirk, P. D., Seshan, K. R., and Röller, H. (1981). The juvenile hormones of *Cecropia*. *In* "Regulation of Insect Development and Behaviour Part I" (F. Sehnal, A. Zabza, J. J. Menn, and B. Cymborowski, eds.), pp. 183–198. Wroclaw Technical Univ. Press, Wroclaw, Poland.

Darjo, A. (1976). Activité des corpora allata et contrôle photoperiodique de la maturation ovarienne chez *Locusta migratoria. J. Insect Physiol.* **22**, 347–355.

Dautry-Varsat, A., and Lodish, H. F. (1983). The Golgi complex and the sorting of membranes and secreted protein. *Trends Neurosci.* **6**, 484–490.

Davey, K. G. (1982). Hormonal integration governing the ovary. *In* "Endocrinology of Insects" (R. G. H. Downer and H. Laufer, eds.), pp. 251–258. Liss, New York.

De Kort, C. A. D., and Granger, N. A. (1981). Regulation of the juvenile hormone titer. *Annu. Rev. Entomol.* **26**, 1–28.

De Kort, C. A. D., Kramer, S. J., and Wieten, M. (1978). Regulation of juvenile hormone titres in the adult Colorado beetle: Interaction with carboxylesterases and carrier proteins. *In* "Comparative Endocrinology" (P. J. Gaillard and H. H. Boer, eds.), pp. 507–510. Elsevier, Amsterdam.

De Kort, C. A. D., Khan, M. A., Bergot, B. J., and Schooley, D. A. (1981). The JH titre in the Colorado Beetle in relation to reproduction in diapause. *In* "Juvenile Hormone Biochemistry" (G. E. Pratt and G. T. Brooks, eds.), pp. 125–134. Elsevier, Amsterdam.

De Kort, C. A. D., Bergot, B. J., and Schooley, D. A. (1982). The nature and titre of juvenile hormone in the Colorado Beetle, *Leptinotarsa decemlineata*. *J. Insect Physiol.* **28**, 471–474.

De Kort, C. A. D., Peter, M. G., and Koopmanschap, A. B. (1983). Binding and degradation of juvenile hormone III by haemolymph proteins of the Colorado potato beetle: A re-examination. *Insect Biochem.* **13**, 481–487.

De Lerma, B. (1956). Corpora cardiaca et neurosecretion protocérébrale chez le Coleoptère *Hydrous piceus L. Ann. Sci. Nat. Zool.* (*11*) **18**, 235–250.

Deleurance, S., and Charpin, P. (1978). Ultrastructural dynamics of the corpus allatum of *Chlovea angustata* Fab. (Coleoptera, Catopidae). *Cell Tissue Res.* **191**, 151–160.

Deleurance, S., Baehr, J. C., Porcheron, P., and Cassier, P. (1980). Diapause et évolution des taux des ecdysteroides et des hormones juveniles chez les nymphes et les imagos de *Leptinotarsa decemlineata* Say. *C.R. Acad. Sci. Paris* **290**, 367–370.

De Wilde, J., and de Boer, J. A. (1969). Humoral and nervous pathways in photoperiodic induction of diapause in *Leptinotarsa decemlineata*. *J. Insect Physiol.* **15**, 661–675.

Dhadialla, T. S., and Wyatt, G. R. (1981). Competence for juvenile hormone-stimulated vitellogenin synthesis in the fat body of female and male *Locusta migratoria*. *In* "Juvenile Hormone Biochemistry" (G. E. Pratt and G. T. Brooks, eds.), pp. 257–270. Elsevier, Amsterdam.

Dhadialla, T. S., and Wyatt, G. R. (1983). Juvenile hormone-dependent vitellogenin synthesis in *Locusta migratoria* fat body: Inducibility related to sex and stage. *Dev. Biol.* **96**, 436–444.

Doane, W. W. (1973). Role of hormones in insect development. *In* "Developmental Systems: Insects" (S. J.Counce and C. H. Waddington, eds.), pp. 291–497. Academic Press, New York.

Dorn, A. (1973). Electron microscopic study of the larval and adult corpus allatum of *Oncopeltus fasciatus* Dallas (Insecta: Heteroptera). *Z. Mikrosk. Anat. Forsch.* **145**, 447–458.

Dorn, A. (1975). Struktur und Funktion des embryonalen Corpus allatum von *Oncopeltus fasciatus* Dallas (Insecta, Heteroptera). *Verh. Dtsch. Zool. Ges. Bochum* **1974**, 85–89.

Dorn, A. (1982). Precocene-induced effects and possible role of juvenile hormone during embryogenesis of the milkweed bug, *Oncopeltus fasciatus*. *Gen. Comp. Endocrinol.* **46**, 42–52.

Eckert, M. (1977). Immunologische Untersuchungen des neuroendokrinen Systems von Insekten. IV. Differenzierte immunohistochemische Darstellung von Neurosekreten des Gehirns und der Corpora cardiaca bei der Schabe *Periplaneta americana*. *Zool. Jb. Physiol.* **81**, 25–41.

Edwards, J. P., and Price, N. R. (1983). Inhibition of juvenile hormone III biosynthesis in *Periplaneta americana* with the fungal metabolite compactin (ML-236B). *Insect Biochem.* **13**, 185–189.

Elliott, H. J. (1976). Structural analysis of the corpus allatum of an aphid, *Aphis craccivora*. *J. Insect Physiol.* **22**, 1275–1279.

El-Salhy, M., Falkner, S., Kramer, K. J., and Speirs, R. D. (1983). Immunohistochemical investigations of neuropeptides in the brain, corpora cardiaca and corpora allata of an adult lepidopteran insect, *Manduca sexta* (L.). *Cell Tissue Res.* **232**, 295–317.

Endo, A. (1981). Biological and pharmacological activity of inhibitors of 3-hydroxy-3-methylglutaryl coenzyme A reductase. *Trends Biochem. Sci.* **6**, 10–12.

Engelmann, F. (1957). Die Steuerung der Ovarfunktion bei der ovoviviparen Schabe *Leucophaea maderae* (Fabr.). *J. Insect. Physiol.* **1**, 257–278.

Engelmann, F. (1959). The control of reproduction in *Diploptera punctata* (Blattaria). *Biol. Bull. Mar. Biol. Lab. Woods Hole* **116**, 406–419.

Engelmann, F. (1970). "The Physiology of Insect Reproduction." Pergamon, Oxford.

Engelmann, F. (1978). Synthesis of vitellogenin after long-term ovariectomy in a cockroach. *Insect Biochem.* **8**, 149–154.

Engelmann, F. (1979). Insect vitellogenin: Identification, biosynthesis, and role in vitellogenesis. *In* "Advances in Insect Physiology" (J. E. Treherne, M. J. Berridge, and V. B. Wigglesworth, eds.), pp. 49–108. Academic Press, New York.

Engelmann, F. (1981). Induction of the insect vitellogenin *in vivo* and *in vitro*. *In* "Juvenile Hormone Biochemistry" (G. E. Pratt and G. T. Brooks, eds.), pp. 470–485. Elsevier, Amsterdam.

Engelmann, F. (1983). Vitellogenesis controlled by juvenile hormone. *In* "Endocrinology of Insects" (R. G. H. Downer and H. Laufer, eds.), pp. 259–270. Liss, New York.

Engelmann, F. (1984a). Juvenile hormone binding compounds in hemolymph and tissues of an insect: The functional significance. *In* "Advances in Invertebrate Reproduction" (W. Engels, ed.), pp. 177–187. Elsevier, Amsterdam.

Engelmann, F. (1984b). Regulation of vitellogenesis in insects: The pleiotropic role of juvenile hormone. *In* "Biosynthesis, Metabolism and Mode of Action of Invertebrate Hormones" (J. A. Hoffmann and M. Porchet, eds.), 444–453. Springer-Verlag, Berlin and New York.

Engelmann, F., and Lüscher, M. (1957). Die hemmende Wirkung des Gehirns auf die Corpora allata bei *Leucophaea maderae* (Orthoptera). *Verh. Dtsch. Zool. Ges. Hamburg* **1956**, 215–220.

Erley, D., Southard, S., and Emmerich, H. (1975). Excretion of juvenile hormone and its metabolites in the locust, *Locusta migratoria*. *J. Insect Physiol.* **21**, 61–70.

Ewing, L. L., and Zirkin, B. R. (1983). Leydig cell structure and steroidogenic function. *Recent Prog. Horm. Res.* **39**, 599–635.

Fain, M. J., and Riddiford, L. M. (1975). Juvenile hormone titers in the hemolymph during late larval development of the tobacco hornworm, *Manduca sexta* (L.). *Biol. Bull. Mar. Biol. Lab. Woods Hole* **149**, 506–521.

Fain-Maurel, M. A., and Cassier, P. (1969a). Etude infrastructurale des corpora allata de *Locusta migratoria migratorioides* (R. et F.), phase solitaire, au cours de la maturation sexuelle et des cycles ovariens. *C. R. Hebd. Seances Acad. Sci. Paris* **268**, 2721–2723.

Fain-Maurel, M. A., and Cassier, P. (1969b). Pléomorphisme mitochondrial dans les corpora allata de *Locusta migratoria migratoria* (R. et F.) au cours de la vie imaginale. *Z. Zellforsch.* **102**, 543–553.

Faulkner, D. J., and Peterson, M. R. (1971). Synthesis of C_{18} cecropia JH to obtain optically active forms of known absolute configuration. *J. Am. Chem. Soc.* **93**, 3766–3767.

Ferenz, H.-J. (1984). Isolation of an allatotropic factor in *Locusta migratoria* and its effect on corpus allatum activity *in vitro*. *In* "Biosynthesis, Metabolism and Mode of Actions of Invertebrate Hormones" (J. A. Hoffman and M. Porchet, eds.), pp. 92–96. Springer-Verlag, Berlin and New York.

Ferenz, H.-J., and Diehl, I. (1983). Stimulation of juvenile hormone biosynthesis *in vitro* by locust allatotropin. *Z. Naturforsch.* **38c**, 856–858.

Ferenz, H.-J., and Kaufner, I. (1981). Juvenile hormone synthesis in relation to oogenesis in *Locusta migratoria*. *In* "Juvenile Hormone Biochemistry" (G. E. Pratt and G. T. Brooks, eds.), pp. 135–145. Elsevier, Amsterdam.

Feyereisen, R. (1984). Regulation of juvenile hormone titer: Synthesis. *In* "Comprehensive Insect Physiology, Biochemistry and Pharmacology" (G. A. Kerkut and L. I. Gilbert, eds.), Vol. 7. Pergamon, Oxford.

Feyereisen, R., and Tobe, S. S. (1981). A rapid partition assay for routine analysis of juvenile hormone release in insect corpora allata. *Anal. Biochem.* **111**, 372–375.

Feyereisen, R., Koener, J., and Tobe, S. S. (1981a). *In vitro* studies with C_2, C_6 and C_{15} precursors of C_{16}JH biosynthesis in the corpora allata of adult female *Diploptera punctata*. *In* "Juvenile Hormone Biochemistry: Action, Agonism and Antagonism" (G. E. Pratt and G. T. Brooks, eds.), pp. 81–92. Elsevier, Amsterdam.

Feyereisen, R., Friedel, T., and Tobe, S. S. (1981b). Farnesoic acid stimulation of C_{16} juvenile hormone biosynthesis by corpora allata of adult female *Diploptera punctata*. *Insect Biochem.* **11**, 401–409.

Feyereisen, R., Pratt, G. E., and Hamnett, A. F. (1981c). Enzymic synthesis of juvenile hormone in locust corpora allata: Evidence for a microsomal cytochrome P-450 linked methyl farnesoate epoxidase. *Eur. J. Biochem.* **118**, 231–238.

Feyereisen, R., Johnson, G., Koener, J., Stay, B., and Tobe, S. S. (1981d). Precocenes as pro-allatocidins in adult female *Diploptera punctata*: A functional and ultrastructural study. *J. Insect Physiol.* **27**, 855–868.

Feyereisen, R., Ruegg, R. P., and Tobe, S. S. (1984). Juvenile hormone III biosynthesis; Stoichiometric incorporation of [2-^{14}C]acetate and effects of exogenous farnesol and farnesoic acid. *Insect Biochem.* **14**, 657–661.

Fraser, J., and Pipa, R. (1977). Corpus allatum regulation during the metamorphosis of *Periplaneta americana*: Axon pathways. *J. Insect Physiol.* **23**, 975–984.

Friedel, T. (1974). Endocrine control of vitellogenesis in the harlequin bug, *Dindymus versicolor*. *J. Insect Physiol.* **20**, 717–733.

Friedel, T., Loughton, B. G., and Andrew, R. D. (1980a). A neurosecretory protein from *Locusta migratoria*. *Gen. Comp. Endocrinol.* **41**, 487–498.

Friedel, T., Feyereisen, R., Mundall, E. C., and Tobe, S. S. (1980b). The allatostatic effect of 20-hydroxyecdysone on the adult viviparous cockroach, *Diploptera punctata*. *J. Insect Physiol.* **26**, 665–670.

Gadot, M., and Applebaum, S. W. (1984). *In vitro* activation of locust corpora allata. *In* "Advances in Invertebrate Reproduction" (W. Engels, ed.), p. 580. Elsevier, Amsterdam.

Gibbs, D., and Riddiford, L. M. (1977). Prothoracicotropic hormone in *Manduca sexta*: Localization by a larval assay. *J. Exp. Biol.* **66**, 255–266.

Giese, Ch., Spindler, K. D., and Emmerich, H. (1977). The solubility of insect juvenile hormone in aqueous solutions and its adsorption by glassware and plastics. *Z. Naturforsch.* **32c**, 158–160.

Gilbert, L. I., and King, D. S. (1973). Physiology of growth and development: Endocrine aspects. *In* "The Physiology of Insects" (M. Rockstein, ed.), pp. 249–370. Academic Press, New York.

Gilbert, L. I., and Schneiderman, H. A. (1961). The content of juvenile hormone and lipid in Lepidoptera: Sexual differences and developmental changes. *Gen. Comp. Endocrinol.* **1**, 453–472.

Gilbert, L. I., Goodman, W., and Granger, N. (1978). Regulation of juvenile hormone titer in Lepidoptera. *In* "Comparative Endocrinology" (P. J. Gaillard and H. H. Boer, eds.), pp. 471–486. Elsevier, Amsterdam.

Girardie, A. (1966). Contrôle de l'activité genitale chez *Locusta migratoria*. Mise en evidence d'un facteur gonadotrope et d'un facteur allatotrope dans la pars intercérébralis. *Bull. Soc. Zool. Fr.* **91**, 423–439.

Girardie, A. (1967). Contrôle neuro-hormonal de la métamorphose et de la pigmentation chez *Locusta migratoria cinerascens* (Orthoptere). *Bull. Biol. Fr. Belg.* **101**, 79–114.

Girardie, A. (1974). Recherches sur le rôle physiologique des cellules neurosecretrices laterales du protocerebron de *Locusta migratoria migratoroides* (Insecte Orthoptere). *Zool. Jb. Physiol.* **78,** 310–326.

Girardie, A. (1983). Neurosecretion and reproduction. *In* "Endocrinology of Insects" (R. G. H. Downer, and H. Laufer, eds.), pp. 305–317. Liss, New York.

Girardie, J., and Granier, S. (1974). Ultrastructure des corps allates d'*Anacridium aegyptium* (Insecte Orthoptere) à l'avant dernier stade larvaire et durant la vie imaginale. *Arch. Anat. Microsc. Morphol. Exp.* **63**, 251–267.

Girardie, J., Tobe, S. S., and Girardie, A. (1981). Biosynthèse de l'hormone juvenile C_{16} (JH-III) et maturation ovarienne chez le criquet migrateur. *C. R. Acad. Sci. Paris* **293**, 443–446.

Girardie, J., Allard, M., Faddoul, A., and Neuzil, E. (1983). Separation of proteins on locust (*Locusta migratoria*) neurosecretory endings with high performance liquid chromatography. *J. Liq. Chromatogr.* **6**, 2501–2512.

Goldsworthy, G. J., Robinson, J., and Mordue, W. (1981). "Endocrinology," pp. 184. Wiley, New York.

Goltzene, F., and Porte, A. (1978). Endocrine control by neurosecretory cells of pars interecerebralis and the corpora allata during the earlier phases of vitellogenesis in *Locusta migratoria migratoroides* R. and D. (Orthoptera). *Gen. Comp. Endocrinol.* **35**, 35–45.

Goodman, W. G. (1983). Hemolymph transport of ecdysteroids and juvenile hormone. *In* "Endocrinology of Insects" (R. G. H. Downer and H. Laufer, eds.), pp. 147–159. Liss, New York.

Goodman, W. G., Schooley, D. A., and Gilbert, L. I. (1978). Specificity of the juvenile hormone binding protein: The geometrical isomers of juvenile hormone I. *Proc. Natl. Acad. Sci. U.S.A.* **75**, 185–189.

Granger, N. A., and Borg, T. K. (1976). The allatotropic activity of the larval brain of *Galleria mellonella* cultured *in vitro*. *Gen. Comp. Endocrinol.* **29,** 349–359.

Granger, N. A., and Goodman, W. G. (1983). Juvenile hormone radioimmunoassays: Theory and practice. *Insect Biochem.* **13**, 333–340.

Granger, N. A., and Sehnal, F. (1974). Regulation of larval corpora allata in *Galleria mellonella*. *Nature (London)* **251**, 415–417.

Granger, N. A., Bollenbacher, W. E., Vince, R., Gilbert, L. I., Baehr, J. C., and Dray, F. (1979). *In vitro* biosynthesis of juvenile hormone by the larval corpora allata of *Manduca sexta*: Quantification by radioimmunoassay. *Mol. Cell. Endocrinol.* **16**, 1–17.

Granger. N. A., Bollenbacher, W. E., and Gilbert, L. I. (1981). A *in vitro* approach for investigating the regulation of the corpora allata during larval-pupal metamorphosis. *In* "Current Topics in Insect Endocrinology and Nutrition" (G. Bhaskaran, S. Friedman, and J. G. Rodriguez, eds.), pp. 84–105. Plenum, New York.

Granger, N. A., Niemiec, S. M., Gilbert, L. I., and Bollenbacher, W. E. (1982a). Juvenile hormone III biosynthesis by the larval corpora allata of *Manduca sexta*. *J. Insect. Physiol.* **28,** 385–391.

Granger, N. A., Niemiec, S. M., Gilbert, L. I., and Bollenbacher, W. E. (1982b). Juvenile hormone synthesis *in vitro* by larval and pupal corpora allata of *Manduca sexta*. *Mol. Cell. Endocrinol.* **28**, 587–604.

Granger. N. A., Mitchell, L. J., and Bollenbacher, W. E. (1985). Stimulation of JH III synthesis *in vitro* by an allatotropic factor from the brain of the tobacco hornworm, *Manduca sexta*. *In* "Insect Neurochemistry and Neurophysiology" (A. B. Borkovec and T. J. Kelly, eds.), pp. 365–367. Plenum, New York.

Greenberg, S., and Tobe, S. S. (1985). Adaptation of a radiochemical assay for juvenile hormone biosynthesis to study caste differentiation in a primitive termite. *J. Insect Physiol.* **31**, 347–352.

Guelin, M., and Darjo, M. A. (1974). Etude ultrastructurale des corpora allata en relation avec contrôle photopériodique de leur fonction gonadotrope chez *Locusta migratoria migratoria* L. *C. R. Hebd. Seances Acad. Sci. Paris* **278**, 491–494.

Gunawan, S., and Engelmann, F. (1984). Esterolytic degradation of juvenile hormone in the haemolymph of the adult female of *Leucophaea maderae. Insect Biochem.* **14**, 601–607.

Gundel, M., and Penzlin, H. (1980). Identification of neuronal pathways between the stomatogastric nervous system and the retrocerebral complex of the cockroach *Periplaneta americana* (L.). *Cell Tissue Res.* **208**, 283–297.

Gwadz, R. W., and Spielman, A. (1973). Corpus allatum control of ovarian development in *Aedes aegypti. J. Insect Physiol.* **19**, 1441–1448.

Hagedorn, H. H., O'Connor, J. D., Fuchs, M. S., Sage, B., Schlaeger, D. A., and Bohm, M. K. (1975). The ovary as a source of ecdysone in an adult mosquito. *Proc. Natl. Acad. Sci. U.S.A.* **72**, 3255–3259.

Hagedorn, H. H., Turner, S., Hagedorn, E. A., Pontcorvo, D., Greenbaum, P., Pfeiffer, D., Wheelock, G., and Flanagan, T. R. (1977). Postemergence growth of the ovarian follicles in *Aedes aegypti. J. Insect Physiol.* **23**, 203–206.

Haget, A. (1977). L'embryologie des insectes. *In* "Traité de Zoologie" (P.-P. Grasse, ed.), Vol. 8, pp. 2–387. Masson, Paris.

Haget, A., Ressouches, A., and Rogueda, J. (1981). Chronological and ultrastructural observations on the activities of the embryonic corpus allatum in *Carausius morosus* Br. (Phasmida: Lonchodidae). *Int. J. Insect Morph. Embryol.* **10**, 65–81.

Hammock, B. D. (1975). NADPH-dependent epoxidation of methyl farnesoate to juvenile hormone in the cockroach *Blaberus giganteus* L. *Life Sci.* **17**, 323–328.

Hammock, B. D., and Mumby, S. M. (1978). Inhibition of epoxidation of methyl farnesoate to juvenile hormone III by cockroach corpus allatum homogenates. *Pestic. Biochem. Physiol.* **9**, 39–47.

Hammock, B. D., and Quistad, G. B. (1981). Metabolism and mode of action of juvenile hormone, juvenoids and other insect growth regulators. *In* "Progress in Pesticide Biochemistry" (D. H. Hutson and T. R. Roberts, eds.), Vol. 1, pp. 1–82. Wiley, New York.

Hammock, B. D., Nowock, J., Goodman, W., Stamoudis, V., and Gilbert, L. I. (1975). The influence of hemolymph-binding protein on juvenile hormone stability and distribution in *Manduca sexta* fat body and imaginal discs *in vitro. Mol. Cell. Endocrinol.* **3**, 167–184.

Hammock, B. D., Sparks, T. C., and Mumby, S. M. (1977). Selective inhibition of JH esterases from cockroach hemolymph. *Pestic. Biochem. Physiol.* **7**, 517–530.

Hammock, B. D., Jones, D., Jones, G., Rudnicka, M., Sparks, T. C., and Wing, K. D. (1981). Regulation of juvenile hormone esterase in the cabbage looper, *Trichoplusia ni. In* "Regulation of Insect Development and Behavior" (F. Sehnal, A. Zebza, J. J. Menn, and B. Cymborowski, eds.), pp. 219–235. Wroclaw Technical Univ. Press, Wroclaw, Poland.

Hamnett, A. F., and Pratt, G. E. (1978). Use of automated capillary column radio gas chromatography in the identification of insect juvenile hormones. *J. Chromatogr.* **158**, 387–399.

Hamnett, A. F., Pratt, G. E., Stott, K. M., and Jennings, R. C. (1981). The use of radio HRLC in the identification of the natural substrate of the *O*-methyl transferase and substrate utilization by the enzyme. *In* "Juvenile Hormone Biochemistry" (G. E. Pratt and G. T. Brooks, eds.), pp. 93–105. Elsevier, Amsterdam.

Hanaoka, K., and Hagedorn, H. H. (1980). Brain hormone control of ecdysone secretion by the ovary in a mosquito. *In* "Progress in Ecdysone Research" (J. A. Hoffmann, ed.), pp. 467–480. Elsevier, Amsterdam.

Handler, A. M., and Postlethwaite, J. H. (1978). Regulation of vitellogenin synthesis in *Drosophila* by ecdysterone and juvenile hormone. *J. Exp. Zool.* **206**, 247–254.

Hansen, B. L., Hansen, G. N., and Scharrer, B. (1982). Immunoreactive material resembling vertebrate neuropeptides in the corpus cardiacum and corpus allatum of the insect *Leucophaea maderae*. *Cell Tissue Res.* **225**, 319–329.

Harper, E., Seifter, S., and Scharrer, B. (1967). Electron microscopic and biochemical characterization of collagen in blattarian insects. *J. Cell Biol.* **23**, 385–393.

Hartmann, R. (1978). The juvenile hormone-carrier in the haemolymph of the acridine grasshopper *Gomphocerus rufus* L.: Blocking of the juvenile hormone's action by means of repeated injections of an antibody to the carrier. *Roux' Arch.* **184**, 301–324.

Hayes, T. K., and Keeley, L. L. (1984). The isolation of insect neuropeptides using reverse-phase high performance liquid chromatography. *In* "Insect Neurochemistry and Neurophysiology" (A. B. Borkovec and T. J. Kelly, eds.), pp. 223–250. Plenum, New York.

Hodkova, M. (1977). Size and gonadotropic activity of corpus allatum after different surgical treatments in *Pyrrhocoris apterus* females (Heteroptera). *Vstn. Cesk. Spol. Zool.* **41**, 8–14.

Hodkova, M. (1979a). Hormonal and nervous inhibition of reproduction by brain in diapausing females of *Pyrrhocoris apterus* L. (Hemiptera). *Zool. Jb. Physiol.* **83**, 126–136.

Hodkova, M. (1979b). Does the insect brain count larval instars? *Experientia* **35**, 135–136.

Hoffmann, J. A., Lagueux, M., Hetru, C., Charlet, M., and Goltzene, F. (1980). Ecdysone in reproductively competent female adults and in embryos of insects. *In* "Progress in Ecdysone Research" (J. A. Hoffmann, ed.), pp. 431–465. Elsevier, Amsterdam.

Imboden, H., Lanzrein, B., Delbecque, J. P., and Lüscher, M. (1978). Ecdysteroids and juvenile hormone during embryogenesis in the ovoviviparous cockroach *Nauphoeta cinerea*. *Gen. Comp. Endocrinol.* **36**, 628–635.

Injeyan, H. S., and Tobe, S. S. (1981). Phase polymorphism in *Schistocerca gregaria*: Assessment of juvenile hormone synthesis in relation to vitellogenesis. *J. Insect Physiol.* **27**, 203–210.

Ito, S., and Karnovsky, M. (1968). Formaldehyde-glutaraldehyde fixatives containing trinitro compounds. *J. Cell Biol.* **39**, 168a.

Jennings, R. C., Judy, K. J., Schooley, D. A., Hall, M. S., and Siddall, J. B. (1975a). The identification and biosynthesis of two juvenile hormones from the tobacco budworm moth (*Heliothis virescens*). *Life Sci.* **16**, 1033–1040.

Jennings, R. C., Judy, K. J., and Schooley, D. A. (1975b). Biosynthesis of the homosesquiterpenoid juvenile hormone JH II [Methyl (2E,6E,10Z)-10,11-epoxy-3,7,11-trimethyltridecadienoate] from [5-^{3}H]Homomevalonate in *Manduca sexta*. *J. Chem. Soc. Chem. Commun.* 21–22.

Johansson, A. S. (1958). Relation of nutrition to endocrine-reproductive functions in the milkweed bug *Oncopeltus fasciatus* (Dallas) (Heteroptera: Lygaeidae). *Nytt. Mag. Zool.* (*Oslo*) **7**, 1–132.

Johnson, R. A., and Hill, L. (1973). Quantitative studies on the activity of the corpora allata in adult male *Locusta* and *Schistocerca*. *J. Insect Physiol.* **19**, 2459–2467.

Johnson, G. and Stay, B. (1985). In preparation.

Johnson, G., Stay, B., and Rankin, S. M. (1985). Ultrastructure of corpora allata of known activity during the vitellogenic cycle in the cockroach, *Diploptera punctata*. *Cell Tiss. Res.* **239**, 317–327.

Joly L., Joly, P., Porte, A., and Girardie, A. (1968). Etude physiologique et ultrastructurale des corpora allata de *Locusta migratoria* L. (Orthoptere) en phase grégaire. *Arch. Zool. Exp. Gen.* **109**, 703–727.

Jones, D., Jones, G., and Bhaskaran, G. (1980). Dietary sugars, hemolymph trehalose levels and supernumerary molting of *Manduca sexta* larvae. *Physiol. Zool.* **54**, 260–266.

Jones, G., Wing, K. D., Jones, D., and Hammock, B. (1981). Source and action of head factors regulating juvenile hormone esterase in larvae of the cabbage looper, *Trichoplusia ni*. *J. Insect Physiol.* **27**, 85–91.

Juberthie, C., and Cassagnau, P. (1971). L'évolution du système neurosecreteur chez les insectes. L'importance des Collemboles et des autres Apterygotes. *Rev. Ecol. Biol., Sol.* **8**, 59–80.

Judy, K. J., Schooley, D. A., Dunham, L. L., Hall, M. S., Bergot, B. J., and Siddall, J. B. (1973a). Isolation, structure, and absolute configuration of a new natural insect juvenile hormone from *Manduca sexta. Proc. Natl. Acad. Sci. U.S.A.* **70**, 1509–1513.

Judy, K. J., Schooley, D. A., Hall, M. S., Bergot, B. J., and Siddall, J. B. (1973b). Chemical structure and absolute configuration of a juvenile hormone from grasshopper corpora allata *in vitro. Life Sci.* **13**, 1511–1516.

Judy, K. J., Schooley, D. A., Troetschler, R. G., Jennings, R. C., Bergot, B. J., and Hall, M. S. (1975). Juvenile hormone production by corpora allata of *Tenebrio molitor in vitro. Life Sci.* **16**, 1059–1066.

Kaiser, H. (1980). Licht und electronenmikroskopische Untersuchung der Corpora allata der Eintagfliege *Ephemera danica* Müll. (Ephemeroptera: Ephemeridae) während der Metamorphose. *Int. J. Insect Morphol. Embryol.* **9**, 395–403.

Kandutsch, A. A., Chen, H. W., and Heiniger, H.-J. (1978). Biological activity of some oxygenated sterols. *Science* **201**, 498–501.

Karnovsky, J. J. (1965). A formaldehyde-glutaraldehyde fixative of high osmolality for use in electron microscopy. *J. Cell. Biol.* **27**, 137a.

Kasturi, T. R., and Manithomas, T. (1967). Essential oil of *Ageratum conyzoides*—isolation and structure of two new constituents. *Tetrahedron Lett.* **1967**, 2573–2575.

Keith, M. L., Rodwell, V. W., Rogers, D. H., and Rudney, H. (1979). *In vitro* phosphorylation of 3-hydroxy-3-methylglutaryl coenzyme A reductase: Analysis of ^{32}P-labeled, inactivated enzyme. *Biochem. Biophys. Res. Commun.* **90**, 969–975.

Khan, M. A., Doderer, A., Koopmanschap, A. B., and de Kort, C. A. D. (1982a). Improved assay conditions for measurement of corpus allatum activity *in vitro* in the adult Colorado potato beetle, *Leptinotarsa decemlineata. J. Insect Physiol.* **28**, 279–284.

Khan, M. A., Koopmanschap, A. B., and de Kort, C. A. D. (1982b). The effects of juvenile hormone, 20-hydroxyecdysone and precocene II on activity of corpora allata and the mode of negative-feedback regulation of these glands in the adult Colorado potato beetle. *J.Insect Physiol.* **28**, 995–1001.

Khan, M. A., Koopmanschap, A. B., and de Kort, C. A. D. (1983). The relative importance of nervous and hormonal pathways for the control of corpus allatum activity in the adult Colorado potato beetle, *Leptinotarsa decemlineata* (Say). *Gen. Comp. Endocrinol.* **52**, 214–221.

Khan, M. A., Romberg-Privee, H. M., and Schooneveld, H. (1984). Innervation of the corpus allatum in the Colorado potato beetle as revealed by retrograde diffusion with horse radish peroxidase. *Gen. Comp. Endocrinol.* **55**, 66–73.

Klages, G., Emmerich, H., and Rembold, H. (1981). Identification and titer of juvenile hormones in the fruitfly, *Drosophila hydei. In* "Regulation of Insect Development and Behavior I" (F. Sehnal, A. Zabza, J. J. Menn, and B. Cymborowski, eds.). p. 207. Wroclaw Technical Univ. Press, Wroclaw, Poland.

Koeppe, J. K. (1981). Juvenile hormone regulation of ovarian maturation in *Leucophaea maderae. In* "Regulation of Insect Development and Behavior" (F. Sehnal, A. Zabza, J. J. Menn, and B. Cymborowski, eds.), pp. 505–522. Wroclaw Technical University Press, Wroclaw, Poland.

Koeppe, J. K., Hobson, K., and Wellman, S. E. (1980). Juvenile hormone regulation of structural changes and DNA synthesis in the follicular epithelium of *Leucophaea maderae. J. Insect Physiol.* **26**, 229–240

Koontz, M., and Edwards, J. S. (1980). The projection of neuroendocrine fibers (NCC I and II) in the brains of three orthopteroid insects. *J. Morphol.* **165**, 285–299.

Kovalick, G. E., and Koeppe, J. K. (1983). Assay and identification of juvenile hormone binding proteins in *Leucophaea maderae. Mol. Cell. Endocrinol.* **31**, 271–286.

Kramer, K. J., and Childs, C. N. (1977). Interaction of juvenile hormone with carrier proteins and hydrolases from insect haemolymph. *Insect Biochem.* **7**, 397–403.

Kramer, K. J., Sanburg, L. L., Kezdy, F. J., and Law, J. H. (1974). The juvenile hormone binding protein in the hemolymph of *Manduca sexta* Johannson (Lepidoptera: Sphingidae). *Proc. Natl. Acad. Sci. U.S.A.* **71**, 493–497.

Kramer, K. J., Dunn, P. E., Peterson, R. C., and Law, J. H. (1976). Interaction of juvenile hormone with binding proteins in insect hemolymph. *In* "The Juvenile Hormones" (L. I. Gilbert, ed.), pp. 327–341. Plenum, New York.

Kramer, S. J. (1978). Age-dependent changes in corpus allatum activity *in vitro* in the adult Colorado potato beetle, *Leptinotarsa decemlineata. J. Insect Physiol.* **24**, 461–464.

Kramer, S. J., and de Kort, C. A. D. (1976). Some properties of hemolymph esterases from *Leptinotarsa decemlineata* Say. *Life Sci.* **19**, 211–218.

Kramer, S. J., and de Kort, C. A. D. (1978). Juvenile hormone carrier lipoprotein(s) in the haemolymph of the Colorado potato beetle, *Leptinotarsa decemlineata. Insect Biochem.* **8**, 87–92.

Kramer, S. J., and Kalish, F. (1984). Regulation of the corpora allata in the black mutant of *Manduca sexta. J. Insect Physiol.* **30**, 311–316.

Kramer, S. J., and Law, J. H. (1980a). Control of juvenile hormone production: The relationship between precursor supply and hormone synthesis in the tobacco hornworm, *Manduca sexta. Insect Biochem.* **10**, 569–575.

Kramer, S. J., and Law, J. H. (1980b). Synthesis and transport of juvenile hormones in insects. *Chem. Res.* **13**, 297–303.

Kramer, S. J., and Staal, G. B. (1981). *In vitro* studies on the mechanism of action of anti-juvenile hormone agents in larvae of *Manduca sexta. In* "Juvenile Hormone Biochemistry" (G. E. Pratt and G. T. Brooks, eds.), pp. 425–437. Elsevier, Amsterdam.

Kramer, S. J., Wieten, M., and de Kort, C. A. D. (1977). Metabolism of juvenile hormone in the Colorado potato beetle, *Leptinotarsa decemlineata. Insect Biochem.* **7**, 231–236.

Lanzrein, B. (1979). The activity and stability of injected juvenile hormones (JH I, JH II and JH III) in last-instar larvae and adult females of the cockroach *Nauphoeta cinerea. Gen. Comp. Endocrinol.* **39**, 69–78.

Lanzrein, B., Gentinetta, V., and Lüscher, M. (1977). *In vivo* and *in vitro* studies on the endocrinology of the reproductives of the termite *Macrotermes subhyalinus.* Proc. 8th Int. Congr. International Union for the Study of Social Insects, Wageningen, The Netherlands. pp. 265–268.

Lanzrein, B., Gentinetta, V., Fehr, R., and Lüscher, M. (1978). Correlation between haemolymph juvenile hormone titre, corpus allatum volume, and corpus allatum *in vivo* and *in vitro* activity during oocyte maturation in a cockroach (*Nauphoeta cinerea*). *Gen. Comp. Endocrinol.* **36**, 339–345.

Lanzrein, B., Wilhelm, R., and Gentinetta, V. (1981a). On relations between corpus allatum activity and oocyte maturation in the cockroach *Nauphoeta cinerea. In* "Regulation of Insect Development and Behaviour" (F. Sehnal, A. Zabza, J. J. Menn, and B. Cymborowski, eds.), pp. 523–534. Wroclaw Technical Univ. Press, Wroclaw, Poland.

Lanzrein, B., Wilhelm, R., and Buschor, J. (1981b). On the regulation of the corpora allata activity in adult females of the ovoviviparous cockroach *Nauphoeta cinerea, In* "Juvenile Hormone Biochemistry" (G. E. Pratt and G. T. Brooks, eds.), pp. 147–160. Elsevier, Amsterdam.

Lauer, R. C., Solomon, P. H., Nakanishi, K., and Erlanger, B. F. (1974). Antibodies to insect C_{16}-juvenile hormone. *Experientia* **30**, 558–560.

Laufer, H., and Borst, D. W. (1983). Juvenile hormone and its mechanism of action. *In* "Endocrinology of Insects" (R. G. H. Downer and H. Laufer, eds.), pp. 203–216. Liss, New York.

Law, J. H. (1983). Biosynthesis of juvenile hormones in insects. *In* "Biosynthesis of Isoprenoid Compounds" (J. W. Porter and S. L. Spurgeon, eds.), pp. 507–534. Wiley, New York.

Lazarovici, P., and Pener, M. P. (1978). The relations of the pars intercerebralis, corpora allata, and juvenile hormone to oocyte development and oviposition in the African migratory locust. *Gen. Comp. Endocrinol.* **35**, 375–386.

Lee, E., Schooley, D. A., Hall, M. S., and Judy, K. J. (1978). Juvenile hormone biosynthesis: Homomevalonate and mevalonate synthesis by insect corpus allatum enzymes. *J. Chem. Soc. Chem. Commun.* **7**, 290–292.

Lensky, Y., Baehr, J. C., and Porcheron, P. (1978). Dosages radio-immunolgiques des ecdysones et des hormones juveniles au cours du developpement post-embryonnaire chez les ouvrières et les reines d'Abeille (*Apis mellifica* L. var. *ligustica*). *C.R. Hebd. Seances Acad. Sci. Paris D* **287**, 821–824.

Lococo, D. J., and Tobe, S. S. (1984). Neuroanatomy of the retrocerebral complex, in particular the pars intercerebralis and partes lateralis in the cockroach *Diploptera punctata* (Dictyoptera: Blaberidae). *Int. J. Insect Morphol. Embryol.* **13**, 65–76.

Lococo, D. J., Thompson, C. S., and Tobe, S. S. (1985). Cell-coupling in an insect endocrine gland. (Submitted)

Loher, W., Ruzo, L., Baker, F. C. Miller, C. A., and Schooley, D. A. (1983). Identification of the juvenile hormone from the cricket, *Teleogryllus commodus*, and juvenile hormone titre changes. *J. Insect Physiol.* **29**, 585–589.

Lüscher, M., and Engelmann, F. (1960). Histologische und experimentelle Untersuchungen über die Auslösung der Metamorphose bei *Leucophaea maderae* (Orthoptera). *J. Insect Physiol.* **5**, 240–258.

Lusky, K. L., Faust, J. R., Chin, D. J., Brown, M. S., and Goldstein, J. L. (1983). Amplification of the gene for 3-hydroxy-3-methylglutaryl coenzyme A reductase, but not for the 53-kDa protein, in UT-1 cells. *J. Biol. Chem.* **258**, 8462–8469.

Madhavan, K., Conscience-Egli, M., Sieber, F., and Ursprung, H. (1973). Farnesol metabolism in *Drosophila melanogaster*: Ontogeny and tissue distribution of actanol dehydrogenase and aldehyde oxidase. *J. Insect Physiol.* **19**, 235–241.

McCaffery, A. R. (1976). Effects of electrocoagulation of cerebral neurosecretory cells and implantation of corpora allata on oocyte development in *Locusta migratoria*. *J. Insect Physiol.* **22**, 1081–1092.

McCaffery, A. R., and Highnam, K. C. (1975a). Effects of corpora allata on the activity of the cerebral neurosecretory system of *Locusta migratoria migratoroides* R. and F. *Gen. Comp. Endocrinol.* **25**, 358–372.

McCaffery, A. R., and Highnam, K. C. (1975b). Effects of corpus allatum hormone and its mimics on the activity of cerebral neurosecretory system of *Locusta migratoria migratoroides* R. and F. *Gen. Comp. Endocrinol.* **25**, 373–386.

McCaffery, A. R., and McCaffery, V. A. (1983). Corpus allatum activity during overlapping cycles of oocyte growth in adult female *Melanoplus sanguinipes*. *J. Insect Physiol.* **29**, 259–266.

McCaleb, D. C., Reddy, G., and Kumaran, A. K. (1980). Some properties of the haemolymph juvenile hormone esterases in *Galleria mellonella* larvae and *Tenebrio molitor* pupae. *Insect Biochem.* **10**, 273–277.

Mason, C. (1973). New features of the brain-retrocerebral neuroendocrine complex of the locust *Schistocerca vaga* (Scudder). *Z. Zellforsch.* **141**, 19–32.

Melnikova, E. J., and Panov, A. A. (1975). Ultrastructure of the larval corpus allatum of *Hyphantria cunea* Drury (Insecta, Lepidoptera). *Cell Tissue Res.* **162**, 395–410.

Metzler, M., Dahm, K. H., Meyer, D., and Röller, H. (1971). On the biosynthesis of juvenile hormone in the adult cecropia moth. *Z. Naturforsch.* **26**, 1270–1276.

Metzler, M., Meyer, D., Dahm, K. H, and Röller, H. (1972). Biosynthesis of juvenile hormone from 10-epoxy-7-ethyl-3,11-dimethyl-2,6-tridecadienoic acid in the adult cecropia moth. *Z. Naturforsch.* **27**, 321–322.

Meyer, A. S., Schneiderman, H. A., Hanzman, E., and Ko, J. H. (1968). The two juvenile hormones from the cecropia silk moth. *Proc. Natl. Acad. Sci. U.S.A.* **60**, 853–860.

Meyer, A. S., Hanzmann, E., and Schneiderman, H. A. (1970). The isolation and identification of the two juvenile hormones from the cecropia silk moth. *Arch. Biochem. Biophys.* **137**, 190–213.

Miller, J. P., and Selverston, A. I. (1979). Rapid killing of single neurons by irradiation of intracellularly injected dye. *Science* **206**, 702–704.

Monger, D. J., and Law, J. H. (1982). Control of juvenile hormone biosynthesis. Evidence for phosphorylation of the 3-hydroxy-3-methylglutaryl coenzyme A reductase of insect corpus allatum. *J. Biol. Chem.* **257**, 1921–1923.

Monger, D. J., Lim, W. A., Kezdy, F. J., and Law, J. H. (1982). Compactin inhibits insect HMG-CoA reductase and juvenile hormone biosynthesis. *Biochem. Biophys. Res. Commun.* **105**, 1374–1380.

Moulins, M., Girardie, A., and Girardie, J. (1974). Manipulation of sexual physiology by brain stimulation in insects. *Nature (London)* **250**, 339–340.

Müller, P. J., Masner, P., and Trautmann, K. H. (1974). The isolation and identification of juvenile hormone from cockroach corpora allata *in vitro*. *Life Sci.* **15**, 915–921.

Nakanishi, K., Schooley, D. A., Koreeda, M., and Dillon, J. (1971). Absolute configuration of the C_{18}JH: Application of a new circular dichroism method using Tris(dipivaloylmethanato)praseodymium. *Chem. Commun.* **1971**, 1235–1236.

Neill, J. D., and Frawley, S. (1983). Detection of hormone release from individual cells in mixed populations using a reverse hemolytic plaque assay. *Endocrinology* **112**, 1135–1137.

Nes, W. R., and McKean, M. L. (1977). "Biochemistry of Steroids and Other Isopentenoids." Univ. Park Press, Baltimore.

Nijhout, H. F. (1975a). A threshold size for metamorphosis in the tobacco hornworm, *Manduca sexta* (L.). *Biol. Bull. Woods Hole* **149**, 214–225.

Nijhout, H. F. (1975b). Axonal pathways in the brain-retrocerebral neuroendocrine complex of *Manduca sexta* (L.). (Lepidoptera: Sphingidae). *Int. J. Insect Morphol. Embryol.* **4**, 529–538.

Nijhout, H. F. (1975c). Dynamics of juvenile hormone action in larvae of the tobacco hornworm, *Manduca sexta* (L.). *Biol. Bull. Woods Hole* **149**, 568–579.

Nijhout, H. F. (1979). Stretch-induced moulting in *Oncopeltus fasciatus*. *J. Insect Physiol.* **25**, 277–281.

Nijhout, H. F. (1981). Physiological control of molting in insects. *Am. Zool.* **21**, 631–640.

Nijhout, H. F., and Wheeler, D. E. (1982). Juvenile hormone and the physiological basis of insect polymorphism. *Q. Rev. Biol.* **57**, 109–133.

Nijhout, H. F., and Williams, C. M. (1974). Control of moulting and metamorphosis in the tobacco hornworm, *Manduca sexta* (L): Cessation of juvenile hormone secretion as a trigger for pupation. *J. Exp. Biol.* **61**, 493–501.

Nowock, J., Hammock, B., and Gilbert, L. I. (1976). The binding protein as a modulator of juvenile hormone stability and uptake. *In* "The Juvenile Hormones" (L. I. Gilbert, ed.), pp. 354–373. Plenum, New York.

Odhiambo, T. R. (1966a). The fine structure of the corpus allatum of the sexually mature male of the desert locust. *J. Insect Physiol.* **12**, 819–828.

Odhiambo, T. R. (1966b). Ultrastructure of the development of the corpus allatum in the adult male of the desert locust. *J. Insect Physiol.* **12**, 995–1002.

Palevody, C., and Grimal, A. (1976). Variations cytologiques des corps allates au cours du cycle réproducteur du Collembole *Folsomia candida*. *J. Insect Physiol.* **22**, 63–72.

Panov, A. A., and Bussurmanova, O. K. (1970). Fine structure of the gland cells in inactive and active corpus allatum of the bug, *Eurygaster integriceps*. *J. Insect Physiol.* **16**, 1265–1281.

Papillon, M., Cassier, P., Girardie, J., and Lafon-Cazal, M. (1976). L'influence de la temperature d'élevage sur l'activité endocrine de *Schistocerca gregaria*. *Arch. Biol.* (*Brussels*) **87**, 103–127.

Papillon, M., Porcheron, P., and Baehr, J. C. (1980). Effects of the rearing temperature upon growth and hormonal balance in *Schistocerca gregaria* during the last two larval instars. *Experientia* **36**, 419–422.

Pener, M. P. (1965). On the influence of corpora allata on maturation and sexual behaviour of *Schistocerca gregaria*. *J. Zool.* **147**, 119–136.

Pener, M. P. (1967). Effects of allatectomy and sectioning of the nerves of the corpora allata on oöcyte growth, male sexual behaviour, and colour change in adults of *Schistocerca gregaria*. *J. Insect Physiol.* **13**, 665–684.

Pener, M. P., Orshan, L., and de Wilde, J. (1978). Precocene II causes atrophy of corpora allata in *Locusta migratoria*. *Nature* (*London*) **272**, 350–353.

Peter, M. G., and Dahm, K. H. (1975). Biosynthesis of juvenile hormone in the cecropia moth, Labelling pattern from 1-[^{14}C]-propionate through degradation to single carbon atom derivatives. *Helv. Chim. Acta* **58**, 1037–1048.

Peter, M. G., Dahm, K. H., and Röller, H. (1976). The juvenile hormones in blood of larvae and adults of *Manduca sexta* (Joh.). *Z. Naturforsch.* **31c**, 129–131.

Peter, M. G., Gunawan, S., Gellissen, G., and Emmerich, H. (1979). Differences in hydrolysis and binding of homologous juvenile hormones in *Locusta migratoria* hemolymph. *Z. Naturforsch.* **34c**, 588–598.

Peter, M. G., Shirk, P. D., Dahm, K. H., and Röller, H. (1981). On the specificity of juvenile hormone biosynthesis in the male cecropia moth. *Z. Naturforsch.* **36c**, 579–585.

Peter, M. G., Stupp, H.-P., and Lentes, K.-U. (1983). Reversal of the enantioselectivity in the enzymatic hydrolysis of juvenile hormone as a consequence of a protein fractionation. *Angew. Chem.* **95**, 794.

Peterson, R. C., Reich, M. F., Dunn, P. E., Law, J. H., and Katzenellenbogen, J. A. (1977). Binding specificity of the juvenile hormone carrier protein from the hemolymph of the tobacco hornworm *Manduca sexta* Johannson (Lepidoptera: Sphingidae). *Biochemistry* **16**, 2305–2311.

Peterson, R. C., Dunn, P. E., Seballos, H. L., Barbeau, B. K., Keim, P. S., Riley, C. T., and Heinrikson, R. L. (1982). Juvenile hormone carrier protein of *Manduca sexta* haemolymph. Improved purification procedure; protein modification studies and sequence of the amino terminus of the protein. *Insect Biochem.* **12**, 643–650.

Pipa, R. L. (1977). Do the brains of wax moth larvae secrete an allatotropic hormone? *J. Insect Physiol.* **23**, 103–108.

Pipa, R. L. (1978). Locations and central projections of neurons associated with the retrocerebral neuroendocrine complex of the cockroach *Periplaneta americana* (L.). *Cell Tissue Res.* **193**, 443–455.

Pipa, R. L. (1980). Neural inhibition of corpus allatum activity during metamorphosis of *Periplaneta americana*: A reassessment. *Ann. Entomol. Soc. Am.* **73**, 275–278.

Pipa, R. L. (1982). Neural influence on corpus allatum activity and egg maturation in starved virgin *Periplaneta americana*. *Physiol. Entomol.* **7**, 449–455.

Pipa, R. L. (1983). Morphological considerations in the integration of nervous and endocrine systems. *In* "Endocrinology of Insects" (R. G. H. Downer and H. Laufer, eds.), pp. 39–53. Liss, New York.

Pipa, R. L., and Novak, F. J. (1979). Pathways and fine structure of neurons forming the nervi corporis allati II of the cockroach *Periplaneta americana* (L.). *Cell Tissue Res.* **201**, 227–237.

Polivanova, E. N., and Bocharova, E. V. (1974). Corpora allata in late embryogenesis of *Eurygaster integriceps* Put. (Insecta: Hemiptera). *Gen. Comp. Endocrinol.* **22**, 409.

Polivanova, E. N., and Bocharova, E. V. (1982). The development of the function of corpora allata of *Eurygaster integriceps* embryos under normal conditions and under the effect of juvenile hormone analogue. *Zh. Obshchei. Biol.* **43**, 88–95.

Poras, M., Baehr, J. C., and Cassier, P. (1983). Control of corpus allatum activity during the imaginal diapause in females of *Locusta migratoria* L. *Int J. Invertebr. Reprod.* **6**, 111–122.

Postlethwaite, J. H., and Shirk, P. D. (1981). Genetic and endocrine regulation of vitellogenesis in *Drosophila. Am. Zool.* **21**, 687–700.

Pratt, G. E. (1975). Inhibition of juvenile hormone carboxyesterase of locust haemolymph by organophosphates *in vitro. Insect Biochem.* **5**, 595–607.

Pratt, G. E., and Bowers, W. S. (1977). Precocene II inhibits juvenile hormone biosynthesis by cockroach corpora allata *in vitro. Nature* (*London*) **265**, 548–550.

Pratt, G. E., and Finney, J. R. (1977). Chemical inhibition of juvenile hormone biosynthesis *in vitro. In* "Crop Protection Agents — their Biological Evaluation" (N. R. McFarlane, ed.), pp. 113–132. Academic Press, New York.

Pratt, G. E., and Pener, M. P. (1983). Precocene sensitivity of corpora allata in adult female *Locusta migratoria* after electrocoagulation of the pars intercerebralis neurosecretory cells. *J. Insect Physiol.* **29**, 33–39.

Pratt, G. E., and Tobe, S. S. (1974). Juvenile hormones radiobiosynthesized by corpora allata of adult female locusts *in vitro. Life Sci.* **14**, 575–586.

Pratt, G. E., and Weaver, R. J. (1975). Juvenile hormone biosynthesis by cultured cockroach corpora allata. *J. Endocrinol.* **64**, 67.

Pratt, G. E., Tobe, S. S., Weaver, R. J., and Finney, J. R. (1975a). Spontaneous synthesis and release of C_{16} juvenile hormone by isolated corpora allata of female locust *Schistocerca gregaria* and female cockroach *Periplaneta americana. Gen. Comp. Endocrinol.* **26**, 478–484.

Pratt, G. E., Tobe, S. S., and Weaver, R. J. (1975b). Relative oxygenase activities in juvenile hormone biosynthesis of corpora allata of an African locust (*Schistocerca gregaria*) and American cockroach (*Periplaneta americana*). *Experientia* **31**, 120–122.

Pratt, G. E., Weaver, R. J., and Hamnett, A. F. (1976). Continuous monitoring of juvenile hormone release by superfused corpora allata of *Periplaneta americana. In* "The Juvenile Hormones" (L. I. Gilbert, ed.), pp. 164–178. Plenum, New York.

Pratt, G. E., Hamnett, A. F., Finney, J. R. and Weaver, R. J. (1978). The biosynthesis of radiolabelled cecropia (C_{18}) juvenile hormone and several analogues by isolated corpora allata of the American cockroach (*Periplaneta americana*) during incubation with methyl-labelled methionine and analogues of farnesenic acid. *Gen. Comp. Endocrinol.* **34**, 113.

Pratt, G. E., Jennings, R. C., Hamnett, A. F., and Brooks, G. T. (1980). Lethal metabolism of prococene-I to a reactive epoxide by locust corpora allata. *Nature* (*London*) **284**, 320–323.

Pratt, G. E., Stott, K. M., Brooks, G. T., Jennings, R. C., Hamnett, A. F., and Alexander, B. A. J. (1981). Utilisation of farnesoic acid analogues by the *O*-methyl transferase from corpora allata of adult female *Locusta migratoria. In* "Juvenile Hormone Biochemistry" (G. E. Pratt and G. T. Brooks, eds.), pp. 107–121. Elsevier, Amsterdam.

Quistad, C. B., Cerf, D. C., Schooley, D. A., and Staal, G. (1981). Fluoromevalonate acts as an inhibitor of insect juvenile hormone biosynthesis. *Nature* (*London*) **289**, 176–177.

Raabe, M. (1982). "Insect Neurohormones." Plenum, New York.

Rademakers, L. H. P. M. (1977). Identification of a secretomotor center in the brain of *Locusta migratoria*, controlling the secretory activity of the adipokinetic hormone producing cells of the corpus cardiacum. *Cell Tissue Res.* **184**, 381–395.

Rankin, S. M., and Stay, B. (1983). Effects of decapitation and ovariectomy on the regulation of juvenile hormone synthesis in the cockroach, *Diploptera punctata. J. Insect Physiol.* **29**, 839–845.

Rankin, S. M., and Stay, B. (1984). The changing effect of the ovary on rates of juvenile hormone synthesis in *Diploptera punctata. Gen. Comp. Endocrinol.* **54**, 382–388.

Rankin, S. M., and Stay, B. (1985a). Regulation of juvenile hormone synthesis during pregnancy in the cockroach, *Diploptera punctata. J. Insect Physiol.* **31**, 145–157.

Rankin, S. M., and Stay, B. (1985b). Ovarian inhibition of juvenile hormone synthesis in the viviparous cockroach, *Diploptera punctata Gen. Comp. Endocrinol.* (in press).

Reddy, G., and Krishnakumaran, A. (1972). Relationship between morphogenetic activity and metabolic stability of insect juvenile hormone analogues. *J. Insect Physiol.* **18**, 2019–2028.

Reddy, G., Hwang-Hsu, K., and Kumaran, A. K. (1979). Factors influencing juvenile hormone esterase activity in the wax moth, *Galleria mellonella. J. Insect Physiol.* **25**, 65–71.

Reibstein, D., and Law, J. H. (1973). Enzymatic synthesis of insect juvenile hormones. *Biochem. Biophys. Res. Commun.* **55**, 266–272.

Reibstein, D., Law, J. H., Bowlus, S. B., and Katzenellenbogen, J. A. (1976). Enzymatic synthesis of juvenile hormone in *Manduca sexta. In* "The Juvenile Hormones" (L. I. Gilbert, ed.), pp. 131–146. Plenum, New York.

Rembold, H. (1981). Modulation of JH III-titer during the gonotrophic cycle of *Locusta migratoria*, measured by gas chromatography-selected ion monitoring mass spectrometry. *In* "Juvenile Hormone Biochemistry" (G. E. Pratt and G. T. Brooks, eds.), pp. 11–20. Elsevier, Amsterdam.

Rembold, H., and Hagenguth, H. (1981). Modulation of hormone pools during postembryonic development of the female honey bee castes. *In* "Regulation of Insect Development and Behavior" (F. Sehnal, A. Zabza, J. J. Menn, and B. Cymborowski, eds.), pp. 427–440. Wroclaw Technical Univ. Press, Wroclaw, Poland.

Rembold, H., Eder, J., and Ulrich, G. M. (1980). Inhibition of allatotropic activity and ovary development in *Locusta migratoria* by anti-brain antibodies. *Z. Naturforsch. Sect. C Biosci.* **35**, 1117–1119.

Renucci, M., and Strambi, A. (1981). Evolution des ecdysteroides ovariens et hémolymphatiques au cours de la maturation ovarienne chez *Acheta domesticus* L. (Orthoptere). *C.R. Hebd. Seances Acad. Sci. Paris* **293**, 825–830.

Riddiford, L. M. (1980). Interaction of ecdysteroids and juvenile hormone in the regulation of larval growth and metamorphosis of the tobacco hornworm. *In* "Progress in Ecdysone Research" (J. A. Hoffmann, ed.), pp. 409–430. Elsevier, Amsterdam.

Riddiford, L. M., and Hori, M. (1981). Control of larval cuticle formation and pigmentation by juvenile hormone. *In* "Juvenile Hormone Biochemistry" (G. E. Pratt and G. T. Brooks, eds.), pp. 241–250. Elsevier, Amsterdam.

Rode-Gowal, H., Abbott, S., Meyer, D., Röller, H., and Dahm, K. H. (1975). Propionate as a precursor of juvenile hormone in the cecropia moth. *Z. Naturforsch.* **30c**, 392–397.

Rodwell, V. W., Nordstrom, J. L., and Mitschlen, J. J. (1976). Regulation of HMG-CoA reductase. *Adv. Lipid Res.* **14**, 1–74.

Roe, R. M., Hammond, A. M., and Sparks, T. C. (1982). Characterization of the plasma juvenile hormone esterase in synchronous last instar female larvae of the sugarcane borer, *Diatraea saccharalis. Insect Biochem.* **13**, 163–170.

Röller, H., and Dahm, K. H. (1970). The identity of juvenile hormone produced by corpora allata *in vitro. Naturwissenschaften* **57**, 454–455.

Röller, H., Dahm, K. H., Sweely, C. C., and Trost, B. M. (1967). The structure of juvenile hormone. *Angew. Chem.* **6**, 179–180.

Röseler, P.-F. (1977). Juvenile hormone control of oögenesis in bumblebee workers, *Bombus terrestris. J. Insect Physiol.* **23**, 985–992.

Röseler, P.-F., and Röseler, I. (1978). Studies on the regulation of the juvenile hormone titre in bumblebee workers, *Bombus terrestris. J. Insect Physiol.* **24**, 707–713.

Röseler, P.-F., Röseler, I., and Strambi, A. (1980). The activity of corpora allata in dominant and subordinated females of the wasp *Polistes gallicus. Insectes Sociaux* **27**, 97–107.

Rossignol, P. A., Feinsod, F. M., and Spielman, A. (1981). Inhibitory regulation of corpus allatum activity in mosquitoes. *J. Insect Physiol.* **27**, 651–654.

Roth, L. M., and Stay, B. (1961). Oocyte development in *Diploptera punctata* (Eschscholtz) (Blattaria). *J. Insect Physiol.* **7**, 186–202.

Rotin, D., and Tobe, S. S. (1983). The possible role of juvenile hormone esterase in the regulation of juvenile hormone titre in the female cockroach *Diploptera punctata. Can. J. Biochem. Cell Biol.* **61**, 811–817.

Rotin, D., Feyereisen, R., Koener, J., and Tobe, S. S. (1982). Haemolymph juvenile hormone esterase activity during the reproductive cycle of the viviparous cockroach *Diploptera punctata. Insect Biochem.* **12**, 263–268.

Rudnicka, M., and Hammock, B. D. (1981). Approaches to the purification of the juvenile hormone esterase from the cabbage looper, *Trichoplusia ni. Insect Biochem.* **11**, 437–444.

Rudnicka, M., and Kochman, M. (1984). Purification of the juvenile hormone esterase from the haemolymph of the wax moth *Galleria mellonella* (*Lepidoptera*). *Insect Biochem.* **14**, 189–198.

Ruegg, R. P., Kriger, F. L., Davey, K. G., and Steel, C. G. H. (1981). Ovarian ecdysone elicits release of a myotropic ovulation hormone in *Rhodnius* (Insecta: Hemiptera). *Int. J. Invertebr. Reprod.* **3**, 357–361.

Ruegg, R. P., Lococo, D. J., and Tobe, S. S. (1983). Control of corpus allatum activity in *Diploptera punctata*: Roles of the pars intercerebralis and pars lateralis. *Experientia* **39**, 1329–1334.

Sanburg, L. L., Kramer, K. J., Kezdy, F. J., Law, J. H., and Oberlander, H. (1975a). Role of juvenile hormone esterases and carrier proteins in insect development. *Nature* (*London*) **253**, 266–267.

Sanburg, L. L., Kramer, K. J., Kezdy, F. J., and Law, J. H. (1975b). Juvenile hormone-specific esterases in the haemolymph of the tobacco horn-worm, *Manduca sexta. J. Insect Physiol.*. **21**, 873–887.

Scharrer, B. (1946). Section of the nervi corporis cardiaci in *Leucophaea maderae* (Orthoptera). *Anat. Rec.* **96**, 577.

Scharrer, B. (1952). Neurosecretion. XI. The effects of nerve section on the intercerebralis-cardiacum-allatum system of the insect *Leucophaea maderae. Biol. Bull. Woods Hole* **102**, 261–272.

Scharrer, B. (1958). Neuro-endocrine mechanisms in insects. *Proc. Int. Symp. Neurosecretion, 2nd* pp. 79–84.

Scharrer, B. (1964). Histophysiological studies on the corpus allatum of *Leucophaea maderae.* IV. Ultrastructure during normal activity cycle. *Z. Zellforsch.* **62**, 125–148.

Scharrer, B. (1968). Neurosecretion. XIV. Ultrastructure study of sites of release of neurosecretory material in blattarian insects. *Z. Zellforsch. Mikrosk. Anat.* **89**, 1–16.

Scharrer, B. (1971). Histophysiological studies on the corpus allatum of *Leucophaea maderae.* V. Ultrastructure of sites of origin and release of a distinctive cellular product. *Z. Zellforsch.* **120**, 1–16.

Scharrer, B. (1978). Histophysiological studies on the corpus allatum of *Leucophaea maderae.* VI. Ultrastructural characteristics in gonadectomized females. *Cell Tissue Res.* **194**, 533–545.

Scharrer, B., and von Harnack, M. (1958). Histophysiological studies on the corpus allatum of *Leucophaea maderae.* I. Normal life cycle in male and female adults. *Biol. Bull Mar. Biol. Lab. Woods Hole* **115**, 508–517.

Schooley, D. A. (1977). Analysis of the naturally occurring juvenile hormones, their isolation, identification, and titer determination at physiological levels. *In* "Analytical Biochemistry of Insects" (R. B. Turner, ed.), pp. 241–287. Elsevier, Amsterdam.

Schooley, D. A., and Baker, F. C. (1985). Juvenile hormone biosynthesis. *In* "Comprehensive Insect Physiology, Biochemistry and Pharmacology" (G. A. Kerkut and L. I. Gilbert, eds.), Vol. 7. Pergamon, Oxford.

Schooley, D. A., Judy, K. J., Bergot, B. J., Hall, M. S., and Siddall, J. B. (1973). Biosynthesis of the juvenile hormone of *Manduca sexta*: Labeling pattern from mevalonate, propionate, and acetate. *Proc. Natl. Acad. Sci. U.S.A.* **70**, 2921–2925.

Schooley, D. A., Judy, K. J., Bergot, B. J., Hall, M. S., and Jennings, R. C. (1976). Determination of the physiological levels of juvenile hormones in several insects and biosynthesis of the carbon skeletons of the juvenile hormones. *In* "The Juvenile Hormones" (L. I. Gilbert, ed.), pp. 101–117. Plenum, New York.

Schooley, D. A., Bergot, B. J., Goodman, W., and Gilbert, L. (1978). Synthesis of both optical isomers of insect juvenile hormone III and their affinity for the juvenile hormone-specific binding protein of *Manduca sexta*. *Biochem. Biophys. Res. Commun.* **81**, 743–749.

Schooley, D. A., Baker, F. C., Tsai, L. W., Miller, C. A., and Jamieson, G. C. (1984). 3 juvenile hormones 0, I, and II exist only in Lepidoptera. *In* "Biosynthesis, Metabolism and Mode of Action of Invertebrate Hormones" (J. A. Hoffmann and M. Porchet, eds.), pp. 373–383. Springer-Verlag, Berlin and New York.

Schooneveld, H. (1970). Structural aspects of neurosecretory and corpus allatum activity in the adult Colorado beetle, *Leptinotarsa decemlineata* (Say), as a function of day-length. *Neth. J. Zool.* **20**, 151–237.

Schooneveld, H., Kramer, S. J., Privee, H., and van Huis, A. (1979). Evidence of controlled corpus allatum activity in the adult Colorado potato beetle. *J. Insect Physiol.* **25**, 449–453.

Sedlak, B. J., Marchione, L., Devorkin, B., and Davino, R. (1983). Correlations between endocrine gland ultrastructure and hormone titres in the fifth larval instar of *Manduca sexta*. *Gen. Comp. Endocrinol.* **52**, 291–310.

Sehnal, F. (1976). Action of juvenoids on different groups of insects. *In* "The Juvenile Hormones" (L. I. Gilbert, ed.), pp. 301–322. Plenum, New York.

Sehnal, F. (1981). Action of juvenile hormone on tissue and cell differentiation. *In* "Regulation of Insect Development and Behavior" (F. Sehnal, A. Zabza, J. J. Menn, and B. Cymborowski, eds.), pp. 463–482. Wroclaw Technical Univ. Press, Wroclaw, Poland.

Sehnal, F., and Granger, N. A. (1975). Control of corpora allata function in larvae of *Galleria mellonella*. *Biol. Bull. Woods Hole* **148**, 106–116.

Shirk, P. D., Dahm, K. H., and Röller, H. (1976). The accessory sex glands as the repository for juvenile hormone in male cecropia moths. *Z. Naturforsch.* **31c**, 199–200.

Shirk, P. D., Bhaskaran, G., and Röller, H. (1983). Developmental physiology of corpora allata and accessory sex glands in the cecropia silkmoth. *J. Exp. Zool.* **227**, 69–79.

Siew, Y. C., and Gilbert, L. I. (1971). Effects of moulting hormone and juvenile hormone on insect endocrine gland activity. *J. Insect Physiol.* **17**, 2095–2104.

Slade, M., and Zibitt, C. H. (1972). Metabolism of cecropia juvenile hormone in insects and mammals. *In* "Insect Juvenile Hormones—Chemistry and Action" (J. J. Menn and M. Beroza, eds.), pp. 155–176. Academic Press, New York.

Slama, K., and Jarolim, V. (1981). Hydrolysis of 2(E) and 2(Z) isomers of juvenoid esters related to their juvenile hormone activity in insects. *In* "Juvenile Hormone Biochemistry" (G. E. Pratt and G. T. Brooks, eds.), pp. 185–192. Elsevier, Amsterdam.

Slama, K., Romanuk, M., and Sorm, F. (1974). "Insect Hormones and Bioanalogues." Springer-Verlag, Berlin and New York.

Sparagana, S. P., Bhaskaran, G., Dahm, K. H., and Riddle, V. (1984). Juvenile hormone production, juvenile hormone esterase, and juvenile hormone acid methyltransferase in corpora allata of *Manduca sexta*. *J. Exp. Zool.* **230**, 309–313.

Sparks, T. C., and Hammock, B. D. (1979a). A comparison of the induced and naturally occurring juvenile hormone esterases from the last instar larvae of *Trichoplusia ni*. *Insect Biochem.* **9**, 411–421.

Sparks, T. C., and Hammock, B. D. (1979b). Induction and regulation of juvenile hormone esterases during the last larval instar of the cabbage looper, *Trichoplusia ni*. *J. Insect Physiol.* **25**, 551–560.

Sparks, T. C., and Hammock, B. D. (1980). Comparative inhibition of the juvenile hormone esterases from *Trichoplusia ni*, *Musca domestica* and *Tenebrio molitor*. *Pestic. Biochem. Physiol.* **14**, 290–302.

Sparks, T. C., and Rose, R. L. (1983). Inhibition and substrate specificity of the haemolymph juvenile hormone esterase of the cabbage looper, *Trichoplusia ni* (Hübner). *Insect Biochem.* **13**, 633–640.

Sparks, T. C., Willis, W. S., Shorey, H. H., and Hammock, B. D. (1979). Haemolymph juvenile hormone esterase activity in synchronous last instar larvae of the cabbage looper, *Trichoplusia ni*. *J. Insect Physiol.* **25**, 125–132.

Sparks, T. C., Hammock, B. D., and Riddiford, L. M. (1983). The haemolymph juvenile hormone esterase of *Manduca sexta* (L.). Inhibition and regulation. *Insect Biochem.* **13**, 529–541.

Staal, G. B. (1975). Insect growth regulators with juvenile hormone activity. *Annu. Rev. Entomol.* **20**, 417–460.

Stay, B., and Tobe, S. S. (1977). Control of juvenile hormone biosynthesis during the reproductive cycle of a viviparous cockroach. I. Activation and inhibition of corpora allata. *Gen. Comp. Endocrinol.* **33**, 531–540.

Stay, B., and Tobe, S. S. (1978). Control of juvenile hormone biosynthesis during the reproductive cycle of a viviparous cockroach. II. Effects of unilateral allatectomy, implementation of supernumerary corpora allata and ovariectomy. *Gen. Comp. Endocrinol.* **34**, 276–286.

Stay, B., and Tobe, S. S. (1981). Control of the corpora allata during a reproductive cycle in a viviparous cockroach. *Am. Zool.* **21**, 663–674.

Stay, B., Friedel, T., Tobe, S. S., and Mundall, E. C. (1980). Feedback control of juvenile hormone synthesis in cockroaches: Possible role for ecdysterone. *Science* **207**, 898–900.

Stay, B., Tobe, S. S., Mundall, E. C., and Rankin, S. (1983). Ovarian stimulation of juvenile hormone synthesis in the viviparous cockroach, *Diploptera punctata*. *Gen. Comp. Endocrinol.* **52**, 341–349.

Stay, B., Ostegaard, L. S., Tobe, S. S., Strambi, A., and Spaziani, E. (1984). Ovarian and haemolymph titres of ecdysteroid during the gonadotropic cycle in *Diploptera punctata*. *J. Insect Physiol.* **30**, 643–651.

Steele, J. E. (1976). Hormonal control of metabolism in insects. *In* "Advances in Insect Physiology" (J. E. Treherne, M. J. Berridge, and V. B. Wigglesworth, eds.), pp. 239–323. Academic Press, New York.

Stefano, G. B., and Scharrer, B. (1981). High affinity binding of an enkephalin analog in the cerebral ganglion of the insect *Leucophaea maderae* (Blattaria). *Brain Res.* **225**, 107–114.

Strambi, C. (1981). Some data obtained by radioimmunoassay of juvenile hormone. *In* "Juvenile Hormone Biochemistry" (G. E. Pratt and G. T. Brooks, eds.), pp. 59–63. Elsevier, Amsterdam.

Strambi, C., Strambi, A., de Reggi, M. L., Hirn, M. H., and Delaage, M. A. (1981). Radioimmunoassay of insect juvenile hormones and of their diol derivatives. *Eur. J. Biochem.* **118**, 401–406.

Strong, L. (1965a). The relationships between the brain, corpora allata, and oöcyte growth in the Central American locust, *Schistocerca* sp. I. The cerebral neurosecretory system, the corpora allata, and oöcyte growth. *J. Insect Physiol.* **11**, 135–146.

Strong, L. (1965b). The relationship between the brain, corpora allata, and oöcyte growth in the Central American locust, *Schistocerca* sp. II. The innervation of corpora allata, the lateral neurosecretory complex, and oöcyte growth. *J. Insect Physiol.* **11**, 271–280.

Szibbo, C. M., and Tobe, S. S. (1981a). Cellular and volumetric changes in relation to the activity cycle in the corpora allata of *Diploptera punctata*. *J. Insect Physiol.* **27**, 655–665.

Szibbo, C. M., and Tobe, S. S. (1981b). The mechanism of compensation in juvenile hormone synthesis following unilateral allatectomy in *Diploptera punctata*. *J. Insect Physiol.* **27**, 609–613.

Szibbo, C. M., and Tobe, S. S. (1982). Intrinsic differences in juvenile hormone synthetic ability between corpora allata of males and females of the cockroach *Diploptera punctata*. *Gen. Comp. Endocrinol.* **46**, 533–540.

Szibbo, C. M., and Tobe, S. S. (1983). Nervous humoral inhibition of C_{16} juvenile hormone synthesis in last instar females of the viviparous cockroach, *Diploptera punctata*. *Gen. Comp. Endocrinol.* **49**, 437–445.

Szibbo, C. M., Rotin, D., Feyereisen, R., and Tobe, S. S. (1982). Synthesis and degradation of C_{16} juvenile hormone (JH III) during the final two stadia of the cockroach, *Diploptera punctata*. *Gen. Comp. Endocrinol.* **48**, 25–32.

Thomsen, E. (1948). Effect of removal of neurosecretory cells in the brain of adult *Calliphora erythrocephala* Meig. *Nature (London)* **161**, 439–440.

Thomsen, E., and Lea, A. O. (1969). Control of the medial neurosecretory cells by the corpus allatum in *Calliphora erythrocephala*. *Gen. Comp. Endocrinol.* **12**, 51–57.

Tobe, S. S. (1977). Asymmetry in hormone biosynthesis by insect endocrine glands. *Can. J. Zool.* **55**, 1509–1514.

Tobe, S. S. (1980). Regulation of the corpora allata in adult female insects. *In* "Insect Biology of the Future: VBW 80" (M. Locke and D. S. Smith, eds.), pp. 345–367. Academic Press, New York.

Tobe, S. S. (1985). In preparation.

Tobe, S. S., and Chapman, C. S. (1979). The effects of starvation and subsequent feeding on juvenile hormone synthesis and oocyte growth in *Schistocerca americana gregaria*. *J. Insect Physiol.* **25**, 701–708.

Tobe, S. S., and Clarke, N. (1985). The effect of L-methionine concentration on juvenile hormone biosynthesis by corpora allata of the cockroach *Diploptera punctata*. *Insect Biochem.* **15**, 175–179.

Tobe, S. S., and Feyereisen, R. (1983). Juvenile hormone biosynthesis: Regulation and assay. *In* "Endocrinology of Insects" (R. G. H. Downer and H. Laufer, eds.), pp. 161–178. Liss, New York.

Tobe, S. S., and Pratt, G. E. (1974a). The influence of substrate concentrations on the rate of insect juvenile hormone biosynthesis by corpora allata of the desert locust *in vitro*. *Biochem. J.* **144**, 107–113.

Tobe, S. S., and Pratt, G. E. (1974b). Dependence of juvenile hormone release from corpus allatum on intraglandular content. *Nature (London)* **252**, 474–476.

Tobe, S. S., and Pratt, Ge. E. (1975a). Corpus allatum activity *in vitro* during ovarian maturation in the desert locust, *Schistocerca gregaria*. *J. Exp. Biol.* **62**, 611–627.

Tobe, S. S., and Pratt, G. E. (1975b). The synthetic activity and glandular volume of the corpus allatum during ovarian maturation in the desert locust *Schistocerca gregaria*. *Life Sci.* **17**, 417–422.

Tobe, S. S., and Pratt, G. E. (1976). Farnesenic acid stimulation of juvenile hormone synthesis as an experimental probe in corpus allatum physiology. *In* "The Juvenile Hormones" (L. I. Gilbert, ed.), pp. 147–163. Plenum, New York.

Tobe, S. S., and Saleuddin, A. S. M. (1977). Ultrastructural localization of juvenile hormone biosynthesis by insect corpora allata. *Cell Tissue Res.* **183**, 25–32.

Tobe, S. S., and Stay, B. (1977). Corpus allatum activity *in vitro* during the reproductive cycle of the viviparous cockroach, *Diploptera punctata* (Eschscholtz). *Gen. Comp. Endocrinol.* **31**, 138–147.

Tobe, S. S., and Stay, B. (1979). Modulation of juvenile hormone synthesis by an analogue in the cockroach. *Nature (London)* **281**, 481–482.

Tobe, S. S., and Stay, B. (1980). Control of juvenile hormone biosynthesis during the reproductive cycle in a viviparous cockroach. III. Effects of denervation and age on compensation with unilateral allatectomy and supernumerary corpora allata. *Gen. Comp. Endocrinol.* **40**, 89–98.

Tobe, S. S., Pratt, G. E., and Saleuddin, A. S. M. (1976). Intracellular visualization of C_{16} juvenile hormone biosynthesis in corpora allata of the adult female desert locust *Schistocerca gregaria. Colloq. Int. C.N.R.S., Paris* **251**, 441–449.

Tobe, S. S., Chapman, C. S., and Pratt, G. E. (1977). Decay in juvenile hormone synthesis by insect corpus allatum after nerve transection. *Nature (London)* **286**, 728–730.

Tobe, S. S., Musters, A., and Stay, B. (1979). Corpus allatum function during sexual maturation of male *Diploptera punctata. Physiol. Entomol.* **4**, 79–86.

Tobe, S. S., Stay, B., Friedel, T., Feyereisen, R., and Paulson, P. (1981). The role of the brain in regulation of the corpora allata in female *Diploptera punctata. In* "Juvenile Hormone Biochemistry" (G. E. Pratt and G. T. Brooks, eds.), pp. 161–174. Elsevier, Amsterdam.

Tobe, S. S., Girardie, J., and Girardie, A. (1982). Enhancement of juvenile hormone biosynthesis in locusts following electrostimulation of cerebral neurosecretory cells. *J. Insect Physiol.* **28**, 867–872.

Tobe, S. S., B. Stay, F. C. Baker, and D. A. Schooley. (1984a). Regulation of juvenile hormone titre in the adult female cockroach *Diploptera punctata. In* "Biosynthesis, Metabolism and Mode of Action of Invertebrate Hormones" (J. A. Hoffmann and M. Porchet, eds.), pp. 397–406. Springer-Verlag, Berlin and New York.

Tobe, S. S., Clarke, N., Stay, B., and Ruegg, R. P. (1984b). Changes in cell number and activity of the corpora allata in the cockroach *Diploptera punctata*: A role for mating and the ovary. *Can. J. Zool.* **62**, 2178–2182.

Tobe, S. S., Ruegg, R. P., Stay, B. A., Baker, F. C., Miller, C. A., and Schooley, D. A. (1985). Juvenile hormone titer and regulation in the cockroach *Diploptera punctata. Experientia* (in press).

Trautmann, K. H., Suchy, M., Masner, P., Wipf, H.-K., and Schuler, A. (1976). Isolation and identification of juvenile hormones by means of a radioactive isotope dilution method: Evidence for JH III in eight species from four orders. *In* "The Juvenile Hormones" (L. I. Gilbert, ed.), pp. 118–130. Plenum, New York.

Ulrich, G. M., Schlagintweit, B., Eder, J., and Rembold, H. (1984). Elimination of allatotropic hormone in locusts by microsurgical and immunological methods. *In* "Advances in Invertebrate Reproduction" (W. Engels, ed.), p. 647. Elsevier, Amsterdam.

Vince, R. K., and Gilbert, L. I. (1977). Juvenile hormone esterase activity in precisely timed last instar larvae and pharate pupae of *Manduca sexta. Insect Biochem.* **7**, 115–120

Waku, Y., and Gilbert, L. I. (1964). The corpora allata of the silkmoth, *Hyalophora cecropia*: An ultrastructural study. *J. Morphol.* **115**, 69–96.

Weaver, R. J. (1979). Comparable activities of left and right corpora allata consistent with humoral control of juvenile hormone biosynthesis in the cockroach. *Can. J. Zool.* **57**, 343–345.

Weaver, R. J. (1981a). Radiochemical assays of corpus allatum activity in adult female cockroaches following ovariectomy in the last nymphal instar. *Experientia* **37**, 435–436.

Weaver, R. J., (1981b). Juvenile hormone regulation of left colleterial gland function in the cockroach *Periplaneta americana*: Oothecin synthesis. *In* "Juvenile Hormone Biochemistry" (G. E. Pratt and G. T. Brooks, eds.). pp. 271–283. Elsevier, Amsterdam.

Weaver, R. J., and Pratt, G. E. (1977). The effect of enforced virginity and subsequent mating on the activity of the corpus allatum of *Periplaneta americana* measured *in vitro*, as related to changes in the rate of ovarian maturation. *Physiol. Entomol* **2**, 59–76.

Weaver, R. J., and Pratt, G. E. (1981). Effects of starvation and feeding upon corpus allatum activity and oöcyte growth in adult female *Periplaneta americana*. *J. Insect Physiol.* **27**, 75–83.

Weaver, R. J., Pratt, G. E., and Finney, J. R. (1975). Cyclic activity of the corpus allatum related to gonotrophic cycles in adult female *Periplaneta americana*. *Experientia* **31**, 597–598.

Weaver, R. J., Pratt, G. E., Hamnett, A. F., and Jennings, R. C. (1980). The influence of incubation conditions on the rates of juvenile hormone biosynthesis by corpora allata isolated from adult females of the beetle *Tenebrio molitor*. *Insect Biochem.* **10**, 245–254.

Weibel, E. R. (1979). "Stereological Methods." Academic Press, New York.

Weirich, G. F., and Culver, M. G. (1979). S-adenosylmethionine—Juvenile hormone acid methyltransferase in male accessory reproductive glands of *Hyalophora cecropia* (L.). *Arch. Biochem. Biophys.* **198**, 175–181.

Weirich, G., and Wren, J. (1973). The substrate specificity of juvenile hormone esterase from *Manduca sexta*. *Life Sci.* **13**, 213–226.

Weirich, G., and Wren, J. (1976). Juvenile hormone esterase in insect development—A comparative study. *Physiol. Zool.* **49**, 341–350.

Whitmore, D., and Gilbert, L. I. (1972). Haemolymph lipoprotein transport of juvenile hormone. *J. Insect Physiol.* **18**, 1153–1167.

Whitmore, D., Whitmore, E., and Gilbert, L. I. (1972). Juvenile hormone induction of esterases: A mechanism for the regulation of juvenile hormone titer. *Proc. Natl. Acad. Sci. U.S.A.* **69**, 1592–1595.

Whitmore, D., Gilbert, L. I., and Ittycheriah, P. J. (1974). The origin of haemolymph carboxylesterases "induced" by the insect juvenile hormone. *Mol. Cell. Endocrinol.* **1**, 37–54.

Wigglesworth, V. B. (1934). The physiology of ecdysis in *Rhodnius prolixus* (Hemiptera). II. Factors controlling moulting and "metamorphosis." *Q. J. Microsc. Sci.* **77**, 191–222.

Wigglesworth, V. B. (1948). Functions of the corpus allatum in *Rhodnius prolixus* (Hemiptera). *J. Exp. Biol.* **25**, 1–14.

Wigglesworth, V. B. (1964). The hormonal regulation of growth and reproduction in insects. *Adv. Insect Physiol.* **2**, 247–336.

Wigglesworth, V. B. (1970). "Insect Hormones." Oliver & Boyd. Edinburgh.

Williams, C. M. (1976). Juvenile hormone ... In retrospect and in prospect. *In* "The Juvenile Hormones" (L. I. Gilbert, ed.), pp. 1–14. Plenum, New York.

Wing, K. D., Sparks, T. C., Lovell, V. M., Levinson, S. D., and Hammock, B. D. (1981). The distribution of juvenile hormone esterase and its interrelationship with other proteins influencing juvenile hormone metabolism in the cabbage looper, *Trichoplusia ni*. *Insect Biochem.* **11**, 473–485.

Wirtz, P. (1973). Differentiation in the honeybee larva. *Meded. Landbouwhogesch. Wageningen* **73**, 1–155.

Yin, C.-M., and Chippendale, G. M. (1979a). Diapause of the southwestern corn borer, *Diatrea grandiosella*: Further evidence showing juvenile hormone to be the regulator. *J. Insect Physiol.* **25**, 513–523.

Yin, C.-M., and Chippendale, G. M. (1979b). Ultrastructural characteristics of insect corpora allata in relation to larval diapause. *Cell Tissue Res.* **197**, 453–461.

Yuhas, D. A., Roe, R. M., Sparks, T. C., and Hammond, A. J., Jr. (1983). Purification and kinetics of juvenile hormone esterase from the cabbage looper, *Trichoplusia ni* (Aubner). *Insect Biochem.* **13**, 129–136.

Zaretsky, M., and Loher, W. (1983). Anatomy and electrophysiology of individual neurosecretory cells of an insect brain. *J. Comp. Neurol.* **216**, 253–263.

Nonspiking Interneurons and Motor Control in Insects

Melody V. S. Siegler

Note added in proof

In the time since this review was submitted for publication further papers have appeared concerning nonspiking interneurons in arthropods:

Hisada, M., Takahata, M., and Nagayama, T. (1984). Structure and output connection of local non-spiking interneurons in crayfish. *Zool. Sci.* **1**, 41–49.
Hisada, M., Takahata, M., and Nagayama, T. (1984). Local non-spiking interneurons in the arthropod motor control systems: Characterization and their functional significance. *Zool. Sci.* **1**, 681–700.
Nagayama, T., Takahata, M., and Hisada, M. (1984). Functional characteristics of local non-spiking interneurons as the pre-motor elements in crayfish. *J. Comp. Physiol.* **154**, 499–510.
Paul, D. H., and Mulloney, B. (1985). Local interneurons in the swimmeret system of the crayfish. *J. Comp. Physiol.* **156**, 489–502.

In addition, the following papers are in press:

Paul, D. H., and Mulloney, B. Nonspiking local interneuron in the motor pattern generator for the crayfish swimmeret. *J. Neurophysiol.* (in press).
Watkins, B. L., Burrows, M., and Siegler, M. V. S. The structure of locust non-spiking interneurones in relation to the anatomy of their segmental ganglion. *J. Comp. Neurol.* (in press).

Subject Index

Page numbers in italics indicate references to figures or tables.

D

Cumulative List of Authors

Numbers in bold face indicate the volume numbers of the series.

Cumulative List of Chapter Titles

Numbers in bold face indicate the volume numbers of the series.